Lecture Notes
in Business Information Processing   440

Series Editors

Wil van der Aalst
RWTH Aachen University, Aachen, Germany

John Mylopoulos
University of Trento, Trento, Italy

Sudha Ram
University of Arizona, Tucson, AZ, USA

Michael Rosemann
Queensland University of Technology, Brisbane, QLD, Australia

Clemens Szyperski
Microsoft Research, Redmond, WA, USA

More information about this series at https://link.springer.com/bookseries/7911

Sander J. J. Leemans

Robust Process Mining with Guarantees

Process Discovery, Conformance Checking and Enhancement

 Springer

Sander J. J. Leemans (iD)
Queensland University of Technology
Brisbane, QLD, Australia

ISSN 1865-1348   ISSN 1865-1356   (electronic)
Lecture Notes in Business Information Processing
ISBN 978-3-030-96654-6   ISBN 978-3-030-96655-3   (eBook)
https://doi.org/10.1007/978-3-030-96655-3

This book is a revised version of the PhD dissertation written by the author at Eindhoven University of
Technology, Netherlands. The original PhD dissertation is available from the university.

This Springer imprint is published by the registered company Springer Nature Switzerland AG
The registered company address is: Gewerbestrasse 11, 6330 Cham, Switzerland

Facilius per partes in cognitionem totius adducimur

Lucius Annaeus Seneca

For Shiva and Aryan

Preface

Due to the omnipresence of digitally supported processes and storage facilities nowadays, plenty of organisational data is available in the typical organisation. Process mining aims to aid business process management initiatives, by enabling analysts to gain insights from organisational behaviour recorded in event logs. The number of process mining techniques that analysts can use to study the processes in an organisation has been steadily growing since the late nineties, and has exploded in recent years, in particular the commercially available tools. The cornerstone of any process mining effort is a process model: *process discovery* aims to discover a process model from recorded behaviour such that the control flow of a process can be studied, *conformance checking* studies the relation between a model and recorded behaviour to study deviations, and *enhancement* projects further information such as time and costs on process models.

Due to the central role that process models play in process mining, good quality models are essential: models without deadlocks or other anomalies are much more likely to describe a process well. Furthermore, process discovery techniques ideally provide guarantees on the circumstances they require in order to discover models that represent the underlying processes exactly.

This book presents techniques for process discovery, conformance checking and enhancement. For process discovery, this book introduces the Inductive Miner framework: a recursive skeleton for discovery techniques that in itself provides several guarantees. The framework is instantiated in several concrete discovery techniques, each of which targets a specific challenge of process discovery, such as incompleteness of information or noisy behaviour. For conformance checking, this book introduces the Projected Conformance Checking framework, which focuses on speed, but nevertheless provides several guarantees, such as that for certain classes of models, it can decide language equivalence. For enhancement, this book introduces the Inductive visual Miner, a well-polished end-user focused tool that includes process discovery, conformance checking and that can visualise performance on a discovered model, all without any user input.

From introducing these algorithms and proving their guarantees, a detailed study of several language abstractions results: using the abstractions, such as the commonly used directly follows abstraction, we formally explore the boundaries of what *any* process discovery technique could do.

Acknowledgements

This book largely stems from my doctoral thesis, defended at the Eindhoven University of Technology in May 2017. I would like to thank my supervisors, Wil van der Aalst and Dirk Fahland, for their ongoing support and for their neverending encouragement, no matter the craziness of ideas I came up with. Valueable feedback on the thesis was provided by the PhD committee members Josep Carmona, Jan-Friso Groote, Javier Esparza, Uzay Kaymak and Wim Nuijten.

The research culture of the group in Eindhoven provided an inspirational environment for my research, in particular my office mates Meng Dou, Murat Firat, Shengnan Guo, Rafal Kocielnik, Cong Liu, Felix Mannhardt, Richard Müller, Marcella Rovani and Alifah Syamsiyah. In particular, I would like to thank Ine van der Ligt and Riet van Buul, whose support of the IS group was indispensable. Furthermore, I'd like to thank Eric Verbeek for his support on technical matters. I had the pleasure to work with/argue with/have lunch with/get stopped by police together with/visit Christmas markets with/visit steam trains with the following colleagues: Han van der Aa, Arya Adriansyah, Nour Assy, Alfredo Bolt, JC Bose, Paul de Bra, Rémi Brochenin, Joos Buijs, Toon Calders, Marcus Dees, Alok Dixit, Boudewijn van Dongen, Maikel van Eck, George Fletcher, Eduardo Gonzáles Lopéz de Murillas, Dennis Schunselaar, Christian Günther, Kees van Hee, Farideh Heidari, Bart Hompes, Julia Kiseleva, Maikel Leemans, Massimiliano de Leoni, Guangming Li, Xixi Lu, Fabrizio Maggi, Joyce Nakatumba, Mykola Pechenizkiy, Elham Ramezani, Hajo Reijers, Anne Rozinat, Natalia Sidorova, Christian Stahl, Natasha Stash, Niek Tax, Jan Martijn van der Werf, Michael Westergaard and Bas van Zelst. Futhermore, I had the pleasure of meeting several visitors: Claudio Di Ciccio, Laura Genga, Anna Kalenkova, Jan Mendling, Jorge Muñoz-Gama, David Redlich, Andreas Solti, Tonatiuh Tapia-Flores and Matthias Weidlich for several inspiring discussions and collaborations.

During my PhD, I had the opportunity to visit and collaborate with the group of Avishai Mandelbaum at the Technion in Haifa, Israel. During my visit, increasing temperatures kept me inside, but this yielded a boost in productivity. Avishai, Avigdor Gal, Shahar Harel and Arik Senderovich: it was a pleasure to work with you on discovering queues from event logs.

At the Queensland University of Technology, I was first welcomed for a collaboration visit and second as a new colleague. I'd like to thank Robert Andrews, Kevin Burrage, Abel Armas Cervantes, Raffaele Conforti, Arthur ter Hofstede, Wei Lai, Alireza Ostovar, Chun Ouyang, Artem Polyvyanyy, Erik Poppe, Marcello La Rosa, Suriadi Suriadi, Ilya Verenich, Moe Wynn and Jingxin Xu for welcoming me to this sunny side of the world.

Finally, I'd like to thank my family: Ilse (guess which discovery technique is named after her?), Lida, Martie and especially Shiva, who had to carry the burden of both handling me and, by transitivity, me writing this thesis.

Brisbane,   *Sander J.J. Leemans*
January 2022

Summary

Organisations store a lot of data from their business processes nowadays. This execution data is often stored in an event log. In many cases, the inner workings of the process are not known to management, or only vaguely resemble their three-year old PowerPoint design. Gaining insights into the process as it is executed using only an event log is the aim of process mining. In this work, we address three aspects of process mining: process discovery, conformance checking and enhancement.

Process Discovery.

Given an event log, process discovery is often the first step of a process mining project. Process discovery aims to automatically discover a process model from the event log, such that this model describes the underlying (unknown) process well. As we do not want to assume that all possible behaviour is included in the event log, process discovery algorithms inherently make a trade-off between several quality criteria. For instance, fitness denotes the fraction of the event log is described in the model, and log precision denotes the fraction of the model is described in the event log. Simplicity denotes whether the model needs few and simple constructs to express its behaviour. For some event logs, process models that score well on fitness, log precision and simplicity might not exist in the representational bias of the algorithm. Other quality criteria of algorithms include how well a discovered model resembles the running process that underlies the event log. Even though the process is typically unknown, studying under which conditions a discovery algorithm rediscovers the process allows us to compare discovery algorithms.

In this work, we argue that process discovery algorithms should provide several guarantees, such as that the discovered models are free of deadlocks and other anomalies, that fitness is perfect, and that the processes underlying the event logs are rediscovered. We conducted a systematic study of abstractions of logs and models to determine which classes of models can be rediscovered, such that the language of a discovered model is equivalent to the language of the process underlying the event log. We proved this property for several abstractions, such as the directly-follows relation (which activity follows which directly), the minimum self-distance relation (which activities may be executed between two as-close-as-possible-occurrences of another activity), the activity relations (which characteristic determines the relation between two activities), and the co-occurrence relation.

These relations were used in several new algorithms that we propose: a basic algorithm, an algorithm that handles deviating and infrequent behaviour, and another that handles incompleteness. Deviating behaviour should not be possible in the process but appears in the event log, while infrequent behaviour denotes little-used parts of the process. Such behaviour needs to be filtered to avoid complex models. An event log that does not fully witness a language abstraction is incomplete, and this challenges discovery algorithms. However, we show that in some cases guarantees can still be given.

We introduce a process discovery framework that provides several guarantees, such as deadlock freedom. Using the framework, we introduce a family of discovery algorithms to handle the mentioned challenges, as well as non-atomic event logs (i.e. in which activity executions take time), large event logs and complex event logs. We evaluated the algorithms and found that they perform well on real-life event logs, and that they are robust to logs with deviations, logs with little-used parts and incomplete logs.

Conformance Checking.

Due to the trade-offs involved in process discovery, discovery algorithms might leave out certain behaviour from an event log, or include behaviour that was not recorded in the event log. Therefore, a discovered model should be evaluated, for instance using a conformance checking technique, before conclusions can be drawn on the absence or presence of behaviour. A conformance checking technique compares a model to either an event log or another model and provides information on their differences. Existing conformance checking techniques often take exponential time in the length of traces in the event log. We propose a conformance checking framework to address this by using the language abstractions, thereby reducing the problem size while keeping certain guarantees, for instance that perfect fitness is reported if and only if a log and a model are perfectly fitting.

Enhancement & Tool Support.

More insights can be gained from a discovered process model by projecting additional information on the model. In this work, we study four types of enhancements: frequency information, performance information, deviations of log and model, and animation. For instance, to measure time spent waiting for activities reliably, we showed that it is important that concurrency is taken into account, because if two activities are concurrent, their waiting time is independent of one another. Therefore, such measures should be based on a process model.

Finally, we introduce a software tool that combines the benefits of commercial and academic tools: given an event log, the Inductive visual Miner discovers a process model, applies a conformance checking technique and enhances the process model with the four types of information. By zooming in and out of the process and the event log by changing parameters of the techniques, users can iteratively explore the event log. The Inductive visual Miner combines the ease-of-use of commercial tools with the reliability and robustness of academic tools, and is being used in several process mining projects.

Contents

1 Introduction   **1**

1.1 Abstractions in Process Mining 6

1.2 Process Discovery 6

1.3 Conformance Checking 11

1.4 Enhancement & Tool Support 15

1.5 Contributions and Structure of this Book 17

References 19

2 Preliminaries   **23**

2.1 Multisets, Traces, Regular Expressions 24

2.2 Process Models 25

    2.2.1 Automata 25

    2.2.2 Petri Nets 27

    2.2.3 Yet Another Workflow Language 32

    2.2.4 Business Process Model and Notation 33

    2.2.5 Process Trees 33

2.3 Event Logs 40

    2.3.1 Atomic Event Logs 40

    2.3.2 Non-Atomic Event Logs 41

    2.3.3 Richer Logs 43

2.4 Directly Follows Relation 43

References 45

3 Process Mining   **49**

3.1 Different Use Cases, Different Process Mining Techniques 51

3.2 Formal Key Challenges of Process Mining 55

    3.2.1 Models with Precise Semantics 57

    3.2.2 System - Log - Model Relations 59

    3.2.3 Simplicity & Balancing Log Criteria 66

    3.2.4 An Ideal Technique (1) 69

3.3 Process Discovery 70

    3.3.1 Discovery Algorithms Guaranteeing Soundness 70

    3.3.2 Other Discovery Algorithms 73

    3.3.3 An Ideal Process Discovery Technique (2) 80

3.4   Conformance Checking 82
      3.4.1   Log Conformance Checking 82
      3.4.2   System Conformance Checking 87
      3.4.3   An Ideal Conformance Checking Technique (2) 90
3.5   Enhancement & Tool Support 91
      3.5.1   Enhancements 91
      3.5.2   Process Mining Tools 95
      3.5.3   Requirements for Tool Support Beyond Process Discovery
              and Conformance Checking 101
3.6   Our Approach 101
      3.6.1   A Process Discovery Framework 102
      3.6.2   A Conformance Checking Framework 104
      3.6.3   Enhancement & Tool Support 105
      3.6.4   Future Work 108
References 108

4   Recursive Process Discovery   119
4.1   Recursive Process Discovery 120
      4.1.1   An Example of Recursive Process Discovery 121
      4.1.2   The IM framework 122
      4.1.3   More Technical Examples 123
      4.1.4   Guarantees 127
4.2   Rediscoverability 128
      4.2.1   Rediscoverability using Abstractions 130
      4.2.2   Rediscoverability and the IM framework 133
References 135

5   Abstractions   137
5.1   A Canonical Normal Form for Process Trees 139
      5.1.1   Reduction Rules 140
      5.1.2   Canonicity of the Reduction Rules 144
5.2   Language Uniqueness with Directly Follows Graphs 148
      5.2.1   A Class of Trees: C_B 148
      5.2.2   Footprints 152
      5.2.3   Language Uniqueness 154
5.3   Language Uniqueness with Activity Relations 158
      5.3.1   Activity Relations 159
      5.3.2   Binary Trees 160
      5.3.3   Language Uniqueness 161
5.4   Language Uniqueness with Interleaving 163
      5.4.1   Footprint 164
      5.4.2   A Class of Trees: C_I 164
      5.4.3   Language Uniqueness 166
5.5   Language Uniqueness with Minimum Self-Distance 169
      5.5.1   Minimum Self-Distance 169

      5.5.2   A Class of Trees: C_M   170
      5.5.3   Footprints   171
      5.5.4   LC-Property   173
      5.5.5   Language Uniqueness   174
  5.6   Language Uniqueness with Optionality & Inclusive Choice   177
      5.6.1   Optionality   177
      5.6.2   Optionality in the Directly Follows Graph   178
      5.6.3   A Class of Trees: C_{COO}   183
      5.6.4   Optionality under Sequence   185
      5.6.5   Optionality under Inclusive Choice & Concurrency   187
      5.6.6   Language Uniqueness   197
  5.7   Language Uniqueness with non-Atomic Process Models   201
      5.7.1   Non-Atomic Process Models   202
      5.7.2   Representational Bias of Non-Atomic Models   203
      5.7.3   Non-Atomic Directly Follows Graphs & Footprints   204
      5.7.4   Concurrency Graphs & Footprints   206
      5.7.5   A Class of Trees: C_{LC}   207
      5.7.6   Language Uniqueness   208
  5.8   Classes of Process Trees: Revisited   209
  References   213

6 Discovery Algorithms   215
  6.1   Inductive Miner (IM)   218
      6.1.1   Example   218
      6.1.2   Inductive Miner (IM)   220
      6.1.3   Guarantees   233
  6.2   Handling Deviating & Infrequent Behaviour   237
      6.2.1   Deviating & Infrequent Behaviour   237
      6.2.2   Inductive Miner - infrequent (IM_F)   240
      6.2.3   Example   247
      6.2.4   Guarantees   248
  6.3   Handling Incomplete Behaviour   249
      6.3.1   Incomplete Behaviour   250
      6.3.2   Inductive Miner - incompleteness (IM_C)   251
      6.3.3   Example   259
      6.3.4   Guarantees   260
      6.3.5   Finding Cuts: Translation to SMT   266
  6.4   Handling More Constructs: τ, $\leftrightarrow$ and $\vee$ 268
      6.4.1   Example   268
      6.4.2   Inductive Miner - all operators (IM_A)   272
      6.4.3   Inductive Miner - infrequent - all operators (IM_{FA})   279
      6.4.4   Guarantees   281
  6.5   Handling Non-Atomic Event Logs   286
      6.5.1   Non-Atomic Event Logs   286
      6.5.2   Inductive Miner - life cycle (IM_{LC})   288

      6.5.3   Inductive Miner - infrequent - life cycle (IM_{FLC}) & Inductive
             Miner - incompleteness - life cycle (IM_{CLC}) 294
      6.5.4   Implementation 297
      6.5.5   Guarantees 298
  6.6   Handling Large Event Logs 301
      6.6.1   Example 301
      6.6.2   Inductive Miner - directly follows based framework (IM_{D}
             framework) 304
      6.6.3   Inductive Miner - directly follows (IM_{D}) 304
      6.6.4   Inductive Miner - infrequent - directly follows (IM_{FD}) . . . 310
      6.6.5   Inductive Miner - incompleteness - directly follows (IM_{CD}) . 312
      6.6.6   Guarantees 314
  6.7   Tool Support 318
  6.8   Summary: Choosing a Miner 322
  References 324

7 Conformance Checking   **327**
  7.1   Projected Conformance Checking Framework 329
      7.1.1   Log to Projected Log to DFA 331
      7.1.2   Model to Projected Model to DFA 331
      7.1.3   Comparing DFAs & Measuring 334
      7.1.4   Measuring over All Activities 339
  7.2   An Example of Non-Conformance and Diagnostic Information . . . 340
  7.3   Guarantees 346
  7.4   Tool Support 347
  7.5   Conclusion 349
  7.6   Ideas to Handle Unbounded & Weakly Unsound Petri Nets 352
  References 354

8 Evaluation   **355**
  8.1   Evaluated Process Discovery Algorithms 357
  8.2   Scalability of Discovery Algorithms 358
      8.2.1   Set-up 359
      8.2.2   Results 360
      8.2.3   Discussion 360
  8.3   Log-Quality Dimensions 363
      8.3.1   Event Logs 363
      8.3.2   Quantitative 364
      8.3.3   Qualitative 371
      8.3.4   Conclusion 391
  8.4   Rediscoverability & its Challenges 392
      8.4.1   Incomplete Behaviour 393
      8.4.2   Deviating & Infrequent Behaviour 398
  8.5   Evaluation of Log-Conformance Checking 402
      8.5.1   Set-up 402

	8.5.2	Results	403
	8.5.3	Discussion	403
	8.5.4	Evaluation Using the PCC framework	407
8.6	Non-Atomic Behaviour	408
	8.6.1	Artificial Log	408
	8.6.2	Real-Life Log	410
8.7	Conclusion	418
	References	419

9 Enhancement & Inductive visual Miner   **423**
9.1	Inductive visual Miner (IvM)	425
	9.1.1	Steps & Architecture	426
	9.1.2	Model Visualisation	427
	9.1.3	Controls & Parameters	428
	9.1.4	Adding Extensions	435
9.2	Deviations	436
	9.2.1	Deviations and the PCC framework	436
	9.2.2	Deviations and Alignments	436
9.3	Frequency Information	438
9.4	Projecting Performance Information on Process Trees	439
9.5	Animation	442
9.6	Conclusion	444
	References	446

10 Conclusion   **449**
10.1	Process Discovery	450
10.2	Conformance Checking	453
10.3	Enhancement & Tool Support	454
10.4	Remaining Challenges	454
	10.4.1	Detailed	455
	10.4.2	Future Work	456
	References	458

Index   **461**

Chapter 1
Introduction

Nowadays, considerable amounts of data are recorded by software, machines and organisations. For instance, high tech systems such as MRI-scanners log hardware and software events, web servers log visits to web pages, ERP systems log business transactions, and workflow (case management) systems log activity executions. However, it might be unknown how an MRI-scanner is running, how the website is used, whether business transactions are processed according to the intended process, i.e. how machines, websites, etc. are used in reality. Therefore, the field of *process mining* [1] aims to extract information from these recorded event logs, such that problems can be identified based on facts, and consequently the processes, machines, web systems, etc. (to which we will refer as *systems*) can be improved.

From these event logs, process mining aims to extract information, for instance business process models (a representation of the order of steps taken in a process), performance information (e.g. queueing points in the process), compliance with rules and regulations (e.g. in which cases these rules are not adhered to), and social networks (e.g. the identification of key people in an organisation) [1], in order to enable stakeholders to gain a better understanding of the underlying system. The starting point for process mining is an *event log*. Typically, such an event log consists of *traces*, each of which consists of records of all steps (*activities*) in an end-to-end system execution, such as a product being manufactured by a machine, a visitor navigating through the website and making a purchase, or an insurance claim being submitted and handled. That is, a trace consists of *events*, each of which describes a step in the process such as a machine being switched on, a visitor requesting a web page or a filled-in form being submitted.

Three Process Mining Challenges.

In this book, we focus on the three main challenges in process mining: *process discovery*, *conformance checking* and *enhancement* [1]. Figure 1.1 illustrates these challenges in their context. The blue-filled region denotes a typical process mining project: an unknown *system* is executing, and from the observable behaviour of the system, an event log is recorded. Process mining can be applied when the precise inner workings of the system are unknown and subject of study. Therefore, the first step performed in a process mining project is typically *process discovery*, which aims to automatically discover a process model, e.g. a Petri net [30] or a BPMN [1] model that describes the inner workings of the system, from an event log. Because process discovery is unsupervised learning, we have to evaluate how well the obtained model corresponds to reality. Thus, as a second step, to guarantee the correctness of conclusions drawn from process models, the model should be evaluated using a *log-conformance checking* technique, which compares the model with the event log. Comparing a model to a log requires the model to have valid semantics, thus the process discovery technique needs to guarantee that the discovered model has clear semantics and is free of deadlocks and other anomalies. Further insights can be derived from performance, deviations and resource information (e.g. utilisation), which is projected onto the event log and a process model by *enhancement* tech-

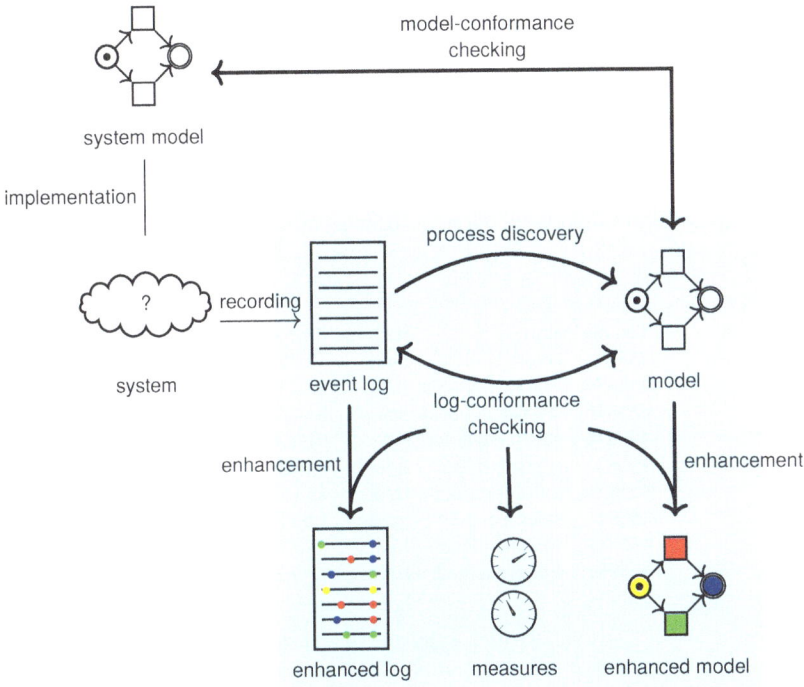

Fig. 1.1: The context of process discovery, log-conformance checking, model-conformance checking and enhancement. The blue-filled region denotes the scope of a typical process mining project.

niques. To do this, enhancement techniques combine the result of log-conformance checking with the event log and the process model, e.g. by indicating frequent and infrequently used steps in the model.

In a process mining project, a model of the system is typically not available. However, in some cases such a *system model* is available, and differences between the system model and its implementation (i.e. the system) can be studied using a *model-conformance checking* technique. Furthermore, in lab settings, such as in our evaluation, we use model-conformance checking to assess the models discovered by process discovery techniques.

First, we give an overview of the three mentioned challenges, then we explain the central theme of this book, after which we review open problems of these challenges in more detail.

Process Discovery.   ·

Process discovery aims to automatically obtain a process model, e.g. a Petri net [30] or a BPMN [1] model, from an event log (in Section 2.2, we introduce these process modelling formalisms). Such a process model describes the order of steps that can be taken for each individual customer and may provide valuable insights, as a model discovered from an event log shows what is actually happening in the system, instead of what management thinks what happens, or what employees or customers say what happens [3, 19].

customer	activity	resource	time stamp
455876	enter claim	Suzie	17-11-2016 10:45
455876	validate request	John	18-11-2016 13:50
455876	secondary check	Stan	01-12-2016 16:55
455876	notify requester	HAL	01-12-2016 16:55
455876	approve refund	John	14-03-2017 11:22
455876	transfer refund	HAL	15-03-2017 00:00
455931	enter request	Famke	12-12-2016 13:00
455931	validate request	John	13-12-2016 15:38
455931	transfer refund	HAL	13-12-2016 15:38
. . .			

(a) Excerpt of an artificial event log.

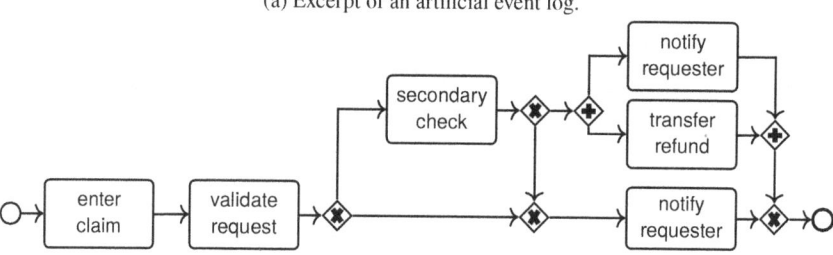

(b) A BPMN model that could be discovered from the event log.

Fig. 1.2: Example of an event log and a discovered process model. The log contains events for two customers (455876) and (455931). The model describes the process for these two customers.

For instance, Figure 1.2a shows an example event log of a fictitious insurance claim process: each line denotes an event, and consists of a customer number, the executed process step (activity), the resource executing the activity and the time at which the event happened. All events that were executed for a particular customer form a trace. Figure 1.2b shows a model that could be discovered from this event log. This BPMN model describes the activities to be performed for each customer. We discuss further challenges of process discovery and the limitations of existing approaches in Section 1.2.

Conformance Checking.

Finding deviations between an event log and a model is the aim of *log-conformance checking* techniques [24]. Log-conformance checking techniques take an event log and a process model as input and return diagnostic information as output (see Figure 1.1). These conformance checking techniques can be used to evaluate discovered process models (notice that the model should be machine-readable), e.g. assess the balance between including and excluding behaviour that the discovery algorithm chose. However, log-conformance checking techniques are not limited to using discovered process models: logs can also be compared to reference implementations (i.e. *system model*s) or specially designed models to verify compliance with rules and regulations (given a model that describes the rule or regulation, a log-conformance checking technique will diagnose and report violations) [28]. In our example (Figure 1.2), the skipping of the secondary check could be reported by log-conformance checking techniques and if company policy requires all transfers of refunds to have completed a secondary check, analysts could investigate the matter further.

Where a log-conformance checking technique compares an event log to a process model, a *model-conformance checking* technique compares two process models. For instance, in case a reference implementation or specification (a *system model*) is available, a model-conformance checking technique can be used to verify that the system implements the system model, and to study the differences between design and reality in the form of a discovered process model. Furthermore, if the system changes over time, i.e. the system exhibits concept drift [2], a model-conformance checking technique could be used to diagnose the differences between the process models discovered on different periods of event data in the event log. Similar strategies can be applied to compare different instances of the same process, for instance different variants of the same process in different organisations or geographical areas [10, 15]. We discuss challenges and limitations of existing log- and model-conformance checking techniques in Section 1.3.

Enhancement & Tool Support.

The third challenge that we address in this book is *enhancement*. That is, additional information of the event log is projected on the model and event log to provide more insight into e.g. the performance, frequencies or social aspects of the system. The discovered process model is enhanced with extra information of the event log, such as time stamps, resource information or other data [23]. For instance, Figure 1.2a shows an event log with time stamps.

We introduce a software tool that automates quick exploration of event logs by applying process discovery, conformance checking and enhancement techniques, to combine strong points of commercial products (e.g. ease-of-use, practical applicability) with strong points of academic software (e.g. reliability of results, ability to evaluate). We discuss challenges of enhancement and our approach in Section 1.4.

1.1 Abstractions in Process Mining

As event logs are finite and thus often incomplete observations of system behaviour, process mining techniques have to compensate for this missing information.
Usually, to handle this incompleteness, process discovery and conformance checking techniques use abstractions of the behaviour of event logs and process models:
process discovery techniques use behavioural abstractions such as directly follows
graphs [1, p.130][33] to avoid having to make the assumption that all behaviour of
the system is in the event log (and thereby disabling itself to include non-observed
behaviour). Conformance checking techniques use abstractions such as directly follows graphs [31], features [25], graphs [26], causal footprints [13], weak order
relations [35] and behavioural profiles [18, 32] to avoid having to consider all behaviour of process models, which might be infeasible for larger models. Currently,
there is not a good understanding of either how these abstractions influence process discovery or conformance checking, nor whether their influence is positive or
negative.

In this book, we introduce systematic approaches to process discovery and conformance checking that can cover various forms of inputs/use cases, are extensible for
new operators, and allow to prove and derive quality guarantees systematically. Furthermore, we perform a systematic investigation of abstractions in process discovery
and conformance checking and we introduce a tool that makes process discovery,
conformance checking and enhancements easily accessible to analysts.

In the remainder of this chapter, we discuss the challenges of process discovery,
conformance checking and enhancement in more detail, as well as our approach to
these challenges. Furthermore, we introduce the guarantees that we seek to provide
and the abstractions we seek to use to achieve this. Finally, we summarise the
contributions and discuss the structure of this book.

1.2 Process Discovery

Process discovery techniques face several challenges. On the one hand, techniques
should be robust in handling real-life event logs and should provide guarantees. For
instance, techniques should guarantee to return models with clear semantics that free
of deadlocks and other anomalies, and should guarantee to rediscover the system.
On the other hand, techniques face trade-offs between the event log and the system
and between including and excluding behaviour, and these trade-offs might depend
on the use case at hand. In this section, we introduce challenges that we identified
from existing techniques and we introduce our approach.

Sound Semantics.

A first challenge is that machines and software have difficulties interpreting models that do not have well-defined semantics, thus to such models conformance checking techniques cannot be reliably applied. Such models can neither be reliably interpreted nor evaluated and therefore conclusions drawn from them cannot be guaranteed to be correct. Furthermore, the discovered model should be *sound*, i.e. free of deadlocks, of unclear behaviour and of other anomalies. Even though an unsound model might be useful for human interpretation, one should be careful with drawing conclusions from such models, as these models might introduce ambiguity. Furthermore, we consider a system by its behaviour, and each system can be described by a sound model that describes this behaviour[1], thus it should not be represented by models having such soundness issues. Therefore, all process discovery algorithms should guarantee to return sound models at all times. Unfortunately, many existing discovery techniques, e.g. [1, 5, 8, 33, 34], do not guarantee to return sound models.

Rediscover the System.

Second, as the main subject of study is the unknown system, ideally, the discovered model expresses the same set of traces as the system. If a discovery algorithm is able to discover such a model, the algorithm possesses *rediscoverability* [6, 20, 22]. Rediscoverability is a formal property that is usually proven using assumptions on the system and the event log, but provides a baseline to compare discovery algorithms. Furthermore, it establishes confidence in discovery algorithms: they are not just working on a best-effort basis, but actually guarantee to return a model similar to the system under the right (known) conditions [20].

Ensure Fitness and Precision.

Third, process discovery algorithms operate in a force field of goals, related to both the event log and the system underlying the event log. We explain this force field using an example use case: the assessment whether a system adheres to certain rules and regulations. This assessment can be performed using two objectives: (1) the rules have not been violated, and (2) the system disallows violations in the future. For (1), the discovered model should represent the event log well, while for (2), the discovered model should represent the system well [1, Section 10.1][29].

To assess whether rules have been violated, the discovered model should represent the event log well. That is, if enough behaviour of the event log is included in the discovered model (the model has a high *fitness* [1, Section 6.4]), conclusions can be drawn about the absence of behaviour, including violations. In fact, a few existing discovery algorithms guarantee to return models with perfect fitness, e.g. [34]. The

---

[1] All systems can be described by for instance a Petri net with inhibitor arcs, as these are Turing complete [27].

model shown in Figure 1.2b does not have a perfect fitness with respect to the log in Figure 1.2a, as for customer 455931 a refund is transferred while no secondary check took place.

However, in order to draw conclusions about the absence of behaviour, the discovered model should not allow for too much more behaviour than the behaviour recorded in the event log. Therefore, the discovered model should describe little more behaviour than the event log (the model should have a high *log precision* [1, Section 4.6]). The model shown in Figure 1.2b does not have a perfect log precision with respect to the log in Figure 1.2a, as there is no customer for which a transfer is made before the customer is notified.

If the goal of the analysis is to describe the future behaviour, e.g. to ensure that deviations cannot occur, then the discovery algorithm needs to discover a model that resembles the system, instead of the event log. That is, algorithms try to discover models that include most of the behaviour of the system (the model should have a high *recall* [1, Section 4.6]). However, not too much behaviour unrelated to the system should be included in the model (the model should have a high *system precision* [1, Section 4.6]).

Thus, discovery techniques typically aim to discover models with as high fitness, log precision, recall and system precision as possible. Ideally, techniques guarantee perfection in one or more of these measures. However, in Chapter 3, we will show that optimising for these four concepts often involves trade-offs, i.e. there might be event logs and systems for which no discovered model with a high fitness, log precision, recall and system precision exists [11].

Exclude Abnormal Behaviour.

Fourth, not all use cases might require strict optimisation on the aforementioned concepts. That is, if the aim of the analysis is to analyse the majority of behaviour or the "happy flow" of the system, it makes little sense to enforce a fitness guarantee if this results in an unreadable and incomprehensible model. Furthermore, for such use cases, typically only the most occurring behaviour should be included in the model. For instance, if an insurance company suddenly receives an abnormally high number of claims due to a severe storm, parts of the normal procedure could be temporarily disabled to speed up service to customers. If the goal of the process mining project is to explore and gain a better understanding of the process, it could be beneficial to exclude such abnormal behaviour and not include it in the discovered model.

Include Missing Behaviour.

Fifth, as process discovery techniques aim to provide new information to analysts, they should not simply represent the event log, as that would not induce any new information (in contrast to Petri net synthesis techniques [7]). That is, process discovery techniques deliberately do not assume that all possible behaviour of the system is present in the event log. Such an assumption could pose infeasibly strong requirements on event logs: a behaviourally complete event log of a system that consists of 10 concurrent activities has 3,628,800 possibilities of execution, and the probability that a reasonably sized event log of such a real-life system would contain every possibility at least once is negligible. Thus, discovery algorithms need to generalise over the behaviour in the event log to deduce the behaviour of the system. Many discovery algorithms approach this by using an abstraction of the event log instead of the event log itself, such as a directly follows graph. In Chapter 5, we conduct a systematic study into several of these abstractions, and the implications the abstractions have on discovering the behaviour of the system.

Thus, discovery algorithms should be robust to too little and too much behaviour in the event log by making trade-offs to include or exclude behaviour.

Handle More Types of Event Logs.

Finally, several non-standard types of event logs can be distinguished; process discovery techniques might benefit from considering these types. For instance, in our evaluation, we will show that current discovery techniques are well able to handle event logs with up to a million events and 100 activities when given limited RAM. However, much larger event logs could be extracted, for instance from the detector control software of the Large Hadron Collider, in which over 25,000 independent control systems collaborate to control e.g. power and airconditioning, resulting in *very* complicated behaviour [17]. Obviously, any representable event log would contain much more than a millon events and discovery techniques need to be adapted to handle such complexity.

In the lion's share of existing process discovery techniques, it is assumed that activity executions are instantaneous (*atomic*). However, in some event logs, the duration of activity executions takes time, i.e. are *non-atomic*. Process discovery techniques should be aware of non-atomic event logs in order to benefit from the extra information it provides (we will show this in Section 5.7).

Our Approach.

Balancing rediscoverability, fitness, precision, excluding abnormal behaviour and including missing behaviour depends on the goal of process mining in a particular situation. In Chapter 3 we will argue that a single process discovery algorithm that always achieves the perfect trade-off cannot exist. Therefore, in this book, we present a family of process discovery techniques. That is, we introduce a framework, the *Inductive Miner framework* (IM framework), that constructs process models. To guarantee soundness, the IM framework limits itself to recursive process models, i.e. process trees, which will be described in Chapter 2. As process trees are sound by definition, all algorithms that implement the IM framework guarantee soundness. The IM framework discovers process models recursively, starting with the identification of the most important behaviour in an event log, splitting the event log into smaller sublogs and recursing until a base case is found. A process discovery technique can fully implement the IM framework by providing parameter functions for each of these steps. Furthermore, the IM framework aids algorithms in providing guarantees such as termination, perfect fitness (i.e. all behaviour of the event log is in the discovered model), perfect log precision (i.e. all behaviour of the discovered model is in the event log) and rediscoverability.

To aid in rediscoverability proofs, we provide a general proof framework for rediscoverability. This proof framework expresses rediscoverability in terms of abstractions, such that it aids abstraction-based algorithms. Furthermore, we linked the proof framework to the IM framework by expressing proof obligations in terms of the parameter functions of the IM framework. Consequently, we use the proof framework to prove rediscoverability for all algorithms introduced in this book.

As described before, many discovery algorithms use abstractions in order to avoid the assumption that all behaviour of the system is present in the event log. However, this implies that models with different behaviour but equivalent abstractions exist, and thus that discovery algorithms are insensitive to some behaviour. We perform a systematic study to these abstractions and their influence on rediscoverability, to better understand the capabilities and limitations of the algorithms introduced in this book and existing algorithms. That is, we describe the abstractions, and explore the boundaries of the systems that can be uniquely identified by the abstraction, by defining classes of systems such that provably no two systems of the class have the same abstraction.

Using the results of the abstractions study, we introduce several discovery algorithms that implement the IM framework, as illustrated in Table 1.1. Using the IM framework and the proof framework, we prove that all of these algorithms guarantee soundness, termination, and rediscoverability. These algorithms illustrate the flexibility of the IM framework: for each algorithm, large parts of earlier algorithms are reused, and nevertheless algorithms with different focus and strategies emerge. That is, we introduce algorithms to handle event logs with deviations, to handle logs with little-used parts and to handle logs in which the abstractions are not fully covered, i.e. incomplete logs. The flexibility of the IM framework is exploited further by the introduction of algorithms that handle different types of event logs, i.e. logs in which events take time (*non-atomic event logs*). Furthermore, we adapt the IM framework slightly to handle large event logs, i.e. with tens of millions of events and thousands of activities, by introducing the IMd framework and corresponding algorithms.

The IM framework and the discovery algorithm that use it have been implemented in the ProM framework [14]. We argue that these algorithms are robust: they return a sound model at all times, they offer several guarantees and, as shown in our evaluation, they handle logs with deviations, logs with little-used parts and incomplete logs, they perform well on real-life event logs and they are scalable.

---

[2] Future work.

[3] We chose not to guarantee fitness for IMd (see Section 6.6.6).

Table 1.1: The family of discovery algorithms, and their guarantees and purposes. Due to the frameworks, all algorithms guarantee soundness, termination and rediscoverability. The algorithms will be introduced in Chapter 6.

use cases	framework	fitness guaranteed	infrequent & deviating behaviour	incomplete behaviour
	IM framework	IM	IM_F	IM_C
discover more behaviour	IM framework	IM_A	IM_{FA}	-[2]
handle non-atomic event logs	IM framework	IM_{LC}	IM_{FLC}	IM_{CLC}
handle larger logs	IM_D framework	IM_D[3]	IM_{FD}	IM_{CD}

For instance, Figure 1.3 shows the results of an existing discovery algorithm (α [1, p.130]) and an algorithm introduced in this book (IM_{FA}, see Section 6.4.3), on the same real-life event log of a mortgage application process of a financial institution [12]. Both models are shown here as Petri nets, which will be introduced in Section 2.2.2. The model discovered by α contains unconnected (i.e. unrestricted) activities, does not contain a clear end state and therefore, the set of traces that this model represents is unclear. Little information can be derived from this model. In contrast, the model returned by IM_{FA} is structured and sound, thus contains no unconstrained activities, deadlocks or other anomalies, and the set of traces that this model represents is clear, which makes it suitable for further analysis. As different use cases might require different discovery techniques, we present a family of discovery techniques, all of which return models that are guaranteed sound.

1.3 Conformance Checking

As process discovery algorithms introduce absent and exclude present behaviour of the event log, an essential step after discovering a model is to evaluate this model. A *conformance checking* technique can be used to perform this evaluation. We consider two types of conformance checking in this book: log-conformance checking and model-conformance checking. A *log-conformance checking* technique compares a process model and an event log, and advises on their differences. If a particular part of the model and the event log deviate strongly, then the model might not represent that part of the system well, conclusions about that part might not be valid for the underlying system, and these conclusions should be drawn with care. A *model-conformance checking* technique compares two process models with one another. Even though in typical process mining projects the system is unknown, a reference model from which the system was implemented might be available (the *system model* in Figure 1.1). Furthermore, process models based on different subsets of the event log can be compared. Event data from e.g. different periods or

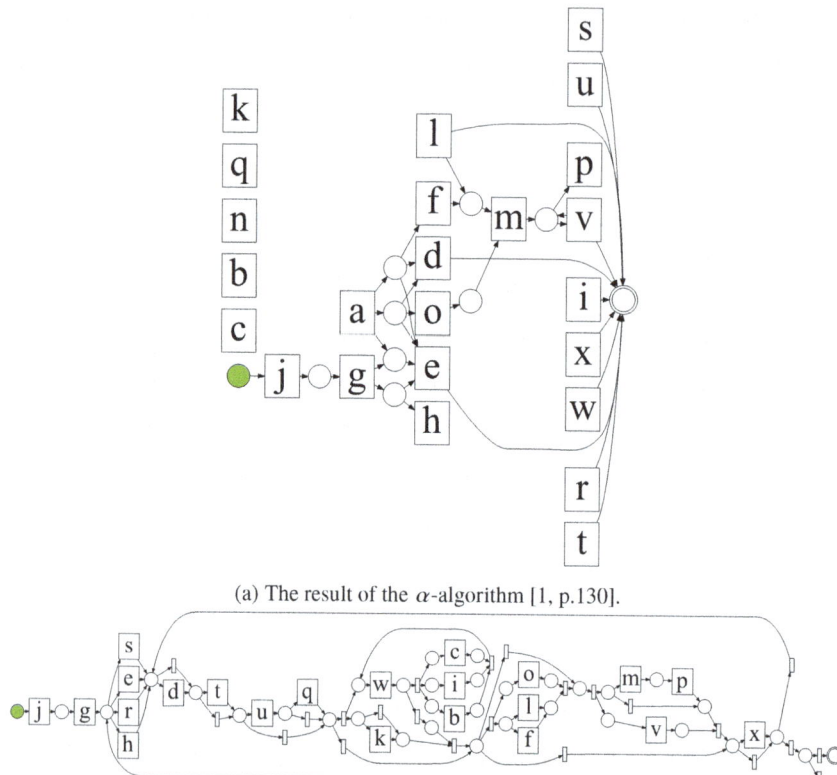

(a) The result of the α-algorithm [1, p.130].

(b) The result of an algorithm introduced in this book (IM$_{FA}$).

Fig. 1.3: Two discovery techniques applied to a real-life event log [12]. The activity names have been replaced with letters.

geographic regions may be used to construct multiple models of the same system and subsequently, the systems of these periods or geographic regions may be compared using model-conformance checking.

Typically, conformance checking techniques express the 'amount' of behaviour that two models or a log and a model have in common, what 'part' of behaviour in the log/system model is represented in the discovered model, or vice versa. Two major challenges of conformance checking techniques are to (1) quantify this amount or part, as models might contain unbounded behaviour, and (2) avoid the state-space explosion problem, as models might contain much or unbounded behaviour. To solve both challenges, also conformance checking techniques often use an abstraction of behaviour, such as the directly follows graph or other behavioural relations [32, 13, 35, 18] (as described earlier in this chapter). However, a downside, shared with process discovery techniques that use abstractions, applies: the measures become insensitive to certain differences in behaviour: if two types of behaviour have the same abstraction, the conformance checking technique cannot distinguish them. Techniques that do not use such abstractions, such as [4], tend to have issues dealing with large event logs, as we will show in our evaluation in Chapter 8.

In this book, we aim to avoid the state-space explosion problem by using an abstraction, while avoiding the insensitivity by choosing an abstraction that is sensitive to a large class of models.

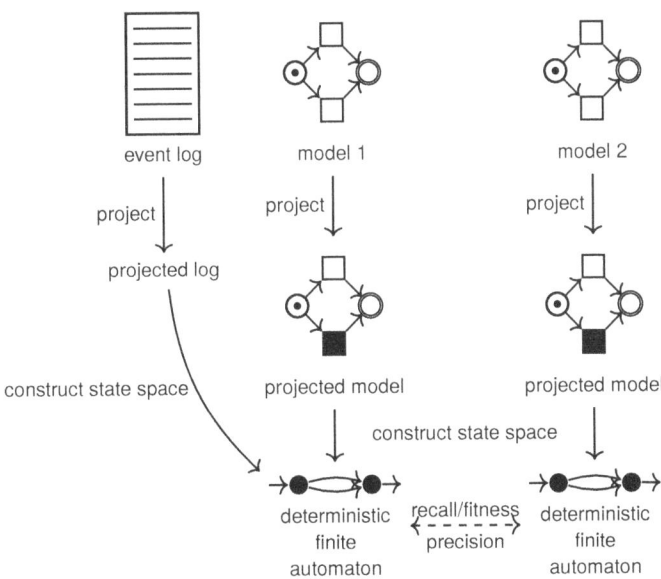

Fig. 1.4: The PCC framework.

Our Approach.

In this book, we introduce the Projected Conformance Checking framework (PCC framework), which supports both log- and model-conformance checking and aims to support large models and event logs, i.e. with over a hundred thousand events and hundreds of activities, which existing techniques cannot handle. The PCC framework combines ideas from existing techniques: it uses automata as abstraction to capture all possible behaviour. That is, model, log and system model are played out [1] and all their behaviour is recorded in automata. However, to improve scalability, the PCC framework considers subsets of activities. That is, for each subset of a fixed size (e.g., all pairs or triplets of activities), the event log, system and/or model are projected onto the activities of the subset, and all projected behaviour is compared to compute fitness, recall, log precision and/or system precision. Figure 1.4 shows the approach of the PCC framework (we will discuss its details in Chapter 7). These measures on subsets of activities provide information on two levels: as summative measures (when averaged over all subsets) and on the parts of the model that deviate from the event log (when averaged over activities, see for instance Figure 1.5). Using the result of our study of abstractions in Chapter 5, we show classes of models for which the PCC framework can reliably decide language equivalence.

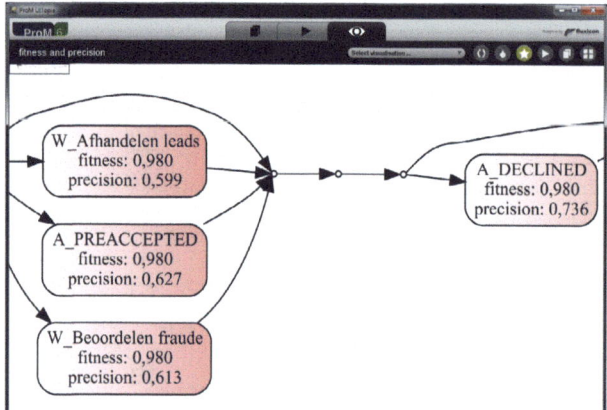

Fig. 1.5: A screenshot of the results of the PCC framework, projected on a process model.

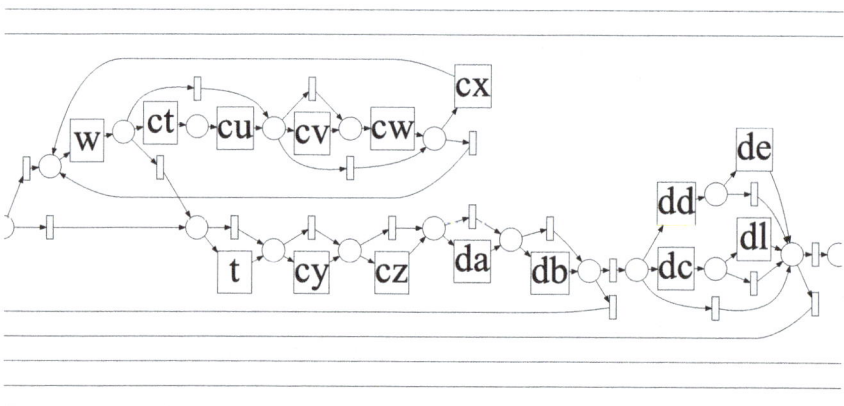

Fig. 1.6: Excerpt of a real-life event log [9]. The complete model contains 151 activities; the activity names have been replaced with letters.

We compare the PCC framework to existing techniques, and find that the PCC framework can handle real-life event logs and models discovered from these event logs that contain hundreds of activities, which the existing techniques cannot handle.

Furthermore, the experiments suggest that the PCC framework needs less computation time. For instance, on the model of which an excerpt is shown in Figure 1.6, the
PCC framework computed fitness and log-precision in less than a second, while the
existing approach [4] could not compute an answer. The PCC framework handled
all real-life event logs of the experiment, whereas current state-of-the-art techniques
could not handle logs with more than 100 different activities.

1.4 Enhancement & Tool Support

Commercially available process mining tools offer many enhancements, and these
enhancements give analysts many more insights into the process than plain process
models. However, enhancements projected on unsound models or on models without
clear semantics can be unreliable. Therefore, we considered these enhancements
offered by commercially available process mining tools, and selected the enhancements that benefit from sound models with clear semantics. For these enhancements,
we describe universally applicable techniques, challenges and concepts. Furthermore, we develop a process mining tool that makes process discovery, conformance
checking and the identified enhancements available in an easy-to-use package. In
this section, we first describe the four enhancements we consider in this book, after
which we describe the process mining tool.

First, Figure 1.7a shows a model enhanced with frequency information. That is,
the activities (i.e. the boxes) are annotated with the number of times the activities
were executed and the edges between the activities are annotated with how often
the edge was used in the routing of cases through the process model. Second, Figure 1.7b shows a model enhanced with performance information. That is, activities
are annotated and coloured with the duration of the activities. Third, the availability
of conformance checking results allows for the visualisation of deviations on the
model. For instance, in Figure 1.7c, it is visualised that in 57 + 4749 cases, an event
happened that was not described by the model. The location of these deviations in
the model is denoted by red-dashed edges. Finally, Figure 1.7d shows a still from
a model enhanced with animation. In this animation, the tokens, which represent
traces, flow over the model. Whenever an event was executed in the trace, a token
flows over the activity that belongs to that event. Tokens of all traces combined
provide insights into e.g. bottlenecks, changes in the process and seasonal effects.
Animation and deviations can only be computed and visualised if the model is sound
and has clear semantics, which illustrates the need for process discovery techniques
that guarantee such models and robust conformance checking techniques that can
handle these models.

Insights gained from enhancement techniques may often lead to new research
questions, and thus explorative process mining projects are typically iterative. For
instance, a model is discovered, which is evaluated and enhanced, after which e.g.
the event log is filtered (to zoom in) on a particular part of the process. After
this, the analysis may be repeated [15]. To support this iterative process, easy-to-

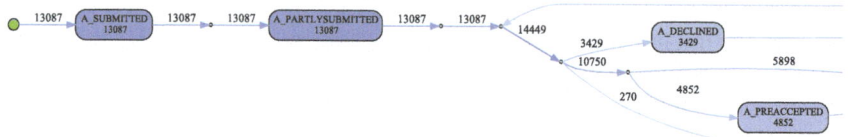

(a) Frequency: the numbers on the edges and in the activities denote how often that edge/activity was executed.

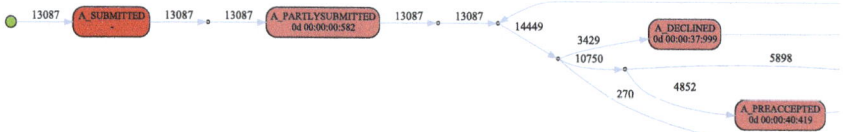

(b) Performance: the digits in the activities denote the average duration of that activity. For instance, A_DECLINED took on average 0 days, 37 seconds and 999 milliseconds.

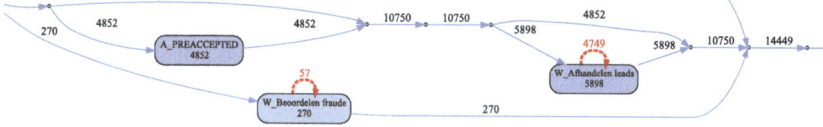

(c) Deviations: the red-dashed edges denote points in the model where the log and the model disagree.

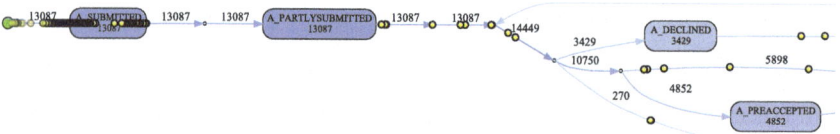

(d) Animation: the yellow dots flow over the model according to the model and indicate e.g. bottlenecks.

Fig. 1.7: Enhancements in Inductive visual Miner.

use software support is necessary [23]. Commercial tools such as Celonis Process Mining and Fluxicon Disco provide the ease-of-use, but do not provide the semantics and conformance checking, which is necessary to evaluate a discovered model and to validate the conclusions drawn from a model [23]. Academic tools provide semantics and conformance checking, however lack the ease-of-use and robustness (e.g. if a model like in Figure 1.3a is discovered, conformance checking might not give useful answers [21]) required for software support [23].

Therefore, we introduce a software tool, the Inductive visual Miner (IvM), which combines the process discovery techniques described in this book with existing conformance techniques (alignments) and the enhancements described in this book. That is, IvM takes an event log as input and discovers a process model using the process discovery techniques described in this book, aligns the event log and the model such that they agree (i.e. computes an *alignment*), and enhances the model and event log using performance, deviations, animation, and frequency information (the enhancements shown in Figure 3.28 were computed and visualised by IvM). All of these steps are performed automatically, and the user gets a result without further interaction necessary. Based on the given results, a user can influence each step or apply filters, after which IvM automatically recomputes all necessary steps. Due to the quick interaction and visualisation, IvM enables users to explore the process as it was recorded in the event log.

1.5 Contributions and Structure of this Book

In this book, we address the process mining challenges process discovery, conformance checking and enhancement. To summarise, this book contains the following contributions:

1. A *framework for process discovery algorithms* (the IM framework, Chapter 4). The IM framework guarantees soundness by its use of process trees, and aids algorithms in guaranteeing fitness, log-precision and rediscoverability. The framework enables discovery algorithms to focus on the *most important* behaviour in an event log, instead of on all behaviour, and enables the design of efficient algorithms that are robust to too little and too much behaviour in the event log.
2. A *systematic study of language abstractions* used in process discovery and conformance checking (Chapter 5). We perform a systematic study to these abstractions and their influence on rediscoverability to better understand the capabilities and limitations of the algorithms presented in this book and existing algorithms. For each abstraction, we study its expressive power, i.e. the class of models that can be represented by the abstraction such that no two models of the class with a different language have the same abstraction.
3. A *family of discovery algorithms* (see Table 1.1, Chapter 6). These algorithms implement the IM framework and therefore guarantee soundness. Each algorithm targets different types of event logs and addresses different challenges of process discovery. Furthermore, all algorithms guarantee rediscoverability and some guarantee perfect fitness. No existing (set of) algorithm(s) possesses this combination of properties.
4. A *conformance checking framework* (the PCC framework, Chapter 7). The PCC framework supports both log-conformance and model-conformance checking and is able to handle larger event logs and models than existing techniques, works faster and supports multiple process modelling formalisms. Furthermore, for certain classes of models, the PCC framework guarantees that the returned measures are perfect if and only if the two models are language equivalent.
5. A *discussion of enhancements* and a *process mining tool* (the IvM, Chapter 9). The IvM combines process discovery, conformance checking and enhancements in an easy-to-use package.

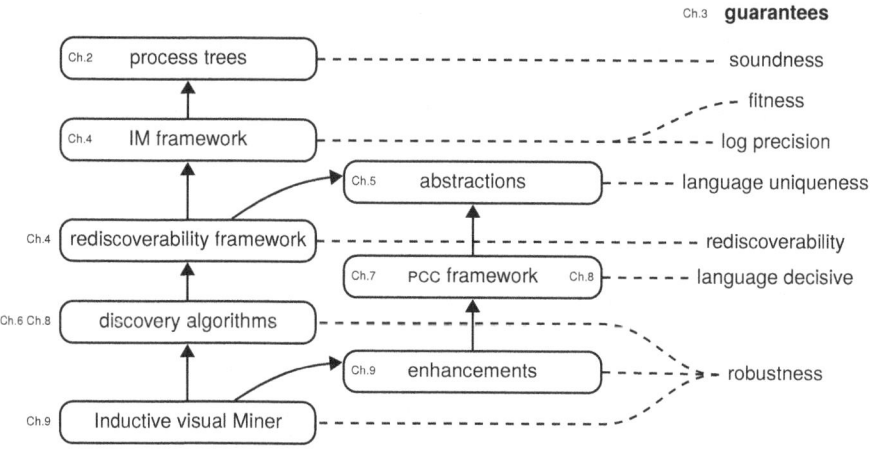

Fig. 1.8: Structure of this book.

The structure of the remainder of this book is shown in Figure 1.8. In this figure, the techniques, concepts and frameworks of this book are shown in boxes, as well as the dependency relations between them.

We first introduce some basic notation such as Petri nets, automata, languages, event logs (not in the figure) and process trees in Chapter 2. Furthermore, the figure shows the guarantees that are provided or enabled for each technique, concept or framework. We elicit guarantees and other requirements of process discovery, conformance checking and enhancement in more detail by considering existing techniques in Chapter 3. In Chapter 4, we study rediscoverability in more detail, introduce the formal rediscoverability proof framework and introduce the IM framework for process discovery.

We perform a systematic study towards the abstractions that are used in both process mining and conformance checking in Chapter 5. In Chapter 6, we introduce concrete discovery algorithms that use these abstractions, i.e. we introduce several algorithms that implement the IM framework, and prove guarantees such as fitness and rediscoverability for these algorithms. We describe the PCC framework, i.e. our conformance checking framework, which supports both log- and model-conformance checking, in Chapter 7. In Chapter 8, we evaluate both process discovery algorithms and conformance checking techniques (this is not denoted in the figure). For discovery algorithms, including the ones introduced in this book, we test their scalability, balancing of log criteria, and robustness to abnormal and missing behaviour. Furthermore, we compare the PCC framework to conformance checking techniques on scalability and their returned measures.

In Chapter 9, we discuss several challenges and solutions of model and event log enhancement. Finally, Chapter 10 concludes the book.

References

1. van der Aalst, W.M.P.: Process mining - discovery, conformance and enhancement of business processes. Springer (2011). DOI 10.1007/978-3-642-19345-3. URL http://dx.doi.org/10.1007/978-3-642-19345-3
2. van der Aalst, W.M.P.: Process Mining - Data Science in Action, Second Edition. Springer (2016)
3. van der Aalst, W.M.P., Adriansyah, A., de Medeiros, A.K.A., Arcieri, F., Baier, T., Blickle, T., Bose, R.P.J.C., van den Brand, P., Brandtjen, R., Buijs, J.C.A.M., Burattin, A., Carmona, J., Castellanos, M., Claes, J., Cook, J., Costantini, N., Curbera, F., Damiani, E., de Leoni, M., Delias, P., van Dongen, B.F., Dumas, M., Dustdar, S., Fahland, D., Ferreira, D.R., Gaaloul, W., van Geffen, F., Goel, S., Günther, C.W., Guzzo, A., Harmon, P., ter Hofstede, A.H.M., Hoogland, J., Ingvaldsen, J.E., Kato, K., Kuhn, R., Kumar, A., Rosa, M.L., Maggi, F.M., Malerba, D., Mans, R.S., Manuel, A., McCreesh, M., Mello, P., Mendling, J., Montali, M., Nezhad, H.R.M., zur Muehlen, M., Munoz-Gama, J., Pontieri, L., Ribeiro, J., Rozinat, A., Pérez, H.S., Pérez, R.S., Sepúlveda, M., Sinur, J., Soffer, P., Song, M., Sperduti, A., Stilo, G., Stoel, C., Swenson, K.D., Talamo, M., Tan, W., Turner, C., Vanthienen, J., Varvaressos, G., Verbeek, E., Verdonk, M., Vigo, R., Wang, J., Weber, B., Weidlich, M., Weijters, T., Wen, L., Westergaard, M., Wynn, M.T.: Process mining manifesto. In: Business Process Management Workshops - BPM 2011 International Workshops, Clermont-Ferrand, France, August 29, 2011, Revised Selected Papers, Part I, *Lecture Notes in Business Information Processing*, vol. 99, pp. 169–194. Springer (2011). DOI 10.1007/978-3-642-28108-2_19. URL http://dx.doi.org/10.1007/978-3-642-28108-2_19
4. Adriansyah, A.: Aligning Observed and Modeled Behavior. Ph.D. thesis, Eindhoven University of Technology (2014)
5. Augusto, A., Conforti, R., Dumas, M., Rosa, M.L., Bruno, G.: Automated discovery of structured process models: Discover structured vs. discover and structure. In: I. Comyn-Wattiau, K. Tanaka, I. Song, S. Yamamoto, M. Saeki (eds.) Conceptual Modeling - 35th International Conference, ER 2016, Gifu, Japan, November 14-17, 2016, Proceedings, *Lecture Notes in Computer Science*, vol. 9974, pp. 313–329 (2016). DOI 10.1007/978-3-319-46397-1_25. URL http://dx.doi.org/10.1007/978-3-319-46397-1_25
6. Badouel, E.: On the α-reconstructibility of workflow nets. In: S. Haddad, L. Pomello (eds.) Application and Theory of Petri Nets - 33rd International Conference, PETRI NETS 2012, Hamburg, Germany, June 25-29, 2012. Proceedings, *Lecture Notes in Computer Science*, vol. 7347, pp. 128–147. Springer (2012). DOI 10.1007/978-3-642-31131-4_8. URL http://dx.doi.org/10.1007/978-3-642-31131-4_8
7. Badouel, E., Bernardinello, L., Darondeau, P.: Petri Net Synthesis. Texts in Theoretical Computer Science. An EATCS Series. Springer (2015). DOI 10.1007/978-3-662-47967-4. URL http://dx.doi.org/10.1007/978-3-662-47967-4
8. vanden Broucke, S.K.L.M.: Advances in process mining: Artificial negative events and other techniques. Ph.D. thesis, KU Leuven (2014)
9. Buijs, J.C.A.M.: Flexible evolutionary algorithms for mining structured process models. Ph.D. thesis, Eindhoven University of Technology (2014)
10. Buijs, J.C.A.M., van Dongen, B.F., van der Aalst, W.M.P.: Towards cross-organizational process mining in collections of process models and their executions. In: F. Daniel, K. Barkaoui, S. Dustdar (eds.) Business Process Management Workshops - BPM 2011 International Workshops, Clermont-Ferrand, France, August 29, 2011, Revised Selected Papers, Part II, *Lecture Notes in Business Information Processing*, vol. 100, pp. 2–13. Springer (2011). DOI 10.1007/978-3-642-28115-0_2. URL http://dx.doi.org/10.1007/978-3-642-28115-0_2
11. Buijs, J.C.A.M., van Dongen, B.F., van der Aalst, W.M.P.: On the role of fitness, precision, generalization and simplicity in process discovery. In: R. Meersman, H. Panetto, T.S. Dillon, S. Rinderle-Ma, P. Dadam, X. Zhou, S. Pearson, A. Ferscha, S. Bergamaschi, I.F. Cruz (eds.) On the Move to Meaningful Internet Systems: OTM 2012, Confederated International Conferences: CoopIS, DOA-SVI, and ODBASE 2012, Rome, Italy, September 10-14, 2012. Proceedings, Part

I, *Lecture Notes in Computer Science*, vol. 7565, pp. 305–322. Springer (2012). DOI 10.1007/
978-3-642-33606-5_19. URL http://dx.doi.org/10.1007/978-3-642-33606-5_19

12. van Dongen, B.: BPI challenge 2012 dataset (2012).   DOI 10.4121/uuid:
3926db30-f712-4394-aebc-75976070e91f.   URL http://dx.doi.org/10.4121/uuid:
3926db30-f712-4394-aebc-75976070e91f

13. van Dongen, B.F., Dijkman, R.M., Mendling, J.: Measuring similarity between business process
models. In: J.A. Bubenko Jr., J. Krogstie, O. Pastor, B. Pernici, C. Rolland, A. Sølvberg (eds.)
Seminal Contributions to Information Systems Engineering, 25 Years of CAiSE, pp. 405–419.
Springer (2013).   DOI 10.1007/978-3-642-36926-1_33.   URL http://dx.doi.org/10.
1007/978-3-642-36926-1_33

14. van Dongen, B.F., de Medeiros, A.K.A., Verbeek, H.M.W., Weijters, A.J.M.M., van der Aalst,
W.M.P.: The prom framework: A new era in process mining tool support. In: G. Ciardo,
P. Darondeau (eds.) Applications and Theory of Petri Nets 2005, 26th International Conference,
ICATPN 2005, Miami, USA, June 20-25, 2005, Proceedings, *Lecture Notes in Computer
Science*, vol. 3536, pp. 444–454. Springer (2005). DOI 10.1007/11494744_25. URL http:
//dx.doi.org/10.1007/11494744_25

15. van Eck, M.L., Lu, X., Leemans, S.J.J., van der Aalst, W.M.P.: PM^2 : A process mining project methodology. In: J. Zdravkovic, M. Kirikova, P. Johannesson (eds.) Advanced Information Systems Engineering - 27th International Conference, CAiSE 2015,
Stockholm, Sweden, June 8-12, 2015, Proceedings, *Lecture Notes in Computer Science*,
vol. 9097, pp. 297–313. Springer (2015).   DOI 10.1007/978-3-319-19069-3_19.   URL
http://dx.doi.org/10.1007/978-3-319-19069-3_19

16. Fournier, F., Mendling, J. (eds.): Business Process Management Workshops - BPM 2014
International Workshops, Eindhoven, The Netherlands, September 7-8, 2014, Revised Papers,
Lecture Notes in Business Information Processing, vol. 202. Springer (2015). DOI 10.1007/
978-3-319-15895-2. URL http://dx.doi.org/10.1007/978-3-319-15895-2

17. Hwong, Y., Keiren, J.J.A., Kusters, V.J.J., Leemans, S.J.J., Willemse, T.A.C.: Formalising and
analysing the control software of the compact muon solenoid experiment at the large hadron
collider. Sci. Comput. Program. **78**(12), 2435–2452 (2013). DOI 10.1016/j.scico.2012.11.009.
URL http://dx.doi.org/10.1016/j.scico.2012.11.009

18. Kunze, M., Weidlich, M., Weske, M.: Querying process models by behavior inclusion. Software
and System Modeling **14**(3), 1105–1125 (2015). DOI 10.1007/s10270-013-0389-6. URL
http://dx.doi.org/10.1007/s10270-013-0389-6

19. Leemans, S.J.J.: Process discovery and exploration. In: Fournier and Mendling [16], pp.
582–585. DOI 10.1007/978-3-319-15895-2_52. URL http://dx.doi.org/10.1007/
978-3-319-15895-2_52

20. Leemans, S.J.J., Fahland, D., van der Aalst, W.M.P.: Discovering block-structured process
models from event logs - a constructive approach. In: J.M. Colom, J. Desel (eds.) Application
and Theory of Petri Nets and Concurrency - 34th International Conference, PETRI NETS
2013, Milan, Italy, June 24-28, 2013. Proceedings, *Lecture Notes in Computer Science*, vol.
7927, pp. 311–329. Springer (2013). DOI 10.1007/978-3-642-38697-8_17. URL http:
//dx.doi.org/10.1007/978-3-642-38697-8_17

21. Leemans, S.J.J., Fahland, D., van der Aalst, W.M.P.: Discovering block-structured process
models from event logs containing infrequent behaviour. In: N. Lohmann, M. Song, P. Wohed (eds.) Business Process Management Workshops - BPM 2013 International Workshops,
Beijing, China, August 26, 2013, Revised Papers, *Lecture Notes in Business Information Processing*, vol. 171, pp. 66–78. Springer (2013). DOI 10.1007/978-3-319-06257-0_6. URL
http://dx.doi.org/10.1007/978-3-319-06257-0_6

22. Leemans, S.J.J., Fahland, D., van der Aalst, W.M.P.: Discovering block-structured process
models from incomplete event logs. In: G. Ciardo, E. Kindler (eds.) Application and Theory
of Petri Nets and Concurrency - 35th International Conference, PETRI NETS 2014, Tunis,
Tunisia, June 23-27, 2014. Proceedings, *Lecture Notes in Computer Science*, vol. 8489, pp.
91–110. Springer (2014). DOI 10.1007/978-3-319-07734-5_6. URL http://dx.doi.org/
10.1007/978-3-319-07734-5_6

23. Leemans, S.J.J., Fahland, D., van der Aalst, W.M.P.: Exploring processes and deviations. In: Fournier and Mendling [16], pp. 304–316. DOI 10.1007/978-3-319-15895-2_26. URL http://dx.doi.org/10.1007/978-3-319-15895-2_26
24. Leemans, S.J.J., Fahland, D., van der Aalst, W.M.P.: Scalable process discovery and conformance checking. Software & Systems Modeling **special issue**, 1–33 (2016). DOI 10.1007/s10270-016-0545-x. URL http://dx.doi.org/10.1007/s10270-016-0545-x
25. Lu, R., Sadiq, S.W.: On the discovery of preferred work practice through business process variants. In: C. Parent, K. Schewe, V.C. Storey, B. Thalheim (eds.) Conceptual Modeling - ER 2007, 26th International Conference on Conceptual Modeling, Auckland, New Zealand, November 5-9, 2007, Proceedings, *Lecture Notes in Computer Science*, vol. 4801, pp. 165–180. Springer (2007). DOI 10.1007/978-3-540-75563-0_13. URL http://dx.doi.org/10.1007/978-3-540-75563-0_13
26. Minor, M., Tartakovski, A., Bergmann, R.: Representation and structure-based similarity assessment for agile workflows. In: R. Weber, M.M. Richter (eds.) Case-Based Reasoning Research and Development, 7th International Conference on Case-Based Reasoning, ICCBR 2007, Belfast, Northern Ireland, UK, August 13-16, 2007, Proceedings, *Lecture Notes in Computer Science*, vol. 4626, pp. 224–238. Springer (2007). DOI 10.1007/978-3-540-74141-1_16. URL http://dx.doi.org/10.1007/978-3-540-74141-1_16
27. Peterson, J.L.: Petri nets. ACM Comput. Surv. **9**(3), 223–252 (1977). DOI 10.1145/356698.356702. URL http://doi.acm.org/10.1145/356698.356702
28. Ramezani, E.: Understanding non-compliance. Ph.D. thesis, Eindhoven University of Technology (2017)
29. Ramezani, E., Fahland, D., van der Aalst, W.: Where did I misbehave? Diagnostic information in compliance checking. In: BPM, *Lecture Notes in Computer Science*, vol. 7481, pp. 262–278. Springer (2012)
30. Reisig, W.: A primer in Petri net design. Springer Compass International. Springer (1992)
31. Rozinat, A., van der Aalst, W.M.P.: Conformance checking of processes based on monitoring real behavior. Inf. Syst. **33**(1), 64–95 (2008). DOI 10.1016/j.is.2007.07.001. URL http://dx.doi.org/10.1016/j.is.2007.07.001
32. Weidlich, M., Polyvyanyy, A., Mendling, J., Weske, M.: Causal behavioural profiles - efficient computation, applications, and evaluation. Fundam. Inform. **113**(3-4), 399–435 (2011)
33. Weijters, A.J.M.M., Ribeiro, J.T.S.: Flexible heuristics miner (FHM). In: Proceedings of the IEEE Symposium on Computational Intelligence and Data Mining, CIDM 2011, part of the IEEE Symposium Series on Computational Intelligence 2011, April 11-15, 2011, Paris, France, pp. 310–317. IEEE (2011). DOI 10.1109/CIDM.2011.5949453. URL http://dx.doi.org/10.1109/CIDM.2011.5949453
34. van der Werf, J.M.E.M., van Dongen, B.F., Hurkens, C.A.J., Serebrenik, A.: Process discovery using integer linear programming. Fundam. Inform. **94**(3-4), 387–412 (2009). DOI 10.3233/FI-2009-136. URL http://dx.doi.org/10.3233/FI-2009-136
35. Zha, H., Wang, J., Wen, L., Wang, C., Sun, J.: A workflow net similarity measure based on transition adjacency relations. Computers in Industry **61**(5), 463–471 (2010). DOI 10.1016/j.compind.2010.01.001. URL http://dx.doi.org/10.1016/j.compind.2010.01.001

Chapter 2
Preliminaries

Sander J. J. Leemans: Robust Process Mining with Guarantees, LNBIP 440, pp. 23–47, 2022
https://doi.org/10.1007/978-3-030-96655-3_2

Abstract In this chapter, we introduce some basic concepts that will be used extensively in the remaining chapters. Introduced concepts include multisets, regular expressions, process automata, Petri nets, workflow nets, soundness, free-choice Petri nets, YAWL, BPMN, process trees, event logs and the directly follows language abstraction.

In Section 2.2 we discuss process models and event logs. Furthermore, we elaborate on a model and event log abstraction, the directly follows relation, in Section 2.4. This notion is used in many process mining techniques [3, 13, 22, 5, 19].

2.1 Multisets, Traces, Regular Expressions

A *multiset* is a set in which elements may occur multiple times, i.e. a multiset A is a function of the elements of A to natural numbers, such that for an element a, $A(a)$ denotes how often a is included in A. For elements not occurring in the multiset, A returns 0. For instance, let a, b and c be different elements, then the multiset $A = [a^2, b]$ is the multiset in which $A(a) = 2$, $A(b) = 1$ and $A(c) = 0$.

- The expression $a \in A$ expresses that element a is in multiset A, i.e. that $A(a) \geq 1$.
- To define multisets using formulas, we use a bracket notation. For instance, for a multiset X, $[a^2 | a \in X]$ denotes the multiset in which every element of X is included twice as often, e.g. $[a^2 | a \in [x^3, y^2]] = [x^6, y^4]$.
- The union of two multisets A and B, denoted with $A \uplus B$, is the sum function, i.e. let $C = A \uplus B$, then for each element a it holds that $C(a) = A(a) + B(a)$.
- Similarly, in the multiset difference $A \setminus B$, for each element a, $(A \setminus B)(a) = \max(0, A(a) - B(a))$.
- A multiset A is a subset of multiset B, denoted with $A \subseteq B$, if and only if for all elements a it holds that $A(a) \leq B(a)$.
- A set can be seen as a special case of a multiset, for which the returned value for each element is bounded by 1. The function *set* transforms a multiset into a set: $set(L) = \{t | t \in L\}$.
- For a given multiset A, the function $\mathbb{M}(A)$ returns the multisets in which all elements of A occur infinitely often, and no other elements occur, i.e. for all multisets A and A' such that $set(A) = set(A')$, it holds that $A' \subseteq \mathbb{M}(A)$ and $set(\mathbb{M}(A)) = set(A)$.

A *trace* is a sequence of elements. For instance, $\langle a, b, a \rangle$ denotes the trace consisting of an a followed by a b followed by an a again. Traces can be concatenated, e.g. $\langle a, b \rangle \cdot \langle c, d \rangle = \langle a, b, c, d \rangle$. The trace without elements is denoted with ϵ. To access the ith element of a trace t, we write $t(i)$, and we denote the length of a trace t with $|t|$.

A *regular expression* expresses sets of traces using three operators: choice $|$, concatenation $\cdot$ and Kleene-star * [15]. In this notation, an expression describes a set of traces, i.e. a for $a \in \Sigma$ describes the set $\{\langle a \rangle\}$, and if A and B are expressions, then $A \mid B = A \cup B$, $A \cdot B = \{a \cdot b \mid a \in A \wedge b \in B\}$, and $A^* = \{a, a \cdot a, a \cdot a \cdot a, \ldots\}$.

2.2 Process Models

In this section, we introduce some process modelling formalisms: automata (Section 2.2.1), Petri nets (Section 2.2.2), Yet Another Workflow Language (Section 2.2.3), Business Process Modelling and Notation (Section 2.2.4), and process trees (Section 2.2.5). The formalism of Petri nets is a well studied formalism that provides executable semantics to many other formalisms. Furthermore, many process discovery and conformance checking techniques use Petri nets, such as [3, 13, 22, 5, 19]. The Yet Another Workflow Language (YAWL) formalism extends Petri nets with several constructs. The Business Process Model and Notation (BPMN) language notation is used in many end-user tools, and inspired the tooling described in Section 9.1.

However, Petri nets, YAWL models and BPMN models might suffer from certain anomalies, such as the presence of deadlocks (see Chapter 1). To avoid such problems, the process discovery and conformance checking techniques that we introduced will use a formalism that is well-behaved by design: *process trees*.

2.2.1 Automata

An *automaton* is one of the most basic ways to represent processes. In this section, we present two types of automata: deterministic finite automata and nondeterministic finite automata.

In the following, let Σ be an alphabet, i.e. the set of activities in the automaton.

Definition 2.1 (deterministic finite automaton) A deterministic finite automaton (DFA) over an alphabet Σ is a tuple (S, s_0, F, A) which consists of:

- a finite set of states S;
- an initial state $s_0 \in S$;
- a set of final states $F \subseteq S$;
- a transition relation $A: S \times \Sigma \rightarrow S$.

Graphically, a state is represented by a circle, the initial state has an unconnected incoming edge, and the final state is represented by a circle with a double border. An element of A, e.g. $A(s, a) = s'$, with $a \in \Sigma$, $s \in S$, $s' \in S$, is represented by a direct edge from s to s', annotated with a. For instance, Figure 2.1 shows an example of a DFA in which $\Sigma = \{a, b\}$, $S = \{s_0, s_1, s_2, s_3\}$ is the set of states, s_0 is the initial state, $F = \{s_3\}$ is the set of final states, and $A(s_0, a) = s_1$, $A(s_0, b) = s_2$, $A(s_1, b) = s_3$, $A(s_2, a) = s_3$ is the transition relation.

Let $s_i, s_j \in S$ be states in a DFA and let $a \in \Sigma$, such that $A(s_i, a) = s_j$, i.e. from state s_i there is an a-edge to s_j. We denote this with $s_i \xrightarrow{a} s_j$, and semantically, this means that the DFA starts in state s_i, executes activity a and ends up in state s_j. Consider a sequence of executions $s_0 \xrightarrow{a_1} s_1 \xrightarrow{a_2} s_2 \ldots s_n$, such that s_0 is the initial state of the DFA, $\forall_{1 \leq i \leq n} a_i \in \Sigma$ and $\forall_{0 \leq i \leq n} s_i \in S$. We refer to such a sequence

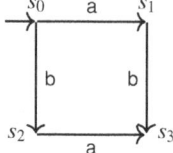

Fig. 2.1: Example of an automaton, NFA and DFA.

$a_1 \ldots a_n$ as a *trace*. If $s_n \in F$, i.e., s_n is a final state, then the trace is *accepted* by the DFA. The sets of all traces that are accepted by a DFA is the *language* of the DFA. For instance, the language of the DFA shown in Figure 2.1 is $ab \mid ba$, using the notation of regular language expressions.

Several DFAs could express the same language, e.g. the two automata of Figure 2.2 have the same language, however Figure 2.2 has fewer states. As proven in [15], for each DFA there is a *minimal DFA*, i.e. a DFA with the same language for which there is no DFA with the same language but fewer states. This minimal DFA is unique, i.e. there is just one DFA with a minimum number of states [15]. Therefore, language equivalence of two DFAs can be determined by examining their minimal DFAs, which we will use for our conformance checking framework in Chapter 7. Figure 2.2b is the minimal DFA of our example.

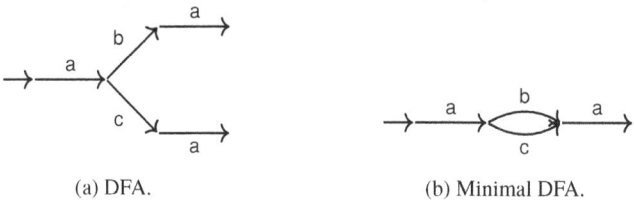

(a) DFA.   (b) Minimal DFA.

Fig. 2.2: Two DFAs with the same language.

A language that can be represented by a DFA is a *regular language*. Languages that are not representable by DFAs are for instance $\{a^n b^n \mid n \in \mathbb{N}\}$, i.e. any number of a-s followed by the same number of b-s. For this language, the automaton would need to count the number of a-s. As the number of states in a DFA is bounded, for each DFA there is trace of the language not properly recognised by the DFA [15]. We denote the set of all regular languages and all process models with regular languages regardless of process modeling formalism with $\mathbb{L}$.

In another class of automata, an automaton has an initial state, but no final states, and its traces are therefore assumed to have no end. Process discovery has been studied on such automata, e.g. on reactive continuously running systems [21], and in the context of Linear Time Logic (LTL) [16]. However, given the context of business processes, we limit ourselves to languages with termination.

Many more types of automata have been defined, e.g. non-deterministic finite automata and infinite automata. For more information, please refer to [15].

For most practical process models, DFAs are not well suited, as each possible state the process can be in needs an explicit state in a DFA: DFAs of real-life processes can be prohibitively large and therefore may be difficult to understand by human analysts. Therefore, in the next sections, we discuss process modelling formalisms and notations that can represent state spaces more compactly.

2.2.2 Petri Nets

In this section, we introduce Petri nets [18], two subclasses of Petri nets, and several extensions. Petri nets can denote some state spaces more compact, especially in presence of concurrent behaviour, i.e. independent executions do not lead to a state space explosion. We first introduce general Petri nets. Second, we introduce two subclasses of Petri nets that are used in many process discovery techniques. We illustrate the limitations of Petri nets by recalling two extensions that make Petri nets Turing complete.

Definition 2.2 (Petri net, unlabelled Petri net) A *Petri net* over a given alphabet Σ is a tuple (P, T, A, M_0, F, l) consisting of:

- a set of places P;
- a set of transitions T, such that $P \cap T = \emptyset$;
- a multiset arc relation $A \subseteq \mathbb{M}((P \times T) \cup (T \times P))$;
- an initial marking $m_0 \subseteq \mathbb{M}(P)$;
- a set of final markings F, being a set of multisets over P;
- a partial labelling function $l : T \to \Sigma$.

In case l is an bijective function, i.e. l is a one-to-one correspondence between T and Σ, then the net is an *unlabelled Petri net*.

Graphically, places are represented by circles, transitions are denoted by rectangles, and arcs are denoted by directed arcs between places and transitions. The initial marking is denoted by a black dot (a *token*). In case the net has one final marking, the places that are part of this final marking are denoted with a doubly bordered circle. Figure 2.3 shows an example.

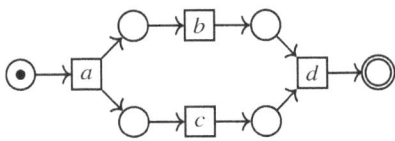

Fig. 2.3: A sound workflow net.

For a transition or place e, let $e^\bullet$ denote the *post set* of e, i.e. all places/transitions to which e has outgoing arcs: $e^\bullet = \{j \mid (e, j) \in A\}$. Similarly, $^\bullet e$ denotes the *pre set* of e: $^\bullet t = \{j \mid (j, e) \in A\}$. We canonically extend pre and post sets to multisets if A is a multiset. The marking of a Petri net denotes the state of the net, and consists of possibly multiple *tokens* in places, i.e. a marking is a multiset of places. A transition t is *enabled* in a marking m if all places from which t has incoming arcs contain at least one token: $^\bullet t \subseteq m$. If transition t is enabled in marking m, it can *fire*, thereby changing the marking of the Petri net to a new marking m', such that it removes tokens from the places connected to its incoming arcs, and produces tokens on places connected with its outgoing arcs: $m' = m \uplus t^\bullet \uplus {}^\bullet t$. We extend this to sequences of transition firings: the Petri net starts in the initial marking m_0, and a sequence of transition firings brings the Petri net to a final marking $m_n \in F$.

The firing of a transition t semantically denotes the execution of its labelled activity, i.e. the firing of t denotes the execution of $l(t)$. If l does not map t, then t does not denote the execution of an activity: t is a *silent transition*. The sequence of labelled transitions in a firing sequence that brings the Petri net from the initial marking m_0 to a final marking $m_n \in F$ denotes a trace of the Petri net. We denote such a firing sequence using $m_0 \rightsquigarrow m_n$. The set of all such labelled firing sequences is the language of the Petri net, which for a net M is denoted with $\mathcal{L}(M)$.

Alternatively, Petri nets could be defined without final markings: the language of such a Petri net would contain every labelled firing sequence that starts in the initial marking m_0. That is, in Definition 2.2, F would consist of all possible markings. Such Petri nets could for instance be used to model continuous processes, or when in case only events are logged without a categorisation into traces [21]. However, the business processes we consider have a clear ending, and the traces in event logs have a clear end as well. Therefore, we limit ourselves to *workflow nets*.

Workflow nets.

Workflow nets are a subclass of Petri nets, that is used by many process discovery algorithms [3, 13, 22], due to their rather natural representation of the final markings by a single place without any outgoing arcs [3, 1].

Definition 2.3 (workflow net) A *workflow net* is a Petri net (P, T, A, m_0, F, l) such that

- there is a single $i \in P$ such that $^\bullet i = \emptyset$;
- there is a single $o \in P$ such that $o^\bullet = \emptyset$;
- all places (P) and transitions (T) are on a path from i to o;
- the initial marking consists of i: $m_0 = [i]$;
- the only final marking consists of o: $F = \{[o]\}$.

For instance, Figure 2.3 shows a workflow net: its language is $\{\langle a, b, c, d \rangle, \langle a, c, b, d \rangle\}$.

Workflow nets might suffer from several behavioural anomalies: the final marking might not be reachable, or there might be tokens remaining in the net after a token reaches the sink place. Figure 2.4a shows an example of a workflow net in which the final marking is not reachable: after firing either a or b, the net is in a deadlock because c would require both a and b to be fired. Figure 2.4b shows an example of a workflow net in which there can be remaining tokens after the sink place is reached: after firing a, b and c, d can fire, which puts a token in the sink place, but there is a remaining token in the place left of d.

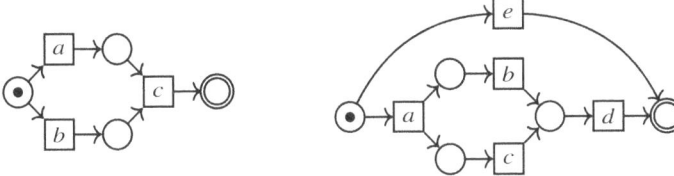

(a) A workflow net with an unreachable final marking.

(b) A workflow net with remaining tokens.

Fig. 2.4: Two workflow nets with soundness issues.

Workflow nets that do not have such issues are sound, i.e. a workflow net has *soundness* if each transition can fire from the initial marking and from each reachable marking it is possible to reach the final marking [3]. Formally:

Definition 2.4 (soundness) Let $W = (P, T, A, [i], [o], l)$ be a workflow net, in which i is the source place and o is the sink place. W is *sound* if and only if:

- every transition can be fired, i.e. $\forall_{t \in T} \exists_{[o] \rightsquigarrow m} {}^\bullet t \subseteq m$;
- from every marking, reachable from $[i]$, it is possible to reach $[o]$, i.e. $\forall_{[i] \rightsquigarrow m} m \rightsquigarrow [o]$;
- every marking, reachable from $[i]$, that puts a token in o has no other tokens, i.e. $\forall_{[i] \rightsquigarrow m} o \in m \Rightarrow [o] = m$.

The soundness property can be summarised in three requirements: (1) every transition can be fired in some marking that is reachable from the initial marking m_0, (2) the final marking (i.e. with one token in the sink place) is reachable from m_0, and (3) once a token is put in the sink place, the rest of the net is empty.

In [2], it was proven that the problem of deciding soundness for a workflow net is equivalent to deciding whether its short-circuited net (i.e. connecting the sink place to the source place using a silent transition) is live and bounded, i.e. from every reachable marking, it is possible to eventually fire every transition, and there exists a number k such that no place in the net has more than k tokens in any reachable marking.

In the next chapter, we will show the importance of soundness for process mining. Here, we continue with a class of process models that are guaranteed sound: *block-structured workflow nets*. A workflow net is *block structured* if for every place or transitions with multiple outgoing arcs, there is a corresponding place or transition with multiple incoming arcs. The parts of the net between the outgoing and incoming arcs form regions, and no arcs can exist between regions, i.e. the regions have a single entry and a single exit. In Figure 2.5, the transitions bound the two dashed regions, and there cannot be an arc between the regions. Due to this structure, block-structured workflow nets are inherently sound.

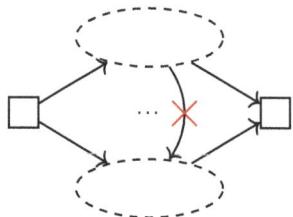

Fig. 2.5: Single-entry-single-exit regions: the dashed regions have only one incoming and one outgoing arc.

An example is shown in Figure 2.6: the filled regions denote the blocks of the block structure. Block-structured workflow nets are sound by definition, and therefore we will use block-structured workflow nets in the techniques presented.

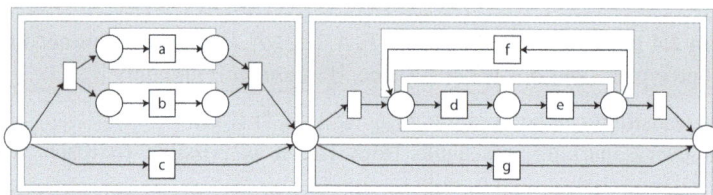

Fig. 2.6: A block-structured workflow net; filled regions denote the block structure.

Free-Choice Petri Nets.

Another class of Petri nets are *free-choice Petri nets*. In such nets, two transitions that share places from which they have incoming arcs, share all such places [11]. For instance, Figure 2.7 shows free choice and a non-free choice constructs. Free choice nets are a well-studied subclass of Petri nets, as several properties, such as reachability of a marking, liveness and boundedness, are decidable [11]. Furthermore, several process discovery algorithms provide guarantees if the system model can be expressed as a free choice Petri net, e.g. the α algorithm [3, p.130] (see Section 3.3.2).

(a) A free choice construct.   (b) A non-free choice construct.

Fig. 2.7: Free choice Petri nets.

Extensions.

Several extensions to Petri nets have been proposed. For instance, a *reset arc* between a place and a transition removes all tokens from the place on execution of the transition, while not influencing the precondition of the transition. Figure 2.8a contains an example: the language of this sound workflow net consists of traces that start with an *a* followed by a *b* and any combination *b* and *c* such that there are never more *c*'s executed than *b*'s, and end with a *d*.

Another extension is the *inhibitor arc*. An inhibitor arc between a place and a transition alters the precondition of the transition, which can only fire if the place is free of tokens. Figure 2.8b shows an example: the language of this sound workflow net consists of traces that start with an *a* followed by any combination *b* and *c* such that there are never more *c*'s executed than *b*'s, and end with a *d* when the number of *b*'s and *c*'s that were executed is equal. A reset arc can be transformed into an inhibitor arc, but the reverse is not necessarily true [3]. The addition of inhibitor arcs makes Petri nets Turing complete [17].

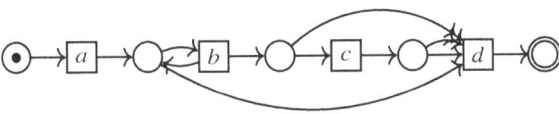

(a) A Petri net with a reset arc.

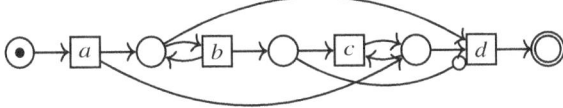

(b) A Petri net with an inhibitor arc.

Fig. 2.8: Examples of Petri net extensions.

2.2.3 Yet Another Workflow Language

The Yet Another Workflow Language (YAWL) was designed to provide an illustrative implementation of a set of typical constructs to represent how cases flow through the process: the *workflow patterns* [4]. YAWL is both a process modelling language and an engine that implements workflow systems by interpreting models written in the YAWL language [14]. We used parts of the YAWL language in the main formalism used here, process trees, which will be introduced in Section 2.2.5.

The YAWL language extends Petri nets using syntactic sugaring, and several constructs that extend the expressibility of Petri nets:

- Data. Control-flow in YAWL can be constrained using data elements. These data elements influence routing of cases. For instance, a gold customer (and this fact is present as a data element in the case) might be routed differently than a silver customer. Using data elements, YAWL is Turing complete.
- Cancellation regions. A transition in YAWL can have an associated *cancellation region* that includes some elements of the model: upon execution of the transition, all tokens that are present in the elements in the cancellation region are removed. Cancellation regions closely resemble reset arcs of Petri nets.
- Or-splits and or-joins. A YAWL transition can be annotated as being an or-split. That is, upon execution of the transition, a selection of the output arcs is made. Only via the selected arcs a token is produced. The selection must consist of at least one arc, and in YAWL, the selection is determined by data elements.

  A transition annotated as an or-join performs the opposite task: it waits for tokens on all incoming arcs as long as a via an arc a token *could* still arrive. As soon as no token can arrive anymore, the transition fires. These or-joins have complicated semantics, which can even lead to paradoxes [10].
- Multiple instance subprocesses. YAWL models can be nested, and such a nested subprocess can be instantiated multiple times using a *multiple instance* YAWL element. For instance, in an insurance claim, a subprocess could be 'send a form to all witnesses and wait for a reply'. As soon as a sufficient number of witness forms have been received, the multiple instance subprocess finishes and the process continues.

  The multiple instance notion is convenient for repeating subprocesses and the YAWL construct offers flexibility: it is possible to model that 10 witness forms must be sent, and after 3 received forms the process should continue.

  Furthermore, it is possible to model that an unspecified (unbounded) number of subprocesses can be started, and all must finish before the multiple instance finishes and the process can continue. This construct brings the YAWL language outside the class of regular languages (see Section 2.2.1), as the model needs to keep track of an unbounded number of started subprocesses, and this cannot be modelled in a finite automaton. (Note that other YAWL constructs and data have the same effect.)

2.2.4 Business Process Model and Notation

Recently, the Business Process Model and Notation (BPMN) has become one of the most widely used languages to model business processes [3]. Given its wide use in industry and the rather intuitive semantics of basic routing constructs, we took inspiration from the BPMN notation for our software tools, which will be introduced in Section 9.1.

A BPMN model is similar to a Petri net, however uses *gateways* as routing constructs, a job served by the tasks (transitions) in Petri nets. The constructs most relevant for us are summarised in Figure 2.9. An example BPMN model is shown in Figure 2.10a, which has the language $(ab \mid ba)(ab \mid ba)^*$. BPMN models might have similar issues as Petri nets, e.g. Figure 2.10b shows a model which Petri net translation would not be sound. For more information, please refer to [8].

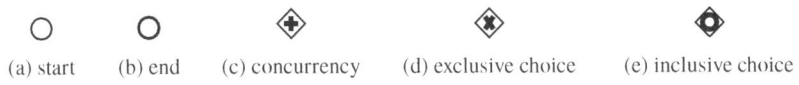

(a) start   (b) end   (c) concurrency   (d) exclusive choice   (e) inclusive choice

Fig. 2.9: BPMN routing constructs.

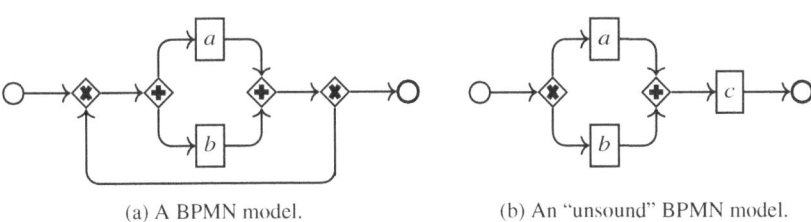

(a) A BPMN model.   (b) An "unsound" BPMN model.

Fig. 2.10: Two BPMN models.

2.2.5 Process Trees

Both Petri nets and BPMN models might suffer from soundness issues. As described in Chapter 1, in process mining, models with soundness issues might be challenging for techniques, and for many use cases, unsound models should be discarded. Therefore, we focus on a modelling formalism that is guaranteed to be free of soundness issues: the *process tree*. By using process trees, discovery algorithms and conformance checking techniques need not to worry about soundness, as soundness is guaranteed by construction.

A *process tree* is an abstract hierarchical representation of a block-structured workflow net: a rooted tree in which the leaves are annotated with activities or the silent activity τ and all other nodes are annotated with operators. We assume a finite alphabet Σ to be given, then we define the syntax of proces trees recursively as follows:

Definition 2.5 (process trees syntax) Let Σ be an alphabet of activities, then

- activity $a \in \Sigma$ is a process tree;
- the silent activity τ ($\tau \notin \Sigma$) is a process tree;
- let $M_1 \ldots M_n$ with $n > 0$ be process trees and let $\oplus$ be a process tree operator, then $\oplus(M_1, \ldots M_n)$ is a process tree, which we sometimes write as
$$\oplus \atop \overbrace{M_1 \ldots M_n}$$

To define the semantics of process trees, we again assume a finite set of activities Σ to be given. The language of an activity a is the trace $\langle a \rangle$, representing the execution of that activity (a process step). The language of the *silent activity* τ contains only the empty trace ϵ. The language of a *process tree operator* is a combination of the languages of its children.

Formally, the language of a process tree is defined recursively as follows, in which $\oplus$ denotes any process tree operator (for us $\oplus \in \{\times, \rightarrow, \wedge, \leftrightarrow, \circlearrowleft, \vee\}$), and in which $\oplus_{\mathcal{L}}$ denotes an operator-specific function that combines the languages of its children:

Definition 2.6 (process tree semantics) Let Σ be an alphabet of activities, then

$$\mathcal{L}(\tau) = \{\epsilon\}$$
$$\mathcal{L}(a) = \{\langle a \rangle\} \text{ for } a \in \Sigma$$
$$\mathcal{L}(\oplus(M_1, \ldots M_n)) = \oplus_{\mathcal{L}}(\mathcal{L}(M_1), \ldots \mathcal{L}(M_n))$$

We refer to such functions $\oplus_{\mathcal{L}}$ as *language-join functions*, and each process tree operator has a different language-join function.

We use six process tree operators: $\times$, $\rightarrow$, $\wedge$, $\circlearrowleft$, $\leftrightarrow$ and $\vee$. The $\times$ operator describes the exclusive choice between its children, $\rightarrow$ the sequential composition, $\wedge$ the concurrent composition, $\leftrightarrow$ the interleaved (i.e. non-overlapping) composition, $\circlearrowleft$ the repetitive composition and $\vee$ the optional concurrent composition.

In the remainder of this section, we finish the definition of the semantics of process trees by giving the language-join function for each operator, and showing a corresponding Petri net.

2.2.5.1 Exclusive Choice

The exclusive choice operator ($\times$) expresses that a trace of one of the children of the operator must be included, i.e.:

Definition 2.7 (exclusive choice semantics) Let $L_1 \ldots L_n$ be languages, such that $n \geq 1$. Then,

$$\times_{\mathcal{L}}(L_1, \ldots, L_n) = L_1 \cup L_2 \cup \ldots \cup L_n$$

For instance, the language of the process tree $\times(a, b)$ is $\{\langle a \rangle, \langle b \rangle\}$. An exclusive choice operator can be translated to a block-structured workflow net as follows, in which $M_1 \ldots M_n$ are process trees. In this translation, the dotted boxes denote the translations of the subtrees $M_1 \ldots M_n$.

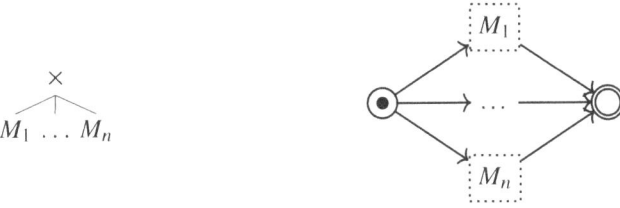

2.2.5.2 Sequence

The sequence operator ($\rightarrow$) expresses that a trace of all children must be included in order, i.e.:

Definition 2.8 (sequence semantics) Let $L_1 \ldots L_n$ be languages, such that $n \geq 1$. Then,

$$\rightarrow_{\mathcal{L}}(L_1, \ldots, L_n) = L_1 \cdot L_2 \cdots L_n$$

For instance, the language of the process tree $\rightarrow(a, b)$ is $\{\langle a, b \rangle\}$. A sequence operator can be translated to a block-structured workflow net as follows, in which $M_1 \ldots M_n$ are process trees:

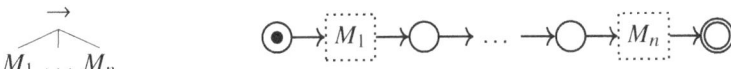

2.2.5.3 Concurrency

The concurrency operator ($\wedge$) expresses that a trace of all children must be included, and these traces may overlap. The following definition uses the shuffle-product operator $\sqcup$, which we will define afterwards.

Definition 2.9 (concurrent semantics) Let $L_1 \ldots L_n$ be languages, such that $n \geq 1$. Then,

$$\wedge_{\mathcal{L}}(L_1, \ldots, L_n) = L_1 \sqcup\!\sqcup L_2 \sqcup\!\sqcup \ldots L_n$$

For instance, the language of the process tree $\wedge(a, b)$ is $\{\langle a, b \rangle, \langle b, a \rangle\}$.

The shuffle product $S_1 \sqcup\!\sqcup \ldots S_n$ takes sets of traces from $S_1 \ldots S_n$ and interleaves their traces $\forall_{1 \leq i \leq n} \, t_i \in S_i$ while maintaining the order within each subtrace t_i. Let t be a trace, and let f be a bijective function that maps each event of t to a subtrace t_j and to a position in that subtrace, i.e. $f: \{1 \ldots |t|\} \rightarrow \{(j, k) \mid 1 \leq j \leq n \wedge k \leq |t_j|\}$. Then,

$$t \in t_1 \sqcup\!\sqcup \ldots \sqcup\!\sqcup t_n \Leftrightarrow \exists_{\text{function } f}$$
$$\forall_{1 \leq i_1 < i_2 \leq |t| \wedge f(i_1) = (t, k_1) \wedge f(i_2) = (t_j, k_2)} \, k_1 < k_2 \wedge$$
$$\forall_{1 \leq i \leq n \wedge f(i) = (t, k)} \, t(i) = t_j(k)$$

where f is a bijective function mapping each event of t to an event in one of the t_i [6].

For instance,

$$\{\langle a, b \rangle\} \sqcup\!\sqcup \{\langle c, d \rangle\} = \{\langle a, b, c, d \rangle, \langle a, c, b, d \rangle, \langle a, c, d, b \rangle,$$
$$\langle c, d, a, b \rangle, \langle c, a, d, b \rangle, \langle c, a, b, d \rangle\}$$

A concurrent operator can be translated to a block-structured workflow net as follows, in which $M_1 \ldots M_n$ are process trees:

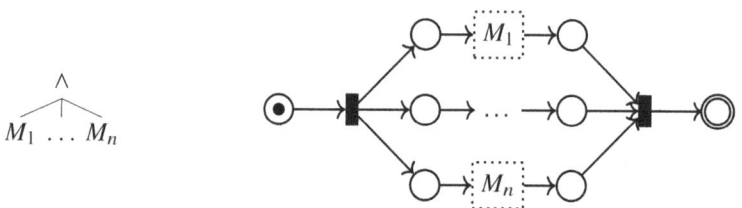

2.2.5.4 Interleaving

The interleaving operator ($\leftrightarrow$) expresses that a trace from all children must be included, and that these traces cannot overlap. The following definition uses the permutation function $p(n)$, which returns all permutations of the numbers $\{1 \ldots n\}$.

Definition 2.10 (interleaved semantics) Let $L_1 \ldots L_n$ be languages, such that $n \geq 1$. Then,

$$\leftrightarrow_{\mathcal{L}}(L_1, \ldots, L_n) = \bigcup_{(i_1 \ldots i_n) \in p(n)} \rightarrow_{\mathcal{L}}(L_{i_1}, \ldots, L_{i_n})$$

For instance, the language of the process tree $\leftrightarrow(a, \rightarrow(b, c))$ is $\{\langle a, b, c\rangle, \langle b, c, a\rangle\}$. Notice that we only define the language here: in a real-life system, the choice in which order the children are executed might be made incrementally, even though the definition does not explicitly express this. An interleaved operator can be translated to a block-structured workflow net as follows, in which $M_1 \ldots M_n$ are process trees.

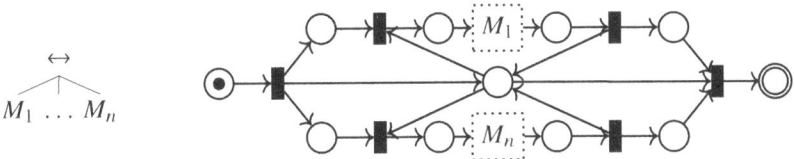

Compared with concurrency, for interleaving a critical section place is added, which ensures that at any time, only one subtree can be executing.

2.2.5.5 Loop

The loop operator ($\circlearrowleft$) expresses that each trace first contains a trace from the first child (the *loop body*). Then, the trace either ends or contains a number of times a trace from a non-first child (a *loop redo*) followed by a trace from the loop body again.

Definition 2.11 (loop semantics) Let $L_1 \ldots L_n$ be languages, such that $n \geq 2$. Then,

$$\circlearrowleft_{\mathcal{L}}(L_1, \ldots, L_n) = L_1 (\times_{\mathcal{L}}(L_2, \ldots, L_n) L_1)^*$$

For instance, $\circlearrowleft(a, b, c)$ is the composition of a trace of the body a, then zero-or-more times a trace from a redo part (b or c) and a body a again: $a((b \mid c)a)^*$, i.e. $\{\langle a\rangle, \langle a, b, a\rangle, \langle a, c, a\rangle, \langle a, b, a, c, a\rangle, \langle a, c, a, b, a\rangle, \langle a, b, a, b, a\rangle \ldots\}$. A loop operator can be translated to a block-structured workflow net as follows, in which $M_1 \ldots M_n$ are process trees:

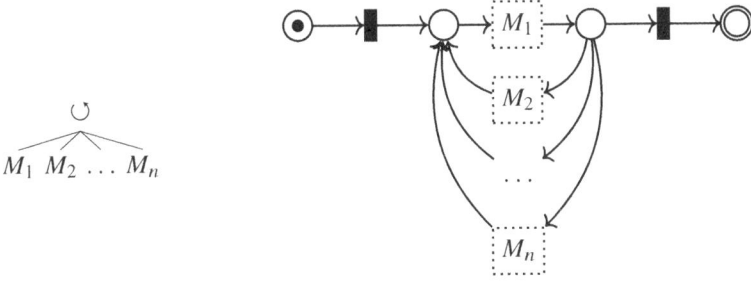

2.2.5.6 Inclusive choice

The inclusive choice operator ($\vee$) expresses that a trace from at least one child is included, and that these traces may overlap. The following definition uses the subset-function $q(n)$, which returns the set of all subsets of the numbers $\{1 \ldots n\}$, without the empty subset.

Definition 2.12 (inclusive choice semantics) Let $L_1 \ldots L_n$ be languages. Then,

$$\vee_{\mathcal{L}}(L_1, \ldots, L_n) = \bigcup_{\{i_1 \ldots i_m\} \in q(n)} \wedge_{\mathcal{L}}(L_{i_1}, \ldots, L_{i_m})$$

For instance, the language of the process tree $\vee(a, b)$ is $\{\langle a \rangle, \langle b \rangle, \langle a, b \rangle, \langle b, a \rangle\}$. An inclusive choice operator can be translated to a block-structured workflow net as follows, in which $M_1 \ldots M_n$ are process trees:

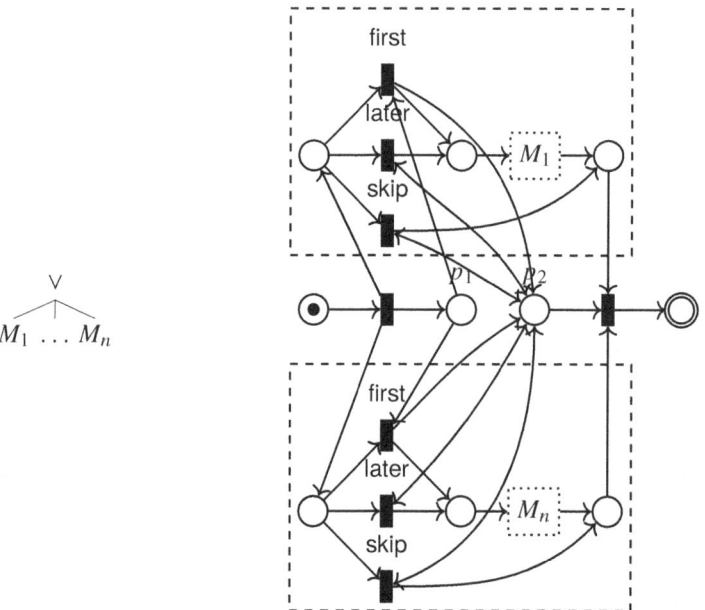

Place p_1 denotes that no subtree has been scheduled for execution yet, while place p_2 denotes that a child has been scheduled. We use this construct to ensure that at least one subtree is executed, which corresponds to the semantics of the $\vee$-operator. We added dashed regions to clarify the Petri net constructs that are added for each subtree. The constructs that are necessary for each subtree include three silent transitions: one to skip execution of the subtree (this can only be done if p_2 denotes that another subtree has been scheduled), one to schedule the subtree as the first child to be executed (this can only be done if p_1 denotes that no other subtree has been scheduled), and finally one to schedule the subtree if another subtree has already been scheduled (this can only be done if p_2 denotes that another subtree has been scheduled). Once a subtree is scheduled, it must be executed, however the order in which the scheduled subtrees are executed is arbitrary. Once all subtrees have been executed or skipped, the rightmost silent transition joins the inclusive choice.

In YAWL models, or-joins might have complex semantics, even leading to para-doxes [10], but such issues do not appear in block-structured models, as for each or-join there is a corresponding or-split, and such structures (local synchronising merge workflow pattern [4]) do not risk expressing paradoxical behaviour.

2.2.5.7 Example

For instance, consider the process tree $M_E = $ 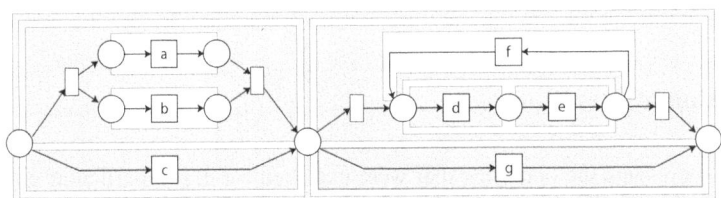 . The language of M_E

is $(ab \mid ba \mid c)(de(fde)^* \mid g)$. Each process tree is easily translatable to a sound block-structured workflow net. For example, the workflow net corresponding to M_E is shown in Figure 2.11.

Fig. 2.11: A Petri net corresponding to the process tree $\to(\times(\wedge(a,b),c),\times(\circlearrowleft(\to(d,e)),f),g)$. The filled regions denote the block structure derived from the process tree operators.

2.2.5.8 Furthermore

Based on the language-join functions, the order of children for $\times$, $\wedge$, $\leftrightarrow$, $\vee$, and the order of non-first children of $\circlearrowleft$ are arbitrary. For instance, $\times(a,b) = \times(b,a)$, $\wedge(a,b) = \wedge(b,a)$ and $\circlearrowleft(a,b,c) = \circlearrowleft(a,c,b)$ but $\to(a,b) \neq \to(b,a)$ and $\circlearrowleft(a,b,c) \neq \circlearrowleft(b,a,c)$.

The *lowest common ancestor* of two nodes in a tree is the node connecting the two nodes, e.g. in $\circlearrowleft$, the lowest common ancestor of a and c is the $\circlearrowleft$ node. We will use

$$
\begin{array}{c}
\wedge \\
\rightarrow\ c \\
\wedge \\
a\ b
\end{array}
$$

Σ to denote the set of activities of a process tree explicitly, e.g. $\Sigma(\times(a,b)) = \{a,b\}$. We refer to the set of all process trees as $\mathbb{T}$.

The process trees introduced here differ slightly from the process trees as defined in [7, 20]. In [7], a loop has precisely three children: a body, a redo and an exit subtree. The semantics are similar: the body is always executed, followed by a repeated redo and body. However, the exit child is executed last. We decided to opt for n-ary loop nodes, as these provide a more natural fit to the cut detection algorithms: these algorithms, which will be introduced in Chapter 6, discover n-ary loops naturally, while an exit-node would be indistinguishable from a sequence-node. Furthermore, we added the interleaved operator.

2.3 Event Logs

In this section, we introduce the input format of discovery algorithms and many other process mining techniques: event logs. We first introduce the most commonly used type of event logs and their notation, i.e. in which activities are atomic. Second, we introduce a variant of event logs in which activities take time. Third, we consider rich event logs, i.e. event logs in which events have meta information attached.

2.3.1 Atomic Event Logs

A trace represents the activities that were executed for e.g. a particular customer, patient, file or claim. An activity is a step in the process, e.g. the recording of an order or treatment of a patient. Each such an execution is an *event*. Formally, a trace is a finite sequence of events, e.g. $\langle a,a,b\rangle$ denotes a trace in which first an event of activity a occurred, then an event of activity a again and finally an event of activity b.

An *atomic event log* is a multiset of traces. For ease of notation, we will refer to an atomic event log simply as an *event log*. For instance, $[\langle a,a,b\rangle^3, \langle b,b\rangle^2]$ denotes an event log in which the trace $\langle a,a,b\rangle$ happened 3 times and $\langle b,b\rangle$ happened twice. We refer to the set of all atomic event logs as $\mathbb{E}$. We denote $\Sigma(L)$ for the alphabet of a log L, i.e., the activities used in L, $|L|$ for the number of traces in L and $||L||$ for the number of events in L.

2.3.2 Non-Atomic Event Logs

In atomic event logs, an event denotes the execution of an activity, but this execution is assumed to be atomic and instantaneous. In this section, we introduce a concept of event logs in which the executions of activities can take time.

In a *non-atomic event log*, executions of activities are represented by two events instead of one: one event denotes the start of the execution, while the other denotes the completion of the execution. Thus, every event is marked as being either a completion or a start event. Notation wise, we write $\tilde{L}$ to denote a non-atomic event log. For instance, the trace $\langle a_s, b_s, b_c, a_c \rangle$ denotes the trace in which first non-atomic activity $\tilde{a}$ was started, then $\tilde{b}$ was started, after which $\tilde{b}$ completed and finally $\tilde{a}$ completed.

The atomic event logs introduced earlier can be mapped to the new notion by transforming each event into both a start and completion event, e.g. $\langle a, b \rangle$ is transformed to $\langle a_s, a_c, b_s, b_c \rangle$ (notice that this keeps the semantics of the atomic trace intact). We do not assume events to contain any information on which start event belongs to which completion event, e.g. in the trace $\langle a_s, a_s, a_c, a_c \rangle$, two explanations are possible (see Figure 2.12). We do not have to make this assumption as the discovery techniques presented here do not need to know which a_s belongs to which a_c.

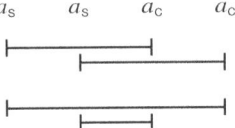

Fig. 2.12: Two ways to explain the trace $\langle a_s, a_s, a_c, a_c \rangle$.

We refer to a set of non-atomic traces as a *non-atomic language*.

Event logs adhering to the XES standard [12] might provide start and completion information. For instance, the BPI Challenge log of 2012 (BPIC12) [9] contains this information. The XES standard describes several extensions, which allow information to be added events. The *lifecycle:transition* extension adds life cycle transitions to events, i.e. events can denote state changes in a state machine that describes their life cycle. Several life cycles with corresponding state machines have been proposed, we will use only use the *start* and *complete* transitions. Here (and in our implementations) we ignore events with other life cycle transitions.

2.3.2.1 Non-Atomic Trace Consistency

Our notion of non-atomic traces inherently introduces a challenge: the start and completion events must not appear out of order to make sense, i.e. the number of 'running' activity executions must never be negative. For instance, the trace $\langle a_c, a_s \rangle$

makes little sense, as activity a was completed before it started. We refer to a non-atomic trace without such issues as a *consistent trace*, i.e. a trace for which one could construct a one-to-one mapping between start and completion events, in which the start event appears before its mapped completion event.

Definition 2.13 (consistent non-atomic trace) A non-atomic trace $t = \langle a_1, a_2, \ldots a_n \rangle$ is *consistent* if there is a mapping $X \subseteq \mathbb{N} \times \mathbb{N}$ such that

- all mapped events are of the same activity:
  $\forall X(a_i, a_j) \Sigma([\langle a_i \rangle]) = \Sigma([\langle a_j \rangle])$
- each completion event is mapped to one preceding start event:
  $\forall_{a_i \in t, a_i \text{ is a completion event}} |\{j \mid X(a_j, a_i) \wedge 1 \leq j < i\}| = 1$
- each start event is mapped to one following completion event:
  $\forall_{a_i \in t, a_i \text{ is a start event}} |\{j \mid X(a_i, a_j) \wedge i < j \leq n\}| = 1$
- no other mappings are present:
  $\forall_{X(a_i, a_j)} a_i$ is a start event $\wedge a_j$ is a completion event

However, even though we assume there is such a mapping, the discovery techniques presented do not need to know an actual mapping.

As inconsistent traces make little sense and to keep the algorithms and proofs simple, we assume that all non-atomic traces we encounter are consistent. In practice however, inconsistent traces might be encountered, e.g. the BPIC12 log [9] has three such start events that cannot be mapped to a corresponding completion event. Therefore, we introduce a pre-processing step to solve inconsistencies.

We explain this step using an inconsistent example trace, $\langle a_s, b_s, b_c \rangle$. This trace could be made consistent in several ways:

$\langle b_s, b_c \rangle$	remove a_s
$\langle a_s, b_s, b_c, \mathbf{a_c} \rangle$	add a_c at the end
$\langle a_s, b_s, \mathbf{a_c}, b_c \rangle$	add a_c in the middle
$\langle a_s, \mathbf{a_c}, b_s, b_c \rangle$	add a_c in the beginning

Without further information, there is no reason why one of these options would prevail over the others. Therefore, we pragmatically choose the last one, as that keeps the impact of the inconsistency as small as possible by inserting a completion event right after the unmatched start event. This is applied to all inconsistent traces in our tools as a pre-processing step. Therefore, in the remainder of this chapter, we can assume that all traces are consistent. In case more information is available, such as when the events are linked to one another with the *concept:instance* extension of XES [12], our pre-processing step could easily be preceded by a such a custom inconsistency resolver.

2.3.3 Richer Logs

In real-life event logs, events may be annotated with additional data elements such as time stamps (when the event was executed), resources (by whom or what the event was executed), the type of customer for which the event was executed, what decision information was available when the event was executed, etc.

For instance, consider the following trace:

$$\Big\langle a_{\text{life cycle: START}} \atop \text{time: 29-02-1900 12:32} \atop \text{resource: Sue} \quad , \quad b_{\text{life cycle: START}} \atop \text{time: 29-02-1900 12:36} \atop \text{resource: Bert} \quad ,$$

$$a_{\text{life cycle: COMPLETE}} \atop \text{time: 29-02-1900 13:52} \atop \text{resource: Sue} \quad , \quad b_{\text{life cycle: START}} \atop \text{time: 29-02-1900 14:00} \atop \text{resource: Bert} \Big\rangle$$

This trace consists of four events, and each event is annotated with a life-cycle transition, a time stamp and a staff member who executed the event. For more information, please refer to [12].

In most techniques presented, we will not consider this extra event information. However, in Chapter 9, we will present techniques to use this data, e.g. by summarising the data and projecting it onto process models.

In the previous sections, we discussed process models and event logs. In the remainder of this section, we introduce an important abstraction of both: the directly follows relation.

2.4 Directly Follows Relation

In this section, we introduce a language abstraction that is used in many process discovery algorithms and implicitly in many conformance checking techniques: the *directly follows relation*. Before we introduce the relation, we first establish some terminology on relations and graphs.

A *graph* is a set of nodes combined with a set of edges, such that each edge connects two nodes. The edges might be annotated, i.e. an *edge weight* attached. If the edges of a graph have no direction, i.e. just connect two nodes without providing an ordering on them, the graph is an *undirected graph*, which corresponds to a commutative relation on the nodes. In a *directed graph*, the edges have a direction, i.e. go from a node to another node, thereby establishing a non-commutative relation between the nodes of the graph. A directed graph can be projected onto an undirected graph by ignoring the direction of the edges.

A (directed) *path* is a sequence of nodes such that each sequential pair of nodes on the path is connected by an edge (in the correct direction). Note that a path contains at least one node. We define an *undirected path* of a directed graph (↔) to be a path of which the direction of the edges is ignored.

A *connected component* in an undirected graph is a non-empty set of nodes such that there is a path between all pairs of nodes in the set. A *strongly connected component* in a directed graph is a non-empty set of nodes such that between each pair of nodes in the set, there is a directed path forth and a directed path back in the graph.

A *directly follows relation* $\rightarrow\!\!\!\rightarrow$ is a combination of a graph and two annotations. The graph (*directly follows graph*) is a directed graph: its nodes are activities, its edges denote which activities can directly follow one another. The two annotations $\top$ and $\bot$ denote the start and the end of a trace. Notice that these annotations are not part of the graph.

Definition 2.14 (directly follows relation) Let Σ be an alphabet such that $\top \notin \Sigma$ and $\bot \notin \Sigma$ and let L be a language over Σ. We define the following relations:

$$a \rightarrow\!\!\!\rightarrow b \Leftrightarrow \exists_{t \in L}\ t = \langle \ldots a, b, \ldots \rangle$$
$$\top \rightarrow\!\!\!\rightarrow a \Leftrightarrow \exists_{t \in L}\ t = \langle a, \ldots \rangle$$
$$a \rightarrow\!\!\!\rightarrow \bot \Leftrightarrow \exists_{t \in L}\ t = \langle \ldots, a \rangle$$
$$\top \rightarrow\!\!\!\rightarrow \bot \Leftrightarrow \exists_{t \in L}\ t = \epsilon$$

For readability, we often write $\text{Start}(L)$ for the set of all start activities ($\{a \in \Sigma \mid \top \rightarrow\!\!\!\rightarrow a\}$), and $\text{End}(L)$ for the end activities ($\{a \in \Sigma \mid a \rightarrow\!\!\!\rightarrow \bot\}$). Directly follows relations can be derived from event logs and models. Therefore, for a log L we denote the directly follows graph of L with $\rightarrow\!\!\!\rightarrow(L)$. Similarly, for a process model M, we use the shorthand $\rightarrow\!\!\!\rightarrow(M)$ to denote $\rightarrow\!\!\!\rightarrow(\mathcal{L}(M))$, and $\text{Start}(M)$ and $\text{End}(M)$ for $\text{Start}(\mathcal{L}(M))$ and $\text{End}(\mathcal{L}(M))$.

Let d be a directly follows relation, then we denote $\Sigma(d)$ for the alphabet over which d was defined. We denote $\text{Start}(d)$ for the start activities of d and $\text{End}(d)$ for the end activities of d.

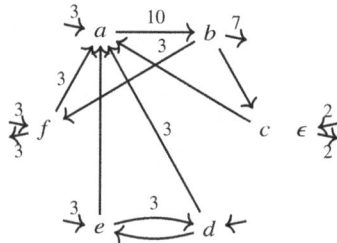

Fig. 2.13: Example of a directly follows graph.

In the directly follows graph that is derived of a process model, frequency information (i.e. how often each edge in the directly follows graph was observed), is not available. In event logs, such information is available and useful for discovery algorithms to assess which behaviour occurred more frequently than other behaviour.

Therefore, we extend the directly follows relation with frequency information for event logs: the sets of start and end activities become multisets, and the edges of the directly follows graph become annotated with a weight, denoting how often some relation happened in the event log. For instance, consider the following event log:

$$L = [\epsilon^2, \quad \langle a, b, f \rangle^3, \quad \langle f, a, b \rangle^2, \quad \langle e, d, a, b \rangle^3, \quad \langle d, e, a, b \rangle, \quad \langle f, a, b, c, a, b \rangle]$$

then, its activities are

$$\Sigma(L) = \{a, b, c, d, e, f\}$$

its start and end actvities are

$$\text{Start}(L) = [a^3, d, e^3, f^2]$$
$$\text{End}(L) = [b^7, f^3]$$

and its directly follows graph is

$$\twoheadrightarrow(L) = [\top \twoheadrightarrow \bot^2, \quad \top \twoheadrightarrow a^3, \quad \top \twoheadrightarrow d, \quad \top \twoheadrightarrow e^3, \quad a \twoheadrightarrow b^{10}, \quad b \twoheadrightarrow c, \quad b \twoheadrightarrow f^3,$$
$$c \twoheadrightarrow a, \quad d \twoheadrightarrow a^3, \quad d \twoheadrightarrow e, \quad e \twoheadrightarrow a, \quad e \twoheadrightarrow d^3, \quad f \twoheadrightarrow a^3, \quad b \twoheadrightarrow \bot^7,$$
$$f \twoheadrightarrow \bot^3]$$

A graphical representation of this relation is shown in Figure 2.13, in which the annotations $\top$ and $\bot$ are visualised by open-ended edges, and the empty trace is denoted with ϵ.

Furthermore, let $\twoheadrightarrow^+$ denote the transitive closure of a directly follows graph, i.e. for two activities a and c

$$a \twoheadrightarrow^+ c \Leftrightarrow \exists_{0 \le n, b_1 \ldots b_n \in \Sigma(L)} \; a \twoheadrightarrow b_1 \twoheadrightarrow b_2 \ldots b_n \twoheadrightarrow c$$

Notice that $a \twoheadrightarrow^+ a$ is not a tautology.

In this chapter, we introduced event logs and process models, and we discussed the directly follows relation, which is a language abstraction used by many process discovery techniques. We will use this relation extensively, especially in Chapter 5. In the next chapter, we will describe the process mining problems that are addressed: process discovery, conformance checking and model enhancement.

References

1. van der Aalst, W., Weijters, A., Maruster, L.: Workflow mining: Discovering process models from event logs. IEEE Trans. Knowl. Data Eng. 16(9), 1128–1142 (2004)
2. van der Aalst, W.M.P.: The application of Petri nets to workflow management. Journal of Circuits, Systems, and Computers 8(1), 21–66 (1998). DOI 10.1142/S0218126698000043. URL http://dx.doi.org/10.1142/S0218126698000043

3. van der Aalst, W.M.P.: Process mining - discovery, conformance and enhancement of business processes. Springer (2011). DOI 10.1007/978-3-642-19345-3. URL http://dx.doi.org/ 10.1007/978-3-642-19345-3

4. van der Aalst, W.M.P., ter Hofstede, A.H.M., Kiepuszewski, B., Barros, A.P.: Workflow patterns. Distributed and Parallel Databases 14(1), 5–51 (2003). DOI 10.1023/A:1022883727209. URL http://dx.doi.org/10.1023/A:1022883727209

5. Adriansyah, A.: Aligning Observed and Modeled Behavior. Ph.D. thesis, Eindhoven University of Technology (2014)

6. Bloom, S.L., Ésik, Z.: Free shuffle algebras in language varieties. Theor. Comput. Sci. 163(1&2), 55–98 (1996)

7. Buijs, J.C.A.M.: Flexible evolutionary algorithms for mining structured process models. Ph.D. thesis, Eindhoven University of Technology (2014)

8. Dijkman, R.M., Dumas, M., Ouyang, C.: Semantics and analysis of business process models in BPMN. Information & Software Technology 50(12), 1281–1294 (2008). DOI 10.1016/j. infsof.2008.02.006. URL http://dx.doi.org/10.1016/j.infsof.2008.02.006

9. van Dongen, B.: BPI challenge 2012 dataset (2012). DOI 10.4121/uuid: 3926db30-f712-4394-aebc-75976070e91f. URL http://dx.doi.org/10.4121/uuid: 3926db30-f712-4394-aebc-75976070e91f

10. Dumas, M., Großkopf, A., Hettel, T., Wynn, M.T.: Semantics of standard process models with or-joins. In: R. Meersman, Z. Tari (eds.) On the Move to Meaningful Internet Systems 2007: CoopIS, DOA, ODBASE, GADA, and IS, OTM Confederated International Conferences CoopIS, DOA, ODBASE, GADA, and IS 2007, Vilamoura, Portugal, November 25-30, 2007, Proceedings, Part I, *Lecture Notes in Computer Science*, vol. 4803, pp. 41–58. Springer (2007). DOI 10.1007/978-3-540-76848-7_5. URL http://dx.doi.org/10. 1007/978-3-540-76848-7_5

11. Esparza, J., Silva, M.: On the analysis and synthesis of free choice systems. In: G. Rozenberg (ed.) Advances in Petri Nets 1990. 10th International Conference on Applications and Theory of Petri Nets, Bonn, Germany, June 1989, Proceedings, *Lecture Notes in Computer Science*, vol. 483, pp. 243–286. Springer (1989). DOI 10.1007/3-540-53863-1_28. URL http: //dx.doi.org/10.1007/3-540-53863-1_28

12. Günther, C., Verbeek, H.: XES v2.0 (2014). URL http://www.xes-standard.org/

13. Guo, Q., Wen, L., Wang, J., Yan, Z., Yu, P.S.: Mining invisible tasks in non-free-choice constructs. In: H.R. Motahari-Nezhad, J. Recker, M. Weidlich (eds.) Business Process Management - 13th International Conference, BPM 2015, Innsbruck, Austria, August 31 - September 3, 2015, Proceedings, *Lecture Notes in Computer Science*, vol. 9253, pp. 109–125. Springer (2015). DOI 10.1007/978-3-319-23063-4_7. URL http://dx.doi.org/10. 1007/978-3-319-23063-4_7

14. ter Hofstede, A.H.M., van der Aalst, W.M.P., Adams, M., Russell, N. (eds.): Modern Business Process Automation - YAWL and its Support Environment. Springer (2010). URL http: //www.yawlbook.com/home/

15. Linz, P.: An introduction to formal languages and automata (4. ed.). Jones and Bartlett Publishers (2006)

16. Maggi, F.M., Burattin, A., Cimitile, M., Sperduti, A.: Online process discovery to detect concept drifts in ltl-based declarative process models. In: R. Meersman, H. Panetto, T.S. Dillon, J. Eder, Z. Bellahsene, N. Ritter, P.D. Leenheer, D. Dou (eds.) On the Move to Meaningful Internet Systems: OTM 2013 Conferences - Confederated International Conferences: CoopIS, DOA-Trusted Cloud, and ODBASE 2013, Graz, Austria, September 9-13, 2013. Proceedings, *Lecture Notes in Computer Science*, vol. 8185, pp. 94–111. Springer (2013). DOI 10.1007/ 978-3-642-41030-7_7. URL http://dx.doi.org/10.1007/978-3-642-41030-7_7

17. Peterson, J.L.: Petri nets. ACM Comput. Surv. 9(3), 223–252 (1977). DOI 10.1145/356698. 356702. URL http://doi.acm.org/10.1145/356698.356702

18. Reisig, W.: A primer in Petri net design. Springer Compass International. Springer (1992)

19. Rozinat, A., van der Aalst, W.M.P.: Conformance testing: Measuring the fit and appropriateness of event logs and process models. In: C. Bussler, A. Haller (eds.) Business Process Management

Workshops, BPM 2005 International Workshops, BPI, BPD, ENEI, BPRM, WSCOBPM, BPS, Nancy, France, September 5, 2005, Revised Selected Papers, vol. 3812, pp. 163–176 (2005). DOI 10.1007/11678564_15. URL http://dx.doi.org/10.1007/11678564_15

20. Schimm, G.: Generic linear business process modeling. In: ER (Workshops), *LNCS*, vol. 1921, pp. 31–39. Springer (2000)

21. Tapia-Flores, T., López-Mellado, E., Estrada-Vargas, A.P., Lesage, J.: Petri net discovery of discrete event processes by computing t-invariants. In: A. Grau, H. Martínez (eds.) Proceedings of the 2014 IEEE Emerging Technology and Factory Automation, ETFA 2014, Barcelona, Spain, September 16-19, 2014, pp. 1–8. IEEE (2014). DOI 10.1109/ETFA.2014.7005080. URL http://dx.doi.org/10.1109/ETFA.2014.7005080

22. van der Werf, J.M.E.M., van Dongen, B.F., Hurkens, C.A.J., Serebrenik, A.: Process discovery using integer linear programming. Fundam. Inform. **94**(3-4), 387–412 (2009). DOI 10.3233/FI-2009-136. URL http://dx.doi.org/10.3233/FI-2009-136

Chapter 3
Process Mining

Sander J. J. Leemans: Robust Process Mining with Guarantees, LNBIP 440, pp. 49–117, 2022
https://doi.org/10.1007/978-3-030-96655-3_3

Abstract In Chapter 1, we introduced three challenges of process mining: *process discovery*, *conformance checking* and *model enhancement*. In this chapter, we elaborate on these challenges, discuss related work and gather requirements for process mining techniques. We first discuss several use cases, and how these might need to be addressed using different process mining techniques. Second, we discuss key challenges of process mining, including the importance of precise semantics of process models, equivalence classes of behaviour, the relation between system, log and model, and the necessity to trade off log-model quality criteria. Third, we discuss related existing techniques for process discovery, conformance checking and enhancement, while gathering requirements for the "ideal" process mining technique. Finally, we introduce our approaches to these three challenges: the Inductive Miner framework, the Projected Conformance Checking framework and the Inductive visual Miner.

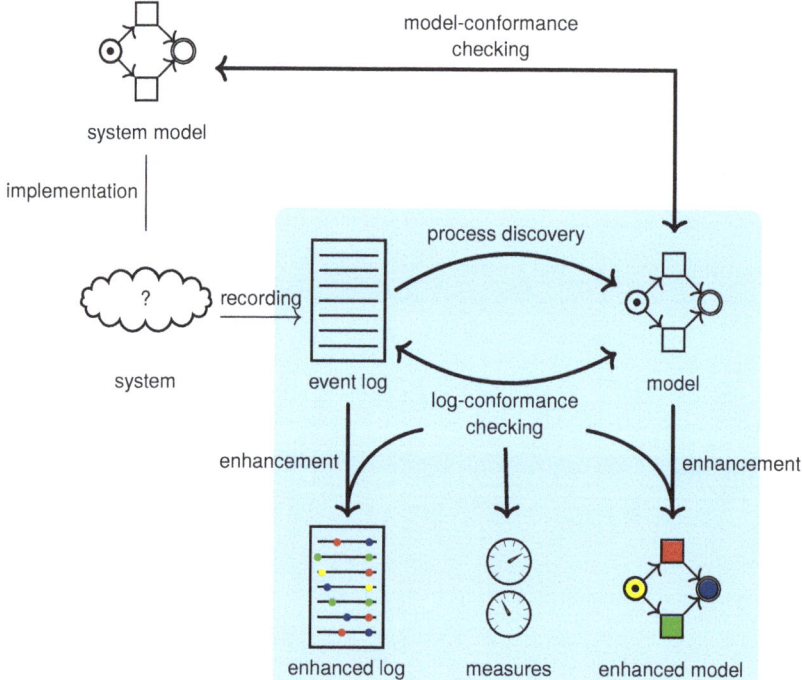

Fig. 3.1: Process discovery, conformance checking and enhancement in their context. The coloured region denotes the scope of a typical process mining project.

In the previous chapter, we discussed some of the core concepts of process mining techniques: event logs, process modelling notations and directly follows graphs. We explained these challenges using Figure 3.1: a system implements a system model, the system executes and these executions are recorded in an event log. To gain insight into the behaviour of the system, a *process discovery* technique uses the event log to obtain a process model that describes the system. To provide insight into the match between a log and the model, a *log-conformance checking* technique measures the correspondence between a log and a model. Notice that the model could be discovered by a discovery technique, but could also have been made by hand. Process discovery and conformance checking techniques are typically concerned with the control flow of a process, i.e. the conditions on and the order in which the process steps can be performed. Besides control flow, process models can express more information, for instance on process performance. A *model-enhancement* technique annotates a process model with aggregated information of the event log, such as time, life cycle or resources, thereby enabling analysts to assess the process on these perspectives. To gain insight into the system as it is running, instead of how the organisation thinks it runs, in typical process mining projects these steps are used. These steps are contained in the blue coloured region in Figure 3.1. We will describe an example process mining project in Section 3.1.

In a typical exploratory process mining project, the system is subject of study but unknown. However, in evaluative process mining projects, a reference system model might be available, using which the system is implemented. To gain insight into the differences between two process models, e.g. the system and the system model, a *model-conformance checking* technique measures the correspondence between two process models, e.g. the system model and the discovered model, to provide insight in their differences. Another application of model-conformance checking is to compare models discovered from event logs from different scopes (e.g. time periods or geographical locations) of the same system to detect differences between these scopes.

In this chapter, we introduce these three process mining problems in more detail: we start with a description of common use cases (Section 3.1), after which we discuss formal challenges of process discovery and conformance checking, and gather formal requirements for both in Section 3.2. In Section 3.3, we discuss existing process discovery techniques and gather practical requirements. In Section 3.4, we do the same for conformance checking techniques, and in Section 3.5 for model enhancement techniques. We finish with a high-level description of our approach and how it addresses the identified challenges in Section 3.6.

3.1 Different Use Cases, Different Process Mining Techniques

Process mining projects (denoted by the blue-coloured region in Figure 3.1) can have several use cases, each of which might require different characteristics of the techniques used. Moreover, use cases may change during the project. We illustrate several use cases using a case study, after which we discuss more common use cases.

Case Study.

In [44], we described a real-life process mining project, performed at IBM, a leading multinational technology and consulting corporation, in a hardware service department. In particular, a spare parts purchasing process was analysed, which is performed independently at several locations around the world. The event log extracted from this process contained hundreds of thousands of events related to thousands of orders of spare parts.

The initial goal of this process mining project was to gain insight into the process and see what would stand out. Therefore, as a first step, several event logs were extracted, each covering a different perspective of the data. As a second step, a process model was discovered using the algorithms that will be described in Chapter 6 and tools that will be discussed in Chapter 9, and this model was discussed with process experts. The main purpose of this process model was to narrow the discussion to points of interest: it was not important that this model showed all behaviour of the event log, as the most occurring behaviour was sufficient. This allowed the process experts to identify several areas of interest in the process model, and the analysis was continued iteratively on these areas by filtering the event log.

On one such filtered event log, a more detailed analysis revealed that in order to cancel an order, users were required to perform two steps, i.e. do double work (see Figure 3.2). This conclusion was validated in the SAP system that supported the process, and changes were proposed to improve the process by making the second step automated. As this analysis went into more detail, a more detailed model was necessary.

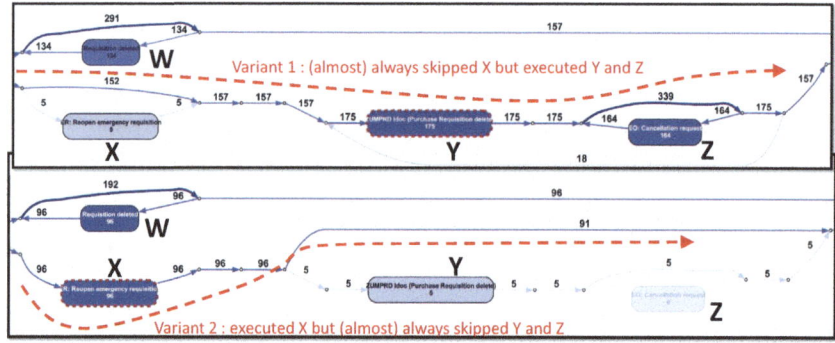

Fig. 3.2: Repetitive filtering of the event logs. The bottom process shows that activity X was executed before activity Y sometimes, while these should be mutually exclusive (image: [44]).

In another analysis, the process was compared over the four locations worldwide. In this analysis, as a first step, the event log was split into four sublogs, each covering the cases handled in one location. Second, all sublogs were filtered to represent around 80% of the most-occurring behaviour (i.e. traces). For one of these sublogs, say location 1, a process model was discovered, which described the "happy flow", i.e. the majority of normal behaviour. Second, this model of location 1 was compared to the sublogs of locations 2, 3 and 4 using log-conformance checking techniques, to spot differences in their executions of the process (see Figure 3.3). Here, the filtering was applied on the event logs themselves, instead of in the discovery algorithms. Therefore, the discovery algorithm was not expected to filter infrequent behaviour. This analysis revealed different bottlenecks in the process at different geographical locations.

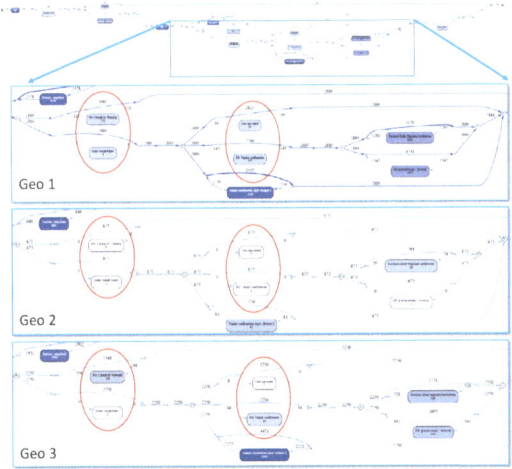

Fig. 3.3: A process executed in four locations. A "happy flow" model of a location is compared to the sublogs of three other locations. Light-coloured activities are less often executed.

These differences were further investigated in a further analysis, revealing that these differences were likely caused by a difference in pricing models, i.e. in some locations, suppliers were paid for unsuccessful repairs, while in other locations the suppliers were not paid in these cases. This led to a proposal to implement order confirmations, and equalise the process over the four locations.

In this case study, the goals changed during the project, and the process mining techniques had to be flexible to deal with the changing requirements. Furthermore, both process discovery, conformance checking and log filtering were used in an iterative fashion. Each time a model was discovered, it led to new insights and new questions. Therefore, an iterative process was used in which event logs were repeatedly filtered.

Other Use Cases.

Another common use case is to get a better understanding of the process as it is
running, instead of how the organisation thinks it runs. For such a use case, it is
important that the model is understandable (the model has a high *simplicity*, i.e. uses
few constructs to express its behaviour), and less important that the model describes
all behaviour of the event log (the model has a high *fitness*, i.e. the model describes
most behaviour of the event log) or describes little more behaviour than the event log
(the model has a high *log precision*, i.e. the event log contains most of the behaviour
of the model) (we will explain fitness and log precision in more detail in Section 3.3,
and simplicity in Section 3.2.3). Such models are sometimes called *80% models*
or *happy flow models*, i.e. as a rule of thumb 80% of the behaviour can often be
expressed using only 20% of the model complexity [2]. However, conclusions should
be drawn with great care: if fitness or log precision is too low, conclusions cannot
be drawn reliably as the model does not fully capture the behaviour in the event log.
For instance, Figure 3.4a shows a Petri net which is an 80% model of the event log
$[\langle a, b \rangle^8, \langle b, a \rangle^2]$. Based on the model, one could conclude that a is always followed
by b. However, the event log and the model do not fit one another perfectly: in two
traces, b and a are reversed. Thus, one should conclude that in *most* of the traces,
a is followed by b, instead of that this is *always* the case. In contrast, Figure 3.4b
shows another Petri net, which allows for any behaviour. Based on this model, one
could conclude that a and b can be executed repeatedly. However, the model is not
very precise: it allows for much more behaviour than the behaviour that was seen in
the event log, thus even though the model allows for repeated execution, this never
happened in reality.

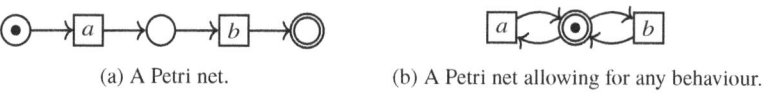

(a) A Petri net.   (b) A Petri net allowing for any behaviour.

Fig. 3.4

Another use case is to find a process model that can be used for *enactment*, i.e.
form the basis from which a process engine manages a process. Then, it is important
that the model represents most of the behaviour of the system (the model has a high
recall). Depending on the context of the enactment, it might be important that the
model does not allow for too much extra behaviour (the model has a high *system
precision*), e.g. in security settings. For enactment, *rediscoverability*, i.e. the ability
of a discovery algorithm to rediscover the behaviour, i.e. the language, of a system,
is a vital property, as the enacted model should support the system well (we will
introduce rediscoverability in more detail in Section 3.2.2.3).

If the use case of the project is to perform auditing-related tasks, e.g. to ensure
behaviour is absent or present, the discovered model should have well-defined semantics, such that log conformance checking techniques can show where deviations

occurred. The number of deviations should not be so high to become unmanageable, which implies that the model should describe most of the event log (fitness). Furthermore, if the model describes lots of behaviour that is not in the event log, conclusions that can be drawn could be weakened (log precision). For instance, Figure 3.5a shows a process model based on directly follows semantics, and was generated by Fluxicon Disco (FD) [56]. Behaviour was filtered from this model, i.e. it does not represent all behaviour in the event log, but it is not visible in the model *which* behaviour has been filtered out, which makes it challenging to spot where the model deviates from the event log and to assess the quality of this model. Therefore, conclusions on the absence of behaviour should not be drawn based on this model. (In Section 3.3.2, we will show that such models have ambiguities as well.) In contrast, Figure 3.5b shows a Petri net to which a log-conformance checking technique has been applied, and this Petri net has been annotated with the result of a log-conformance technique, e.g. the yellow large places and red-bordered transitions denote deviations (In Section 3.4.1.1, we will explain these deviations in more detail). Using these deviations, conclusions about absence of behaviour can be drawn, as the annotated model contains all information that is present in the event log.

Finally, the certain types of analyses may require reasoning about all behaviour, which requires all behaviour to be present (perfect fitness). In the next section, we will show that such a perfectly fitting model is always achievable, but often not desirable. If the process does not allow for an understandable perfectly fitting model, alternatively one could obtain an 80% model and study deviations using a log conformance checking techniques.

In the next section, we formalise challenges related to process discovery and conformance checking. Thereafter, we discuss discovery techniques, in Section 3.4, we discuss conformance checking techniques and we discuss enhancement techniques in Section 3.5.

3.2 Formal Key Challenges of Process Mining

In the previous section, we described a few typical use cases for process mining. In this section, we provide an introduction of formal challenges to two of the process mining problems that are addressed here, i.e. the problems of process discovery and conformance checking. The problem of *process discovery* is to, given an event log, return a process model, which ideally is "good". The problem of *conformance checking* is to, given a process model and either another model or an event log as input, return "good" measurements and information on the correspondence between them. In this section, we discuss what makes a "good" process model or conformance measurement, i.e. we elaborate on challenges of these two process mining subfields.

Ideally, a discovery algorithm returns a process model with well-defined semantics that is free of deadlocks, that is equal to the system, and is readable by both human analysts and machines. As illustrated in Figure 3.1, two entities are relevant in assessing the quality of a discovered process model or conformance measurement:

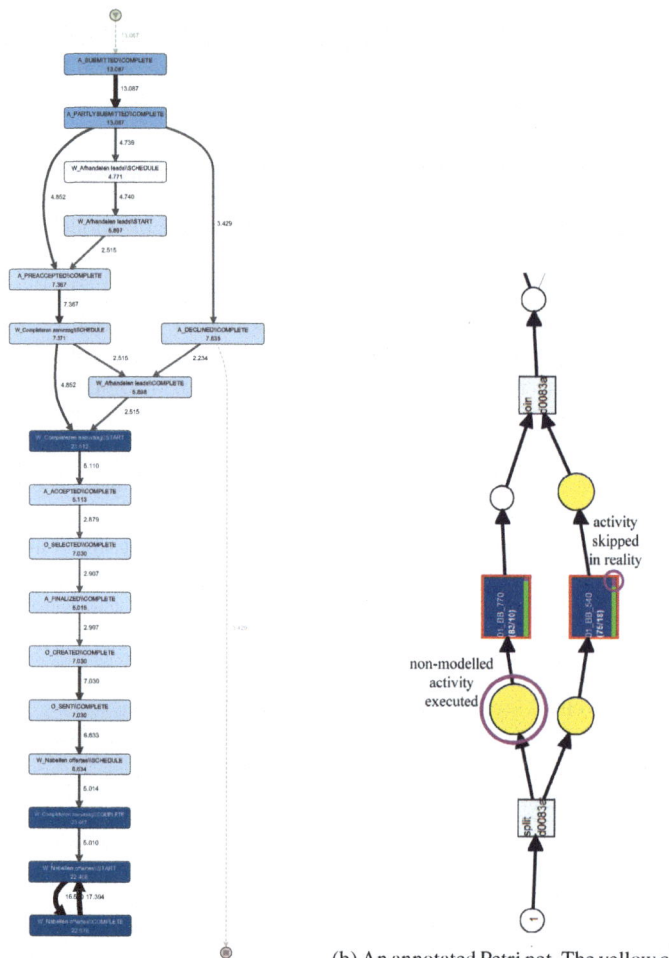

(b) An annotated Petri net. The yellow circles
and purple parts of bars in the blue boxes
(a) An FD model. Deviations are not visible.   denote deviations.

Fig. 3.5: Two models that do not correspond to their event logs.

the event log and the system. Therefore, process discovery can be hampered by
e.g. unclear semantics, the system being difficult to capture neatly in the output
formalism, the event log containing too little behaviour, or the event log containing
erroneous behaviour. The ideal conformance checking technique expresses, using
a few measures, how well a discovered model satisfies these challenges, compared
to a system or an event log, and the values of these measures compare well with
values obtained from other measures. Thereby, conformance checking techniques
face similar challenges as discovery techniques: the model might not have well-

defined semantics, the event log might contain too little information or too much false information to give reliable measurements on the underlying system, the model might be complex, and the measures might be biased by little-relevant features of the models and the event logs. In this section, we discuss these challenges of process discovery and conformance checking in more detail. We will recall that not all of them can necessarily be solved together: in real-life cases, different use cases of process mining projects might influence which challenges prevail.

Discovery techniques enjoy many degrees of freedom: the formalism of the process model, which behaviour to include in the process model, what behaviour to exclude, and how neat and readable the process model will be. Some quality dimensions are independent of the event log and universal to process models, such as producing sound models and producing models that can be layout such that they can be understood by human analysts. In this section, we gather formal requirements for process discovery and conformance checking. We first argue in favour of process models having precise semantics in Section 3.2.1, after which we discuss the relation between system, event log and discovered model in Section 3.2.2. We finish with a summary of the desirable properties of discovery algorithms and conformance checking techniques in Section 3.2.4. After this section, in sections 3.3 and 3.4, we will study existing techniques and gather practical requirements.

3.2.1 Models with Precise Semantics

Regardless of the use case at hand (see Section 3.1), all discovered process models need to adhere to some universal quality criteria. Most importantly, for most use cases, the model should have executable semantics, i.e. a language: for each trace it should be clear whether it is described by the model. Even though a model without such semantics can be useful for human interpretation, one should be careful with drawing conclusions from such models, as they might impose ambiguity. Obviously, computers have difficulties with models that do not have well-defined semantics, thus conformance checking techniques cannot be reliably applied. For instance, reconsider the process model shown in Figure 3.5a: in this model, the boxes denote activities, and the edges denote which activities denote the process flow, i.e. which activity can be executed after which activity. However, in this model it is not clear what the splits and joins represent: intuitively, all splits and joins represent exclusive choices, however they might also represent inclusive choices, interleavings and concurrencies [87]. Therefore, process discovery techniques should return models with executable semantics, yielding a new Requirement DR1:

Requirement DR1. *The model has a well-defined unambiguous language, and this model should be guaranteed to be sound.*

In order to determine whether a trace is represented by a process model, the model needs two key ingredients: an initial state and a (possibly unlimited) set of

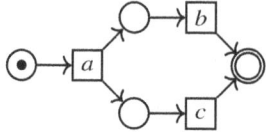

Fig. 3.6: An unsound workflow net.

final states. In a workflow net, these ingredients are present by construction, and therefore many process discovery techniques focus on workflow nets.

However, workflow nets might suffer from other issues. For instance, the workflow net shown in Figure 3.6 has an issue: the final state, i.e. a token in the sink place and no other tokens elsewhere, is not reachable. As a final marking is not reachable, this model has no traces and thus has an empty language. Even though a human analyst might be able to derive information from this model, conclusions should be drawn carefully as the perceived language of this model can be ambiguous. Applying automated analysis such as conformance checking to such models may lead to counterintuitive results. Furthermore, unsound models are obviously undesirable, as in reality there should not be unexecutable process steps, or customers waiting in a deadlock. Therefore, ideally a process discovery algorithm guarantees that every process model it discovers is sound, and as a first requirement for process discovery (Requirement DR1), any discovered model should be sound.

A weaker notion of soundness is *weak soundness*, i.e. a workflow net is weakly sound if it is possible to reach the final state from the initial state. If a model is not weakly sound, i.e. the final state of the model cannot be reached, then the language of that model is by definition empty, i.e. does not contain any trace. In the case a language is empty, language-based measures do not make much sense. A weakly sound model suffices for several log conformance checking techniques, such as [14] and the techniques we will introduce in Chapter 7 to get sensible measurements. Therefore, we add a new requirement for conformance checking techniques, being that any weakly sound model should be accepted. As weakly unsound models have empty languages, some techniques may apply heuristics to derive more information from such models, i.e. make certain assumptions such that the language of such models is not empty. If such heuristics are applied, then the measurements on weakly unsound models should still be comparable to measures on weakly sound models.

Requirement CR1. *The technique should return measures for all weakly sound models. If unsound models are accepted by the technique by applying heuristics, these measurements should be comparable to measures on weakly sound models. Log conformance techniques should only take language into account.*

Besides a language, systems have more properties that determine its behaviour. For instance, Figure 3.7 shows three *language equivalent* Petri nets. In the models of figures 3.7a and 3.7b, the choice for transition b and c is made at a different moment. These models can be distinguished using *bisimilarity*, which entails that one model can "mimic" all moves of the second or vice versa [2, Section 5.3]. That is, that a mapping exists between the states of the first model and the states of the second model, such that for every pair of mapped states, executing the same step in both models leads to two states in both models that are mapped [2, Section 5.3]. The model in Figure 3.7c contains silent transitions, which change the state of the system without a corresponding activity execution. The models of figures 3.7a and 3.7c are distinguished by bisimilarity, as the silent steps of Figure 3.7c cannot be mimicked in Figure 3.7a. However, intuitively, these models are perfectly capable of mimicking one another's visible steps and moments of choice: the choice between *b* and *c* remains unclear unless either is executed. The equivalence relations *branching bisimilarity* and *weak bisimilarity* capture this intuitive notion by taking silent transitions into account. That is, the models of figures 3.7a and 3.7c are branching and weakly bisimilar [51]. As both these bisimilarity notions are weaker than bisimilarity, the models of figures 3.7a and 3.7b are branching and weak bisimilar as well.

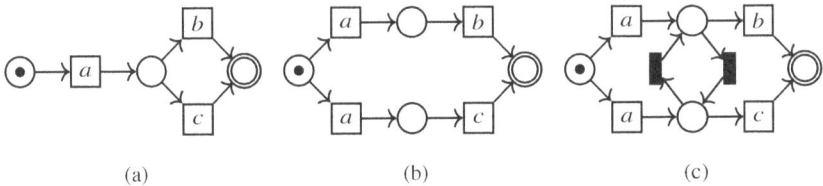

(a)   (b)   (c)

Fig. 3.7: Three Petri nets with the same language.

As moments of choice and unobservable behaviour are not recorded in an event log, and the models of Figure 3.7 have the same language, all of these models could produce the same event logs. Therefore, without further information, process discovery and conformance checking algorithms could not make a reliable choice between any of these three models or assess one as "better" over the others based on the event log. That is, we argue that conformance checking techniques should only take the language of a model into account (Requirement DR1). Using extra information, discovery algorithms and conformance checking techniques could guarantee stronger notions (which could be useful for analysts), but we limit ourselves to languages.

3.2.2 System - Log - Model Relations

Other quality dimensions depend on the system and the event log: we discuss the overlap in behaviour between these entities using a Venn diagram (figures 3.8, 3.9 and 3.10): in this diagram, each circle denotes the behaviour of either the system, the log, or a discovered model. The concepts explained using this diagram are independent of a particular notion of behaviour (branching behaviour, language, directly follows graph, activities, . . .), however for our purposes, we limit ourselves to languages. That is, we only assume that we can clearly distinguish whether a trace belongs to system, log and/or model. In [26], all 8 overlapping areas are discussed; here, we limit ourselves to the areas most relevant for our purposes. Furthermore, we do not elaborate on the relation of system models to systems, event logs and models, as we assume that the system model perfectly represents the system.

3.2.2.1 Model - Log Relation

First, we discuss how the discovered model can be positioned with respect to the event log, yielding *log conformance* concepts. In Figure 3.8a the green filled area denotes *fitting behaviour*, i.e. behaviour of the event log that is present in the model [30]. The blue filled area denotes the opposite, i.e. unfitting behaviour. For several use cases, it is important that the model contains a lot of fitting behaviour. For instance, in auditing, conclusions on the absence of real behaviour would be wrong when drawn from a model with lots of unfitting behaviour, as such behaviour happened in reality but is not described in the event log. However, if the model contains lots of imprecise behaviour, then conclusions on the absence of behaviour could not be drawn in the first place. If a general overview of the system is the aim of the analysis, a model with more unfitting behaviour might be desirable (see Section 3.1).

In Figure 3.8b the green filled area denotes *log-precise* behaviour, i.e. behaviour of the model that is present in the event log [1]. Conversely, the blue area denotes imprecise behaviour, i.e. behaviour of the model that is not in the event log. Log-precise behaviour is important for similar use cases as fitting behaviour, e.g. in auditing, conclusions about presence of behaviour should be drawn from a model with little log-imprecise behaviour, as log-imprecise behaviour is behaviour that is included in the model but was not observed in reality.

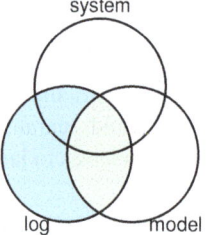

(a) Fitting & unfitting behaviour.

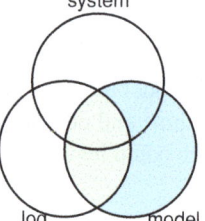

(b) Log-precise & -imprecise behaviour.

Fig. 3.8: Log conformance concepts.

---

[1] In contrast to literature we use the term log precise (instead of "precise") to distinguish log and system precision.

If the log and the model contain neither unfitting nor log imprecise behaviour, then the model is equivalent to the log according to the behavioural notion of the Venn diagram.

Measures.

The log-conformance measure *fitness* describes the balance of fitting and unfitting behaviour, i.e. in typical measures, fitness is 1 if all behaviour is fitting, i.e. all behaviour of the event log is represented in the model, and 0 if all behaviour is unfitting [30].

The log-conformance measure *log precision* denotes the part of behaviour in the event log that is precise: if all behaviour of the model is present in the event log, then precision is 1 [30]. In process discovery, the assumption is made that the event log does not contain all behaviour, i.e. the event log only contains examples. If all possible behaviour is assumed to have been recorded (which is not very realistic), one could just use the event log as a prediction of future behaviour. Therefore, in many use cases, perfect log precision is not necessary. Notice that models and systems might have an unbounded number of traces, e.g. in case of loops, while event logs are always bounded. Thus, similar to [83], an event log can never contain all behaviour of a system that allows for indefinite execution. Therefore, a model with a loop can conceptually never be perfectly log precise. This challenges log measures, as these measures need to quantify how much of the unbounded behaviour in the model is used in the bounded event log as well.

Ideally, log conformance measures provide guarantees, e.g. if fitness and log conformance are both perfect, the log conformance measure should guarantee that the event log and model are language equivalent (as discussed in Section 3.2.1, a stronger equivalence notion is not feasible as the event log only contains a language). This yields a new requirement:

Requirement CR2. *The technique measures fitness and log precision between an event log and a (system) model, and provides guarantees that these measures are perfect if and only if the event log and the (system) model have the same language.*

3.2.2.2 System - Model Relation

Second, we discuss how the discovered model can be positioned with respect to the system, yielding *system conformance* concepts: in Figure 3.9a, the green coloured area denotes *recalled* behaviour, i.e. behaviour of the system that is in the model as well. Conversely, the blue coloured region denotes behaviour that is possible in the system, but not in the discovered model. Process discovery algorithms aim to minimise the blue and maximise the green region, as this ensures that the model describes all behaviour that is possible in reality. In Figure 3.9b, the green coloured region denotes *system-precise* behaviour, i.e. behaviour of the model that is also present in the system [63]. Conversely, the blue coloured region denotes behaviour that is in the model, but not in the system, i.e. superfluous behaviour that should be minimised. Recalled and system precise behaviour are relevant for the same use cases as fitting and log precise behaviour, however taken from the viewpoint of the system: if conclusions about the impossibility of behaviour have to be drawn, the model should not contain (much) non-recalled behaviour. For the possibility of behaviour, there should not be (much) system-imprecise behaviour.

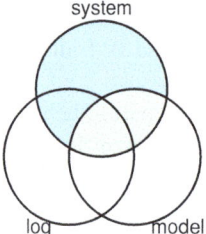

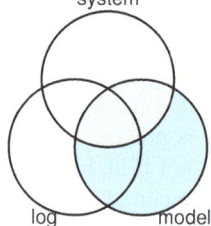

(a) Recalled & non-recalled behaviour.   (b) System-precise & -imprecise behaviour.

Fig. 3.9: System conformance concepts.

Measures.

The system conformance measure *recall* measures the amount of recalled behaviour compared to the unrecalled behaviour, i.e. if all behaviour of the system is in the model, then recall is 1 [63].

As the system is unknown in typical process mining projects, the log measure *generalisation* aims to estimate recall. That is, generalisation aims to indicate the likelihood that future behaviour will be represented by the model [28], and computes this estimate using only the event log. Generalisation techniques typically measure some property of the combination of the event log and the discovered model, for instance the variety in the event log: if the event log repeatedly contains the same behaviour, then generalisation is assumed to be high, as it is likely that a future trace will be supported by the model. However, if each trace in the event log represents different behaviour, then generalisation is assumed to be low, as it is likely that a future trace has not seen before and is therefore unlikely to be supported by the model [30].

In Section 3.4, we will discuss generalisation measures in more detail. However, in the evaluation (Chapter 8), we will not use generalisation estimations but instead measure recall directly using k-fold cross validation.

The system-conformance measure *system precision* measures the opposite of recall, i.e. if all behaviour of the model is in the system, then system precision is typically 1 [63]. Ideally, a system conformance checking technique should guarantee that two models have perfect recall and system precision if and only if they have the same behaviour (up to some notion of behaviour, such as language equivalence or bisimilarity), yielding a new requirement:

Requirement CR3. *The technique measures recall and system precision between process models, and provides guarantees that these measures are perfect if and only if some behavioural notion of equivalence holds.*

Similar to log precision, the measures recall and system precision are challenged by the possible unboundedness of system and model. Furthermore, model-model comparison might hit formal boundaries, e.g. for general Petri nets, language inclusion is undecidable [48], hence guarantees such as Requirement CR3 cannot be given, i.e. there cannot exist a set of system conformance measures that can reliably detect whether the language of a Petri nets includes the language of the other Petri net without making further assumptions on the Petri nets.

Rediscoverability.

If recall and system precision are both conceptually perfect and the circles of the Venn diagram overlap completely, the system is rediscovered. That is, the discovered model is language equivalent the system. A desirable property of discovery techniques is to guarantee language equivalence of models and systems, which is captured by a new requirement:

Requirement DR2. *The technique guarantees rediscoverability.*

We refer to this property as *rediscoverability*. Formally, rediscoverability is the ability of a process discovery algorithm to discover a model that is language equivalent to the system, while making assumptions on the event log and the system. For instance, for the α algorithm, rediscoverability has been proven. In Section 4.2, we will discuss rediscoverability in more detail and introduce a formal framework that aids in proving it for specific algorithms.

3.2.2.3 Log - System Relation

Finally, we describe how the event log and the system relate to one another, i.e. the two remaining Venn diagrams denote the relation between a log and the system (Figure 3.10). We refer to them as *log quality* concepts. In Figure 3.10a, the green coloured area denotes the *observed behaviour*, i.e. the behaviour that is possible in the system that actually happened and was recorded in the event log. Conversely, the blue coloured area denotes the unobserved behaviour, i.e. behaviour that is possible in the system but has not been observed in the event log. The more behaviour is observed, the easier it is for a process discovery algorithm to describe the system correctly. From the log perspective, in Figure 3.10b, the green coloured region denotes *correct* behaviour, i.e. behaviour recorded in the event log that was also possible according to the system. We refer to incorrect behaviour as *deviating* behaviour: this is denoted by the blue coloured region.

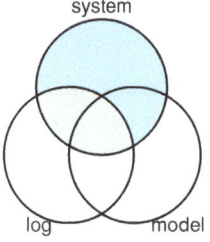
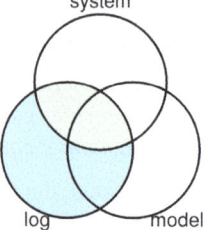

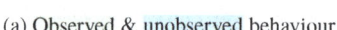

(a) Observed & unobserved behaviour.   (b) Correct & deviating behaviour.

Fig. 3.10: Log quality concepts.

In typical process mining projects (as described in Section 3.1), a model of the system is unavailable, but the event log can be influenced: more data can be extracted from the system if the event log seems to be incomplete, and infrequent behaviour can be filtered. The collection of Venn diagrams depicted in figures 3.8, 3.9 and 3.10 illustrates that if the event log is close to the system, i.e. has little unobserved and little deviating behaviour, then discovering a model that is close to the event log, i.e. with little unfitting or log imprecise behaviour, will result in a model that closely resembles the system, i.e. has little unrecalled and system imprecise behaviour. Conversely, if the event log has lots of unobserved and deviating behaviour, or the discovered model has lots of unfitting or log imprecise behaviour, then the discovered model might have lots of unrecalled and system imprecise behaviour.

Measures.

The log conformance concepts discussed earlier in this section apply to systems as well, as these measures express the differences between logs and models: *completeness* describes the balance of observed and unobserved behaviour, i.e. in typical measures, completeness is 1 if all behaviour of the system is present in the event log. Similarly, *correctness* describes the balance of correct and deviating behaviour.

For instance, in practise, we cannot expect an event log to contain all possible traces of the system, i.e. be *language complete*, as this may require unbounded event logs in case of loops. Furthermore, language complete event logs could become infeasibly large, e.g. there are $10! = 3,628,800$ possibilities to execute 10 activities concurrently, so an event log with all behaviour would need to contain at least $3,628,800$ traces. Furthermore, this would prevent any new information to be gained by discovering a model. Therefore, *instead of considering an event log directly, process discovery algorithms typically construct an abstraction of the behaviour in the event log first, and use that abstraction to discover a model*. This changes the notion of completeness, as the event log no longer needs to contain all behaviour, but just enough traces to "cover" the abstraction, and frees algorithms of the requirement that the log needs to contain every possible trace.

Challenges.

As discussed earlier, rediscoverability relies on making assumptions on event logs: the event log must contain enough correct information about the actual process. Therefore, several glitches in the relation between event log and system might challenge discovery. We discuss three of these challenges: one related to deviating behaviour, one related to *infrequent behaviour* and one related to completeness.

- As described, deviating behaviour is the occurrence of behaviour that is not part of the system but ends up in the event log anyway. For instance, consider an HR hiring process, during which the CEO kindly requests the department to hire a certain person, or an insurance claim handling process that is flooded with claims after a natural disaster and to help customers faster, mandatory checks are skipped. If deviating behaviour ends up in the event log, a discovery algorithm might have to handle it, e.g. exclude the behaviour from the model, depending on the use case. If such abnormal behaviour is the subject of the analysis, it might need to be included, whereas if the goal of the analysis is to visualise the big picture and leave out details, it better be excluded. Deviating behaviour can interfere with rediscoverability, because if the algorithm is unable to detect and exclude the noisy behaviour, it will be included in the model and rediscovery fails as the system did not contain this behaviour. Thus, process discovery algorithms need to be able to handle deviating behaviour, i.e. decide to filter or include it, but also be flexible in doing so, i.e. have the possibility to perform the opposite.
- A slightly different challenge arises from infrequent behaviour: *infrequent behaviour* are the little-used parts of a system, i.e. behaviour that is supposed to happen, but does not happen or happens in just a few cases in the event log. In the Venn diagram of Figure 3.10, infrequent behaviour is part of the observed and correct behaviour. The difference between deviating and infrequent behaviour is intention: if behaviour is not part of the system, the behaviour is deviating and if it is, the behaviour is infrequent. Systems with infrequent behaviour pose challenges to rediscoverability, as discovery algorithms have to decide whether behaviour is infrequent (and include it in the model if the entire system is to be described, or exclude it if only the most frequent behaviour is to be described) or deviating (and exclude it from the model). Obviously, this is difficult given just an event log. Luckily, in some use cases it is not necessary to distinguish deviating and infrequent behaviour, e.g. if the aim is to obtain the big picture of a process, both must be excluded.
- In some cases, event logs might be *incomplete*, i.e. the event log contains too little information for the complexity of the system. For instance, this might happen if the event log was extracted from data spanning a short period, or if the system contains many activities. The notion of completeness depends on the discovery algorithm, and different algorithms have different completeness notions. For instance, for IM (that we will introduce in Chapter 6) and the α algorithm, which both require the directly follows relation (that we introduced in Section 2.4). If an event log is incomplete for the combination of a system and a discovery algorithm, rediscovery fails. Therefore, a third challenge for

algorithms is to have a completeness notion such that small event logs can already be complete.

Notice that the three challenges, i.e. deviating behaviour, infrequent behaviour and incompleteness, might be ambiguous or overlapping. If the event log contains behaviour that occurs little, this behaviour might be deviating (the system does not contain it), infrequent behaviour (the system contains it but it's exceptional), or incomplete behaviour (the system contains it and there's similar behaviour that is unseen). For instance, consider the example log used before:

$$[\langle a, b, c, d \rangle^{18}, \langle a, c, b, d \rangle^{25}, \langle a, d, b, c \rangle]$$

Three systems from which this log could have emerged are shown in Figure 3.11. The trace $\langle a, d, b, c \rangle$ occurs seldom compared to the other traces. In the system shown in Figure 3.11a, this trace would indicate incomplete behaviour, i.e. several other traces are missing from the event log, e.g. $\langle a, c, d, b \rangle$. In Figure 3.11b, the trace would be deviating, as this system does not allow for it. The model shown in Figure 3.11c includes this trace, however the part of the model that supports this trace is rarely executed, thus the model contains infrequent behaviour. It's up to the discovery algorithm to choose, and solving the ambiguity is a though challenge for discovery algorithms, in particular as the choice might depend on the use case, which yields a new requirement:

Requirement DR3. *The technique is able to distinguish deviating behaviour, infrequent behaviour and incomplete behaviour, and guarantees rediscoverability in presence of these challenges.*

We argue that there cannot exist a discovery algorithm that performs this classification correctly in all cases. Therefore, human involvement in algorithms will remain necessary to make the right choice depending on the use case.

In [5], a different view on the relations between system, model and log is presented in which for each trace, a model expresses a probability that that trace occurs, instead of expressing simply whether a trace can happen. In this view, it is assumed that any behaviour can happen, though most behaviour with low probability, which consequently defines process discovery as the problem of finding a model that represents this trace probability distribution well. In further research, discovery algorithms might be adapted to include such distributions.

3.2.3 Simplicity & Balancing Log Criteria

In the previous sections, three log conformance measures were introduced: fitness, log precision and generalisation. In this section, we add a fourth one, *simplicity*, that expresses whether a model is understandable by human analysts. Furthermore, we show that discovery algorithms need to take these measures into account to avoid trivial uninteresting models.

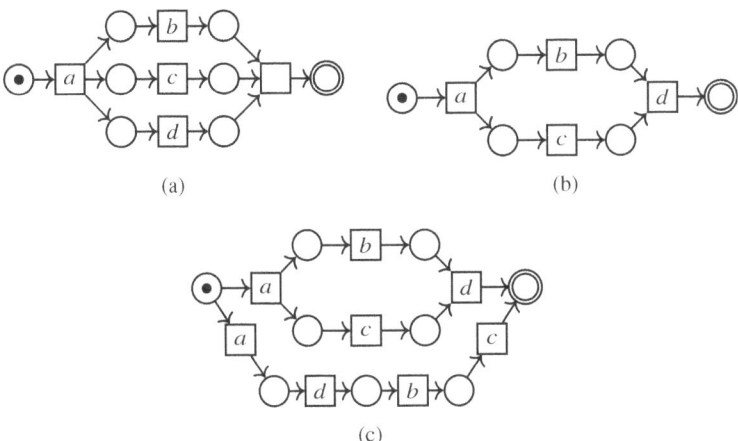

Fig. 3.11: Three systems that could have produced the event log $[\langle a, b, c, d \rangle^{18}, \langle a, c, b, d \rangle^{25}, \langle a, d, b, c \rangle]$.

3.2.3.1 Simplicity

The measure *simplicity* is intuitively defined as the understandability of a process model by human analysts, which is a highly subjective and ambiguous definition. To measure simplicity, typically circumstantial presumably complicating factors can be considered, such as the size of the model, partitionability (e.g. structuredness), cyclicity (e.g. the amount of looping behaviour or structures), concurrency [75] or redundant model elements [30], and all these factors can be measured in several ways. Furthermore, simplicity could be measured with respect to an event log or system, i.e. it could express whether the model is a complex representation of an event log or system. A detailed discussion of simplicity measures is out of scope; we refer to [76] for an overview of simplicity measures defined in literature. For our purposes, we will use a size-based simplicity measure. However, in the next section, we will show that discovery algorithms need to take simplicity into account to avoid trivial uninteresting models.

3.2.3.2 Balancing Log Measures

As process discovery algorithms take event logs as input and produce a model in an output formalism, a challenge discovery techniques face is that they need to represent the event log well in the output formalism [8]. In this section, we illustrate the challenges that arise with respect to event logs using four of the log conformance quality measures: *fitness*, *log precision*, *generalisation* and *simplicity* [2, Section 6.4.3].

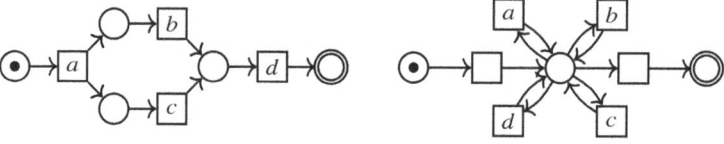

(a) An unsound model:
fitting, log precise, general, simple.

(b) A flower model:
fitting, log precise, general, simple.

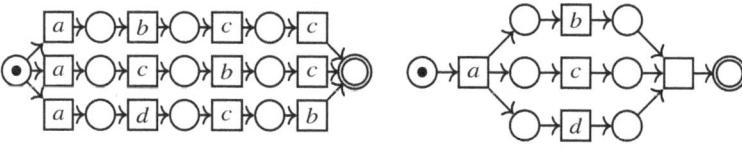

(c) A trace model:
fitting, log precise, general, simple.

(d) A general model:
fitting, log precise, general, simple.

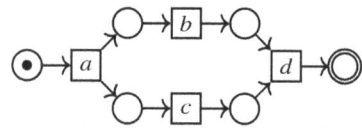

(e) A balanced model:
fitting, log precise, general, simple.

Fig.   3.12:   Competing   log   quality   criteria   for   the   log
$[\langle a, b, c, d\rangle^2, \langle a, c, b, d\rangle^2, \langle a, d, b, c\rangle]$.

Process discovery algorithms are challenged by these measures, as they should ideally all be high, but they sometimes compete. Therefore, discovery algorithms have to strike a balance between them. For instance, consider the following event log:

$$[\langle a, b, c, d\rangle^2, \langle a, c, b, d\rangle^2, \langle a, d, b, c\rangle]$$

Figure 3.12 shows several workflow nets that could describe this event log. The first (Figure 3.12a) is not sound and its language of correctly terminating executions is empty, as the final marking, i.e. a single token in the sink place, cannot be reached. Nevertheless, it is perfectly log precise as all traces of the model, i.e. none, are represented in the event log as well. The second (Figure 3.12b) is a *flower model*, i.e. it allows for any behaviour (i.e. traces) of a, b, c and d. It is therefore perfectly fitting and general, but not log precise as it describes much more behaviour than seen in the event log. The third (Figure 3.12c) is a *trace model*, i.e. it describes all traces of the event log and nothing else. It is therefore fitting and log precise, but not general and not simple, as it is unlikely that future behaviour will be represented in the model, and the model contains much more constructs than the other models. The fourth (Figure 3.12d) is a model that provides more information than the first three. It is fitting, general and simple. However, it allows for more behaviour than seen in the event log and is thus not log precise. The fifth (Figure 3.12e) seems to balance the quality criteria and seems to take the occurrences of traces in the event log into account: only the trace $\langle a, d, b, c \rangle$ has been left out and therefore the model is not fitting, but this left-out trace occurs just once in the log, and it therefore seems reasonable to leave it out in exchange for a higher log precision and simplicity.

In these examples, we illustrated that there might not be a model with a perfect or high score for all four measures in the representational bias of the discovery algorithm. Furthermore, even three of the four measures being perfect may still yield useless trivial models that do not provide any new information (figures 3.12b and 3.12c). Therefore, algorithms should avoid such trivial models and balance all four log measures, which yields a new requirement:

Requirement DR4. *The technique allows to balance log-conformance measures if the behaviour of the event log does not fit the representational bias of the technique well.*

However, this balance might depend on the use case: in our example, either of the models in figures 3.12d (for instance for implementation purposes) or 3.12e (for a view on the most-occurring behaviour) might be preferable (see Section 3.1). The challenge is made more difficult by the fact that for some event logs, there exists no model that scores well on all four log measures [30].

3.2.4 An Ideal Technique (1)

To summarise the previous sections, an ideal discovery technique takes an event log and produces a process model, such that:

DR1 The model has a well-defined unambiguous language, and this model should be guaranteed to be sound.

DR2 The technique guarantees rediscoverability.

DR3 The technique is able to distinguish deviating behaviour, infrequent behaviour and incomplete behaviour, and guarantees rediscoverability in presence of these challenges.

DR4 The technique allows to balance log-conformance measures if the behaviour of the event log does not fit the representational bias of the technique well.

Similarly, an ideal conformance checking technique adheres to the following requirements:

CR1 The technique should return measures for all weakly sound models. If unsound models are accepted by the technique by applying heuristics, these measurements should be comparable to measures on weakly sound models. Log conformance techniques should only take language into account.

CR2 The technique measures fitness and log precision between an event log and a (system) model, and provides guarantees that these measures are perfect if and only if the event log and the (system) model have the same language.

CR3 The technique measures recall and system precision between process models, and provides guarantees that these measures are perfect if and only if some behavioural notion of equivalence holds.

As discussed before, discovery algorithms and conformance checking techniques satisfying all of these requirements are unlikely to exist given the discussed trade-offs. In the remainder of this chapter, we discuss existing process mining techniques and extract practical requirements for conformance checking and process discovery. That is, the lists of requirements will be extended in sections 3.3.3 and 3.4.3.

3.3 Process Discovery

Process discovery aims at discovering a process model from an event log. In the previous chapter, we explored challenges and fundamental requirements of process discovery algorithms. In this section, we explore more practical challenges and illustrate how existing process discovery techniques solve these challenges.

In particular, we will consider existing techniques on rediscoverability (Requirement DR2) and balancing of log requirements (Requirement DR4).

To rediscover a system, a discovery algorithm requires a log to contain enough information about the system. As discussed in Section 3.2, if an algorithm would require all behaviour to be present in the event log, there might be systems for which a language complete event log is infeasible or even impossible (in case of loops). Therefore, many process discovery algorithms use a language abstraction. Typically, abstractions are rather small, e.g. the directly follows graph is quadratic in the number of activities of the process: for a system with 10 concurrent activities, less than $10 \cdot 9 = 90$ traces could suffice instead of $3,628,800$ for language completeness. Therefore, an assumption that an event log resembles an abstraction is much more likely to hold than an assumption on the full behaviour of the system. In Chapter 5, we study such abstractions in more detail.

3.3.1 Discovery Algorithms Guaranteeing Soundness

Many process discovery techniques have been proposed. In this section, we discuss some techniques: we first discuss some algorithms that guarantee to return sound models, after which we discuss some that do not provide this guarantee.

3.3.1.1 Evolutionary Tree Miner & Evolutionary Miner

The Evolutionary Tree Miner (ETM) [26, 27] and the Evolutionary Miner (EM) [78] are process discovery algorithms that uses a genetic approach: they generate an initial population of process models and iteratively alters this population until a predefined criterion is met. Altering steps include cross-over between process models and random changes. Moreover, in each iteration log performance of the population is measured and only the best performing models are kept.

To guarantee soundness, both ETM and EM limit their search space to an abstract hierarchical view of languages: process trees (Requirement DR1).

The ETM is very flexible: it can optimise any combination of any log based measure, e.g. if the use case requires a fitting model over all else, then ETM can be configured to prefer the best fitting models in population iterations, i.e. it satisfies Requirement DR4. Moreover, ETM has proven its flexibility on the used formalism as well: techniques have been proposed that let it discover configurable process trees, i.e. instead of a single process tree, a family of trees is discovered, which has proven useful in comparing similar processes over organisations [29]. EM differs from ETM in the log conformance measures used (in particular, the log precision measure), the flavour of process trees and the altering steps.

However, given their random nature, it is difficult to provide guarantees with respect to rediscoverability (Requirement DR2), and they tend to be intractable for real-life event logs. As we will show in Chapter 8, ETM can be slow on real-life event logs as its state space is huge, even though extensions have been defined [43] to traverse this state space in a smarter way than randomly: using 1 activity at most once, there are 5 process trees with a different language. For 2 activities, there are 46 trees with different languages, but many more with equivalent languages. This yields a new requirement:

Requirement DR5. *The algorithm should work fast on real-life event logs and systems.*

3.3.1.2 Constructs Competition Miner

The Constructs Competition Miner (CCM) [86] uses a process model formalism similar to the process trees used in ETM and EM (which were explained in Section 2.2.5), but discovers a model constructively by recursion.

The first step in CCM is to derive an *eventually follows graph*, i.e. activity a is eventually followed by b if there is a trace in the event log in which a happens somewhere before b. Second, the recursion begins: first the most important behaviour, i.e. the operator of the root of the process tree, is determined and the activities are divided in a binary partition. Based on the operator and partition, the eventually follows graph is split in two smaller sub-graphs, and on these sub-graphs, CCM recurses until a base case, i.e. a graph containing a single activity, is encountered.

CCM guarantees soundness (as it uses process trees), supports deviating behaviour filtering and is quite fast in practice, even though its run time is worst-case exponential in the number of activities. As a witness of the running time, CCM has been applied in streaming environments, in which the event log cannot be stored [85]. However, due to the use of the eventually follows graph, rediscoverability cannot be proven (Requirement AP.5, which will be introduced in Section 4.2.2, does not hold) (Requirement DR2).

In most discovery algorithms (e.g. α, HM, LT, and more discussed below), choices in a process model are considered to be independent, and each activity can only appear once in the resulting model. A *long-distance dependency* between two moments of choice in a process breaks this independence, and *duplicate activities* allow for activities appearing twice in the model. Figure 3.13 shows an example: both workflow nets have the same language, but the first uses a long-distance dependency, while the second uses a duplicated activity. Furthermore, due to the process tree representation in which each activity occurs at most once, CCM cannot represent long-distance dependencies and *non-free choice constructs*, which yields a new requirement:

Requirement DR6. *The algorithm should be able to (re)discover models with non-free choice constructs, duplicate activities and long-distance dependencies.*

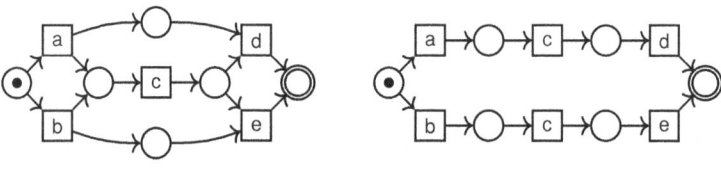

(a) Long-distance dependency: the choice between d and e depends on the choice between a and b.

(b) Duplicate activity c.

Fig. 3.13: Two language-equivalent workflow models: one with long-distance dependencies, one with a duplicate activity.

3.3.1.3 Maximal Pattern Mining

The Maximal Pattern Miner (MPM) [66] uses patterns to cover the entire event log. First, for each trace loops are folded, and second, a prefix tree is constructed of the entire event log. This prefix tree is guaranteed to be sound. On this prefix tree, sub-parts are folded and combined to discover concurrency and choice, thereby the soundness is preserved. MPM is able to handle long-distance dependencies, duplicate activities (except under concurrency) and non-free choice constructs (Requirement DR6). However, this often comes at the price of simplicity: when applied to a typical real-life event log, the resulting model may contain many duplicate activities [66].

The technique Process Miner (PM) [92] performs a similar strategy, i.e. it unfolds loops, merges similar traces based on concurrency, merges these into a process tree, and finally this process tree is minimised using reduction rules. In determining concurrency, PM uses non-atomic event logs (see Section 2.3.2). That is, if two activity executions overlap in time, they are considered concurrent. A downside of PM is that is was specifically designed not to generalise, i.e. all and only behaviour of the event log is to be included in the model [92].

3.3.2 Other Discovery Algorithms

Besides the class of soundness guaranteeing algorithms, many algorithms exist that do not provide the soundness guarantee. In this section, we list a few.

α-algorithm.

The α-algorithm (α) [2, p.130] was one of the first process discovery algorithms. It constructively builds a Petri net from an event log L, using an intermediate abstraction, i.e. a directly follows graph. The directly follows graph denotes for each activity which activities can be executed immediately after it. From such a graph, the α-algorithm first derives causal relations, i.e. for a pair of activities a and b, $a >_L b$ if a is directly follows by b somewhere in the event log. Second, the causal relations are combined into activity relations:

$$a \rightarrow_L b \Leftrightarrow a >_L b \wedge a \not>_L b$$
$$a \#_L b \Leftrightarrow a \not>_L b \wedge b \not>_L a$$
$$a \|_L b \Leftrightarrow a >_L b \wedge b >_L a$$

Informally, $a \rightarrow_L b$ denotes that a and b are in a sequential relation, $a\#_L b$ that a and b are mutually exclusive, and $a\|_L b$ that a and b are concurrent. From these relations, a Petri net is constructed by searching for maximal patterns in the activity relations. For instance, if $a \rightarrow_L b$, then there will be a place connected from a to b. Additionally, if there is a c such that $c \rightarrow_L b$ and $a\#_L c$, then the place also has a connection from c (see Figure 3.14). For more information, please refer to [2, p.130].

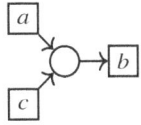

Fig. 3.14: An example pattern of the α-algorithm, based on the relations $a \rightarrow_L b$, $c \rightarrow_L b$ and $a\#_L c$.

As finding maximal patterns in the activity relations can be exponential in the number of activities, the α-algorithm might be slow for event logs with lots of activities. In Chapter 8, we will show that for larger common real-life event logs, we were unable to discover a model using the α algorithm. Nevertheless, the α-algorithm works fast in practice as it builds a Petri net using a constructive approach and is linear in the size of the event log (assuming the number of activities is fixed). Rediscoverability has been proven for the α-algorithm [20], using the assumption that the activity relations in the log precisely match the activity relations in the system, which is equivalent to their directly follows graphs being the same (Requirement DR2).

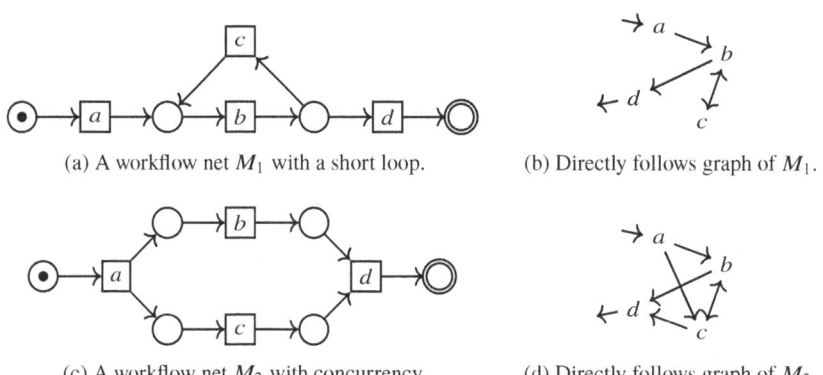

(a) A workflow net M_1 with a short loop.

(b) Directly follows graph of M_1.

(c) A workflow net M_2 with concurrency.

(d) Directly follows graph of M_2.

Fig. 3.15: An example of short loops.

Some of the limitations on the system are that it cannot have *short loops*. Figure 3.15 shows an example of a short loop: activities b and c can directly follow one another in both Figure 3.15a and Figure 3.15c. The pattern in the directly follows graph is that b and c have a back and forth connection, and, on an event log derived from this model, the α algorithm therefore considers these activities to be concurrent, as in Figure 3.15c, even though the directly follows graphs of these models are different. This yields a new requirement:

Requirement DR7. *The algorithm should be able to (re)discover models with short loops.*

The short loop restriction was dropped in the α^+-algorithm [72]. Further restrictions include:

- the system cannot have long-distance dependencies. This was addressed in [110] and is expressed in Requirement DR6;
- the system cannot have non free-choice constructs. This was addressed in α^{++} [107] and is expressed in Requirement DR6;
- the system cannot have silent transitions. This was addressed in $\alpha^{\#}$ [109, 111] and in $\alpha^{\$}$ [57], and yields a new requirement:

Requirement DR8. *The algorithm should be able to (re)discover models with silent steps.*

- the system cannot have duplicate activities. This is expressed in Requirement DR6.

To the best of our knowledge, this last requirement has not been solved yet, and in Section 5.8, we will show that it cannot be completely solved.

Another variant of the α algorithm, called the Tsinghua-α algorithm (Tα) [108], deals with non-atomic event logs, i.e. event logs in which executions of activities take time (see Section 2.3.2). This algorithm uses a variant of the directly follows relation that takes activity instances into account instead of events. Furthermore, a new relation keeps track of activities that overlap in time and hence are concurrent. Using these two relations, Tα constructs a Petri net, similar to the other α variants. Due to the concurrency relation, logs need to contain fewer traces, e.g. the normal α algorithm would need to observe $[\langle a, b \rangle, \langle b, a \rangle]$ to conclude concurrency, while Tα only needs to observe $[\langle a_s, b_s, a_c, b_c \rangle]$.

A downside of the α-algorithm and its derivatives is that they do not guarantee the discovery of sound models (Requirement DR1). As a consequence, neither fitness nor log precision can be guaranteed. Moreover, neither any of these derivatives nor the original algorithm is able to handle deviating behaviour, incompleteness or infrequent behaviour (Requirement DR3).

Causal Nets / Heuristic Miner / Little Thumb / Fodina / Structured Miner.

In response to discovery algorithms having problems returning useful models for deviating behaviour, infrequent behaviour and complex event logs, *causal nets* were introduced in [6]. Causal nets aid soundness-seeking techniques by guaranteeing proper completion and absence of deadlocks and livelocks: by definition, the language of a causal net only consists of traces that reach the final state (Requirement DR1). As a side-effect, syntactical OR-splits and joins are possible in causal nets. However, deadlocks, livelocks and unreachable parts remain possible in causal nets, even though they are not considered to be part of the behaviour of the net. Notice that we apply this strategy of causal nets to Petri nets: only traces that end in a final marking are considered to be part of the language of the net, and traces that end in a deadlock or livelock are not considered to be part of the language. The causal net discovery technique proposed in [95] uses Satisfiability Modulo Theories (SMT) to find the smallest causal net that includes the event log. However, SMT problems become intractable for large event logs, which was alleviated in [93], in which the event log is split in smaller parts recursively before the SMT technique is applied to discover several smaller intermediate causal nets. Afterwards, the intermediate causal nets are combined into the final causal net.

The Flexible Heuristic Miner (HM) [106] and the Little Thumb (LT) [105] apply a strategy similar to α: they construct a directly follows abstraction and use activity relations to construct a causal or workflow net. However, to handle deviating and infrequent behaviour (Requirement DR3), the activity relations are derived probabilistically, e.g. if $a >_L b$ happened once and $a >_L b$ 400 times, then the probability that $b \rightarrow_L a$ is higher than if $b >_L a$ happened twice and $a >_L b$ did not happen at all. These probabilities are considered locally, i.e. no other activities have influence on the relation between a and b. Next, the probabilistic relations are filtered: for each activity a, let b be the activity with the strongest probabilistic $\rightarrow_L$-relation. Then, the outgoing relations of a that are weaker than $a \rightarrow_L b$ by a user-given relative threshold are removed. In LT, from the filtered α activity relations a Petri net is constructed similar to the α-algorithm [105]. In HM, first a causal net is constructed, which can be converted into a Petri net [106].

As LT uses the workflow net construction step of α, it inherits its exponential runtime in the number of activities, while HM is polynomial (Requirement DR5). HM supports long-distance dependencies and has been extended to partially support duplicate activities [23] (we will show limitations to this in Section 5.8, Requirement DR6), LT short loops and non-free-choice constructs. However, both HM and LT do not guarantee the discovery of sound models, and as a consequence fitness nor log precision can be guaranteed.

Fodina (FO) [23] extends HM with support for long-distance dependencies and duplicate activities (Requirement DR6): FO discovers these constructs by considering the context of an activity c, that is which combinations of activities happen right before and right after c. From this information (in our example: $(\{a\}, c, \{d\})$ and $(\{b\}, c, \{d\})$), the long-distance dependency or duplicate activity is derived. A downside of this approach is that it might imply that the completeness notion requires even larger event logs, e.g. a model with a choice at the end of the process depending on a choice at the start of a process might require the log to be language-complete. That is, a technique can only decide that two choices are not dependent if all combinations of the choices have been seen, and if dependent choices might be arbitrarily far apart, then all choices are assumed to be depending on one another until all possible combinations have been seen, thus if a single combination is missing, the model will change.

A shared downside of HM and FO is that they do not guarantee to return sound models, and tend to return even weakly unsound models. To counteract this, the Structured Miner (SM) [19] post-processes the models discovered by HM or FO. In this post-processing step, the model is first translated to BPMN, after which the model is transformed into a block-structured model by applying several transformation rules exhaustively [84]. For instance, one such rule resolves unmatched splits and joins (e.g. a concurrent split and an exclusive choice join), by "pushing" the gateways outwards. Figure 3.16 shows an example: the exclusive choice gateway in the center of the process is pushed out, thereby duplicating c. The set of rules proposed in [19] is not complete, i.e. not every model can be transformed into a fully block-structured model, and consequently soundness is not guaranteed. Furthermore, the run-time complexity in terms of activities and gateways in the BPMN model is $O(n^n)$.

Fig. 3.16: Example of a structuring rule of the Structured Miner.

Theory of Regions & Integer Linear Programming Miner.

The theory of regions aims to find a Petri net which represents the same behaviour as a given specification of a system. We discuss two variants of this theory: state- and language-based region theory. In early state-based region techniques, a Petri net is synthesised from a state space such that the state space of the Petri net is branching bisimilar to the state space of the log, by searching for sets of states such that every activity either enters this set, exits it or never crosses the boundary of the set. These sets are the *regions* and correspond to places in the discovered Petri net [45, 46].

Region-based techniques typically guarantee weak soundness. However, to guarantee soundness, typical region-based techniques depend on certain properties of the state space. For instance, a unique start and end activity might be required to guarantee that the returned Petri net will be a workflow net, and the state space needs to be "elementary" to guarantee liveness and absence of deadlocks [11]. The approach proposed in [36] drops this last restriction and guarantees soundness for arbitrary bounded state spaces, however it duplicates activities in the Petri net.

To obtain a state space from an event log, in [11] a two-stage approach is proposed that first constructs a state space of the event log and generalises this state space by using abstractions. Second, this approach uses the approach of [36] to generate a corresponding Petri net, which might be forced to be a sound workflow net if each trace in the event log is appended with an artificial start and an artificial end event beforehand and the state space abstraction method is chosen carefully.

However, such techniques have difficulties generalising, to filter infrequent or deviating behaviour [112], and might have long computation times.

Later techniques such as [33] adapt these techniques to generalise over the behaviour of the event log, i.e. only require that the language of the model includes the event log (i.e. perfect fitness). However, the inability to filter infrequent or deviating behaviour remains. As a way to avoid lengthy computations, in [94] it is shown that state spaces could be split before computing regions, thereby saving time. In [32, 31], heuristics are proposed that split the event log in concurrent parts and continue computation on the smaller split logs, putting their results concurrently in the final result (a strategy similar to [93] for causal nets). Our approach will extend this divide-and-conquer idea to avoid the further theory-of-regions computations altogether.

In language-based region theory, a Petri net is derived from a language, such that the resulting net has the smallest behaviour possible while still containing the language. Language-based region theory starts with a Petri net containing all transitions, but no places. In this net, all behaviour is possible. Second, places are added that do not prohibit any trace from the language. Thus, the main challenge is to choose the places such that the behaviour is limited as much as possible, without excluding any behaviour from the language. This principle is applied to process discovery by the Integer Linear Programming Miner (ILP) [112]. The challenge is solved by translating the problem of adding places to an ILP problem. An initial marking is always provided, and an optional extension ensures that the net has no remaining tokens at the end of a trace from the log. With this optional extension, ILP guarantees weak soundness, though for traces not in the log, the net may reach a deadlock with tokens remaining in the net, or it might be impossible to reach a final marking, thus soundness is not guaranteed.

Given this optional extension, one could assume the final marking to be the empty marking. With that extension and assumption, the models returned by ILP have a defined language, i.e. we can determine the traces that are included in the language of the model (Requirement DR1), and then ILP guarantees perfect fitness and, for its representational bias, best log precision. Moreover, due to its global nature, in contrast to the locally working HM and α, ILP is able to discover workflow patterns such as the milestone, the parallel interleaved routing, the critical section and the arbitrary cycle [9], yielding the new requirement:

Requirement DR9. *The algorithm should be able to (re)discover models with workflow patterns such as the milestone, the parallel interleaved routing, the critical section and the arbitrary cycle.*

As a consequence of perfect fitness and best log precision for its representational bias, ILP often suffers from overfitting, i.e. a lack of generalisation and deviating/infrequent behaviour handling. These problems have been addressed in an extension [116]. However, the discovered net is not guaranteed to be a workflow net, and even if a workflow net is returned, it is not guaranteed to be sound. Furthermore, ILP is not a fast algorithm: the ILP algorithm is exponential in the number of events in the log.

Fuzzy Miner & Fluxicon Disco.

The Fuzzy Miner (FM) [55] and Fluxicon Disco (FD) [56] apply the analogy of road maps to directly follows graphs: on such maps, roads connect cities but not every road or city is equally important. Thus, little-used alleys can be removed while highways remain (edge filtering) and suburbs and neighbourhoods can be grouped into cities (activity clustering). This makes such models (fuzzy models) suitable for interactive hierarchical exploration, and allows for balancing log-conformance criteria, and yields a new requirement:

Requirement DR10. *The algorithm should be able to balance log conformance criteria depending on the use case at hand, and this balance should be influenceable by analysts.*

In a fuzzy model, outgoing edges of an activity a mean that these activities might be executed after the execution of a. However, as inherited from directly follows graphs, it is not immediately clear whether this implies an exclusive choice, an inclusive choice, an interleaving or a concurrency [87]. In some commercial tools, e.g. FD, concurrency related edges, i.e. pairs of directly follows edges between two activities that overlap in time, are filtered out in some cases, which makes the language ambiguous. Moreover, fuzzy models might suffer from a problem, which is similar to a problem found in unsound workflow nets: it might not always be possible to reach the end state (Requirement DR1). Thus, these models are less suitable for enactment and concurrency detection.

3.3.2.1 Declare

A different angle to business processes is given by *Declare*. While Petri nets, BPMN models and other formalisms describe what can happen, a Declare model can also express what can *not* happen [71]. Such constraints are expressed between activities, e.g. activity a is always eventually followed by activity b. Declare models can be discovered by algorithms, for instance in [113], all possible constraints are checked on the event log and the ones that hold are returned. This method obviously overfits, thus in [38], constraints are filtered using thresholds for support and confidence, while taking care that the set of constraints remains satisfiable (Requirement DR1). Furthermore, many Declare techniques make sure not to return models with overlapping constraints.

Despite the different angle and the much greater flexibility in modelling, a Declare model still describes a language and all the challenges described before hold for it, with the added complexity that executing a Declare model is challenging, as a next step is only supported by the model if a final state can still be reached, i.e. a state in which no constraints are violated[2]. Therefore, it is essential that discovery techniques guarantee a satisfiable set of constraints. Furthermore, the (extensible) constraints of Declare are based on LTL-formulae, which implies that there are regular languages (non star-free languages) that are not expressible by Declare models [114], for instance the language $\{(aa)^n | n > 0\}$ [91].

3.3.2.2 Negative Events

Process discovery algorithms implicitly make assumptions about which behaviour can and cannot happen. *Negative events* make this assumption explicit, as they show for each prefix trace in an event log what activities can not happen [52]. Several methods have been proposed to infer (discover) such negative events artificially, as

---

[2] Conjectured NP-complete: given a trace, checking whether it is in the Declare model can be done in polynomial time (nondeterministic polynomial time), and the 3-SAT problem can be solved by transforming it in polynomial time to a Declare model, using custom 3-ways constraints, such that the 3-SAT-problem is satisfiable if and only if the Declare model has a trace (NP-hard).

they are not commonly found in event logs [52]. Using negative events, process discovery becomes a binary classification problem, and the generation of negative events provides an extra degree of freedom towards handling deviating behaviour. However, the technique described in [52] does not guarantee soundness, and challenges related to log precision and generalisation are shifted to the introduction of the negative events, i.e. these key problems are not addressed.

3.3.2.3 Further Optimisations

Process discovery on larger or complex event logs, i.e. containing millions of traces or hundreds of activities, can be challenging. Therefore, several techniques have been proposed to decrease the run time complexity of process discovery and the complexity of its resulting models. For instance, in [73] similar traces are clustered, which reduces complexity (horizontal partitioning). In [32] and [12], traces are split in concurrent sub-traces (vertical partitioning).

3.3.3 An Ideal Process Discovery Technique (2)

While discussing the practical challenges of existing algorithms we collected a list of additional requirements that complements the list in Section 3.2.4:

DR5 The algorithm should work fast on real-life event logs and systems.
DR6 The algorithm should be able to (re)discover models with non-free choice constructs, duplicate activities and long-distance dependencies.
DR7 The algorithm should be able to (re)discover models with short loops.
DR8 The algorithm should be able to (re)discover models with silent steps.
DR9 The algorithm should be able to (re)discover models with workflow patterns such as the milestone, the parallel interleaved routing, the critical section and the arbitrary cycle.
DR10 The algorithm should be able to balance log conformance criteria depending on the use case at hand, and this balance should be influenceable by analysts.

Requirement DR10 is an extension of Requirement DR4 and entails that algorithms are robust to deviating, infrequent and incomplete behaviour. This requirement is well supported by various algorithms. However, the "best" support is provided by ETM and EM, that allow fine-grained control of this balance.

Regarding Requirement DR5: the fast and robust algorithms tend to use an abstraction (α, HM, FO, FD, etc.) instead of the full event log (ETM, EM, ILP), and work deterministically (α, HM, FO, FD, etc.) instead of randomly (ETM, EM).

Duplicate activities (Requirement DR6) are not well supported by abstraction-based algorithms (α, HM, FO, etc.) and ILP. They are supported by the current algorithms that discover process trees (ETM, CCM), however these algorithms provide no rediscovery guarantees. To deal with duplicate activities, several techniques have been proposed that are independent from specific discovery techniques. For instance, [18] and [69] pre-process the event log to detect certain types of duplicate activities, annotate the event log (e.g. each occurrence of an activity a is labelled with a sequence number: the first occurrence in a trace is renamed to a_1, the second to a_2 etc.), and continue with a discovery technique.

As shown in the discussion of α, Requirement DR7 is inherent to using a directly follows graph, that cannot distinguish a short loop from two concurrent activities. Algorithms not using the directly follows graph (ETM, EM, ILP, etc.) tend to have little problems with this construct.

Requirement DR8 comes with Petri nets, i.e. silent transitions, and process trees, i.e. τ leaves. These silent steps do not cause an event to be recorded in a trace on execution, and therefore cannot be observed directly, which makes them hard to detect. The algorithms that do not use an abstraction (ETM, EM) support this construct, unless it is not in the representational bias of the algorithm (ILP).

The only algorithm in our study that supports workflow patterns such as the milestone (Requirement DR9) is ILP. This is due to the abstractions used by other algorithms that are often activity-based, e.g. in a directly follows relation, these constructs are hard to detect. Moreover, these constructs often do not fit the representational bias of algorithms. For instance, the ETM does not use an abstraction, but its use of process trees limits support for these constructs, as they cannot be easily expressed in process trees.

Up till now, there is no discovery algorithm that satisfies all of these requirements. We would prioritise balancing and speed, however we argue that a desirable algorithm (re)discovering all the mentioned constructs cannot exist, as the more models a technique can (re)discover, the more information it needs in an event log to distinguish all of these models. In Section 5.8, we elaborate on this.

Most of the algorithms we considered use a language abstraction: we encountered directly follows graphs and eventually follows graphs as typical abstractions. Algorithms that do not use an abstraction, such as ETM or ILP, tend to be able to (re)discover more constructs, such as short loops, milestones and long-distance dependencies. These constructs are challenging for other, abstraction-based, algorithms, as these constructs express more complex behaviour than what can be captured with only the relation between two activities. A downside of these techniques is that they tend to be slow on real-life logs, and that that they may yield overfitting models.

ILP can guarantee to return weakly sound models, MPM and the approach described in [11] can guarantee soundness and all discovery algorithms that use process trees guarantee soundness. In the approach presented here, we will fix the output formalism to process trees, and leave implementing algorithms free to use abstractions or not. In Section 3.6, we introduce our approach, we formally define it in Chapter 4, elaborate on abstractions in Chapter 5 and introduce process discovery techniques in Chapter 6.

In the next section, we consider existing conformance checking techniques.

3.4 Conformance Checking

Conformance checking aims to derive information from the difference between a process model and either a system or an event log (see Figure 3.17). Furthermore, conformance checking aims to measure the quality of an, e.g. discovered, model with respect to the event log or the system. In Section 3.2, we introduced conformance checking and its challenges. In this section, we describe how existing techniques measure the quality of a model, and what information can be derived from these measures. Furthermore, we extend the list of requirements started in Section 3.2.4. We first discuss log conformance, after which we discuss system conformance.

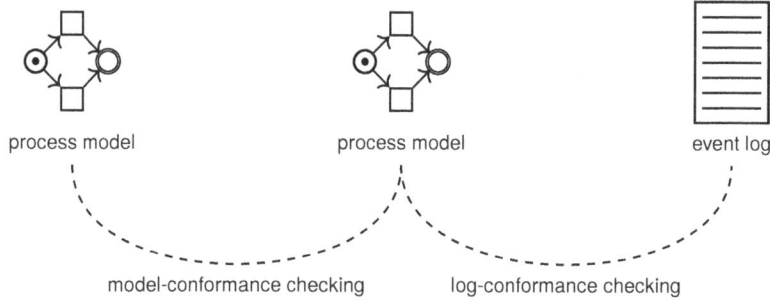

Fig. 3.17: The context of two types of conformance checking.

3.4.1 Log Conformance Checking

Log conformance checking provides insight in the relation between an event log and a (system) model, for instance to verify that the model is describing the behaviour in the event log well, and that therefore the model resembles reality. In many specific use cases of log conformance checking, it is useful to have an estimate of which path a given trace took through the model, i.e. how the trace can be projected onto the model, as this provides information about how model and log are alike or differ, and where they differ. For instance, this could show that certain parts of a model are skipped in reality, or that extra activities are executed: this information is essential for computing animations, measuring frequency and measuring performance (see Section 3.5). Projecting a trace on a model can be done unambiguously if the model does not contain duplicate activities and the trace is part of the language of the model. However, if the trace is not part of the language of the model or the model contains duplicate activities, then projecting might be more difficult.

We discuss several techniques to perform log conformance checking: we first elaborate on techniques that use such projections, and second we discuss other techniques.

3.4.1.1 Alignment-based Approaches

An *alignment* is the result of a projection of a trace on a model: an alignment contains the trace and a path through the model, interleaved to match, e.g. pairing events in the log with executions of transitions in the model while preserving execution order of log and model [14].

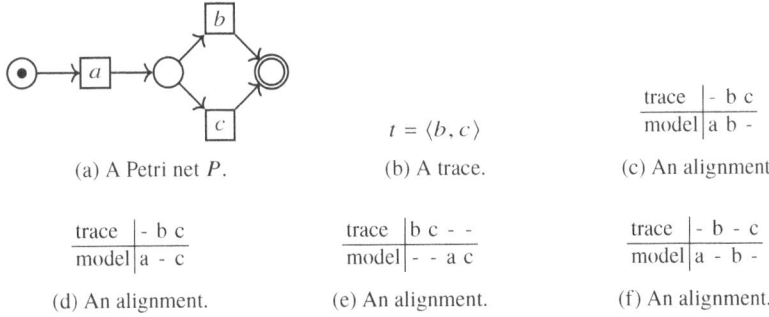

<table>
<tr><td></td><td></td><td>trace</td><td>- b c</td></tr>
<tr><td></td><td></td><td>model</td><td>a b -</td></tr>
</table>

(a) A Petri net P. $t = \langle b, c \rangle$ (c) An alignment.

(b) A trace.

trace	- b c
model	a - c

(d) An alignment.

trace	b c - -
model	- - a c

(e) An alignment.

trace	- b - c
model	a - b -

(f) An alignment.

Fig. 3.18: An example of a Petri net, a trace and some alignments of them.

For instance, Figure 3.18a contains a Petri net P, Figure 3.18b shows a trace t, and Figure 3.18c shows an alignment between P and t. In this representation of an alignment, the top row represents t and the bottom row represents a trace of the language of P. Trace t is not in the language of P, i.e. $t \notin \mathcal{L}(P)$, as a is missing from t. Such an omission is a *model move*, which is represented by a - on the top row. Moreover, b and c are exclusive but t contains both. In this alignment, only b corresponds to the model; c is omitted. Such an omission is a *log move*, which is represented by a - on the bottom row. In this alignment, t and the trace of P agree on b, which makes b a *synchronous move*.

Notice that the alignment given in Figure 3.18c is not the only possible alignment: figures 3.18d - 3.18f show a few more alignments of trace $\langle b, c \rangle$ to model P. An *optimal alignment* is an alignment with a minimum number of log and model moves. In our example, the alignments of figures 3.18c and 3.18d are optimal, as there is no alignment of t and P with less than two log/model moves. Optimal alignments can be computed, but a major challenge for alignment-based techniques is their complexity. The complexity of the optimal alignment computation is worse than exponential. Even optimised implementations of these techniques cannot deal with millions of traces or hundreds of activities [63, 81]. Therefore, we add a new requirement:

Requirement CR4. *The algorithm should work fast on real-life models and event logs.*

To alleviate this problem, decomposition techniques have been proposed, for instance using passages [3, 12], single-entry-single-exit decompositions [81], or even more general constructs [4]. However, in practice alignment computations might be slow up to the point that they become infeasible to compute on large models encountered in practice (as will be shown in Chapter 8).

The example illustrates a limitation of alignments as well: even though it's tempting to consider an optimal alignment to be the most likely explanation of the path that a trace in reality took through a model, this might not be correct: in our example, both optimal alignments provide an equally likely explanation whether b or c was executed. Obviously, alignments cannot provide guarantees about what happened in reality as well. Nevertheless, they are useful in spotting problems in models, computing animations and measuring performance, given the limitations described.

An alignment can exist if and only if the model can reach a final state (weak soundness [14], Requirement CR1).

Three Levels of Insights.

Alignments may provide insights into the conformance of log and model on three levels. First, fitness and log precision measures summarise the conformance in a number. We will discuss these summarative measures later in this section.

Second, the information of the alignment, i.e. model moves, log moves and synchronous moves, can be aggregated and projected on the model, possibly leading to insights into which parts of the model deviate from the event log. Figure 3.19 shows an example: on this workflow net, places coloured yellow denote that log moves occurred when the place was marked, and their sizes increase with the number of log moves. Similarly, a red border around a transition denotes that a model move happened, i.e. that activity was skipped, and a green/purple bar denotes how often this happened.

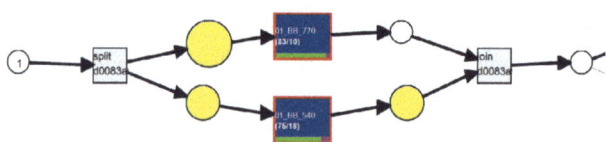

Fig. 3.19: Example of an alignment projected to a Petri net: a yellow-filled circle denotes the location of log moves and purple parts of activities denote model moves.

Third, the information of the alignment can be projected on the event log. Figure 3.20 shows an example: the sequence of wedges represents the events in a trace.

The colour denotes the type of model move: green for synchronous moves, purple for model moves and yellow for log moves. Using this information, deviations between model and log can be traced back to individual traces in the event log.

Log-conformance checking techniques ideally support all three levels, thus we add a new requirement:

Requirement CR5. *Log-conformance techniques should provide insights at three levels: summarative numbers, projections on models and projections on event logs. Similarly, system conformance techniques should provide two levels: summarative numbers and projections on models.*

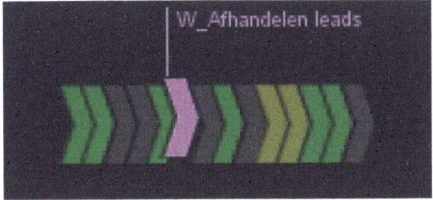

Fig. 3.20: Example of an alignment projected to a trace. Each wedge represents an event in a trace: a green wedge represents a synchronous move, a grey wedge a move on a silent transition, a purple wedge a model move and a yellow wedge a log move.

Measuring Log Conformance Using Alignments.

Alignments yield a straightforward measure for fitness: the number of log moves and model moves. In order to make this measure comparable over models and traces, it is normalised to a number between 0 and 1, where 0 denotes the worst possible fitness and 1 that the trace is in the language of the model. This makes measures of different models and event logs comparable, which we add as a new requirement:

Requirement CR6. *Both measures should be normalisable, i.e. a scaled version should give a result between 0 and 1. Two models having nothing in common should result in a recall and system precision of 0, while two equivalent models should result in a recall and system precision of 1 (depending on the equivalence notion).*

The normalisation is performed using a worst-case alignment in which each event of the trace is a log move and each execution of an activity in the model is a model move. In our example, Figure 3.18e shows such a worst-case alignment. Based on this worst-case alignment, the fitness of t with respect to P is the fraction of log and model moves, i.e. $\frac{2}{4} = 0.5$. Fitness over the entire event log is then the average fitness over all traces of the event log [14].

The most basic measure for log precision is to enumerate all traces of a model, and to compute the fraction of traces that was seen in the event log [53]. However, this approach fails for models which have infinitely many traces, e.g. models with loops. Another approach to log precision is *behavioural appropriateness* [89] and *footprint matrices* [2, Section 8.4], which compare directly and eventually follows graphs. However, obtaining these graphs might require a full state-space exploration. In practice, state spaces can be huge or unbounded, which makes many of these techniques slow (Requirement CR4). Therefore, several log precision techniques have been proposed that avoid creating a full state space of the model. For instance, instead of creating a full state space of the model, the *ETConformance* [80] technique creates a state space of the event log and counts the possibilities that are enabled in the model but not used in the event log. The initial measure assumed that the event log was fitting; this assumption was dropped in [15, 16], which uses alignments.

Finally, using an alignment, generalisation can be estimated, which aims to prevent overfitting. Generalisation measures use the principle that if the model contains little-used parts, it is likely that there are parts that are not seen yet and thus generalisation is low, and if all parts of the model are frequently used, then probably all future behaviour is captured and generalisation is high [26]. This principle is incorporated in the generalisation measure of [26, p118], that works on process trees: the number of executions of a node is divided by the total number of nodes. To counter the unwanted effect that increasing the number of "useless" nodes increases the measure, executions of such useless nodes are excluded. The measure described in [7] is similar, but uses either events or the state space of a model.

Although generalisation is claimed to balance well against the competing measures of fitness, log precision and simplicity ([26] and Section 3.2.3), in practice generalisation measures tend to heavily skew towards the maximum value 1.0 [7, 26, 60]. Further empirical and formal research needs to show the true predictive value of the concept of generalisation and these measures, hence we cannot add it as a requirement yet. Nevertheless, as discussed using Figure 3.12, besides fitness, log precision and simplicity, a fourth measure is necessary to avoid some trivial useless models. When comparing discovery algorithms (instead of logs & models), we would suggest to use different approaches to measure/avoid over-fitting, such as k-fold cross validation [7, 26, 66]. In the evaluation, i.e. Chapter 8, we decided to use k-fold cross validation.

3.4.1.2 Other Approaches

A basic approach is the *parsing measure* [104]. This measure denotes the part of the traces in the log that are expressed by the model. Although such checks can be fast, it is obviously a too coarse measure: a minor local deviation in the model can have a large impact on this measure, and correctly captured behaviour in the model is unaccounted for if a trace encounters even a single deviation [52].

In the *token-based replay* technique, a trace is replayed on the model (in this case, a Petri net). In case there are not enough tokens available to replay an event, this is recorded as a missing token and the event/transition is executed anyway. At the end of the trace, tokens remaining in the net are recorded as well [88]. The distribution of missing and remaining tokens provides normalised (Requirement CR6) measures of fitness and log precision. This provides insights into where the model deviates from the log [74], and it could provide insights in where the log deviates from the model, thus the three levels of Requirement CR5 are all present. However, this method has difficulties handling ambiguity such as invisible transitions and duplicate activities, and it does not take a final marking into account. Therefore, not all Petri nets are supported by token replay (Requirement CR1). These difficulties can be partially solved, at the cost of speed [14]. If negative events are present, a closely related technique provides fitness, log-precision and generalisation measures [24, 52].

Another approach to find the path through a model that a trace took is to convert this problem into a hidden Markov model [90], from which well established algorithms provide the most likely path [96]. However, this technique does not support concurrency [14] and requires the model to be annotated with probabilities, which restricts its use as a conformance checking technique on general Petri nets.

For more examples of techniques to measure fitness, log precision and generalisation, please refer to [100].

3.4.2 System Conformance Checking

In typical explorative process mining projects [63], as described in Section 3.1, the system is typically not known. However, in some cases the system is known, for example in the form of another process model, therefore we coin the term *model-model comparison*. For instance, when evaluating process discovery algorithms, it can be insightful to take a known system, generate an event log from it, run a discovery algorithm and compare the discovered model to the system (see Chapter 8). Moreover, model-model comparison methods are useful for efficient retrieval of relevant process models from model repositories [42], and two models describing the process in two different periods, e.g. summer and winter, could be compared to detect and explain concept drift [22].

A pre-processing step that might be necessary is to find matching activities between the two process models. For instance, one model might be more abstract than the other or the activities might have slightly different names such as "payment received" and "receive payment". For this problem, many techniques have been proposed [47, 101]. We assume that this step has already been performed and that equivalent activities have equivalent names.

In case the two models are both process trees, a tree-based edit-distance measure can be applied to measure the distance between the trees. For instance, in [21], this problem is characterised as the minimum number of insertion, deletion and renaming steps required to transform one tree into the other. Such measures have been defined on both ordered, i.e. trees in which the order of the children of a node is important [67], and unordered trees [17]. For ordered trees, computing edit-distance is polynomial, while for unordered trees, this is NP-complete and a PTAS cannot exist [67]. These methods are not directly applicable to process trees, as the trees used by us combine orderedness and unorderedness in a single formalism: for some operators ($\times$, $\wedge$, $\leftrightarrow$, $\vee$ and non-first children of $\circlearrowleft$) the children are unordered, while for $\rightarrow$ and the first child of $\circlearrowleft$, the children must be ordered. A larger drawback is that these methods compare process trees by syntax rather than by semantics. For instance, the tree $\rightarrow(a, a)$ does not have a zero edit-distance to $\wedge(a, a)$, even though these trees are language equivalent. In Chapter 5 we introduce a set of reduction rules to equate syntax and semantics. These rules guarantee that two trees with a different syntax also have different semantics. However, this holds only for a subclass process trees, so further research would be necessary to enable the use of process-tree edit-distance techniques for model-model conformance checking.

Similarly, Petri nets can be compared using graph edit distance [50] and in [77] similarity is measured using graph edit-distance. A general downside of graph-based techniques is that they depend on the structure of the graph, which consequently has to be *language unique*, i.e. if there are two such graphs with a different language, two language-equivalent models might be considered non-similar. In Chapter 5, we will study the language uniqueness property in detail. However, similar syntax vs semantics issues arise, especially for Petri nets with silent and duplicate transitions.

Another approach is to convert the process models to a graph of the process behaviour, and applying edit-distance techniques to these graphs [40]. Such graphs have been constructed from BPMN models and EPCs. A lot of complexity of this technique is caused by the need to match the activity names of both process models; we assume this mapping is already present. In [47], a graph of the process behaviour is generated. The similarity of two models is then determined using the context of elements in this graph.

In [65], an edit distance between two process models is computed and the edit steps (change patterns) are reported; the edit distance is normalised to a similarity measure. However, this method is limited to sound block-structured process models without loops, and its underlying problem is NP-hard [65] (Requirement CR1). [68] measures similarity using features of the process models. [70] works on graphs, assigns similarity measures to elements of the graphs and performs a fixpoint computation, which in each iteration adjusts the similarities of elements using the similarities of nearby elements. However, the formalism and structures of the two models must match.

Another approach is to compute the α-relations on both models and to compare these relations. For each pair of activities, an α-relation is determined, which yields a matrix for each model. The similarity between two models is that a normalised value describing on how many pairs of activities the models 'agree'. Such relations can be efficiently computed on a subset of the sound free-choice workflow nets, however for arbitrary workflow nets this requires a full state-space exploration [102].

The difference or similarity between two models can be expressed in several ways: some techniques provide a single similarity or distance measure [35, 97]. However, given the focus on languages and as is common in the data mining field, we propose the use of two measures: *recall* expresses the part of the behaviour of the system that is represented in the model, and *system precision* expresses the part of the behaviour of the model that is in the system. These measures are of course similar to the log conformance measures fitness and log precision. Generalisation is absent as the notion of future behaviour is captured by recall.

Ideally, both measures should correlate with at least language inclusion, which should follow directly from a measure being perfect. Consequently, the combination of perfect recall and system precision should bi-imply language equivalence. Using such a property, these measures can be used to quantify language-rediscovery: given a system, it can be tested how close a discovery algorithm got to rediscovering the language, up to a perfect score, from which rediscovery can be concluded (Requirement CR3).

In [10], several recall and system precision measures are proposed. A basic method is to generate all traces of one model and replay them on the other model, yielding symmetric recall and system precision measures. When comparing models, an asymmetric recall and precision measure makes little sense, i.e. we argue that swapping the models should not change the measure. Therefore, we add a new requirement:

Requirement CR7. *Model-model measures should be symmetrical, i.e. for two models a and b, recall(a, b) = system precision(b, a), and reflexive, i.e. recall(a, a) = 1 [47].*

A problem arises with iterative behaviour (loops), which allows for an unlimited number of traces. In [10], this problem is circumvented by assuming/generating an event log. However, this poses assumptions on loop behaviour and might be prohibitively expensive on complex models (Requirement CR4).

In [97], similarity between process models is measured using *principal transition sequences*, i.e. traces in which loops are unfolded a fixed number of times. Similarity is defined for two such transition sequences, and the similarity between two models is obtained by averaging maximally-similar pairs of sequences. Obviously, this measure is quadratic in the number of possible traces in the models, and therefore infeasible for models containing many activities (Requirement CR4).

[82, 115] quantify similarity on state machines, i.e. models without concurrency: [82] focuses on hierarchical state machines and bi-similarity, while [115] constructs an abstraction of the languages of the state machines and compares these.

As noted in [39], many model-model comparison techniques suffer from exponential complexity due to concurrency and loops in the models. To overcome this problem, several techniques apply an abstraction, for instance using causal footprints [42], weak order relations [117] or behavioural profiles [59, 103]. A downside of using an abstraction is that the comparison inherits the limitations of the abstraction, as an abstraction inherently contains less information than the model itself. Another technique to reduce the state space is to consider parts of the model separately [59]. Even though this technique reduces state space and hence run time, the parts cannot be arbitrarily chosen, and not all models can be divided into parts. Furthermore, the technique only works on Petri nets (Requirement CR1).

3.4.3 An Ideal Conformance Checking Technique (2)

Based on our analysis of existing algorithms we collected additional requirements that extend the list composed in Section 3.2.4:

CR4 The algorithm should work fast on real-life models and event logs.

CR5 Log-conformance techniques should provide insights at three levels: summarative numbers, projections on models and projections on event logs. Similarly, system conformance techniques should provide two levels: summarative numbers and projections on models.

CR6 Both measures should be normalisable, i.e. a scaled version should give a result between 0 and 1. Two models having nothing in common should result in a recall and system precision of 0, while two equivalent models should result in a recall and system precision of 1 (depending on the equivalence notion).

CR7 Model-model measures should be symmetrical, i.e. for two models a and b, recall(a, b) = system precision(b, a), and reflexive, i.e. recall(a, a) = 1 [47].

Techniques that tend to construct or traverse the state space of a model tend to be slower (Requirement CR4). Especially if the technique (needs to) construct the state space in memory, then the technique is bound to run out of memory when facing larger event logs and models, i.e. is less robust to large inputs. Options to counter this are to use an abstraction or to decompose the problem and compute measures on smaller state spaces.

Requirement CR5 is satisfied by the alignment-based techniques and by token-based replay techniques. Techniques that use an abstraction or decompose the problem tend to have difficulties supporting the third level, i.e. projecting their measures on event logs.

Most techniques we discussed seem to support Requirement CR6, even though e.g. alignment seem to be approach a fitness or log precision of 0 only in the limit.

Most techniques we discussed satisfy the second part of Requirement CR7, i.e. comparing a model with itself will yield a measure of 1. As most techniques do not use two measures, we did not encounter a technique that satisfies the symmetry part.

Up till now, we did not encounter a conformance checking technique that satisfies all of these requirements. We would prioritise speed (Requirement CR4) and measures on three levels of detail (Requirement CR5), as we argue that these are the most useful in practise. However, in order to evaluate process discovery algorithms, needed to assess the presented discovery techniques (Chapter 8), it is necessary that the conformance checking technique handles unsound or weakly sound models well (Requirement CR1), and that the measures are symmetric and reflexive (Requirement CR7).

In the evaluation, we applied conformance checking techniques to thousands of event logs and models. Therefore, our conformance checking approach focuses on speed (Requirement CR4). We combine the idea of using an abstraction with the idea of decomposing the problem, which substantially reduces the state space that has to be traversed.

3.5 Enhancement & Tool Support

As discussed in the previous parts of this chapter, often a single model serving all process mining goals does not exist. Therefore, process mining projects tend to be highly iterative, thus easy-to-use software is necessary to support analysts in performing process discovery and conformance checking. Furthermore, process discovery and conformance checking techniques introduced before concern only the control-flow perspective, i.e. the ordering and occurrence of activities, as they assume that the event log contains only information on the order of activities that were executed. However, if the events are annotated with more information, different perspectives can be explored for more insights into the process, which make process mining tools (much) more useful. In Section 1.4, we discussed four types of enhancements: frequencies, performance, deviations and animation. In this section, we first elaborate on these enhancements from the log perspective, i.e. we describe what information in the log is necessary and how the measures corresponding to these enhancements are computed, and we gather requirements. Second, we discuss existing process mining tools, elaborate on how they satisfy the requirements, and devise new ones.

3.5.1 Enhancements

In Section 1.4, we introduced four enhancements: frequencies, performance, deviations and animation. In this section, we discuss three types of data in event logs that enable these enhancements: life cycle, time, and resources and other data, and discuss the enhancements in more detail. Furthermore, we describe challenges in case of missing information.

3.5.1.1 Life Cycle

The life cycle perspective assumes that each execution of an activity is represented by several events, that together describe the *life cycle* of that *activity instance*, i.e. activity execution. For instance, one event denotes the start of an activity instance, while another event denotes its completion. Several life cycle models have been proposed, such as in the XES standard[54] and in [2], and ideally, techniques handle arbitrary life cycle models. However, in order to draw conclusions from arbitrary life cycle models, these models need to have semantics.

For our purposes, we define semantics using and use the life cycle model denoted in Figure 3.21 [62]: this model uses three states and three life-cycle transitions:

enqueue, start and complete. The *enqueue* transition denotes that the execution of an activity is added to a queue of work items, e.g. a new call in call centre is added to the call queue of a group of employees, or in a hospital a patient enters a queue of patients queueing to see a doctor [62]. The *start* transition denotes that the activity instance starts processing, e.g. the patient is called in or the customer is answered by an agent. Finally, the *complete* transition denotes that the activity instance finishes, e.g. the patient leaves the room or the customer call ends. An enqueue event of activity a is denoted with a_E, a start event with a_S and a complete event with a_C.

In Section 5.7, we describe the influence of events of the start transition on process models. In Section 6.6, we introduce an algorithm that uses these events. We did not address the usage of enqueue events for process discovery.

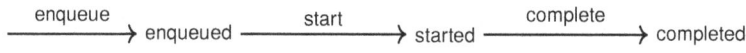

Fig. 3.21: The life cycle model used here.

The process discovery and conformance checking techniques introduced could easily be adapted to use arbitrary life cycle models, i.e. with proper semantics a process model with a life cycle model still describes a language. Defining semantics in a uniform and computer-interpretable way remains part of future research.

Future work 3.1: Investigate semantics for arbitrary life cycle models.

3.5.1.2 Time

A second perspective is the time perspective. This perspective can be considered if (some of) the events carry a timestamp. Analysing the time perspective might give insight into bottlenecks, i.e. parts of the process with undercapacity, and other time-related aspects [2].

The available time measures depend on the life cycle model, as if for instance the enqueue life-cycle transition is missing, there is no information available when a trace entered a queue. In this section, we assume the life cycle model of Figure 3.21; other life cycle models might have comparable measures. Consider the trace $\langle a_E,$ $c_E, c_S, a_S, a_C, b_E, b_S, c_C, b_C \rangle$: Figure 3.22 shows a graphical representation of t. It is obvious that c is concurrent to a and b, as it is executed at the same time. Furthermore, assume that a and b are sequential. Then, the *sojourn time* of b is the time between completes, i.e. a_C and b_C. Intuitively, the sojourn time of b is the time from the moment that the process allowed b to be executed till the moment it finished. The *waiting time* of b is the time between a_C and b_S, i.e. the time between the enabling of b and its start. The *queueing time* of b is the time between b_E and b_S, i.e. the time spent in queue. The *service time* of b is the time between b_S and b_C, i.e. the time spent on actually executing b [62, 14].

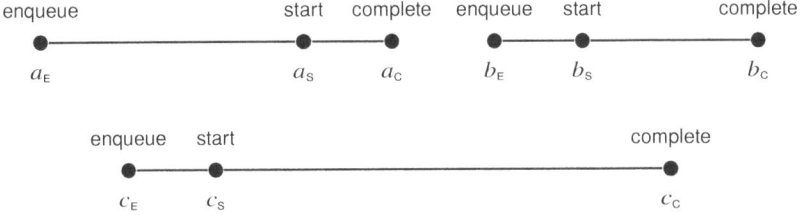

Fig. 3.22: A graphical representation of the trace $\langle a_E, a_S, a_C, b_E, b_S, b_C \rangle$.

Waiting and sojourn time depend on the moment that the process allowed the activity to be started. Thus, for such time measures, it is important to take a process model into account. For instance, consider the sojourn time of activity instance e in the trace

$$\langle a_E, a_S, a_C, b_E, b_S, b_C, c_E, c_S, c_C, d_E, d_S, d_C, e_E, e_S, e_C \rangle$$

In the model of Figure 3.23a, the sojourn time of e is the time between d_C and e_C: from the moment d completed, e could start according to the model. However, in the model of Figure 3.23b, e does not depend on d but on c, and therefore the sojourn time of e is the time between c_C and e_C [62, 89]. Therefore, such performance measures should take a process model into account, which we add as a new requirement:

Requirement ER1. *The tool should provide several performance measures, based on timestamps of the event log, using life cycle semantics. Furthermore, the tool should take a process model into account.*

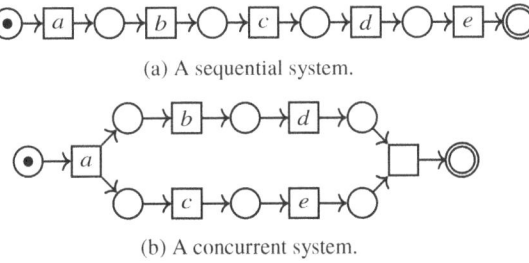

(a) A sequential system.

(b) A concurrent system.

Fig. 3.23: Two systems that could produce the trace $\langle a, b, c, d, e \rangle$. The sojourn time of e depends on the previously executed non-concurrent activity: either d or c.

Similar situations occur when computing the other measures, such as waiting time. Notice that these measures do not require all timestamps and life-cycle transitions, e.g. the sojourn time computation of our example does not need the enqueue event or its timestamp.

Another powerful feature that is enabled by the presence of timestamps is animation: using log animation, a user can inspect this time perspective. The event log is visually replayed on the map by tokens flowing over the process model, and each activity that is encountered is executed. This reveals frequent paths and bottlenecks over time, and makes concept drift explicit. For instance, the animation could highlight that a part of a model is little used in the beginning of the time period that is represented in the event log, but later becomes heavily used. If an animation can be paused, it gives a frozen view of the map with the traces that were in the process at a particular point in time [61]. Furthermore, in our experience, animation increases the confidence of process owners and other stakeholders in the discovered model and increases understandability of the model. This yields a new requirement:

Requirement ER2. *The tool should animate the event log on a process model.*

3.5.1.3 Resources and Other Data

If each event in an event log is enriched with which person, machine or other resource executed the event, the flow of information through an organisation can be visualised. This may lead to more insights in the social aspects of a process, e.g. by showing how cases flow through an organisation. Studies in this field of *sociometry* [98] become much easier by the data mining techniques as described [2]. We do not cover resources in detail; please refer to [2] for more information. However, using XES classifiers [54], one can discover a model from a log where activities are not derived from the activity names of events, but rather the values of the resource attribute, thereby transforming the techniques introduced to discover social process models.

Many other types of data attached to events can be used using such classifiers. Moreover, any data type can be used to enrich a process model with choice information, i.e. for each moment of choice in a process model, a decision model can be discovered based on the event data at that point in the process [64]. We do not cover other data types in detail; please refer to [2] for more information.

3.5.1.4 Missing Information

A challenge common to these perspectives is that they take the event log into account, and therefore the similarity of the event log and process model becomes relevant. Obviously, if the log and model are completely different, few of these perspectives make sense. In many use cases however, log and model can be related, but the model might not represent the event log completely (i.e. is not perfectly fitting). Furthermore, timestamps might be missing from some events, leading to incomplete information. As the event logs we have encountered rarely contained all information, enhancements and measures need to be robust to such inconsistencies between log and model and missing information, which yields a new requirement:

Requirement ER3. *The tool should be robust to missing information and to mismatches between log and model.*

We argue for a pragmatic approach for each type of enhancement. For instance, animation only makes sense if the information is complete: if an event is recorded without a timestamp in a trace, the animation of that trace has to estimate when that event happened in order to keep the animation smooth. However, performance measures might become unreliable with the introduction of estimated information, so we argue that incomplete measures are best omitted.

Finally, as process discovery techniques might exclude behaviour of the event log from the model or include behaviour in the model that is not observed in the event log, the model should be evaluated using a conformance checking technique (see Section 3.4.2). Ideally, the process mining tool should integrate such a technique:

Requirement ER4. *The tool should provide a way to evaluate the discovered process model.*

3.5.2 Process Mining Tools

In this section, we consider several existing process mining tools: we address whether and how these tools satisfy the gathered requirements, and we devise new requirements relevant for such tools. We consider two commercial tools: Fluxicon Disco (FD, [56], version 1.9.7) and Celonis Process Mining (CPM, version 4.0.1, web-based [34], accessed 6-1-2017).

3.5.2.1 Fluxicon Disco

Fluxicon Disco (FD) [56]) is a commercial process mining tool developed by Fluxicon (http://www.fluxicon.com/). In Section 3.3.2, we considered the process discovery of FD, however in this section we focus on features and on usability aspects. After importing an event log, the user interface of FD consists of several views, which are shown in Figure 3.24. The first view (Figure 3.24a) shows the process model. In this view, the process model is visualised, as well as controls to influence the model. That is, the two sliders in the dashed ellipses control the number of activities and the complexity of the discovered model. If the user changes such a slider, a new model is computed automatically, and this is done fast, which allows for quick and iterative process discovery. As described in Section 3.1, filtering the event log allows analysts to drill down and focus on several parts of the event log. FD contains an extensive set of stackable filters, which are applied consistently through the user interface (see Figure 3.24b). That is, a filter is applied once and affects all views on the data. This yields the following requirement:

Requirement ER5. *The tool should allow for quick iterative exploration and log filtering.*

The process model visualisation is enhanced with frequency information, e.g. the activities and edges are annotated with how often they have been executed.

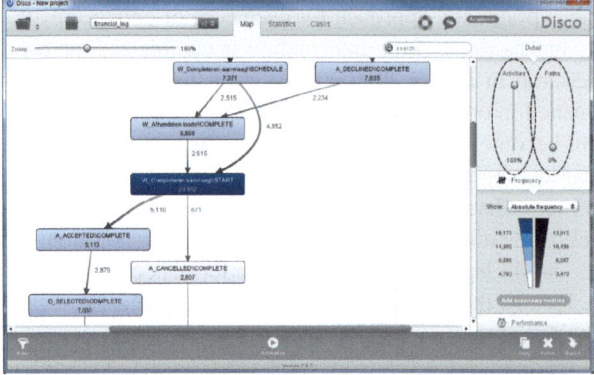

(a) Process model view with frequencies.

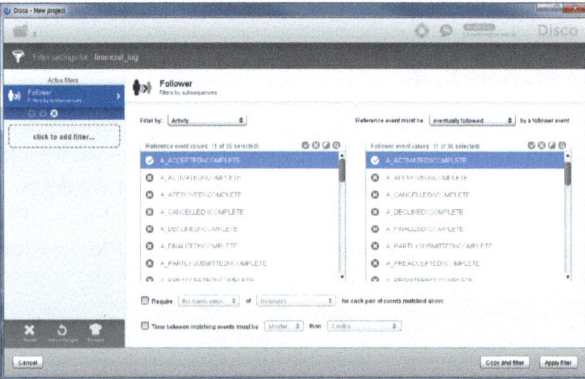

(b) Filters.

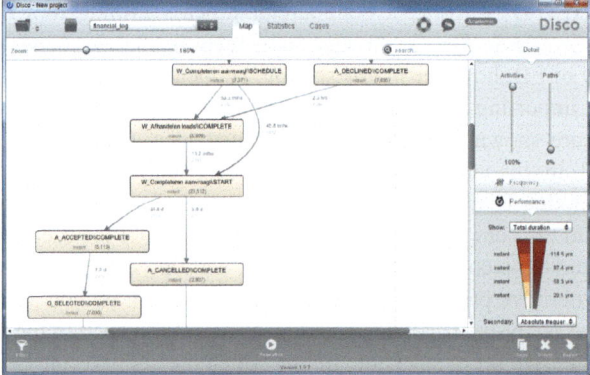

(c) Enhanced process model.

Fig. 3.24: Screenshots of Fluxicon Disco, showing several enhancements.

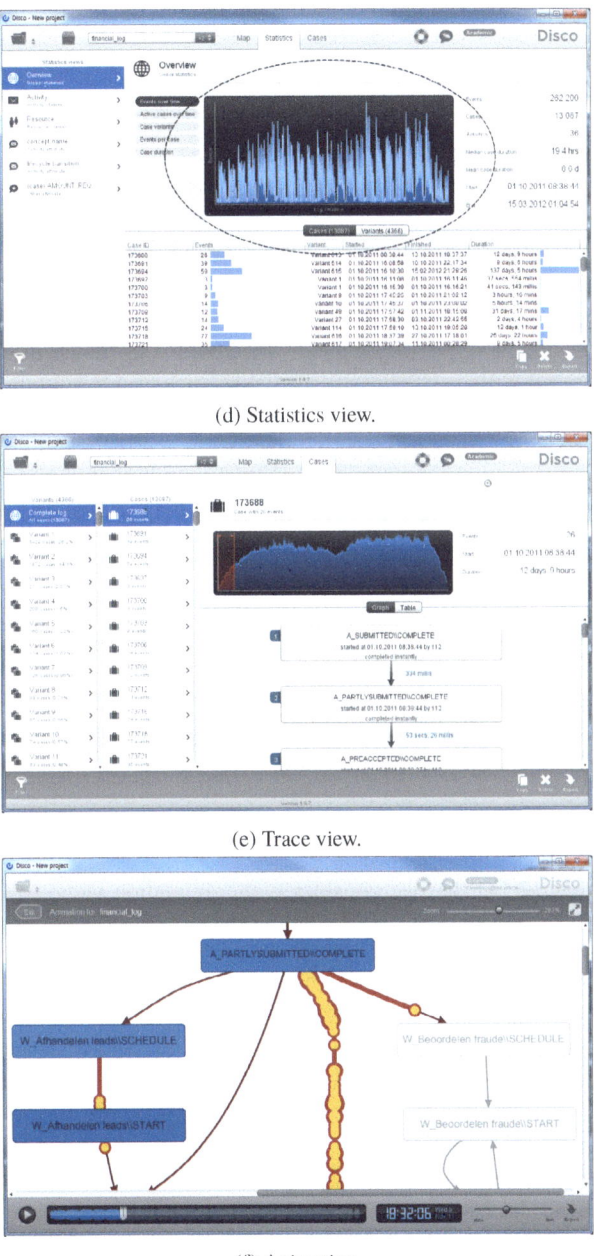

(d) Statistics view.

(e) Trace view.

(f) Animation.

Fig. 3.24: Screenshots of Fluxicon Disco, showing several enhancements.

Furthermore, the activity colouring and edge thickness are adjusted accordingly to highlight the often-used parts of the process. This yields a new requirement:

Requirement ER6. *The tool should be able to visualise frequencies on the model.*

This process model can be enhanced with several performance measures, such as the total duration that an activity took and the total time that was spent between the execution of activities (Figure 3.24c). In Section 3.5.1.2, we showed that a process model is necessary to obtain reliable performance measures. As discussed in Section 3.3.2, the semantics of FD models might be ambiguous, thus the performance measures of FD should be used with care (Requirement ER1).

The second view in FD is the statistics view, which is shown in Figure 3.24d. In this view, several global statistics are available, such as a histogram showing the number of events being executed over time (as denoted by the dotted ellipse). Several groupings are available, such as by activity, resource and by trace information. For instance, choosing resources provides a graph of the number of events each resource executed. This yields a new requirement:

Requirement ER7. *The tool should provide frequency statistics over the entire event log (e.g. histograms) and over sub-groupings of the event log.*

The third view in FD is the traces view (see Figure 3.24e), which enables analysts to study traces in detail. Studying individual traces might be useful at later stages of process mining, i.e. once areas of interest are identified, analysts could drill down to cases that are involved in this area of interest, identify these traces in the trace view and verify the drawn conclusions in the system. Therefore, we add a new requirement:

Requirement ER8. *The tool should provide a visualisation of individual traces.*

The final view of FD that we consider is the animation view, which is shown in Figure 3.24f. In this view, tokens (yellow red-bordered circles) represent traces (Requirement ER2).

We consider the main drawback of FD to be the unclear semantics of the discovered process models, and the lack of guarantees and possibilities to evaluate the models (Requirement ER4). Therefore, all model-related enhancements such as performance measures and animation should be interpreted with care.

3.5.2.2 Celonis Process Mining

Celonis Process Mining (CPM) [34] is a web-based process mining toolkit developed by Celonis (http://celonis/com). The user interface of CPM has a high customisability, as graphs, statistics and other overview elements can be arranged in dashboards and customised, which might make CPM useful in Business Intelligence [99] settings as well (Requirement ER6).

Figure 3.25a shows the process model component of CPM. In this component, there are two sliders that influence discovery of the model: one that controls the

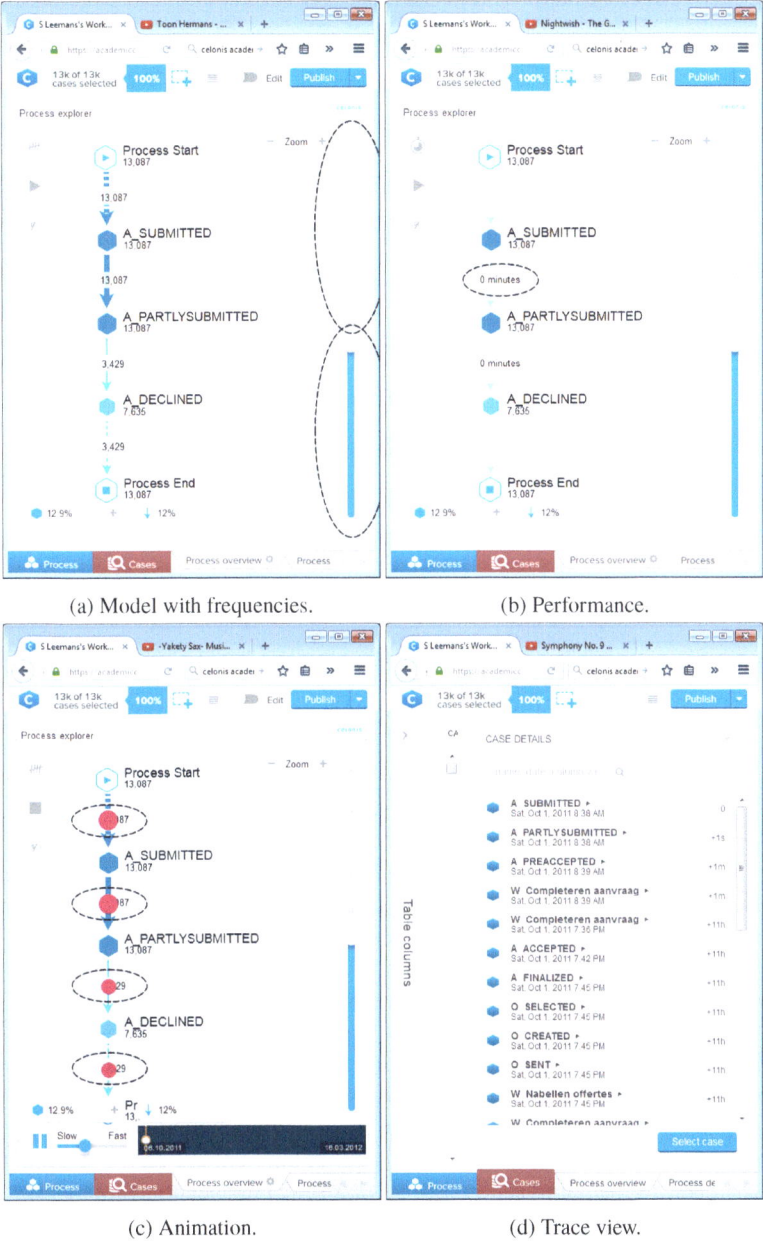

(a) Model with frequencies.

(b) Performance.

(c) Animation.

(d) Trace view.

Fig. 3.25: Screenshots of Celonis Process Mining showing several supported enhancements.

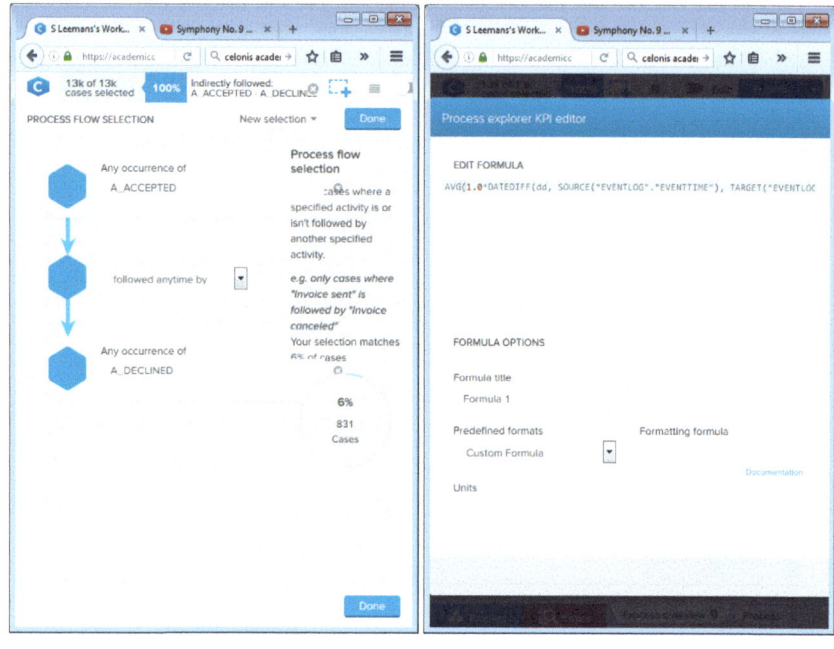

(e) Filters.   (f) Custom enhancements.

Fig. 3.25: Screenshots of Celonis Process Mining showing several supported enhancements.

fraction activities that is included and one that controls the connections. Using these sliders changes the model immediately. Furthermore, CPM offers extensive filtering options, e.g. one can click on an edge or activity to filter the event log, such that the log only contains cases that use the edge or activity (Requirement ER5).

Even though we believe that CPM and FD should be able to discover similar models, there is a notable difference in the model of CPM and FD (Figure 3.24f): both screenshots were obtained at their default settings and on the same event log. FD shows the entire process, and consequently the animation clearly suggests a bottleneck where the concentration of tokens is high. In contrast, CPM seems to show the most-occurring path through the process, thereby ignoring the remainder of the process. Both approaches may be useful in different use cases: this event log was recorded in a loan-application process of a Dutch financial institution [41]. The traces contained in the model of CPM are probably the least interesting, as they are automatically rejected before any employee gets involved (hence the short time between activities). However, a good second step in the analysis would be to filter these traces and repeat the analysis.

The performance measures computed by CPM are illustrated by Figure 3.25b (Requirement ER1). CPM supports animation (Requirement ER2): Figure 3.25c shows a screenshot (we've circled the tokens in the screenshot, these tokens are pink). The animation of CPM seems to round the timestamps to seconds, which seems the cause of the tokens being grouped, making the animation far less useful in such a short process. Figure 3.25d shows the trace view of CPM (Requirement ER8), and Figure 3.25e one of the filters of CPM.

Unlike FD that does not provide any way to evaluate the model, CPM shows some basic measures, i.e. the percentage of included activities and edges between activities. Unfortunately, there is no link to the included or excluded behaviour of the event log (Requirement ER4).

We finish this discussion of Celonis Process Mining with a feature not found in FD: the ability to add custom enhancements. That is, by typing a formula, any derived piece of data can be visualised on the edges and/or the activities: Figure 3.25f shows the user interface. We add this as a new requirement:

Requirement ER9. *The tool should be extensible with custom enhancements.*

3.5.3 Requirements for Tool Support Beyond Process Discovery and Conformance Checking

To summarise the previous sections, an ideal process mining tool with enhancements supports the following:

ER1   The tool should provide several performance measures, based on timestamps of the event log, using life cycle semantics. Furthermore, the tool should take a process model into account.

ER2   The tool should animate the event log on a process model.

ER3   The tool should be robust to missing information and to mismatches between log and model.

ER4   The tool should provide a way to evaluate the discovered process model.

ER5   The tool should allow for quick iterative exploration and log filtering.

ER6   The tool should be able to visualise frequencies on the model.

ER7   The tool should provide frequency statistics over the entire event log (e.g. histograms) and over sub-groupings of the event log.

ER8   The tool should provide a visualisation of individual traces.

ER9   The tool should be extensible with custom enhancements.

3.6 Our Approach

Our approach consists of three parts: (1) a framework and algorithms for process discovery, (2) a framework and algorithms for conformance checking and (3) a process mining software tool. For each part, we first describe our approach, after which we discuss how it addresses the identified requirements.

3.6.1 A Process Discovery Framework

From the list of requirements on process discovery, it is clear that a perfect process discovery algorithm cannot exist. Therefore, we introduce a modular and flexible approach, that always guarantees soundness and optionally guarantees fitness and log precision. Our approach contains a framework, the *Inductive Miner framework* (IM framework), that builds process trees constructively. The framework, like CCM, determines the most important behaviour in an event log, i.e. the root operator of the process tree. However, instead of splitting an abstraction, the entire event log is split by the IM framework. The IM framework uses both horizontal and vertical partitioning, and thus combines [73] and [32].

As guarantees are our focus, we first study three guarantees that can be provided by algorithms that implement the IM framework: rediscoverability, perfect fitness and perfect log precision. To ease formal proofs for these guarantees and to avoid duplication, we express these guarantees in terms of the IM framework. That is, we introduce formal frameworks for them. In the proof framework for rediscoverability, the language abstractions used by most process discovery algorithms play a major role. Therefore, we perform a systematic study into the expressive power of these abstractions, i.e. we study classes of models that can be uniquely represented by several abstractions, such that no two models with different languages have the same abstraction. Using this systematic study and the formal frameworks, we prove rediscoverability for all algorithms and perfect fitness for some algorithms (see Table 1.1).

The framework is instantiated in several algorithms, each focusing on a different aspect: we'll introduce a rediscoverability-focused version, a deviating- and infrequent-behaviour-filtering focused version and an incompleteness-handling focused version. These three algorithms each strike a particular balance, but a big advantage of the IM framework is that this balance can be easily customised by replacing particular interchangeable steps. Thus, an algorithm developer might adjust an algorithm without losing any guarantees given by the framework. Furthermore, we introduce sets of algorithms to handle non-atomic event logs, to handle more process tree operators, and to provide an even better scalability, i.e. handle event logs with thousands of activities and billions of events.

The IM framework and the algorithms introduced have been implemented as part of the ProM framework. Figure 3.26 shows the user interface, in which users can choose a discovery algorithm and, if applicable, a deviating behaviour filtering threshold.

Of the identified requirements for discovery algorithms, the IM framework, by its use of process trees, guarantees sound models with clear semantics (Requirement DR1). Rediscoverability (Requirement DR2) depends on the particular discovery algorithm, so the IM framework cannot provide rediscoverability guarantees by itself. However, we provide proof obligations in a formal framework for the algorithms that implement the IM framework, and use this formal framework to prove that every discovery algorithm that was introduced guarantees rediscoverability. The IM framework provides the flexibility for particular algorithms to focus on distin-

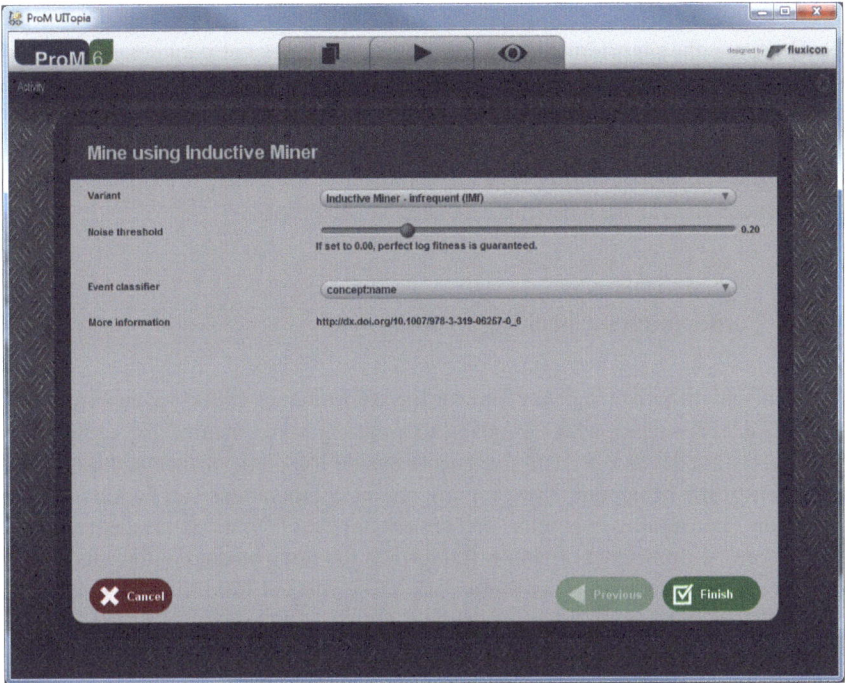

Fig. 3.26: User interface of the IM framework.

guishing deviating, infrequent and incomplete behaviour (Requirement DR3), as we will show in our evaluation. Similarly, algorithms can focus on different balances in log-conformance measures (Requirement DR4). The IM framework aids algorithms to guarantee fitness and log precision, and several algorithms that we introduced guarantee to return a perfectly fitting model.

Requirement DR5 entails that a discovery technique should work fast on real-life event logs and systems. The IM framework enables fast discovery algorithms: we introduce algorithms that are faster than existing discovery algorithms and that work on real-life event logs. However, the speed depends on the specific algorithms: some introduced algorithms apply exponential steps and become intractable on large real-life event logs containing hundreds of activities. The IM framework in itself only prevents rediscovery of some non-free-choice constructs by the choice for process trees (a part of Requirement DR6), all other problematic constructs could in principle be handled, even though we do not have algorithms for all of them at the time of writing (requirements DR6, DR7 and DR8). Similarly, milestones and other more advanced workflow patterns could in principle be handled in specific cases (Requirement DR9). For instance, the ↔ contains a critical section.

The final Requirement DR10 entails that the balance of log-conformance criteria should be influenceable by a user. The user can influence the IM framework by choosing its modules (or a combination of modules, i.e. an algorithm), which allows users to select the focus. This has not been implemented yet in a user-accessible way. However, the algorithms themselves might have several parameters, such as deviating and infrequent behaviour thresholds, that allow the user to easily influence certain aspects of the discovery.

The IM framework is described in more detail in Chapter 4, as well as the formal framework for rediscoverability. The algorithms that instantiate the IM framework are introduced in Chapter 6 and are evaluated in Chapter 8.

3.6.2 A Conformance Checking Framework

Alignments are the current state-of-art for log conformance checking: an alignment provides a decision about which events in an event log were executed "in reality" and which events are deviations, and alignments enable log-conformance techniques to ignore non-fitting behaviour. However, alignment computations can be too lengthy and memory-consuming in practise and are not applicable to model-model settings.

Therefore, we introduce a single framework for both model-conformance and log-conformance checking. This framework, the Projected Conformance Checking framework (PCC framework), approximates recall, system precision, fitness and log precision by projecting the behaviour of a system/log/model on all subsets of k activities (for a given k). For each such k-subset, a *deterministic finite automaton* is constructed from both the system/log and the model, the requested measure is computed, and this is repeated for and averaged over all such k-subsets. The PCC framework resembles many techniques that use language abstractions, e.g. has similarities with [115, 102], but abstracts from activities instead of from types of behaviour.

Due to its projections on subsets of activities, the PCC framework can give clues about where in a process model deviations occur. Figure 3.27 shows an example, in which each activity (the boxes) is enriched with fitness and log-precision information. Furthermore, the activities are coloured: the more problematic, the more red the activities are visualised.

The properties of the PCC framework are immediately clear: any log can be handled, as well as any model with a finite state space. Furthermore, due to the projection, it's an approximation. We propose "hacks" for weakly unsound models, however we do not understand them well enough yet to implement them and to consider them to be part of the PCC framework. Nevertheless, we prove that for a certain class of systems (rediscoverable by the IM framework) and $k = 2$, recall/fitness and model/log precision are 1 if and only if system and model are language-equivalent.

Of the identified requirements for conformance checking techniques, the PCC framework handles any weakly unsound model, thus the PCC framework satisfies Requirement CR1. The language-equivalence proofs show that the PCC framework satisfies requirements CR2 and CR3. The main advantage of the projection steps is a much lower complexity than alignments: instead of traversing the complete state space, only the much smaller projected state spaces are to be traversed. In our

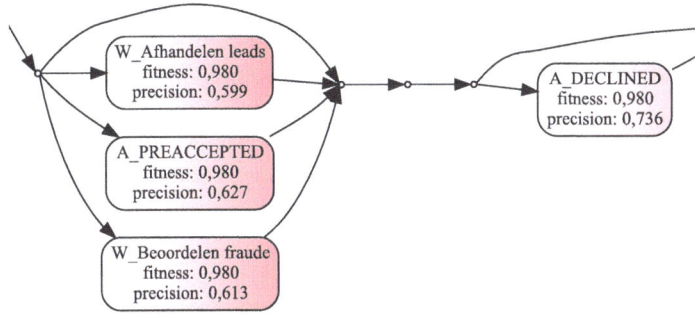

Fig. 3.27: Result of the PCC framework projected on a process model.

evaluation, we show that not only the PCC framework is faster than alignments, but it is able to handle much larger real-life event logs and models (Requirement CR4) as well. Alignments have been made faster using divide-and-conquer as well. For instance using passages [3, 4] or single-entry-single-exit decompositions [81]. The PCC framework can be seen as a generalisation of these techniques, that takes the context of decomposed nets into account. In the future, it might be interesting to extend the PCC framework to measure stronger equivalence notions, such as bisimilarity. For the log model part, this is of course impossible without further information in the event log.

Requirement CR5 entails that conformance techniques provide insights on two (model-model conformance checking) or three (log-model conformance checking) levels. The PCC framework provides insights on most of these levels: summarative measures are available, as well as a projection on process models (see Figure 3.27). The measures of the PCC framework are all numbers between 0 and 1, and the model-model measures are symmetrical and reflexive, which satisfies requirements CR6 and CR7.

The PCC framework is introduced in Chapter 7 and is evaluated in Chapter 8.

3.6.3 Enhancement & Tool Support

To support analysts in process mining projects, we developed a tool Inductive visual Miner (IvM) that makes iteration easy: it takes an event log as input and performs several steps automatically: it discovers a model, computes an alignment on the discovered model and the log, animates the log on top of the model, and computes performance. Intermediate results are shown and the user can interact with the tool at all times, e.g. if the user changes a discovery parameter, computations are restarted automatically. Furthermore, the event log can be filtered, which makes iteration seamless. Figure 3.28a shows a screenshot of the user interface. IvM supports and shows deviations (Figure 3.28b) and performance measures (Figure 3.28c). To solve inconsistencies between log and model, IvM uses alignments, as they provide a convenient conceptual abstraction layer: each event is classified as fitting (synchronous move), non-fitting in the log (model move) or non-fitting in the model (log move). Due to this classification, deviations, frequencies, animation and performance enhancements do not have to deal with non-fitting behaviour. Finally, IvM provides histograms of all traces in the system (Figure 3.29a) and of the executions of activities (Figure 3.29b), and provides a deviation-showing trace view (Figure 3.29c).

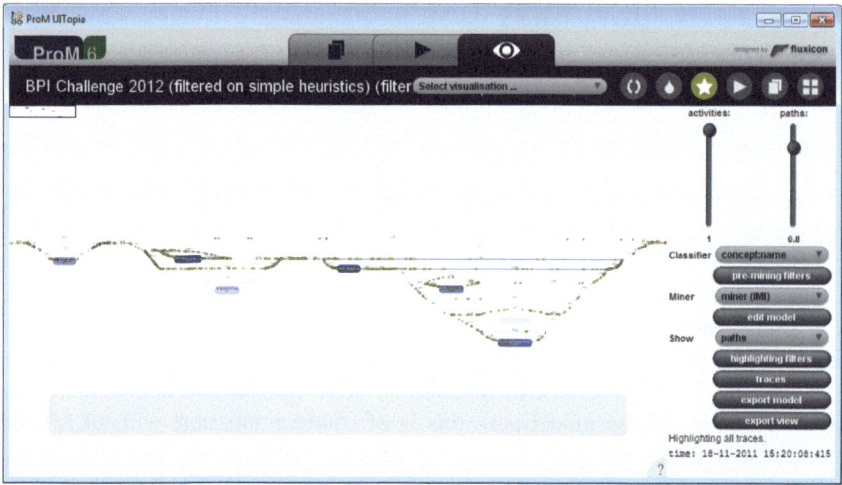

(a) Screenshot of the user interface of IvM. The yellow dots flowing over the model visualise the traces with animation, and the model is enhanced with frequency information.

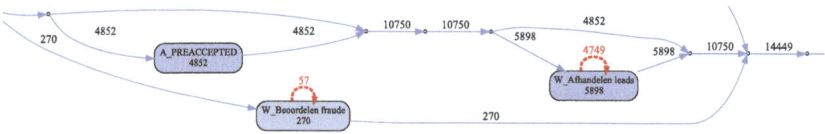

(b) Deviations: the red-dashed edges denote points in the model where the log and the model disagree.

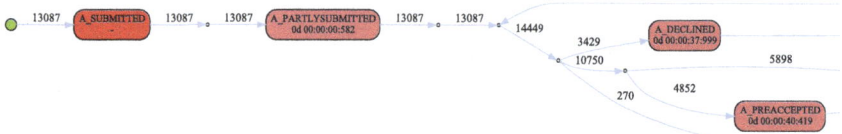

(c) Performance: the digits in the activities denote the average duration of that activity. For instance, A_DECLINED took on average 37 seconds and 999 milliseconds.

Fig. 3.28: The Inductive visual Miner (IvM).

Next, we consider the identified requirements for enhancements and process mining tools. Corresponding to Requirement ER1, IvM provides several performance

(a) Frequency statistics: when the user puts the mouse pointer on the animation controls, a histogram of traces in the system appears.

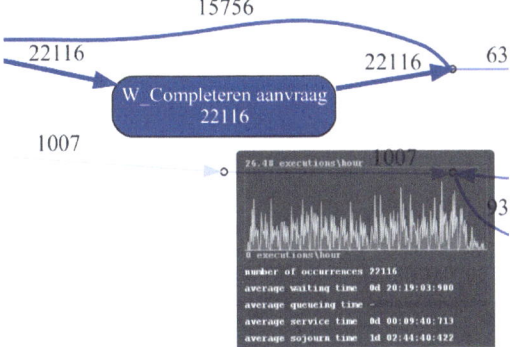

(b) Frequency statistics: when the user puts the mouse pointer on an activity, a pop-up appears that shows performance measures and a histogram of the number of executions for that activity.

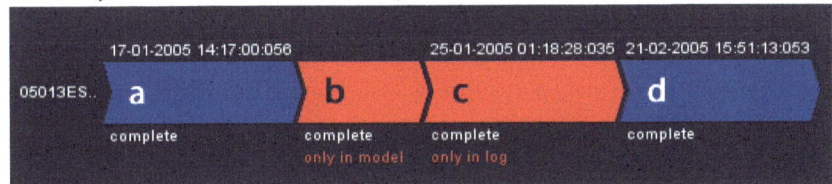

(c) A view of the traces of the event log: each line is one trace (one trace is shown here), each wedge is an event, and red-coloured events deviate from the discovered process model.

Fig. 3.29: Features of IvM.

measurements, and takes both the discovered process model and life-cycle information into account while computing them. Furthermore, IvM supports animation
(Requirement ER2).

Due to the use of alignments IvM is robust against missing information and
inconsistencies between log and model (Requirement ER3). Furthermore, due to the
deviation enhancements, users can evaluate the discovered model in detail. Enabling
evaluation of models brings IvM beyond the commercial tools.

Iteration is quick using the two sliders that immediately update the model (Requirement ER5), and several filters are provided. The provided filters of IvM are not
as extensive as those in FD and CPM, however IvM is open source and additional
log filters can easily be added. Requirement ER6 entails that frequency information
can be projected on the event log. Even though IvM provides further frequency information in log and activity histograms (Requirement ER7), this is not comparable
to the plethora of options offered by CPM and FD.

Finally, IvM contains a view of traces like CPM and FD, such that the traces that
remain after filtering can be traced back to the system for verification and further

analysis (Requirement ER8). However, in the trace view of IvM, each event is enhanced with deviation information, i.e. if the event log deviates from the discovered process model, it is coloured red. Notice that the current implementation of alignments offers a similar feature (see Figure 3.20 [14]. As IvM is open source, custom enhancements could be added by developers, however this will not be easier than CPM, which allows enhancements to be added by end users.

In Chapter 9, we discuss the concepts of enhancements and IvM in more detail. In the next chapter, we introduce our process discovery framework, i.e. the IM framework.

3.6.4 Future Work

In Section 3.2.1, we limited our scope to languages, as event logs typically do not contain information about choices. However, in future research, meta-information or richer event logs could be used to enable process mining techniques to consider stronger notions than language equivalence.

Future work 3.2: Use other information next to event logs in process discovery and conformance checking, and apply ideas of PCC framework to similarity measures stronger than language-equivalence.

The PCC framework does not provide information on the log level currently, which remains future work:

Future work 3.3: Investigate whether it's possible to extend the PCC framework to provide information on the log level (Requirement CR5).

The IvM uses alignments, which might take a long time to compute on larger and more complex real-life event logs. We would like to use techniques like the PCC framework without the presence of alignments.

Future work 3.4: Obtain and visualise deviations and performance measures without alignments.

Finally, we will not address the usage of enqueue events for process discovery, this remains part of future work:

Future work 3.5: Study what enqueue events can contribute to process discovery.

References

1. Proceedings of the IEEE Symposium on Computational Intelligence and Data Mining, CIDM 2011, part of the IEEE Symposium Series on Computational Intelligence 2011, April 11-15, 2011, Paris, France. IEEE (2011). URL http://ieeexplore.ieee.org/xpl/mostRecentIssue.jsp?punumber=5937059

2. van der Aalst, W.M.P.: Process mining - discovery, conformance and enhancement of business processes. Springer (2011). DOI 10.1007/978-3-642-19345-3. URL http://dx.doi.org/10.1007/978-3-642-19345-3

3. van der Aalst, W.M.P.: Decomposing process mining problems using passages. In: Haddad and Pomello [58], pp. 72–91. DOI 10.1007/978-3-642-31131-4_5. URL http://dx.doi.org/10.1007/978-3-642-31131-4_5

4. van der Aalst, W.M.P.: Decomposing Petri nets for process mining: A generic approach. Distributed and Parallel Databases 31(4), 471–507 (2013). DOI 10.1007/s10619-013-7127-5. URL http://dx.doi.org/10.1007/s10619-013-7127-5

5. van der Aalst, W.M.P.: Mediating between modeled and observed behavior: The quest for the "right" process: Keynote. In: R. Wieringa, S. Nurcan, C. Rolland, J. Cavarero (eds.) IEEE 7th International Conference on Research Challenges in Information Science, RCIS 2013, Paris, France, May 29-31, 2013, pp. 1–12. IEEE (2013). DOI 10.1109/RCIS.2013.6577675. URL http://dx.doi.org/10.1109/RCIS.2013.6577675

6. van der Aalst, W.M.P., Adriansyah, A., van Dongen, B.F.: Causal nets: A modeling language tailored towards process discovery. In: J. Katoen, B. König (eds.) CONCUR 2011 - Concurrency Theory - 22nd International Conference, CONCUR 2011, Aachen, Germany, September 6-9, 2011. Proceedings, Lecture Notes in Computer Science, vol. 6901, pp. 28–42. Springer (2011). DOI 10.1007/978-3-642-23217-6_3. URL http://dx.doi.org/10.1007/978-3-642-23217-6_3

7. van der Aalst, W.M.P., Adriansyah, A., van Dongen, B.F.: Replaying history on process models for conformance checking and performance analysis. Wiley Interdisc. Rew.: Data Mining and Knowledge Discovery 2(2), 182–192 (2012). DOI 10.1002/widm.1045. URL http://dx.doi.org/10.1002/widm.1045

8. van der Aalst, W.M.P., Adriansyah, A., de Medeiros, A.K.A., Arcieri, F., Baier, T., Blickle, T., Bose, R.P.J.C., van den Brand, P., Brandtjen, R., Buijs, J.C.A.M., Burattin, A., Carmona, J., Castellanos, M., Claes, J., Cook, J., Costantini, N., Curbera, F., Damiani, E., de Leoni, M., Delias, P., van Dongen, B.F., Dumas, M., Dustdar, S., Fahland, D., Ferreira, D.R., Gaaloul, W., van Geffen, F., Goel, S., Günther, C.W., Guzzo, A., Harmon, P., ter Hofstede, A.H.M., Hoogland, J., Ingvaldsen, J.E., Kato, K., Kuhn, R., Kumar, A., Rosa, M.L., Maggi, F.M., Malerba, D., Mans, R.S., Manuel, A., McCreesh, M., Mello, P., Mendling, J., Montali, M., Nezhad, H.R.M., zur Muehlen, M., Munoz-Gama, J., Pontieri, L., Ribeiro, J., Rozinat, A., Pérez, H.S., Pérez, R.S., Sepúlveda, M., Sinur, J., Soffer, P., Song, M., Sperduti, A., Stilo, G., Stoel, C., Swenson, K.D., Talamo, M., Tan, W., Turner, C., Vanthienen, J., Varvaressos, G., Verbeek, E., Verdonk, M., Vigo, R., Wang, J., Weber, B., Weidlich, M., Weijters, T., Wen, L., Westergaard, M., Wynn, M.T.: Process mining manifesto. In: Business Process Management Workshops - BPM 2011 International Workshops, Clermont-Ferrand, France, August 29, 2011, Revised Selected Papers, Part I, Lecture Notes in Business Information Processing, vol. 99, pp. 169–194. Springer (2011). DOI 10.1007/978-3-642-28108-2_19. URL http://dx.doi.org/10.1007/978-3-642-28108-2_19

9. van der Aalst, W.M.P., ter Hofstede, A.H.M., Kiepuszewski, B., Barros, A.P.: Workflow patterns. Distributed and Parallel Databases 14(1), 5–51 (2003). DOI 10.1023/A:1022883727209. URL http://dx.doi.org/10.1023/A:1022883727209

10. van der Aalst, W.M.P., de Medeiros, A.K.A., Weijters, A.J.M.M.: Process equivalence: Comparing two process models based on observed behavior. In: S. Dustdar, J.L. Fiadeiro, A.P. Sheth (eds.) Business Process Management, 4th International Conference, BPM 2006, Vienna, Austria, September 5-7, 2006, Proceedings, Lecture Notes in Computer Science, vol. 4102, pp. 129–144. Springer (2006). DOI 10.1007/11841760_10. URL http://dx.doi.org/10.1007/11841760_10

11. van der Aalst, W.M.P., Rubin, V.A., Verbeek, H.M.W., van Dongen, B.F., Kindler, E., Günther, C.W.: Process mining: a two-step approach to balance between underfitting and overfitting. Software and System Modeling 9(1), 87–111 (2010). DOI 10.1007/s10270-008-0106-z. URL http://dx.doi.org/10.1007/s10270-008-0106-z

12. van der Aalst, W.M.P., Verbeek, H.M.W.: Process discovery and conformance checking using passages. Fundam. Inform. **131**(1), 103–138 (2014). DOI 10.3233/FI-2014-1006. URL http://dx.doi.org/10.3233/FI-2014-1006

13. Abramowicz, W. (ed.): Business Information Systems - 18th International Conference, BIS 2015, Poznań, Poland, June 24-26, 2015, Proceedings, *Lecture Notes in Business Information Processing*, vol. 208. Springer (2015). DOI 10.1007/978-3-319-19027-3. URL http://dx.doi.org/10.1007/978-3-319-19027-3

14. Adriansyah, A.: Aligning Observed and Modeled Behavior. Ph.D. thesis, Eindhoven University of Technology (2014)

15. Adriansyah, A., Munoz-Gama, J., Carmona, J., van Dongen, B.F., van der Aalst, W.M.P.: Alignment based precision checking. In: M.L. Rosa, P. Soffer (eds.) Business Process Management Workshops - BPM 2012 International Workshops, Tallinn, Estonia, September 3, 2012. Revised Papers, *Lecture Notes in Business Information Processing*, vol. 132, pp. 137–149. Springer (2012). DOI 10.1007/978-3-642-36285-9_15. URL http://dx.doi.org/10.1007/978-3-642-36285-9_15

16. Adriansyah, A., Munoz-Gama, J., Carmona, J., van Dongen, B.F., van der Aalst, W.M.P.: Measuring precision of modeled behavior. Inf. Syst. E-Business Management **13**(1), 37–67 (2015). DOI 10.1007/s10257-014-0234-7. URL http://dx.doi.org/10.1007/s10257-014-0234-7

17. Akutsu, T., Fukagawa, D., Takasu, A., Tamura, T.: Exact algorithms for computing the tree edit distance between unordered trees. Theor. Comput. Sci. **412**(4-5), 352–364 (2011). DOI 10.1016/j.tcs.2010.10.002. URL http://dx.doi.org/10.1016/j.tcs.2010.10.002

18. Armas-Cervantes, A., Dumas, M., Rosa, M.L.: Discovering local concurrency relations in business process event logs (2016). URL http://eprints.qut.edu.au/97615/

19. Augusto, A., Conforti, R., Dumas, M., Rosa, M.L., Bruno, G.: Automated discovery of structured process models: Discover structured vs. discover and structure. In: I. Comyn-Wattiau, K. Tanaka, I. Song, S. Yamamoto, M. Saeki (eds.) Conceptual Modeling - 35th International Conference, ER 2016, Gifu, Japan, November 14-17, 2016, Proceedings, *Lecture Notes in Computer Science*, vol. 9974, pp. 313–329 (2016). DOI 10.1007/978-3-319-46397-1_25. URL http://dx.doi.org/10.1007/978-3-319-46397-1_25

20. Badouel, E.: On the α-reconstructibility of workflow nets. In: Haddad and Pomello [58], pp. 128–147. DOI 10.1007/978-3-642-31131-4_8. URL http://dx.doi.org/10.1007/978-3-642-31131-4_8

21. Bille, P.: A survey on tree edit distance and related problems. Theor. Comput. Sci. **337**(1-3), 217–239 (2005). DOI 10.1016/j.tcs.2004.12.030. URL http://dx.doi.org/10.1016/j.tcs.2004.12.030

22. Bose, R.P.J.C., van der Aalst, W.M.P., Zliobaite, I., Pechenizkiy, M.: Dealing with concept drifts in process mining. IEEE Transactions on Neural Networks and Learning Systems **25**(1), 154–171 (2014). DOI 10.1109/TNNLS.2013.2278313. URL http://dx.doi.org/10.1109/TNNLS.2013.2278313

23. vanden Broucke, S.K.L.M.: Advances in process mining: Artificial negative events and other techniques. Ph.D. thesis, KU Leuven (2014)

24. vanden Broucke, S.K.L.M., Weerdt, J.D., Vanthienen, J., Baesens, B.: Determining process model precision and generalization with weighted artificial negative events. IEEE Trans. Knowl. Data Eng. **26**(8), 1877–1889 (2014). DOI 10.1109/TKDE.2013.130. URL http://dx.doi.org/10.1109/TKDE.2013.130

25. Bubenko Jr., J.A., Krogstie, J., Pastor, O., Pernici, B., Rolland, C., Sølvberg, A. (eds.): Seminal Contributions to Information Systems Engineering, 25 Years of CAiSE. Springer (2013). DOI 10.1007/978-3-642-36926-1. URL http://dx.doi.org/10.1007/978-3-642-36926-1

26. Buijs, J.C.A.M.: Flexible evolutionary algorithms for mining structured process models. Ph.D. thesis, Eindhoven University of Technology (2014)

27. Buijs, J.C.A.M., van Dongen, B.F., van der Aalst, W.M.P.: A genetic algorithm for discovering process trees. In: Proceedings of the IEEE Congress on Evolutionary Computation, CEC

2012, Brisbane, Australia, June 10-15, 2012, pp. 1–8. IEEE (2012). DOI 10.1109/CEC.2012. 6256458. URL http://dx.doi.org/10.1109/CEC.2012.6256458

28. Buijs, J.C.A.M., van Dongen, B.F., van der Aalst, W.M.P.: On the role of fitness, precision, generalization and simplicity in process discovery. In: R. Meersman, H. Panetto, T.S. Dillon, S. Rinderle-Ma, P. Dadam, X. Zhou, S. Pearson, A. Ferscha, S. Bergamaschi, I.F. Cruz (eds.) On the Move to Meaningful Internet Systems: OTM 2012, Confederated International Conferences: CoopIS, DOA-SVI, and ODBASE 2012, Rome, Italy, September 10-14, 2012. Proceedings, Part I, *Lecture Notes in Computer Science*, vol. 7565, pp. 305–322. Springer (2012). DOI 10.1007/978-3-642-33606-5_19. URL http://dx.doi.org/10.1007/978-3-642-33606-5_19

29. Buijs, J.C.A.M., van Dongen, B.F., van der Aalst, W.M.P.: Mining configurable process models from collections of event logs. In: F. Daniel, J. Wang, B. Weber (eds.) Business Process Management - 11th International Conference, BPM 2013, Beijing, China, August 26-30, 2013. Proceedings, *Lecture Notes in Computer Science*, vol. 8094, pp. 33–48. Springer (2013). DOI 10.1007/978-3-642-40176-3_5. URL http://dx.doi.org/10.1007/978-3-642-40176-3_5

30. Buijs, J.C.A.M., van Dongen, B.F., van der Aalst, W.M.P.: Quality dimensions in process discovery: The importance of fitness, precision, generalization and simplicity. Int. J. Cooperative Inf. Syst. **23**(1) (2014). DOI 10.1142/S0218843014400012. URL http://dx.doi.org/10.1142/S0218843014400012

31. Carmona, J.: Projection approaches to process mining using region-based techniques. Data Min. Knowl. Discov. **24**(1), 218–246 (2012). DOI 10.1007/s10618-011-0226-x. URL http://dx.doi.org/10.1007/s10618-011-0226-x

32. Carmona, J., Cortadella, J., Kishinevsky, M.: Divide-and-conquer strategies for process mining. In: Dayal et al. [37], pp. 327–343. DOI 10.1007/978-3-642-03848-8_22. URL http://dx.doi.org/10.1007/978-3-642-03848-8_22

33. Carmona, J., Cortadella, J., Kishinevsky, M.: New region-based algorithms for deriving bounded petri nets. IEEE Trans. Computers **59**(3), 371–384 (2010). DOI 10.1109/TC.2009. 131. URL http://dx.doi.org/10.1109/TC.2009.131

34. Celonis. https://www.celonis.com/. Accessed: 06-01-2017

35. Cook, J.E., Wolf, A.L.: Software process validation: Quantitatively measuring the correspondence of a process to a model. ACM Trans. Softw. Eng. Methodol. **8**(2), 147–176 (1999). DOI 10.1145/304399.304401. URL http://doi.acm.org/10.1145/304399.304401

36. Cortadella, J., Kishinevsky, M., Lavagno, L., Yakovlev, A.: Deriving petri nets for finite transition systems. IEEE Trans. Computers **47**(8), 859–882 (1998). DOI 10.1109/12.707587. URL http://dx.doi.org/10.1109/12.707587

37. Dayal, U., Eder, J., Koehler, J., Reijers, H.A. (eds.): Business Process Management, 7th International Conference, BPM 2009, Ulm, Germany, September 8-10, 2009. Proceedings, *Lecture Notes in Computer Science*, vol. 5701. Springer (2009). DOI 10.1007/978-3-642-03848-8. URL http://dx.doi.org/10.1007/978-3-642-03848-8

38. Di Ciccio, C., Maggi, F.M., Mendling, J.: Efficient discovery of target-branched declare constraints. Inf. Syst. **56**, 258–283 (2016). DOI 10.1016/j.is.2015.06.009. URL http://dx.doi.org/10.1016/j.is.2015.06.009

39. Dijkman, R.M., van Dongen, B.F., Dumas, M., García-Bañuelos, L., Kunze, M., Leopold, H., Mendling, J., Uba, R., Weidlich, M., Weske, M., Yan, Z.: A short survey on process model similarity. In: Bubenko et al. [25], pp. 421–427. DOI 10.1007/978-3-642-36926-1_34. URL http://dx.doi.org/10.1007/978-3-642-36926-1_34

40. Dijkman, R.M., Dumas, M., García-Bañuelos, L.: Graph matching algorithms for business process model similarity search. In: Dayal et al. [37], pp. 48–63. DOI 10.1007/978-3-642-03848-8_5. URL http://dx.doi.org/10.1007/978-3-642-03848-8_5

41. van Dongen, B.: BPI challenge 2012 dataset (2012). DOI 10.4121/uuid:3926db30-f712-4394-aebc-75976070e91f. URL http://dx.doi.org/10.4121/uuid:3926db30-f712-4394-aebc-75976070e91f

42. van Dongen, B.F., Dijkman, R.M., Mendling, J.: Measuring similarity between business process models. In: Bubenko et al. [25], pp. 405–419. DOI 10.1007/978-3-642-36926-1_33. URL http://dx.doi.org/10.1007/978-3-642-36926-1_33

43. van Eck, M.L., Buijs, J.C.A.M., van Dongen, B.F.: Genetic process mining: Alignment-based process model mutation. In: Fournier and Mendling [49], pp. 291–303. DOI 10.1007/978-3-319-15895-2_25. URL http://dx.doi.org/10.1007/978-3-319-15895-2_25

44. van Eck, M.L., Lu, X., Leemans, S.J.J., van der Aalst, W.M.P.: PM^2 : A process mining project methodology. In: J. Zdravkovic, M. Kirikova, P. Johannesson (eds.) Advanced Information Systems Engineering - 27th International Conference, CAiSE 2015, Stockholm, Sweden, June 8-12, 2015, Proceedings, *Lecture Notes in Computer Science*, vol. 9097, pp. 297–313. Springer (2015). DOI 10.1007/978-3-319-19069-3_19. URL http://dx.doi.org/10.1007/978-3-319-19069-3_19

45. Ehrenfeucht, A., Rozenberg, G.: Partial (set) 2-structures. part I: basic notions and the representation problem. Acta Inf. **27**(4), 315–342 (1990). DOI 10.1007/BF00264611. URL http://dx.doi.org/10.1007/BF00264611

46. Ehrenfeucht, A., Rozenberg, G.: Partial (set) 2-structures. part II: state spaces of concurrent systems. Acta Inf. **27**(4), 343–368 (1990). DOI 10.1007/BF00264612. URL http://dx.doi.org/10.1007/BF00264612

47. Ehrig, M., Koschmider, A., Oberweis, A.: Measuring similarity between semantic business process models. In: J.F. Roddick, A. Hinze (eds.) Conceptual Modelling 2007, Proceedings of the Fourth Asia-Pacific Conference on Conceptual Modelling (APCCM2007), Ballarat, Victoria, Australia, January 30 - February 2, 2007, Proceedings, *CRPIT*, vol. 67, pp. 71–80. Australian Computer Society (2007). URL http://crpit.com/abstracts/CRPITV67Ehrig.html

48. Esparza, J., Nielsen, M.: Decidability issues for Petri nets - A survey. Bulletin of the EATCS **52**, 244–262 (1994)

49. Fournier, F., Mendling, J. (eds.): Business Process Management Workshops - BPM 2014 International Workshops, Eindhoven, The Netherlands, September 7-8, 2014, Revised Papers, *Lecture Notes in Business Information Processing*, vol. 202. Springer (2015). DOI 10.1007/978-3-319-15895-2. URL http://dx.doi.org/10.1007/978-3-319-15895-2

50. Gao, X., Xiao, B., Tao, D., Li, X.: A survey of graph edit distance. Pattern Anal. Appl. **13**(1), 113–129 (2010). DOI 10.1007/s10044-008-0141-y. URL http://dx.doi.org/10.1007/s10044-008-0141-y

51. van Glabbeek, R.J., Weijland, W.P.: Branching time and abstraction in bisimulation semantics. J. ACM **43**(3), 555–600 (1996). DOI 10.1145/233551.233556. URL http://doi.acm.org/10.1145/233551.233556

52. Goedertier, S., Martens, D., Vanthienen, J., Baesens, B.: Robust process discovery with artificial negative events. Journal of Machine Learning Research **10**, 1305–1340 (2009). DOI 10.1145/1577069.1577113. URL http://doi.acm.org/10.1145/1577069.1577113

53. Greco, G., Guzzo, A., Pontieri, L., Saccà, D.: Discovering expressive process models by clustering log traces. IEEE Trans. Knowl. Data Eng. **18**(8), 1010–1027 (2006). DOI 10.1109/TKDE.2006.123. URL http://dx.doi.org/10.1109/TKDE.2006.123

54. Günther, C., Verbeek, H.: XES v2.0 (2014). URL http://www.xes-standard.org/

55. Günther, C.W., van der Aalst, W.M.P.: Fuzzy mining - adaptive process simplification based on multi-perspective metrics. In: G. Alonso, P. Dadam, M. Rosemann (eds.) Business Process Management, 5th International Conference, BPM 2007, Brisbane, Australia, September 24-28, 2007, Proceedings, *Lecture Notes in Computer Science*, vol. 4714, pp. 328–343. Springer (2007). DOI 10.1007/978-3-540-75183-0_24. URL http://dx.doi.org/10.1007/978-3-540-75183-0_24

56. Günther, C.W., Rozinat, A.: Disco: Discover your processes. In: N. Lohmann, S. Moser (eds.) Proceedings of the Demonstration Track of the 10th International Conference on Business Process Management (BPM 2012), Tallinn, Estonia, September 4, 2012, *CEUR Workshop Proceedings*, vol. 940, pp. 40–44. CEUR-WS.org (2012). URL http://ceur-ws.org/Vol-940/paper8.pdf

57. Guo, Q., Wen, L., Wang, J., Yan, Z., Yu, P.S.: Mining invisible tasks in non-free-choice constructs. In: Motahari-Nezhad et al. [79], pp. 109–125. DOI 10.1007/978-3-319-23063-4_7. URL http://dx.doi.org/10.1007/978-3-319-23063-4_7

58. Haddad, S., Pomello, L. (eds.): Application and Theory of Petri Nets - 33rd International Conference, PETRI NETS 2012, Hamburg, Germany, June 25-29, 2012. Proceedings, *Lecture Notes in Computer Science*, vol. 7347. Springer (2012). DOI 10.1007/978-3-642-31131-4. URL http://dx.doi.org/10.1007/978-3-642-31131-4

59. Kunze, M., Weidlich, M., Weske, M.: Querying process models by behavior inclusion. Software and System Modeling **14**(3), 1105–1125 (2015). DOI 10.1007/s10270-013-0389-6. URL http://dx.doi.org/10.1007/s10270-013-0389-6

60. Leemans, S.J.J., Fahland, D., van der Aalst, W.M.P.: Discovering block-structured process models from event logs containing infrequent behaviour. In: N. Lohmann, M. Song, P. Wohed (eds.) Business Process Management Workshops - BPM 2013 International Workshops, Beijing, China, August 26, 2013, Revised Papers, *Lecture Notes in Business Information Processing*, vol. 171, pp. 66–78. Springer (2013). DOI 10.1007/978-3-319-06257-0_6. URL http://dx.doi.org/10.1007/978-3-319-06257-0_6

61. Leemans, S.J.J., Fahland, D., van der Aalst, W.M.P.: Exploring processes and deviations. In: Fournier and Mendling [49], pp. 304–316. DOI 10.1007/978-3-319-15895-2_26. URL http://dx.doi.org/10.1007/978-3-319-15895-2_26

62. Leemans, S.J.J., Fahland, D., van der Aalst, W.M.P.: Using life cycle information in process discovery. In: M. Reichert, H.A. Reijers (eds.) Business Process Management Workshops - BPM 2015, 13th International Workshops, Innsbruck, Austria, August 31 - September 3, 2015, Revised Papers, *Lecture Notes in Business Information Processing*, vol. 256, pp. 204–217. Springer (2015). DOI 10.1007/978-3-319-42887-1_17. URL http://dx.doi.org/10.1007/978-3-319-42887-1_17

63. Leemans, S.J.J., Fahland, D., van der Aalst, W.M.P.: Scalable process discovery and conformance checking. Software & Systems Modeling **special issue**, 1–33 (2016). DOI 10.1007/s10270-016-0545-x. URL http://dx.doi.org/10.1007/s10270-016-0545-x

64. de Leoni, M., van der Aalst, W.: Data-aware process mining: discovering decisions in processes using alignments. In: SAC, pp. 1454–1461. ACM (2013)

65. Li, C., Reichert, M., Wombacher, A.: On measuring process model similarity based on high-level change operations. In: Q. Li, S. Spaccapietra, E.S.K. Yu, A. Olivé (eds.) Conceptual Modeling - ER 2008, 27th International Conference on Conceptual Modeling, Barcelona, Spain, October 20-24, 2008. Proceedings, *Lecture Notes in Computer Science*, vol. 5231, pp. 248–264. Springer (2008). DOI 10.1007/978-3-540-87877-3_19. URL http://dx.doi.org/10.1007/978-3-540-87877-3_19

66. Liesaputra, V., Yongchareon, S., Chaisiri, S.: Efficient process model discovery using maximal pattern mining. In: Motahari-Nezhad et al. [79], pp. 441–456. DOI 10.1007/978-3-319-23063-4_29. URL http://dx.doi.org/10.1007/978-3-319-23063-4_29

67. Lu, C.L., Su, Z., Tang, C.Y.: A new measure of edit distance between labeled trees. In: J. Wang (ed.) Computing and Combinatorics, 7th Annual International Conference, COCOON 2001, Guilin, China, August 20-23, 2001, Proceedings, *Lecture Notes in Computer Science*, vol. 2108, pp. 338–348. Springer (2001). DOI 10.1007/3-540-44679-6_37. URL http://dx.doi.org/10.1007/3-540-44679-6_37

68. Lu, R., Sadiq, S.W.: On the discovery of preferred work practice through business process variants. In: C. Parent, K. Schewe, V.C. Storey, B. Thalheim (eds.) Conceptual Modeling - ER 2007, 26th International Conference on Conceptual Modeling, Auckland, New Zealand, November 5-9, 2007, Proceedings, *Lecture Notes in Computer Science*, vol. 4801, pp. 165–180. Springer (2007). DOI 10.1007/978-3-540-75563-0_13. URL http://dx.doi.org/10.1007/978-3-540-75563-0_13

69. Lu, X., Fahland, D., van den Biggelaar, F.J.H.M., van der Aalst, W.M.P.: Handling duplicated tasks in process discovery by refining event labels. In: M.L. Rosa, P. Loos, O. Pastor (eds.) Business Process Management - 14th International Conference, BPM 2016, Rio de Janeiro, Brazil, September 18-22, 2016. Proceedings, *Lecture Notes in Computer Science*,

vol. 9850, pp. 90–107. Springer (2016). DOI 10.1007/978-3-319-45348-4_6. URL http://dx.doi.org/10.1007/978-3-319-45348-4_6

70. Madhusudan, T., Zhao, J.L., Marshall, B.: A case-based reasoning framework for workflow model management. Data Knowl. Eng. **50**(1), 87–115 (2004). DOI 10.1016/j.datak.2004.01.005. URL http://dx.doi.org/10.1016/j.datak.2004.01.005

71. Maggi, F.M., Mooij, A.J., van der Aalst, W.M.P.: User-guided discovery of declarative process models. In: Proceedings of the IEEE Symposium on Computational Intelligence and Data Mining, CIDM 2011, part of the IEEE Symposium Series on Computational Intelligence 2011, April 11-15, 2011, Paris, France [1], pp. 192–199. DOI 10.1109/CIDM.2011.5949297. URL http://dx.doi.org/10.1109/CIDM.2011.5949297

72. de Medeiros, A.K.A., van Dongen, B.F., van der Aalst, W.M.P., Weijters, A.J.M.M.: Process mining for ubiquitous mobile systems: An overview and a concrete algorithm. In: L. Baresi, S. Dustdar, H.C. Gall, M. Matera (eds.) Ubiquitous Mobile Information and Collaboration Systems, Second CAiSE Workshop, UMICS 2004, Riga, Latvia, June 7-8, 2004, Revised Selected Papers, *Lecture Notes in Computer Science*, vol. 3272, pp. 151–165. Springer (2004). DOI 10.1007/978-3-540-30188-2_12. URL http://dx.doi.org/10.1007/978-3-540-30188-2_12

73. de Medeiros, A.K.A., Guzzo, A., Greco, G., van der Aalst, W.M.P., Weijters, A.J.M.M., van Dongen, B.F., Saccà, D.: Process mining based on clustering: A quest for precision. In: A.H.M. ter Hofstede, B. Benatallah, H. Paik (eds.) Business Process Management Workshops, BPM 2007 International Workshops, BPI, BPD, CBP, ProHealth, RefMod, semantics4ws, Brisbane, Australia, September 24, 2007, Revised Selected Papers, *Lecture Notes in Computer Science*, vol. 4928, pp. 17–29. Springer (2007). DOI 10.1007/978-3-540-78238-4_4. URL http://dx.doi.org/10.1007/978-3-540-78238-4_4

74. de Medeiros, A.K.A., Weijters, A.J.M.M., van der Aalst, W.M.P.: Genetic process mining: an experimental evaluation. Data Min. Knowl. Discov. **14**(2), 245–304 (2007). DOI 10.1007/s10618-006-0061-7. URL http://dx.doi.org/10.1007/s10618-006-0061-7

75. Mendling, J.: Detection and prediction of errors in epc business process models. Ph.D. thesis, Vienna University of Economics and Business Administration (2007)

76. Mendling, J., Neumann, G., van der Aalst, W.M.P.: Understanding the occurrence of errors in process models based on metrics. In: R. Meersman, Z. Tari (eds.) On the Move to Meaningful Internet Systems 2007: CoopIS, DOA, ODBASE, GADA, and IS, OTM Confederated International Conferences CoopIS, DOA, ODBASE, GADA, and IS 2007, Vilamoura, Portugal, November 25-30, 2007, Proceedings, Part I, *Lecture Notes in Computer Science*, vol. 4803, pp. 113–130. Springer (2007). DOI 10.1007/978-3-540-76848-7_9. URL http://dx.doi.org/10.1007/978-3-540-76848-7_9

77. Minor, M., Tartakovski, A., Bergmann, R.: Representation and structure-based similarity assessment for agile workflows. In: R. Weber, M.M. Richter (eds.) Case-Based Reasoning Research and Development, 7th International Conference on Case-Based Reasoning, ICCBR 2007, Belfast, Northern Ireland, UK, August 13-16, 2007, Proceedings, *Lecture Notes in Computer Science*, vol. 4626, pp. 224–238. Springer (2007). DOI 10.1007/978-3-540-74141-1_16. URL http://dx.doi.org/10.1007/978-3-540-74141-1_16

78. Molka, T., Redlich, D., Gilani, W., Zeng, X., Drobek, M.: Evolutionary computation based discovery of hierarchical business process models. In: Abramowicz [13], pp. 191–204. DOI 10.1007/978-3-319-19027-3_16. URL http://dx.doi.org/10.1007/978-3-319-19027-3_16

79. Motahari-Nezhad, H.R., Recker, J., Weidlich, M. (eds.): Business Process Management - 13th International Conference, BPM 2015, Innsbruck, Austria, August 31 - September 3, 2015, Proceedings, *Lecture Notes in Computer Science*, vol. 9253. Springer (2015). DOI 10.1007/978-3-319-23063-4. URL http://dx.doi.org/10.1007/978-3-319-23063-4

80. Munoz-Gama, J., Carmona, J.: A fresh look at precision in process conformance. In: R. Hull, J. Mendling, S. Tai (eds.) Business Process Management - 8th International Conference, BPM 2010, Hoboken, NJ, USA, September 13-16, 2010. Proceedings, *Lecture Notes in Computer Science*, vol. 6336, pp. 211–226. Springer (2010). DOI 10.1007/978-3-642-15618-2_16. URL http://dx.doi.org/10.1007/978-3-642-15618-2_16

81. Munoz-Gama, J., Carmona, J., van der Aalst, W.M.P.: Single-entry single-exit decomposed conformance checking. Inf. Syst. **46**, 102–122 (2014). DOI 10.1016/j.is.2014.04.003. URL http://dx.doi.org/10.1016/j.is.2014.04.003

82. Nejati, S., Sabetzadeh, M., Chechik, M., Easterbrook, S.M., Zave, P.: Matching and merging of statecharts specifications. In: 29th International Conference on Software Engineering (ICSE 2007), Minneapolis, MN, USA, May 20-26, 2007, pp. 54–64. IEEE Computer Society (2007). DOI 10.1109/ICSE.2007.50. URL http://dx.doi.org/10.1109/ICSE.2007.50

83. Polyvyanyy, A., Armas-Cervantes, A., Dumas, M., García-Bañuelos, L.: On the expressive power of behavioral profiles. Formal Asp. Comput. **28**(4), 597–613 (2016). DOI 10.1007/s00165-016-0372-4. URL http://dx.doi.org/10.1007/s00165-016-0372-4

84. Polyvyanyy, A., García-Bañuelos, L., Dumas, M.: Structuring acyclic process models. Inf. Syst. **37**(6), 518–538 (2012). DOI 10.1016/j.is.2011.10.005. URL http://dx.doi.org/10.1016/j.is.2011.10.005

85. Redlich, D., Galushka, M., Molka, T., Gilani, W., Blair, G.S., Rashid, A.: Evaluation of the dynamic construct competition miner for an ehealth system. In: Abramowicz [13], pp. 115–126. DOI 10.1007/978-3-319-19027-3_10. URL http://dx.doi.org/10.1007/978-3-319-19027-3_10

86. Redlich, D., Molka, T., Gilani, W., Blair, G.S., Rashid, A.: Constructs competition miner: Process control-flow discovery of bp-domain constructs. In: S.W. Sadiq, P. Soffer, H. Völzer (eds.) Business Process Management - 12th International Conference, BPM 2014, Haifa, Israel, September 7-11, 2014. Proceedings, *Lecture Notes in Computer Science*, vol. 8659, pp. 134–150. Springer (2014). DOI 10.1007/978-3-319-10172-9_9. URL http://dx.doi.org/10.1007/978-3-319-10172-9_9

87. Rozinat, A.: Process Mining: Conformance and Extension. Ph.D. thesis, Eindhoven University of Technology (2010)

88. Rozinat, A., van der Aalst, W.M.P.: Conformance testing: Measuring the fit and appropriateness of event logs and process models. In: C. Bussler, A. Haller (eds.) Business Process Management Workshops, BPM 2005 International Workshops, BPI, BPD, ENEI, BPRM, WSCOBPM, BPS, Nancy, France, September 5, 2005, Revised Selected Papers, vol. 3812, pp. 163–176 (2005). DOI 10.1007/11678564_15. URL http://dx.doi.org/10.1007/11678564_15

89. Rozinat, A., van der Aalst, W.M.P.: Conformance checking of processes based on monitoring real behavior. Inf. Syst. **33**(1), 64–95 (2008). DOI 10.1016/j.is.2007.07.001. URL http://dx.doi.org/10.1016/j.is.2007.07.001

90. Rozinat, A., Veloso, M., van der Aalst, W.: Using hidden markov models to evaluate the quality of discovered process models. BPM Center Report BPM-08-10, BPMcenter.org (2008)

91. Salomaa, A.: Jewels of formal language theory. Computer Science Press, Oelgeschlager (1981)

92. Schimm, G.: Mining exact models of concurrent workflows. Computers in Industry **53**(3), 265–281 (2004). DOI 10.1016/j.compind.2003.10.003. URL http://dx.doi.org/10.1016/j.compind.2003.10.003

93. Solé, M., Carmona, J.: A high-level strategy for c-net discovery. In: J. Brandt, K. Heljanko (eds.) 12th International Conference on Application of Concurrency to System Design, ACSD 2012, Hamburg, Germany, June 27-29, 2012, pp. 102–111. IEEE Computer Society (2012). DOI 10.1109/ACSD.2012.20. URL http://dx.doi.org/10.1109/ACSD.2012.20

94. Solé, M., Carmona, J.: Incremental process discovery. Trans. Petri Nets and Other Models of Concurrency **5**, 221–242 (2012). DOI 10.1007/978-3-642-29072-5_10. URL http://dx.doi.org/10.1007/978-3-642-29072-5_10

95. Solé, M., Carmona, J.: An smt-based discovery algorithm for c-nets. In: Haddad and Pomello [58], pp. 51–71. DOI 10.1007/978-3-642-31131-4_4. URL http://dx.doi.org/10.1007/978-3-642-31131-4_4

96. Viterbi, A.J.: Error bounds for convolutional codes and an asymptotically optimum decoding algorithm. IEEE Trans. Information Theory **13**(2), 260–269 (1967). DOI 10.1109/TIT.1967.1054010. URL http://dx.doi.org/10.1109/TIT.1967.1054010

97. Wang, J., He, T., Wen, L., Wu, N., ter Hofstede, A.H.M., Su, J.: A behavioral similarity measure between labeled Petri nets based on principal transition sequences - (short paper). In: R. Meersman, T.S. Dillon, P. Herrero (eds.) On the Move to Meaningful Internet Systems: OTM 2010 - Confederated International Conferences: CoopIS, IS, DOA and ODBASE, Hersonissos, Crete, Greece, October 25-29, 2010, Proceedings, Part I, *Lecture Notes in Computer Science*, vol. 6426, pp. 394–401. Springer (2010). DOI 10.1007/978-3-642-16934-2_27. URL http://dx.doi.org/10.1007/978-3-642-16934-2_27

98. Wasserman, S., Faust, K.: Social network analysis: Methods and applications, vol. 8. Cambridge university press (1994)

99. Watson, H.J., Wixom, B.: The current state of business intelligence. IEEE Computer **40**(9), 96–99 (2007). DOI 10.1109/MC.2007.331. URL http://dx.doi.org/10.1109/MC.2007.331

100. Weerdt, J.D., Backer, M.D., Vanthienen, J., Baesens, B.: A multi-dimensional quality assessment of state-of-the-art process discovery algorithms using real-life event logs. Inf. Syst. **37**(7), 654–676 (2012). DOI 10.1016/j.is.2012.02.004. URL http://dx.doi.org/10.1016/j.is.2012.02.004

101. Weidlich, M., Dijkman, R.M., Weske, M.: Behaviour equivalence and compatibility of business process models with complex correspondences. Comput. J. **55**(11), 1398–1418 (2012). DOI 10.1093/comjnl/bxs014. URL http://dx.doi.org/10.1093/comjnl/bxs014

102. Weidlich, M., Polyvyanyy, A., Mendling, J., Weske, M.: Efficient computation of causal behavioural profiles using structural decomposition. In: J. Lilius, W. Penczek (eds.) Applications and Theory of Petri Nets, 31st International Conference, PETRI NETS 2010, Braga, Portugal, June 21-25, 2010. Proceedings, *Lecture Notes in Computer Science*, vol. 6128, pp. 63–83. Springer (2010). DOI 10.1007/978-3-642-13675-7_6. URL http://dx.doi.org/10.1007/978-3-642-13675-7_6

103. Weidlich, M., Polyvyanyy, A., Mendling, J., Weske, M.: Causal behavioural profiles - efficient computation, applications, and evaluation. Fundam. Inform. **113**(3-4), 399–435 (2011)

104. Weijters, A., van der Aalst, W., de Medeiros, A.: Process mining with the heuristics miner-algorithm. BETA Working Paper series 166, Eindhoven University of Technology (2006)

105. Weijters, A.J.M.M., van der Aalst, W.M.P.: Rediscovering workflow models from event-based data using little thumb. Integrated Computer-Aided Engineering **10**(2), 151–162 (2003). URL http://content.iospress.com/articles/integrated-computer-aided-engineering/ica00143

106. Weijters, A.J.M.M., Ribeiro, J.T.S.: Flexible heuristics miner (FHM). In: Proceedings of the IEEE Symposium on Computational Intelligence and Data Mining, CIDM 2011, part of the IEEE Symposium Series on Computational Intelligence 2011, April 11-15, 2011, Paris, France [1], pp. 310–317. DOI 10.1109/CIDM.2011.5949453. URL http://dx.doi.org/10.1109/CIDM.2011.5949453

107. Wen, L., van der Aalst, W.M.P., Wang, J., Sun, J.: Mining process models with non-free-choice constructs. Data Min. Knowl. Discov. **15**(2), 145–180 (2007). DOI 10.1007/s10618-007-0065-y. URL http://dx.doi.org/10.1007/s10618-007-0065-y

108. Wen, L., Wang, J., van der Aalst, W.M.P., Huang, B., Sun, J.: A novel approach for process mining based on event types. J. Intell. Inf. Syst. **32**(2), 163–190 (2009). DOI 10.1007/s10844-007-0052-1. URL http://dx.doi.org/10.1007/s10844-007-0052-1

109. Wen, L., Wang, J., van der Aalst, W.M.P., Huang, B., Sun, J.: Mining process models with prime invisible tasks. Data Knowl. Eng. **69**(10), 999–1021 (2010). DOI 10.1016/j.datak.2010.06.001. URL http://dx.doi.org/10.1016/j.datak.2010.06.001

110. Wen, L., Wang, J., Sun, J.: Detecting implicit dependencies between tasks from event logs. In: X. Zhou, J. Li, H.T. Shen, M. Kitsuregawa, Y. Zhang (eds.) Frontiers of WWW Research and Development - APWeb 2006, 8th Asia-Pacific Web Conference, Harbin, China, January 16-18, 2006, Proceedings, *Lecture Notes in Computer Science*, vol. 3841, pp. 591–603. Springer (2006). DOI 10.1007/11610113_52. URL http://dx.doi.org/10.1007/11610113_52

111. Wen, L., Wang, J., Sun, J.: Mining invisible tasks from event logs. In: G. Dong, X. Lin, W. Wang, Y. Yang, J.X. Yu (eds.) Advances in Data and Web Management, Joint 9th Asia-

Pacific Web Conference, APWeb 2007, and 8th International Conference, on Web-Age Information Management, WAIM 2007, Huang Shan, China, June 16-18, 2007, Proceedings, *Lecture Notes in Computer Science*, vol. 4505, pp. 358–365. Springer (2007). DOI 10.1007/978-3-540-72524-4_38. URL http://dx.doi.org/10.1007/978-3-540-72524-4_38

112. van der Werf, J.M.E.M., van Dongen, B.F., Hurkens, C.A.J., Serebrenik, A.: Process discovery using integer linear programming. Fundam. Inform. **94**(3-4), 387–412 (2009). DOI 10.3233/FI-2009-136. URL http://dx.doi.org/10.3233/FI-2009-136

113. Westergaard, M., Maggi, F.M.: Declare: A tool suite for declarative workflow modeling and enactment. In: H. Ludwig, H.A. Reijers (eds.) Proceedings of the Demo Track of the Nineth Conference on Business Process Management 2011, Clermont-Ferrand, France, August 31st, 2011, *CEUR Workshop Proceedings*, vol. 820. CEUR-WS.org (2011). URL http://ceur-ws.org/Vol-820/Demo3.pdf

114. Wolper, P.: Temporal logic can be more expressive. Information and Control **56**(1/2), 72–99 (1983). DOI 10.1016/S0019-9958(83)80051-5. URL http://dx.doi.org/10.1016/S0019-9958(83)80051-5

115. Wombacher, A.: Evaluation of technical measures for workflow similarity based on a pilot study. In: R. Meersman, Z. Tari (eds.) On the Move to Meaningful Internet Systems 2006: CoopIS, DOA, GADA, and ODBASE, OTM Confederated International Conferences, CoopIS, DOA, GADA, and ODBASE 2006, Montpellier, France, October 29 - November 3, 2006. Proceedings, Part I, *Lecture Notes in Computer Science*, vol. 4275, pp. 255–272. Springer (2006). DOI 10.1007/11914853_16. URL http://dx.doi.org/10.1007/11914853_16

116. van Zelst, S.J., van Dongen, B.F., van der Aalst, W.M.P.: Avoiding over-fitting in ILP-based process discovery. In: Motahari-Nezhad et al. [79], pp. 163–171. DOI 10.1007/978-3-319-23063-4_10. URL http://dx.doi.org/10.1007/978-3-319-23063-4_10

117. Zha, H., Wang, J., Wen, L., Wang, C., Sun, J.: A workflow net similarity measure based on transition adjacency relations. Computers in Industry **61**(5), 463–471 (2010). DOI 10.1016/j.compind.2010.01.001. URL http://dx.doi.org/10.1016/j.compind.2010.01.001

Chapter 4
Recursive Process Discovery

Sander J. J. Leemans: Robust Process Mining with Guarantees, LNBIP 440, pp. 119–136, 2022
https://doi.org/10.1007/978-3-030-96655-3_4

Abstract In the previous chapters, we introduced the input and outputs of process mining techniques, and described challenges of process discovery techniques. Due to the trade offs identified in these challenges, we argued that different algorithms might be necessary in different use cases. In this chapter, we introduce a framework for process discovery in a recursive divide-and-conquer fashion: the Inductive Miner framework, using several illustrative examples. Furthermore, we provide the foundations of the guarantees that this framework can provide: soundness, termination, perfect fitness, perfect precision and rediscoverability (which is the ability for an algorithm to rediscover the language of a system given some assumptions on the system and the log).

Therefore, in Chapter 6, we introduce a family of process discovery techniques to handle different situations. For instance, we introduce an algorithm handling deviating behaviour and an algorithm handling missing information.

All these algorithms need to be robust, e.g. handle real-life logs with ease, and provide several guarantees, such as soundness, rediscoverability, and in some cases fitness and log precision. Proving these guarantees, especially rediscoverability and fitness, can be tedious. Therefore, to enable reuse of code and formal results, in this chapter, we introduce a novel process discovery framework, called the *Inductive Miner framework* (IM framework), which provides some guarantees by itself (e.g. soundness). Furthermore, the IM framework aids algorithms in guaranteeing fitness, log precision, rediscoverability, and a polynomial run time, as we express these properties in terms of the framework, which makes them easier to prove.

The IM framework is *abstract*, i.e., it does not define a complete algorithm. We will illustrate through a number of examples how the framework can be instantiated to yield different kinds of discovery algorithms. As we introduce the framework, we will already point to the elements in the framework that allow to ensure soundness, fitness, log precision and balancing log criteria. In contrast, rediscoverability will be less easy to ensure. Therefore, in Section 4.2, we specifically consider which aspects of models, logs and discovery algorithms contribute to rediscoverability. In the same section, we establish a number of sufficient conditions for rediscovering the original model based on behavioural abstractions. We will investigate these behavioural abstractions in detail in Chapter 5. In Chapter 6, we introduce concrete algorithms that use the IM framework and the behavioural abstractions of Chapter 5, and we prove rediscoverability using the sufficient conditions of Section 4.2.

4.1 Recursive Process Discovery

The main idea of the IM framework is to construct a process tree recursively in a top-down fashion, i.e. it starts with the entire event log and splits it into smaller parts, on which the framework recurses, thereby building up a process tree. In this way, any algorithm using the framework will only return process trees, which are sound by construction (Requirement CR1).

We first introduce the framework using an example in Section 4.1.1. Second, in Section 4.1.2 we introduce the framework formally, and in Section 4.1.4, we discuss the guarantees provided by the framework. In Chapter 5, we strengthen the formal foundations of the framework by studying abstractions in more detail, while in Chapter 6, we instantiate the framework by providing several concrete discovery algorithms.

4.1.1 An Example of Recursive Process Discovery

In this section, we provide an intuitive introduction to the framework using the example event log shown in Figure 4.1a.

On this event log L_3, the IM framework performs several steps, of which the first one is to identify the "most important" behaviour of the event log, i.e. the root of the corresponding process tree, and to divide the activities of the event log into smaller subsets. We refer to the combination of a root operator and a partition of activities as a *cut*. In Figure 4.1a, a cut has been highlighted with a red dashed line. Intuitively, this cut denotes that in every trace, first something with activity a happens, and afterwards something with activities b and c, which can be described as a sequence between a left subprocess involving a and a right subprocess involving b and c. We denote this cut with $(\rightarrow, \{a\}, \{b, c\})$, and the first step of the IM framework is to identify such a cut (we will introduce cut detection in detail in chapters 5 and 6). After detection of the cut, the root operator, i.e. $\rightarrow$ in our example, is noted.

$$L_3 = [\langle a, b \rangle$$
$$\langle a, b \rangle$$
$$\langle a, c \rangle]$$

(a) Initial log. The red dashed line denotes the cut $(\rightarrow, \{a\}, \{b, c\})$.

$$L_4 = [\langle a \rangle \qquad\qquad L_5 = [\langle b \rangle$$
$$\langle a \rangle \qquad\qquad\qquad \langle b \rangle$$
$$\langle a \rangle] \qquad\qquad\qquad \langle c \rangle]$$

(b) Sublog after log splitting. This is a base case.

(c) Sublog after log splitting. The red dashed line denotes the cut $(\times, \{b\}, \{c\})$

$$L_6 = [\langle b \rangle \qquad\qquad L_7 = [\langle c \rangle]$$
$$\langle b \rangle]$$

(d) Sublog after log splitting (2). This is a base case.

(e) Sublog after log splitting (2). This is a base case.

Fig. 4.1: Example run of the IM framework.

Second, the IM framework splits the event log according to this cut. In our example, splitting log L_3 leads to the logs $L_4 = [\langle a\rangle, \langle a\rangle, \langle a\rangle]$ and $L_5 = [\langle b\rangle, \langle b\rangle, \langle c\rangle]$, as shown in figures 4.1b and 4.1c. The process tree discovered up till this point is $\rightarrow$, and the IM framework still needs to process L_4 and L_5. Therefore, as the third step, the IM framework recurses, say on L_4.

"L_4" "L_5"

In L_4, only an activity a is present, i.e. there cannot be any cut, thus the IM framework hits a base case of the recursion. The process tree discovered up till now is $\rightarrow$, and the IM framework still needs to process L_5.

a "L_5"

Then, the IM framework recurses on L_5, which is shown in Figure 4.1c. In L_5, the cut $\{\times, \{b\}, \{c\}\}$ can be found, as each trace has either activity b or activity c, but never both. Therefore, the IM framework splits L_5 into sublogs $L_6 = [\langle b\rangle, \langle b\rangle]$ and $L_7 = [\langle c\rangle]$. The process tree discovered up till now is $\rightarrow$, and the IM

framework still needs to process L_6 and L_7.

Finally, at recursion of L_6 and L_7, the IM framework will discover both of them to be base cases of the recursion, and return the process tree $\rightarrow$.

4.1.2 The IM framework

In this section, we define the framework formally, give another example and briefly discuss its implementation.

To summarise, the IM framework defines four steps: first, a cut is detected, second, the event log is split into smaller sublogs and third, the IM framework recurses on these sublogs until a base case is encountered. Fourth, if no cut can be found, a *fall through* is returned, i.e. a process tree is discovered such that recursion can continue (this was not shown in the example). These four steps are parameters of the IM framework and have to be provided as plug-ins by a process discovery algorithm: each algorithm that implements the IM framework should provide each of these four functions. That is, for a log L:

- The parameter function BASECASE detects base cases of the recursion: BASECASE(L) takes a log L, and if L contains a base case returns a process tree that represents this base case.

- The parameter function FINDCUT searches for a cut c, consisting of a process tree operator and an activity partition i.e. FINDCUT(L) searches for a cut in log L and returns that cut if it exists.
- The parameter function SPLITLOG splits the log into smaller sublogs. SPLITLOG(L, c) splits log L according to cut c and returns the remaining sublogs.
- The parameter function FALLTHROUGH(L) returns a fall through for L, i.e. a process tree that describes L. This function must not fail and always return a process tree. Notice that this function will only be invoked if neither a base case nor a cut can be found.

From the viewpoint of the framework, these functions are independent, e.g. it is possible to interchange FINDCUT functions of different algorithms. Nevertheless, algorithms might pose restrictions on this interchangeability, e.g. for some algorithms introduced in Chapter 6, the SPLITLOG function assumes certain properties on the cuts returned by FINDCUT.

Formally, given four functions BASECASE, FINDCUT, SPLITLOG and FALLTHROUGH, the *Inductive Miner framework* (IM framework) has the type $IM framework : \mathbb{E} \rightarrow \mathbb{T}$ and is defined as follows, using that $\square$ denotes "nothing":

function IM FRAMEWORK$_{\text{BASECASE,FINDCUT,SPLITLOG,FALLTHROUGH}}(L)$

    $bc \leftarrow$ BASECASE(L)

    if $bc \neq \square$ **then**

        return bc

    end if

    $(\oplus, \Sigma_1, \ldots, \Sigma_n) \leftarrow$ FINDCUT(L)

    if $(\oplus, \Sigma_1, \ldots, \Sigma_n) \neq \square$ **then**

        $L_1 \ldots L_n \leftarrow$ SPLITLOG($L, (\oplus, \Sigma_1, \ldots, \Sigma_n)$)

        return $\oplus(IM framework(L_1), \ldots, IM framework(L_n))$

    else

        return FALLTHROUGH(L)

    end if

end function

In the remainder of this chapter, for conciseness, we might omit the parameter functions from IM framework if they are clear from the context. We continue this section with examples of several algorithms that instantiate the IM framework. We will not give definitions or code here, but this will illustrate the flexibility of the framework.

4.1.3 More Technical Examples

In this section, we show examples for three algorithms: one using the best case, i.e. if all information that an algorithm needs is present in the event log, and two to illustrate challenges of rediscoverability, i.e. incompleteness and deviating behaviour.

To illustrate the functions of the IM framework, we discuss a second example using a basic algorithm called *Inductive Miner* (IM). For this example, consider the event log

$$L_8 = [\langle a,b,c,d,e \rangle, \langle a,d,b,e \rangle, \langle a,e,b \rangle, \langle a,c,b \rangle, \langle a,b,d,e,c \rangle]$$

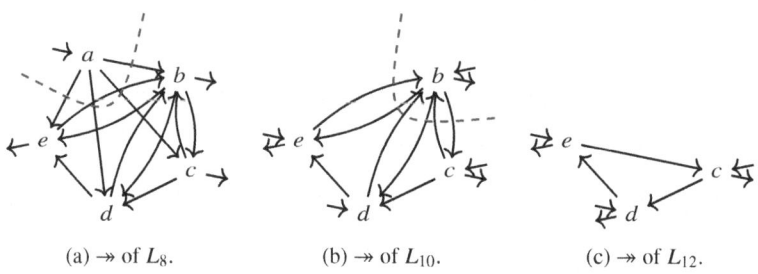

(a) $\twoheadrightarrow$ of L_8.　　　　　(b) $\twoheadrightarrow$ of L_{10}.　　　　　(c) $\twoheadrightarrow$ of L_{12}.

Fig. 4.2: Directly follows graphs of logs used in the recursion. The dashed red curves denote cuts. No cut can be found for L_{12}.

The BASECASE function of IM does not detect a base case in L_8, as multiple activities are present. The FINDCUT function considers the directly follows graph of L_8 (see Figure 6.1a) and searches for characteristic footprints of process tree operators, returning the first footprint found. In this graph, the cut $c_1 = (\rightarrow, \{a\}, \{b,c,d,e\})$ is present, as all edges cross this line in one direction (hence, the sequence). Then, SPLITLOG(L_8, c_1) splits the log in sublogs L_9 and L_{10}:

$$L_9 = [\langle a \rangle^5]$$
$$L_{10} = [\langle b,c,d,e \rangle, \langle d,b,e \rangle, \langle e,b \rangle, \langle c,b \rangle, \langle b,d,e,c \rangle]$$

Furthermore, IM records the choice and recurses, i.e. $\mathrm{IM}(L_8) = \rightarrow(\mathrm{IM}(L_9), \mathrm{IM}(L_{10}))$. We first consider the recursive step on L_9, for which BASECASE(L_9) returns a base case, supposedly being the process tree a:

$$\mathrm{BASECASE}(L_9) = a$$

Next, we give the computation steps taken and the results of the recursive calls:

$$\mathrm{IM}(L_9) = a$$
$$\textsc{baseCase}(L_{10}) = \square$$
$$\textsc{findCut}(L_{10}) = c_3 = (\wedge, \{b\}, \{c, d, e\}), \text{ see Figure 6.1b}$$
$$\textsc{splitLog}(L_{10}, c_3) = L_{11}, L_{12}$$
$$L_{11} = [\langle b \rangle^5]$$
$$L_{12} = [\langle c, d, e \rangle, \langle d, e \rangle, \langle e \rangle, \langle c \rangle, \langle d, e, c \rangle]$$
$$\mathrm{IM}(L_{10}) = \wedge(\mathrm{IM}(L_{11}), \mathrm{IM}(L_{12}))$$
$$\textsc{baseCase}(L_{11}) = b$$
$$\mathrm{IM}(L_{11}) = b$$
$$\textsc{baseCase}(L_{12}) = \square$$
$$\textsc{findCut}(L_{12}) = \square, \text{ see Figure 6.1c}$$
$$\textsc{fallThrough}(L_{12}) = \circlearrowleft(\tau, c, d, e)$$
$$\mathrm{IM}(L_{12}) = \circlearrowleft(\tau, c, d, e)$$

Combining all intermediate steps, IM will discover the process tree $M_{13} = $

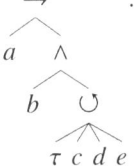

In our next example, we will show how rediscoverability can be challenged by incompleteness and deviating behaviour. For these examples, we assume that the system model from which the event log was derived is indeed M_{13}.

For the incompleteness example, we remove the trace $\langle a, c, b \rangle$ from L_8, i.e.

$$L_{14} = [\langle a, b, c, d, e \rangle, \langle a, d, b, e \rangle, \langle a, e, b \rangle, \langle a, b, d, e, c \rangle]$$

The first cut $(\rightarrow, \{a\}, \{b, c, d, e\})$ is still present in the directly follows graph of L_{14}, so recursion continues as in the previous example, and the following sublog is obtained:

$$L_{15} = [\langle b, c, d, e \rangle, \langle d, b, e \rangle, \langle e, b \rangle, \langle b, d, e, c \rangle]$$

Figure 4.3a shows the directly follows graph of L_{15}. In this graph, there is no cut $(\wedge, \{b\}, \{c, d, e\})$, as the edge $c \twoheadrightarrow b$ is missing. Therefore, the basic algorithm IM would not detect a cut, and instead return the tree $\rightarrow$ 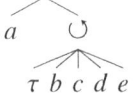, hence would not

rediscover the system M_{13}. In Section 6.3, we will introduce an algorithm that is able

to handle this incompleteness in the directly follows graph and discover the correct
cut despite the missing edge, by searching for a likely cut instead of a perfect cut.
The algorithm thus derives the presence of the missing edge and continues discovery
as IM, and rediscovers M_{13}.

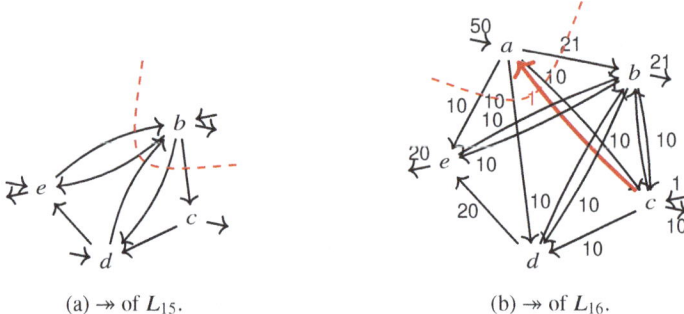

(a) $\twoheadrightarrow$ of L_{15}.   (b) $\twoheadrightarrow$ of L_{16}.

Fig. 4.3: Directly follows graphs of logs used in the recursion. The dashed red curves
are not cuts here.

For the deviating behaviour example, we duplicate the event log 10 times, and
add a deviating trace $\langle c, a, b \rangle$.

$$L_{16} = [\langle a,b,c,d,e \rangle^{10}, \langle a,d,b,e \rangle^{10}, \langle a,e,b \rangle^{10}, \langle a,b,d,e,c \rangle^{10}, \langle c,a,b \rangle]$$

Figure 4.3b shows the directly follows graph of L_{16}. In this figure, we added the
frequencies on the directly follows edges. In the directly follows graph, the deviating
trace adds, amongst other things, the edge $c \twoheadrightarrow a$, and therefore the dashed line is not
a sequence cut, as the edge $c \twoheadrightarrow a$ crosses it in the "wrong" direction. Therefore, the
basic algorithm IM would not detect a cut, and instead return the tree $\circlearrowleft$,

$$\tau\ a\ b\ c\ d\ e$$

hence would not rediscover the system M_{13}. In Section 6.2, we will introduce an
algorithm that spots that of all outgoing edges of c, the edge $a \twoheadrightarrow c$ is 10 times less
frequent than the other edges. The algorithm then filters this edge out and continues
discovery as IM, and rediscovers M_{13}.

In these examples, we illustrated how the IM framework and the directly follows abstraction can be used to (re)discover process trees. We showed a best case,
i.e. when the directly follows abstraction is correct and complete, we showed an
example in which the abstraction was incomplete, and we showed an example in
which the abstraction was erroneous (i.e. the log contained information that did not
correspond to the system). In the remainder of this section, we discuss guarantees
that IM framework can provide, regardless of the abstraction, correctness notion and
incompleteness notion of the specific algorithm.

4.1.4 Guarantees

In the previous sections, we introduced the IM framework as our solution for process discovery. In this section, we address several guarantees that can be provided by the IM framework, possibly requiring some proof obligations on the concrete functions defined by an algorithm, i.e. we address termination, fitness and log precision. Rediscoverability is discussed in Section 4.2.

4.1.4.1 Termination

The IM framework guarantees termination based on the given parameter functions: FINDCUT and SPLITLOG together must guarantee that the size of the event log decreases. In case BASECASE and FALLTHROUGH make a recursive call to the IM framework themselves, the event logs on which these calls are made must be strictly smaller in size than the original log. With these guarantees, IM framework obviously guarantees termination.

4.1.4.2 Fitness & Log Precision

The IM framework is able to provide several guarantees, two of which are perfect fitness and log precision. In this section, we study the conditions under which these guarantees hold. Using the recursive nature of the IM framework, we introduce two local properties. The first property expresses that a step of the framework can never exclude traces from consideration, i.e. locally, fitness is preserved. The second property expresses the reverse, i.e. the model does not represent any trace that was not in the log. Both use a language combining function $\oplus_{\mathcal{L}}$ that combines several languages using its corresponding process tree operator $\oplus$ definition, i.e. $\mathcal{L}(\oplus(K_1, \ldots K_k)) = \oplus_{\mathcal{L}}(\mathcal{L}(K_1), \ldots \mathcal{L}(K_k))$.

Definition 4.1 (Local fitness & log precision preservation) For all event logs L,

- a combination of a cut detection function FINDCUT and a log splitting function SPLITLOG is *locally fitness preserving* if

$$set(L) \subseteq \oplus_{\mathcal{L}}(\text{SPLITLOG}(L, \text{FINDCUT}(L)))$$

  in which $\oplus_{\mathcal{L}}$ is the language combination function corresponding to the operator selected by FINDCUT(L) (if FINDCUT returns a cut);
- a base case function BASECASE is *locally fitness preserving* if

$$set(L) \subseteq \mathcal{L}(\text{BASECASE}(L))$$

  (if BASECASE(L) $\neq \square$, i.e. it applies to L);
- a fall through function FALLTHROUGH is *locally fitness preserving* if

$$set(L) \subseteq \mathcal{L}(\text{FALLTHROUGH}(L))$$

For log precision, the definition is similar, using:

$$\oplus_{\mathcal{L}}(L_1, \dots L_n) \subseteq L$$
$$\mathcal{L}(\text{BASECASE}(L)) \subseteq L$$
$$\mathcal{L}(\text{FALLTHROUGH}(L)) \subseteq L$$

Given this definition, it is not hard to reason that if all steps applied by the IM framework are locally fitness preserving, then the overall result will be perfectly fitting and similarly for log precision. Formally:

Corollary 4.1 *Let* $\Diamond \colon \mathbb{E} \to \mathbb{T}$ *be a discovery technique using the IM framework in which all parameter functions are locally fitness preserving. Then, for every log L it holds that* $\Diamond(L)$ *fits L, i.e.* $set(L) \subseteq \mathcal{L}(\Diamond(L))$. *If all parameter functions are locally log-precision preserving, then* $\Diamond(L)$ *is log precise to L, i.e.* $\mathcal{L}(\Diamond(L)) \subseteq set(L)$.

If the locality is clear from the context, we will omit the word 'local'. This corollary illustrates the modularity of the IM framework: given the use case at hand, one can choose a custom set of operators, cut detection algorithms, base cases and fall throughs. As long as all choices are locally fitness/log precision preserving, the end result will be guaranteed accordingly.

In Chapter 6, we will show local fitness and log precision preservation while introducing concrete functions. Here, we illustrate these concepts using some examples. An example of a fall through is the flower model. A flower model function takes an alphabet $\Sigma(L)$ and returns a model that can generate any behaviour of the alphabet, i.e. $\circlearrowleft(\tau, a_1, \dots a_n)$ where $\{a_1, \dots a_n\} = \Sigma(L)$. This fall through is fitness preserving, as the resulting model allows for any behaviour. However, it is not log precision preserving, as L is bounded, and the model allows for unbounded traces. This argument holds for all $\circlearrowleft$s, i.e. any function that returns a model that contains a $\circlearrowleft$ cannot be log precision preserving.

An example of a log precision preserving function is the fall through that returns a trace model, i.e. a choice between all traces in the event log. For instance, let $L = \{\langle a, b \rangle, \langle a, c \rangle\}$ be an event log. Then, $\times(\to(a, b), \to(a, c))$ is a trace model of L. Obviously, such a trace model is both fitness and log precision preserving. However, the model is not generalising and merely enumerates the log.

4.2 Rediscoverability

As introduced in Section 3.2.2.2, the property rediscoverability entails that a discovery algorithm is able to discover a model that is language equivalent to the system that underlies the given event log. Figure 4.4 illustrates rediscoverability: a system model is implemented by a system, the system executes and of this execution an event log is recorded, and from the event log a model is discovered.

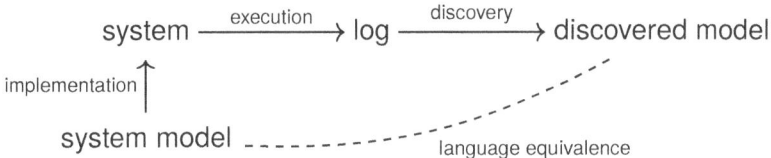

Fig. 4.4: Rediscoverability: rediscover the language of the system model.

The system model is *rediscovered* if it has the same language as the discovered model (as the discovered model is derived from an event log and an event log only contains information about a language, process discovery algorithms cannot rediscover stronger notions of equivalence). An algorithm that guarantees rediscovery possesses *rediscoverability*. Formally:

Definition 4.2 (rediscoverability) Let SM be a system model, S be a system that implements SM and let L be an event log generated from S. Furthermore, let M be a process model that is discovered by a process discovery algorithm. Then, the process discovery algorithm provides *rediscoverability* for SM for L if and only if $\mathcal{L}(SM) = \mathcal{L}(M)$.

Typically, in order to prove rediscoverability for a specific discovery technique, one needs to make assumptions on the system S and the event log L. Obviously, rediscoverability is then only proven for cases in which these assumptions hold. For instance, for the α algorithm, if the system can be represented by an unlabelled sound free-choice workflow net without short loops and without implicit places (see Section 3.3) and the directly follows graph of the event log is equivalent to the directly follows graph of the system, then the discovered model is isomorphic to the system [1, 3]. The restriction on implicit places is necessary for isomorphic rediscovery because these places do not change the language of the system, i.e. removing them preserves language and hence, the α algorithm has no way to discover them. Therefore, allowing implicit places straightforwardly still guarantees language rediscovery.

In this section, we consider how rediscoverability can be achieved in a practical process discovery setting where only partial knowledge about system behaviour (an event log) is available, and we ease the proofs using the abstractions that are used by many process discovery techniques: we introduce a framework that uses these abstractions and makes some assumptions (Section 4.2.1). Furthermore, we discuss how the framework and the IM framework can be combined to prove rediscoverability for actual discovery algorithms in Section 4.2.2.

4.2.1 Rediscoverability using Abstractions

In this framework, let A denote an abstraction function over languages, and let C denote a class of models. For instance, in case of the α algorithm, the abstraction A is a function that takes a language and returns a directly follows graph of the language, and the class of models C is the set of all models that can be represented by unlabelled sound free-choice workflow nets without short loops.

The key of the framework is a property of the combination of the abstraction A and the class of models C: there must not be two models with different languages in C that have the same abstraction. We refer to this property as *language uniqueness*. Formally:

Definition 4.3 (language uniqueness) A class of models C and a language abstraction A: $\mathbb{T} \cup \mathbb{E} \to \mathbb{A}$ are *language unique* if and only if each two models of C with different languages have different abstractions:

$$\forall_{K,M \in C} \; \mathcal{L}(K) = \mathcal{L}(M) \Leftrightarrow A(K) = A(M)$$

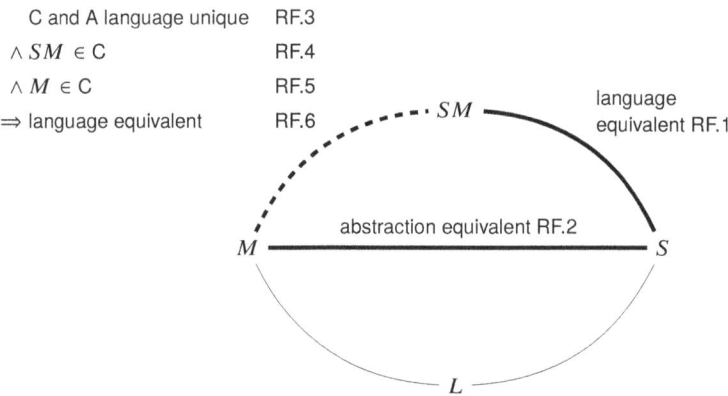

Fig. 4.5: Formal framework for rediscoverability using abstractions: if the system model is of the class C and is language equivalent to the system, the system has the same abstraction A as the discovered model, the discovered model is of class C, and language uniqueness holds for class C and abstraction A, then the system model and the discovered model are language equivalent.

If language uniqueness is proven, then the discovery technique only needs to guarantee to discover a model with the same *abstraction* as the system, instead of a model with the same *language* (as in Definition 4.2). The framework is shown in Figure 4.5. We explain the assumptions and proof obligations, in which RF stands for Rediscoverability Framework:

RF.1 The system model is language equivalent to the system, i.e. the system model is implemented correctly by the system. Although a plethora of issues might challenge the correctness of an implementation of a system model, these issues are outside of our scope, and therefore we assume that the system model and the running system have the same language. Notice that this assumption has little practical influence, as both system model and system are unknown in typical process mining projects.

RF.2 The abstraction of the system is equivalent to the abstraction of the discovered model. This is a proof obligation by the discovery algorithm.

We refer to this property as *abstraction rediscovery*, i.e. the abstraction of the discovered model is equivalent to the abstraction of the system. A process discovery algorithm that guarantees abstraction rediscovery possesses *abstraction rediscoverability*:

Definition 4.4 (abstraction rediscoverability) Let $A: \mathbb{T} \cup \mathbb{E} \rightarrow \mathbb{A}$ be a language abstraction and let C be a class of models. Then, a process discovery algorithm provides *abstraction rediscoverability* for A and C if for each system $S \in$ C, the model M that is discovered by the algorithm has the same abstraction as S, i.e. $A(S) = A(M)$.

A typical proof strategy of discovery algorithms would be to (1) make an assumption on the completeness and correctness of the event log L, e.g. that $A(S) = A(L)$, and (2) prove that the discovered model has the same abstraction as the event log: $A(L) = A(M)$. We chose not to put this assumption in Definition 4.4, as discovery techniques can make arbitrary assumptions on the event log. For instance, the abstraction in the event log may be incomplete or contain erroneous information; in such cases, the assumption $A(S) = A(L)$ could be weakened. Notice that the weaker these assumptions, the more powerful the discovery technique, and more event logs can be handled.

RF.3 The abstraction A and the class of models C are language unique (Definition 4.3). This is a proof obligation that comes with the discovery technique. However, as a property of the abstraction and a class of languages, it is independent of the technique itself and can therefore be reused by several techniques. We will prove this property for several classes of languages and abstractions in Chapter 5.

RF.4 The system model is of class C. This is an assumption on the class of the system model, and restricts rediscoverability accordingly.

RF.5 The discovered model is of class C. This is a proof obligation of the discovery technique: if a technique discovers a model outside the class of C, then language uniqueness should be proven for this model, otherwise the framework does not apply.

From these assumptions and proof obligations, language equivalence between the discovered model and the system model (RF.6) follows directly:

Theorem 4.1 (rediscoverability using abstractions) *Let C be a class of models, let* $A: \mathbb{T} \cup \mathbb{E} \rightarrow \mathbb{A}$ *be a language abstraction, and let SM, S and M be models. Then, if all*

- $\mathcal{L}(SM) = \mathcal{L}(S)$ *(RF.1)*
- $A(S) = A(M)$ *(RF.2)*
- *the combination of C and A is language unique (RF.2)*
- $SM \in C$ *(RF.4)*
- $M \in C$ *(RF.5)*

then $\mathcal{L}(SM) = \mathcal{L}(M)$ *(RF.6).*

Examples.

Rediscoverability has been proven for several discovery algorithms, e.g. α [3, 1, 2] and IM [5]. In both of these proofs, the class of models is limited: a subset of process trees for IM, and a subset of free-choice workflow nets for α, and it is assumed that the event log is directly follows complete and noise free with respect to the system. In terms of the abstraction rediscoverability framework, this assumption means that the event log and the system have the same abstraction (see Figure 4.6).

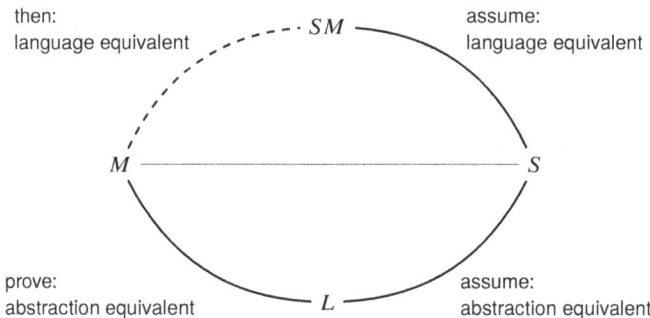

Fig. 4.6: A typical rediscoverability proof mapped onto the abstraction rediscoverability framework.

Formally, this assumption can be mapped onto the abstraction rediscoverability framework:

Corollary 4.2 *Let C be a class of models, let* $A\colon \mathbb{T} \cup \mathbb{E} \to \mathbb{A}$ *be a language abstraction, and let S, L and M be models. Then, if all*

- $A(S) = A(L)$ *(typical assumption)*
- $A(L) = A(M)$ *(typical proof obligation)*

then $A(S) = A(M)$ *RF.2.*

We finish this chapter with a translation of the identified assumptions and proof obligations to the IM framework.

4.2.2 Rediscoverability and the IM framework

In the IM framework, a divide-and-conquer strategy is applied to discover a process tree recursively. In this section, we study the influence of the requirements of abstraction rediscoverability on the IM framework, i.e. we show how discovery algorithms that implement the IM framework can be proven to rediscover the system model. In this proof strategy, we will use four elements.

- First, we use a class of process trees C.
- Second, we use a language abstraction function A that takes an event log or a process tree and returns the abstraction of type $\mathbb{A}$.
- Third, we use a log-assumptions function LA that takes a process model and returns the set of all logs that adhere to the assumptions made by the algorithm. For instance, for the α algorithm, this log-assumptions function would return all event logs that have the same directly follows graph as the model).
- Fourth, we use the algorithm $\Diamond$ itself, which implements the IM framework, using the four parameter functions BASECASE$_\Diamond$, FINDCUT$_\Diamond$, SPLITLOG$_\Diamond$ and FALLTHROUGH$_\Diamond$.

Similar assumptions have been made in rediscoverability proofs. For instance, for e.g. the α algorithm [4], rediscoverability could be proven using C as the class of all unlabelled free-choice Petri nets without short loops, A as the directly follows graph and LA would entail that the log has the same directly follows relation as the system model (although the α algorithm does not consider process trees).

We first introduce a property, i.e. a set of requirements, that expresses the requirements of abstraction rediscoverability using the four parameter functions of the IM framework, i.e. for an actual discovery algorithm, this property should be proven, which we will do in Chapter 6. Second, in Lemma 4.1, we prove that this property is a sufficient condition for abstraction rediscoverability (Definition 4.4). Third, we express RF.5 in terms of C, A and LA (Definition 4.6). Fourth, we prove that these properties are sufficient to conclude rediscoverability.

Definition 4.5 (abstraction preservation) Let C be a class of process trees, let $A: \mathbb{E} \cup \mathbb{T} \to \mathbb{A}$ be a language abstraction, let $LA: C \to 2^{\mathbb{E}}$ be a log assumption function, and let $\Diamond = $ IM FRAMEWORK be a discovery algorithm implementing the IM framework with BASECASE$_\Diamond$, FINDCUT$_\Diamond$, SPLITLOG$_\Diamond$ and FALLTHROUGH$_\Diamond$, i.e. $\Diamond: \mathbb{E} \to \mathbb{T}$. Then, $\Diamond$ is *abstraction preserving* if for every tree $S \in C$ under abstraction A and for any log $L \in LA(S)$:

AP.1 The abstraction of an activity is preserved: for all systems $a \in C$ such that a is an activity, and for all logs $L \in LA(a)$, it holds that $A(\text{BASECASE}_\Diamond(L)) = A(a)$.

AP.2 The abstraction of a τ step is preserved: for the system $\tau \in C$ and for all logs $L \in LA(\tau)$, it holds that $A(\text{BASECASE}_\Diamond(L)) = A(\tau)$.

AP.3 If the algorithm applies a base case, the abstraction is preserved. Let $S \in C$ be a system such that $S = \oplus(S_1, \ldots S_n)$, and let $L \in LA(S)$ be a log adhering to the log-assumptions. Then $A(\text{BASECASE}_\Diamond(L)) = A(S)$ (if BASECASE$_\Diamond(L)$ applies).

To ease proofs, we weaken this requirement using the assumption that it holds for all smaller systems: assume that for all S' such that $|S'| \leq |S|$ and $L' \in LA(S')$, it holds that $A(\text{BASECASE}_\Diamond(L')) = A(S')$.

AP.4 If the algorithm detects a cut, then this cut conforms to the system: for all systems $S = \oplus(S_1, \ldots S_n)$ with $S \in C$ and for all logs $L \in LA(S)$ holds and for which $\text{BASECASE}_\Diamond(L)$ does not apply, it holds that $\text{FINDCUT}_\Diamond(L)$ conforms to S (Definition 5.4).

AP.5 If a conforming cut is found, then the log assumptions hold for the sublogs (for the next recursive step). Let $S = \oplus(S_1, \ldots S_n)$ be a system with $S \in C$, let $c = (\otimes, \Sigma_1, \ldots \Sigma_m)$ be a cut that conforms to S (Definition 5.4), and let $L_1 \ldots L_m = \text{SPLITLOG}(L, c)$, then there exist trees $M_1 \ldots M_m$ such that $A(\oplus(M_1, \ldots M_m)) = A(S)$, and $\forall_{1 \leq i \leq m} L_i \in LA(M_i)$.

AP.6 If the algorithm uses a fall through, the abstraction is preserved: let $S \in C$ be a system such that $S = \oplus(S_1, \ldots S_n)$, and let $L \in LA(S)$ be a log, but neither $\text{BASECASE}_\Diamond(L)$ nor $\text{FINDCUT}_\Diamond(L)$ applies. Assume that for all S' such that $|S'| \leq |S|$ and $L' \in LA(S')$, it holds that $A(\Diamond(L')) = A(S')$. Then, $A(\text{FALLTHROUGH}_\Diamond(L)) = A(S)$.

In the following, we prove that any algorithm that implements the IM framework and satisfies Definition 4.5 will have rediscoverability (according to Definition 4.2). For this, we first prove that any such algorithm will rediscover the *abstraction* of the system (Lemma 4.1). Then, under the assumption that the algorithm preserves the class of its input (i.e., Requirement RF.5), we will be able to conclude that the algorithm also rediscovers the system (Theorem 4.2).

Lemma 4.1 (Abstraction rediscoverability of the IM framework) *Let C be a class of process trees, A be a language abstraction, LA be a log assumption function, and let* $\Diamond = $ *IM* FRAMEWORK$_{\text{BASECASE}_\Diamond, \text{FINDCUT}_\Diamond, \text{SPLITLOG}_\Diamond, \text{FALLTHROUGH}_\Diamond}$, *such that* $\Diamond$ *is abstraction preserving (Definition 4.5). Then, for all systems $S \in C$ and logs $L \in LA(S)$, it holds that $A(\Diamond(L))A(S)$.*

Proof We prove the theorem by induction on process tree sizes, being $|S|$.

- Base case: $S = a$, with $a \in \Sigma$. By Requirement AP.1, $A(\text{BASECASE}_\Diamond(L)) = A(S)$.
- Base case: $S = \tau$. By Requirement AP.2, $A(\text{BASECASE}_\Diamond(L)) = A(S)$.
- Induction step: assume $S = \oplus(S_1, \ldots S_n)$ and that the theorem holds for all models smaller than S. By code inspection, three cases apply:

  - $\text{BASECASE}_\Diamond$ applies. By Requirement AP.3, $A(\text{BASECASE}_\Diamond(L)) = A(S)$.
  - $\text{BASECASE}_\Diamond$ does not apply, then by Requirement AP.4, the cut $c = (\oplus, \Sigma_1, \Sigma2)$ such that $c = \text{FINDCUT}_\Diamond(L)$ conforms to S. Let $L_1 \ldots L_m$ be the sublogs returned by $\text{SPLITLOG}_\Diamond(L, c)$. By Requirement AP.5, there exist trees $M_1, \ldots M_m$ such that $A(\oplus(M_1, \ldots M_m)) = A(S)$, and $\forall_{1 \leq i \leq m} L_i \in LA(M_i)$. By the induction hypothesis, $A(\oplus(\Diamond(L_1), \ldots \Diamond(L_m))) = A(\oplus(M_1, \ldots M_m)) = A(M)$. Hence, $A(\Diamond(L)) = A(S)$.
  - If neither a base case $\text{BASECASE}_\Diamond$ nor a cut $\text{FINDCUT}_\Diamond$ applies, then by Requirement AP.6, $A(\text{FALLTHROUGH}_\Diamond(L)) = A(\Diamond(L)) = A(S)$.

Hence, $A(\Diamond(L)) = A(S)$ (Definition 4.4). $\square$

As a final requirement, we define Requirement RF.5 of Section 4.2.1, i.e. that the discovered model should be of class C, in terms of C and LA:

Definition 4.6 (language-class preservation) A combination of a class of process trees C, a log assumption function LA and an algorithm $\Diamond$ is *language-class preserving* if and only if for all systems $S \in$ C and logs $L \in LA(S)$, it holds that $\Diamond(L) \in$ C.

Finally, we prove the main theorem, i.e. an algorithm that is abstraction preserving and language-class preserving has rediscoverability:

Theorem 4.2 *Let C be a class of process trees, $A \colon \mathbb{T} \cup \mathbb{E} \to \mathbb{A}$ be a language abstraction, LA be a log assumption function, and let $\Diamond = $ IM FRAMEWORK be a discovery algorithm implementing the IM framework with BASECASE$_\Diamond$, FINDCUT$_\Diamond$, SPLITLOG$_\Diamond$ and FALLTHROUGH$_\Diamond$, such that the combination of C, A, LA and $\Diamond$ is abstraction preserving (Definition 4.5), such that the combination of C, LA and $\Diamond$ is language-class preserving (Definition 4.6), and such that the combination of A and the set of languages represented by C is language unique (Definition 4.3).*

Let SM and S be process trees such that $\mathcal{L}(SM) = \mathcal{L}(S)$, $SM \in$ C and $S \in$ C. Then, $\Diamond$ has rediscoverability (Definition 4.2): for each log $L \in LA(S)$, it holds that $\mathcal{L}(SM) = \mathcal{L}(\Diamond(L))$.

Proof We discuss each of the requirements of Theorem 4.1: by assumption, $\mathcal{L}(SM) = \mathcal{L}(S)$ (RF.1). By Lemma 4.1, $A(S) = A(M)$ (RF.2). By assumption, the combination of C and A is language unique (RF.3). By assumption, $SM \in$ C (RF.4). By Definition 4.6, $\Diamond(L) \in$ C (RF.5). Then, by Theorem 4.1, $\mathcal{L}(SM) = \mathcal{L}(\Diamond(L))$. $\square$

In subsequent chapters, we will use the abstraction rediscoverability framework and IM framework to introduce several discovery algorithms. In Chapter 5, we analyse language uniqueness for several combinations of abstractions and classes of process trees, after which we introduce several process discovery algorithms and prove rediscoverability for them in Chapter 6.

References

1. van der Aalst, W., Weijters, A., Maruster, L.: Workflow mining: Discovering process models from event logs. IEEE Trans. Knowl. Data Eng. **16**(9), 1128–1142 (2004)
2. van der Aalst, W.M.P.: Process mining - discovery, conformance and enhancement of business processes. Springer (2011). DOI 10.1007/978-3-642-19345-3. URL http://dx.doi.org/ 10.1007/978-3-642-19345-3
3. Badouel, E.: On the α-reconstructibility of workflow nets. In: S. Haddad, L. Pomello (eds.) Application and Theory of Petri Nets - 33rd International Conference, PETRI NETS 2012, Hamburg, Germany, June 25-29, 2012. Proceedings, *Lecture Notes in Computer Science*, vol. 7347, pp. 128–147. Springer (2012). DOI 10.1007/978-3-642-31131-4_8. URL http://dx. doi.org/10.1007/978-3-642-31131-4_8

4. Badouel, E., Darondeau, P.: Theory of regions. In: W. Reisig, G. Rozenberg (eds.) Lectures on Petri Nets I: Basic Models, Advances in Petri Nets, the volumes are based on the Advanced Course on Petri Nets, held in Dagstuhl, September 1996, *Lecture Notes in Computer Science*, vol. 1491, pp. 529–586. Springer (1996). DOI 10.1007/3-540-65306-6_22. URL `http://dx.doi.org/10.1007/3-540-65306-6_22`

5. Leemans, S.J.J., Fahland, D., van der Aalst, W.M.P.: Discovering block-structured process models from event logs - a constructive approach. In: J.M. Colom, J. Desel (eds.) Application and Theory of Petri Nets and Concurrency - 34th International Conference, PETRI NETS 2013, Milan, Italy, June 24-28, 2013. Proceedings, *Lecture Notes in Computer Science*, vol. 7927, pp. 311–329. Springer (2013). DOI 10.1007/978-3-642-38697-8_17. URL `http://dx.doi.org/10.1007/978-3-642-38697-8_17`

Chapter 5
Abstractions

Abstract In this chapter, we perform an in-depth study of several language abstractions. To this end, we first introduce a set of reduction rules for process trees, and show that this set is confluent. Second, we show for several abstractions that any two trees from a certain class of process trees that have the same language, have the same canonical form. As a side effect, footprints of process tree operators are identified for each abstraction. The abstractions covered are the directly follows relation, the activity relations, the minimum self-distance abstraction, the concurrent-optional-or abstraction, the non-atomic directly follows relation, and the concurrency relation. We finish the chapter with a comparison of the classes of trees involved.

In process discovery, it is typically assumed that not all possible behaviour is actually present in the event log. Therefore, most process discovery algorithms do not use an event log directly, but use an intermediate step, i.e. an abstraction, and instead of assuming that the entire behaviour is present in the event log, it is assumed that the "entire" abstraction has been seen. For instance, the abstraction that the α algorithm uses is the directly follows relation.

In the previous chapter, we introduced the IM framework to discover process models from event logs. Furthermore, we showed how the IM framework aids in guaranteeing rediscoverability, i.e. we introduced a proof framework that poses proof obligations for concrete discovery algorithms. In Chapter 6, we will introduce concrete algorithms that use this framework, and that provide several guarantees.

Abstractions pose limitations to discovery algorithms: if two models have the same abstraction, the discovery algorithm cannot distinguish the models. Therefore, one of the proof obligations of the proof framework in Section 4.2.1 entails that for a class of models, the uniqueness of the abstraction needs to be proven, i.e. the combination of an abstraction and the class of models needs to be *language unique*: all models from this class with different languages should have different abstractions (Definition 4.3). Language uniqueness of an abstraction and a class of models provides a formal basis for discovery algorithms: the discovery algorithm can simply discover a model with the same abstraction to provide rediscoverability.

In this chapter, we systematically study abstractions in combination with process trees, by, for each abstraction, studying the models and languages it can distinguish, and which classes of process trees correspond to these languages. In Chapter 6, we will use these results to prove rediscoverability for the algorithms defined therein.

A challenge to proving language uniqueness is the loose relation between a language and the set of process trees that can represent it, i.e. between semantics and syntax, thereby forbidding reasoning on the structure of process trees directly. For instance, the trees

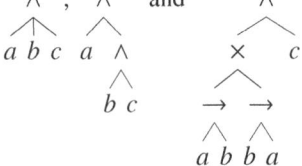

have the same language, and therefore proving two process trees language equivalent has to be performed on behaviour rather than structure, which complicates the language-uniqueness proofs. To address this, we first introduce structural reduction rules on process trees that preserve the language of a tree, and we show that repeated application of these rules leads to a syntactic unique normal form, i.e. the normal form is *canonic*.

Second, using the set of reduction rules, we show the language uniqueness of several abstractions. For each abstraction, the proof strategy is to show that the abstraction of each process tree in normal form is different from the abstractions of all other process trees in normal form in the class of process trees considered (and due to the normal form, we only need to consider syntactically equivalent trees). As all abstractions we use are language based, different abstractions imply different languages. The reduction rules therefore establish a one to one relationship between semantics and syntax of process trees, within the classes considered.

The first abstraction we consider is the directly follows graph, in combination with a simple class of trees, i.e. without duplicate activities and only using four basic operators ($\times$, $\rightarrow$, $\wedge$ and $\circlearrowleft$). We show that with some restrictions on the nesting of operators, the abstraction of the tree onto its directly follows graph suffices to uniquely identify the language of the process tree, and distinguish it from other trees both semantically and syntactically. We study this class of process trees separately as it is similar to a representational bias used in many process discovery algorithms, such as α and its derivatives, Heuristic Miner (HM) and Fodina (FO). Furthermore, we show that different classes of process trees, i.e. having less restrictions, τ steps to allow for skips and other operators such as the inclusive choice, cannot be distinguished by the directly follows graph. In a later section, we show language uniqueness of directly follows graphs and a larger class of process trees, i.e. including $\leftrightarrow$.

In Section 5.1, we introduce the normal form and prove that it is canonical. Second, in each of several sections, we introduce a new abstraction, a class of process trees and prove language uniqueness for this combination. The abstractions covered in this chapter are directly follows graphs (Section 5.2), activity relations (Section 5.3), minimum self-distance (Section 5.5) and concurrent-optional-or relations (Section 5.6).

A limitation of the languages and concepts used before is that all activities are atomic and therefore are limited in their expression of concurrency, and therefore lack an important process modelling feature. Furthermore, it is challenging to measure time and performance on such models, as will be shown in Chapter 9. Therefore, in Section 5.7, the study is repeated on models with non-atomic activities: a different set of reduction rules is introduced, and the directly follows graph is proven to distinguish languages of another class of process trees.

In Section 5.8, we revisit the classes of process trees and compare them to other formalisms and to rediscoverability classes of other algorithms.

5.1 A Canonical Normal Form for Process Trees

There might be multiple process trees with the same language. For instance, the tree $\rightarrow(a, \rightarrow(b, c))$ has the same language as $\rightarrow(\rightarrow(a, b), c)$. We are not interested in the structural difference between these trees, as their behaviour is the same. Therefore, we introduce structural reduction rules on process trees that preserve the language of a tree, and we show that repeated application of these rules leads to a syntactic unique normal form, i.e. for each language, there is at most one process tree in normal form. In our example, the normal form would be $\rightarrow(a, b, c)$. Furthermore, we prove that the repeated application of reduction rules always terminates in finitely many steps.

We first give the rules in Section 5.1.1, after which we prove that their repeated application terminates and that they can be applied in any order (*canonicity*) in Section 5.1.2.

5.1.1 Reduction Rules

A reduction rule applies to a subtree of a process tree and transforms it into another subtree. For instance, the rule $\rightarrow(\ldots_1, \rightarrow(\ldots_2), \ldots_3) \rightarrow \rightarrow(\ldots_1, \ldots_2, \ldots_3)$ transforms trees to remove nested sequence operators. The left hand side denotes that it applies to any sequence operator that is a child of a sequence operator, and the right hand side denotes that after application of the rule, all children of the nested sequence operator are now children of the topmost operator. If this rule would be applied to $\rightarrow(a, \rightarrow(b, c))$, the tree $\rightarrow(a, b, c)$ would result.

We identified four categories of reduction rules: the singularity rule, associativity rules, τ reduction rules and $\wedge$ relation rules. We first give the rules, after which we explain each category and discuss some properties of reduced trees.

Most of these rules apply to a certain patterns of process tree operators, however some rules have additional restrictions. For instance, the trees $S_1 \ldots S_n$ ensure that the length of each trace in the language of each S_i has only traces of at most one event. After the definition, we will elaborate more on the rules and these restrictions.

Definition 5.1 (Reduction rules) Let M, P, Q, Q_1, Q_2 and $S_1 \ldots S_n$ be process trees, and let $\ldots$ be any number of process trees (possibly 0). Then, the reduction rules are as follows:

$$\text{------------------------ singularity rule}$$

(S) $\qquad\qquad\qquad \oplus(M) \Rightarrow M \text{ with } \oplus \in \{\times, \rightarrow, \wedge, \leftrightarrow, \vee\}$

$$\text{------------------------ associativity reduction rules}$$

$(A_\times)$ $\qquad \times(\ldots_1, \times(\ldots_2)) \Rightarrow \times(\ldots_1, \ldots_2)$

$(A_\rightarrow)$ $\quad \rightarrow(\ldots_1, \rightarrow(\ldots_2), \ldots_3) \Rightarrow \rightarrow(\ldots_1, \ldots_2, \ldots_3)$

$(A_\wedge)$ $\qquad \wedge(\ldots_1, \wedge(\ldots_2)) \Rightarrow \wedge(\ldots_1, \ldots_2)$

$(A_\vee)$ $\qquad \vee(\ldots_1, \vee(\ldots_2)) \Rightarrow \vee(\ldots_1, \ldots_2)$

$(A_{\circlearrowleft B})$ $\quad \circlearrowleft(\circlearrowleft(M, \ldots_1), \ldots_2) \Rightarrow \circlearrowleft(M, \ldots_1, \ldots_2)$

$(A_{\circlearrowleft R})$ $\quad \circlearrowleft(M, \ldots_1, \times(\ldots_2)) \Rightarrow \circlearrowleft(M, \ldots_1, \ldots_2)$

$$\text{------------------------ } \tau\text{-reduction rules}$$

$(T_\times)$ $\qquad \times(\ldots, Q, \tau) \Rightarrow \times(\ldots, Q) \text{ with } \epsilon \in \mathcal{L}(Q)$

$(T_\rightarrow)$ $\qquad \rightarrow(\ldots, M, \tau) \Rightarrow \rightarrow(\ldots, M)$

$(T_\wedge)$ $\qquad \wedge(\ldots, M, \tau) \Rightarrow \wedge(\ldots, M)$

$(T_\leftrightarrow)$ $\qquad \leftrightarrow(\ldots, M, \tau) \Rightarrow \leftrightarrow(\ldots, M)$

$(T_\vee)$ $\qquad \vee(\ldots, M, \tau) \Rightarrow \times(\tau, \vee(\ldots, M))$

$(T_{\vee,\times})$ $\quad \vee(\ldots_1, \times(\ldots_2, M, \tau)) \Rightarrow \times(\tau, \vee(\ldots_1, \times(\ldots_2, M)))$

$(T_{\circlearrowleft B})$ $\qquad \circlearrowleft(\tau, \ldots, P) \Rightarrow \times(\tau, \circlearrowleft(\times(\ldots, P), \tau)) \text{ with } \mathcal{L}(P) \neq \{\epsilon\}$

$(T_{\circlearrowleft R})$ $\quad \circlearrowleft(M, \ldots, Q, \tau) \Rightarrow \circlearrowleft(M, \ldots, Q) \text{ with } \epsilon \in \mathcal{L}(Q)$

$(T_{\circlearrowleft BR})$ $\qquad \circlearrowleft(\tau, \tau) \Rightarrow \tau$

$$\text{------------------------ } \wedge\text{-relation rules}$$

$(C_\leftrightarrow)$ $\qquad \leftrightarrow(S_1, \ldots S_n) \Rightarrow \wedge(S_1, \ldots S_n) \text{ with } \forall_{1 \le i \le n, t \in \mathcal{L}(S_i)} |t| \le 1$

$(C_\vee)$ $\qquad \wedge(\ldots, Q_1, Q_2) \Rightarrow \wedge(\ldots, \vee(Q_1, Q_2)) \text{ with } \epsilon \in \mathcal{L}(Q_1) \cap \mathcal{L}(Q_2)$

Notice that none of the rules introduces an operator without children, and a loop operator always keeps at least two children. A process tree to which no rule can be applied is in *normal form*. For simplicity, we will refer to such a tree as a *reduced process tree*. Later, we will prove that this normal form is unique for several classes of process trees.

Notice that we defined the order of children for commutative process tree operators to be irrelevant, e.g. $\times(a, b) = \times(b, a)$ and $\circlearrowleft(a, b, c) = \circlearrowleft(a, c, b) \neq \circlearrowleft(b, a, c)$, and thus rules equalising such process trees are not necessary.

Singularity Rule.

The singularity rule (S) applies to all operators except $\circlearrowleft$, as a $\circlearrowleft$-node always has at least two children. By definition of the process tree operators (Definition 2.6), a node with these operators with a single child has the same behaviour as the child itself.

Associativity Rules.

For all operators, we identified one associativity rule, except for the loop operator which requires two associativity rules due to its asymmetric definition: one for the loop body ($A_{\circlearrowleft B}$) and one for the redo parts ($A_{\circlearrowleft R}$), and except for the $\leftrightarrow$ operator, which is not associative.

We briefly discuss why some seemingly obvious associativity rules are missing from Definition 5.1. An absent and invalid rule would be a rule that reduces nested loops in redo parts, i.e.

$$\circlearrowleft(M, \ldots_1, \circlearrowleft(\ldots_2)) \not\Rightarrow \circlearrowleft(M, \ldots_1, \ldots_2)$$

This rule would not preserve language, e.g. the following example shows two trees with a different language, as witnessed by two example traces:

$$M_{17} = \circlearrowleft(a, b, c) \qquad \langle a, b, c, b, a\rangle \notin \mathcal{L}(M_{17}) \quad \langle a, c, a\rangle \in \mathcal{L}(M_{17})$$

$$M_{18} = \circlearrowleft(a, \circlearrowleft(b, c)) \quad \langle a, b, c, b, a\rangle \in \mathcal{L}(M_{18}) \quad \langle a, c, a\rangle \notin \mathcal{L}(M_{18})$$

Furthermore, the associativity rules do not apply to the $\leftrightarrow$ operator, i.e.

$$\leftrightarrow(\ldots_1 \leftrightarrow(\ldots_2)) \not\Rightarrow \leftrightarrow(\ldots_1, \ldots_2)$$

as witnessed by the following counterexample:

$$M_{19} = \leftrightarrow(\leftrightarrow(a, b), c) \quad \langle a, c, b\rangle \notin \mathcal{L}(M_{19}) \quad \langle b, a, c\rangle \in \mathcal{L}(M_{19})$$

$$M_{20} = \leftrightarrow(a, \leftrightarrow(b, c) \quad \langle a, c, b\rangle \in \mathcal{L}(M_{20}) \quad \langle b, a, c\rangle \notin \mathcal{L}(M_{20})$$

$$M_{21} = \leftrightarrow(a, b, c) \qquad \langle a, c, b\rangle \in \mathcal{L}(M_{21}) \quad \langle b, a, c\rangle \in \mathcal{L}(M_{21})$$

τ-Reduction Rules.

Given the unobservable nature of the τ, we identified several reduction rules targeting this construct. For instance, a τ as a child of an $\times$ is redundant if another child of the $\times$ can already produce the empty trace (rule $T_\times$). Under $\rightarrow$, $\wedge$ and $\leftrightarrow$-operators, τ leafs do not change the language (rules $T_\rightarrow$, $T_\wedge$, $T_\leftrightarrow$). A τ as a child of an $\vee$-node enables the $\vee$-node to produce the empty trace, which we reduce to $\times(\tau, \vee(\ldots))$ ($T_\vee$, $T_{\vee,\times}$). These two rules illustrate the aim of our τ-reduction rules: to minimise the number of τ leafs in a tree and to make the empty trace explicit. For instance, we consider $\times(\tau, \vee(a, b))$ to be more elegant than $\vee(\times(\tau, a), \times(\tau, b))$. In specific discovery algorithms, which will be described in Chapter 6, τ leafs will be detected using a base case (as introduced in Chapter 4). Denoting empty traces explicitly allows discovery algorithms to do this. Furthermore, our conformance checking framework, which will be described in Chapter 7, will use these rules to achieve a speedup.

We identified three $\circlearrowleft$ reduction rules to deal with τ leafs: the first one ($T_{\circlearrowleft B}$) removes τ children from the loop body. This makes the empty trace explicit, which eases rediscovery and ensures discovery does not have to consider $\circlearrowleft(\tau, \ldots)$. The P in this rule (such that $\mathcal{L}(P) \neq \{\epsilon\}$) ensures termination in combination with rules S and $T_\times$:

$$\circlearrowleft(\tau, \tau) \overset{T_{\circlearrowleft B}}{\not\Longrightarrow} \times(\tau, \circlearrowleft(\times(\tau), \tau)) \overset{S}{\Rightarrow} \times(\tau, \circlearrowleft(\tau, \tau)) \overset{T_\times}{\Rightarrow} \circlearrowleft(\tau, \tau)$$

As the redo-parts of a loop are defined using $\times$-semantics, the $T_{\circlearrowleft R}$ rule corresponds to the $T_\times$ rule. Finally, $T_{\circlearrowleft BR}$ covers the base case $\circlearrowleft(\tau, \tau)$ which has the same language as τ.

$\wedge$-Relation Rules.

The final set of reduction rules establishes the connection between the concurrent-like operators $\wedge$, $\leftrightarrow$ and $\vee$. Rule $C_\leftrightarrow$ establishes that if all children have languages of at most one event, executing them concurrently or interleaved does not matter. (In Section 5.7, we will explore alternative semantics in which execution of activities take time, thus invalidating this rule.) Rule $C_\vee$ establishes the relation between $\vee$ and $\wedge$: $\vee$ expresses that at least one of its children must be executed. Therefore, if two children of a $\wedge$-node can be skipped, we consider them to be in an $\vee$-relation that can be skipped. For instance:

$$\wedge(\times(\tau, a), \times(\tau, b)) \overset{\text{rule } C_\vee}{\Longrightarrow} \wedge(\vee(\times(\tau, a), \times(\tau, b)))$$
$$\overset{\text{rule } S}{\Longrightarrow} \vee(\times(\tau, a), \times(\tau, b))$$
$$\overset{\text{rule } T_{\vee, \times} \text{ 2 times}}{\Longrightarrow} \times(\tau, \times(\tau, \vee(\times(a), \times(b))))$$
$$\overset{\text{rule } T_\times}{\Longrightarrow} \times(\tau, \times(\vee(\times(a), \times(b))))$$
$$\overset{\text{rule } S \text{ 3 times}}{\Longrightarrow} \times(\tau, \vee(a, b))$$

Wrap up.

In the remaining chapters, we will often use some properties of reduced process trees, that directly follow from exhaustively applying the reduction rules:

Corollary 5.1 (Properties of reduced trees) *For all subtrees* $\oplus(M_1, \ldots, M_n)$ *in a reduced tree, it holds that*

- $n \geq 2$;
- *If* $\oplus \in \{\times, \rightarrow, \wedge, \vee\}$, *then no direct child is of the same operator:* $\forall_{1 \leq i \leq n} M_i \neq \oplus(\ldots)$;

- *If $\oplus = \circlearrowleft$, then M_1 is not a $\circlearrowleft$ and any non-first child is not an $\times$: $\forall_{2 \leq i \leq n} M_i \neq \times(\ldots)$.*
- *If $\oplus = \leftrightarrow$, then for at least one M_i, it holds that $\exists_{t \in \mathcal{L}(M_i)} |t| > 1$.*
- *τ leafs appear only as children of $\times$-nodes or as non-first children of $\circlearrowleft$-nodes.*

The aim of this set of reduction rules is twofold: it should establish a one to one relationship between syntax and semantics of process trees in normal form, i.e. any two different process trees in normal form have different languages (the process trees are *language unique*). Furthermore, it should establish the same one to one relation between languages/process trees and behavioural abstractions. Then, discovery algorithms can use these behavioural abstractions and provide guarantees.

In the remainder of this chapter, we will first prove that the set of reduction rules always leads to a normal form. Second, we will study several abstractions and show for which classes of process trees they provide language uniqueness.

The current set of reduction rules does not fulfill this aim completely, i.e. the rules are too weak to reduce all language equivalent trees completely. For instance, for the following pairs of trees, both trees have the same language (same semantics), are both in normal form but are not equivalent (same semantics, different syntax):

$$\circlearrowleft(\rightarrow(\times(\tau, a), \circlearrowleft(b, \tau)), \tau) \xleftrightarrow{\text{future work}} \circlearrowleft(\rightarrow(\times(\tau, a), b), \tau)$$

$$\wedge(a, a) \xleftrightarrow{\text{future work}} \rightarrow(a, a)$$

$$\times(a, a) \xleftrightarrow{\text{future work}} a$$

$$\times(\rightarrow(a, b), \rightarrow(b, a)) \xleftrightarrow{\text{future work}} \wedge(a, b)$$

The core challenge of the first pair of trees is that a loop has a τ as a non-first child, while in the second and third pair the challenge is that an activity appears more than once. There might be further reduction rules to equalise these trees, however such rules would need to target several layers deep, or instead of structural use language-based left-hand sides. For now, this remains future work.

Future work 5.1: Extend reduction rules to reduce trees with τ leafs as non-first children and duplicate activities.

5.1.2 Canonicity of the Reduction Rules

Using these structural reduction rules for process trees, we show that the repeated application results in a process tree that is unique, i.e. the order in which the rules are applied is irrelevant (*canonicity*). We do this by first showing that for every process tree, the repeated application of these rules is terminating, i.e. at some point no further rules can be applied (Lemma 5.1). Second, in Lemma 5.2 we show that if two rules (A and B) both apply to a process tree M and yield two different trees

(M_A and M_B), then there is a sequence of reduction rules to make them equal again, i.e. that the rules are *locally confluent*. Figure 5.1 illustrates local confluency. From these two properties and Newman's Lemma [11], canonicity follows. In the sections thereafter, we consider to what extent this normal form can be used to distinguish the languages of process trees.

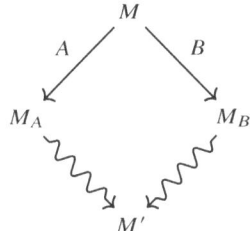

Fig. 5.1: Local confluency: if from a tree M two reduction rules are possible (A and B), then there are sequences of reduction rules possible that converge into an M'.

Lemma 5.1 (Termination of repeated reduction rules application) *Repeated application of the reduction rules of Definition 5.1 is terminating.*

Proof We prove termination by first defining a set of functions that count certain redundancies in a process tree M. The weighted combination of these functions yields an expression that evaluates to a natural number for each process tree. The weighing depends on the process tree at hand, as a rule application might decrease some functions, but increase others. Second, we show that for each process tree there exists a weighing that decreases monotonically with each rule application. From this, termination of rule applications follows.

- Let $LBE(M)$ be the number of $\circlearrowright$ nodes M' of M such that the loop body of M' can produce the empty trace:

$$LBE = |\circlearrowright(M_1, \ldots) \text{ such that } \epsilon \in \mathcal{L}(M_1)|$$

  An upper bound for LBE is the number of nodes in the tree.
- Let $D_\oplus(M)$ denote the number of times a node $M' = \oplus(\ldots)$ has a direct parent $\oplus$. For instance, $D_\times(\times(\times(\times(a))))$ is 2. An upper bound for $D_\oplus$ regardless of $\oplus$ is the number of nodes in the tree.
- Let $T_\vee(M)$ denote the number of times a child $M' = \vee$ has an (in)direct τ child, e.g. $T_\vee(\vee(\vee(\tau, \tau)))$ is 4. An upper bound for $T_\vee$ is n^2, in which n is the number of nodes in the tree.
- Finally, let N denote the number of nodes in a process tree.

Table 5.1 shows how applying each reduction rule influences these functions. Notice that not all functions are monotonically decreasing with each rule application, however the function $LBE(M) * k^5 + D_\wedge(M) * k^4 + D_\vee(M) * k^3 + T_\vee(M) * k + N(M)$ is for a sufficiently high k, i.e. such that in each rule application, $k \geq N(M)$. Therefore, for each process tree there exists a function that decreases with each application of a rule. Hence, the reduction rules are terminating. $\qquad\square$

Table 5.1: Termination of reduction rules. A row denotes, for a reduction rule, the changes in the values of the functions used in the proof of Lemma 5.1 when applying the rule. n is bounded by the number of nodes in the tree.

	LBE	$D_{\hookleftarrow}$	$D_\wedge$	$D_\vee$	$T_\vee$	N
S	$=$	$-1/=$	$-1/=$	$-1/=$	$=$	-1
$A_\times$	$=$	$=$	$=$	$=$	$=$	-1
$A_\rightarrow$	$=$	$=$	$=$	$=$	$=$	-1
$A_\wedge$	$=$	$=$	-1	$=$	$=$	-1
$A_\vee$	$=$	$=$	$=$	-1	$-n$	-1
$A_{\cup B}$	$=$	$=$	$=$	$=$	$=$	-1
$A_{\cup R}$	$=$	$=$	$=$	$=$	$=$	-1
$T_\vee$	$=$	$=$	$=$	-1	-1	$+1$
$T_{\vee,\times}$	$=$	$=$	$=$	$=$	-1	$+1$
$T_{\cup B}$	-1	$=$	$=$	$=$	$+n/=$	$+3$
$T_{\cup R}$	$=$	$=$	$=$	$=$	$-1/=$	-1
$T_{\cup BR}$	-1	$=$	$=$	$=$	$-1/=$	-2
$C_{\leftrightarrow}$	$=$	$-n$	$+n$	$=$	$=$	$=$
$C_\vee$	$=$	$=$	-1	$+2$	$+n$	$+1$

Lemma 5.2 (Reduction rules are locally confluent) *The reduction rules of Definition 5.1 are locally confluent.*

Proof We prove local confluency for each pair of rules of which the left sides overlap, i.e. we prove that after applying one of the rules, the effect can be made equal to applying the other. Table 5.2 shows the left sides of which rules overlap.

Many pairs of rules of which the left sides overlap can be applied independently, i.e. if rule a was applicable before applying b, it is still applicable after applying b (and the other way around). These rule pairs have been denoted in Table 5.2 with I. Left to prove: the remaining rule pairs (+ in the table) are locally confluent.

S   Rule S and $T_{\vee,\times}$ overlap if $\ldots_1$ in Rule $T_{\vee,\times}$ is empty. If Rule $T_{\vee,\times}$ is applied first, then the result of Rule S can be reached by applying rules S and $A_\times$. A similar argument holds for Rule $C_{\leftrightarrow}$, which overlaps if $n = 1$.

$A_\times$ This rule overlaps with Rule $A_{\cup R}$ if $\ldots_2$ contains an $\times$. In case Rule $A_{\cup R}$ is applied first, Rule $A_{\cup R}$ should be applied a second time to obtain the same effect as applying Rule $A_\times$.

$A_\vee$ This rule overlaps with Rule $T_{\vee,\times}$, if either $\ldots_1$ in the left hand side of $A_\vee$ is the left hand side of $T_{\vee,\times}$, or the other way round. Both rule applications let the $\times(\tau,\ldots)$ appear in a different position. Applying rules $T_\times$, S and $T_{\vee,\times}$ repeatedly yields an equivalent result. A similar argument holds for Rule $T_\vee$.

$A_{\cup_R}$ This rule overlaps with Rule $T_{\cup_B}$, if ... of the latter contains an $\times$ child. Applying Rule $T_{\cup_B}$ followed by Rule $A_\times$ yields the same result as first applying Rule $A_{\cup_R}$ followed by Rule $T_{\cup_B}$.

$T_\times$ This rule overlaps with Rule $T_{\vee,\times}$ if $\ldots_2$ of the latter contains a Q. Applying $T_\times$ yields the same results as applying rules $T_{\vee,\times}$, $T_\times$ and S.

$T_\wedge$ This rule overlaps with Rule $C_\vee$ if one of the Q's is a τ. Applying rules $C_\vee$, $T_\vee$ and S yields the same result as applying Rule $T_\wedge$.

$T_\leftrightarrow$ This rule overlaps with Rule $C_\leftrightarrow$. However, Rule $T_\wedge$ provides local confluence in a single step.

$T_{\cup_B}$ This rule overlaps with Rule $T_{\cup_R}$. However, Rule $T_\times$ provides local confluence in a single step.

$C_\vee$ This rule overlaps with itself, but Rule $A_\vee$ provides local confluence.

Hence, the reduction rules of Definition 5.1 are locally confluent.   □

Table 5.2: Overlapping rules. - denotes that the two rules cannot be applied to the same nodes in any tree; I denotes that the left sides might overlap, but execution is independent; + denotes that the rules might apply to the same nodes.

	S	$A_\times$	$A_\to$	$A_\wedge$	$A_\vee$	$A_{\cup_B}$	$A_{\cup_R}$	$T_\times$	$T_\to$	$T_\wedge$	$T_\leftrightarrow$	$T_\vee$	$T_{\vee,\times}$	$T_{\cup_B}$	$T_{\cup_R}$	$T_{\cup_{BR}}$	$C_\leftrightarrow$	$C_\vee$
S	-	I	I	I	I	I	-	-	-	-	-	-	+	-	-	-	+	-
$A_\times$	I	-	-	-	-	-	+	I	-	-	-	-	I	-	-	-	-	-
$A_\to$		I	-	-	-	-	-	-	I	-	-	-	-	-	-	-	-	-
$A_\wedge$			I	-	-	-	-	-	-	I	-	-	-	-	-	-	-	I
$A_\vee$				I	-	-	-	-	-	-	+	+	-	-	-	-	-	-
$A_{\cup_B}$					I	-	-	-	-	-	-	-	-	-	I	-	-	-
$A_{\cup_R}$						-	-	-	-	-	-	-	-	+	I	-	-	-
$T_\times$								I	-	-	-	-	+	-	-	-	-	-
$T_\to$									I	-	-	-	-	-	-	-	-	-
$T_\wedge$										I	-	-	-	-	-	-	-	+
$T_\leftrightarrow$											I	-	-	-	-	-	+	-
$T_\vee$												I	+	-	-	-	-	-
$T_{\vee,\times}$													I	-	-	-	-	-
$T_{\cup_B}$														I	+	-	-	-
$T_{\cup_R}$															I	-	-	-
$T_{\cup_{BR}}$																-	-	-
$C_\leftrightarrow$																	-	-
$C_\vee$																		+

Newman's Lemma [11] states that a set of reduction rules (a *system of rewriting rules*) that is locally confluent and terminating is confluent, thus our reduction rules are confluent and thus canonical.

Corollary 5.2 (Normal form is canonical) *By Lemma 5.2, Lemma 5.1 and Newman's Lemma [11], the reduction rules of Definition 5.1 are confluent. Therefore, the normal form is canonical.*

This means that the end result of applying the reduction rules exhaustively does not depend on the order in which these rules are applied, i.e. the same end result (*normal form*) will be reached. In the next sections, we will prove that this normal form is *language unique*, i.e. each process tree in normal form has a unique language, for several classes of process trees. Language uniqueness will be proven by taking an abstraction, e.g. directly follows graphs, and proving that the abstraction is unique for each normal form.

5.2 Language Uniqueness with Directly Follows Graphs

A first abstraction that we discuss is the *directly follows graph*, which was introduced in Section 2.4, and which is used by many process discovery algorithms [13, 6, 12]. A directly follows graph is an abstraction of a language, and thus also an abstraction of the behaviour of a process tree.

In this section, we study the expressive power of directly follows graphs to represent the behaviour of process trees, with the aim of showing that a different structure (in normal form) implies different behaviour. Studying this property clarifies the boundary of directly follows based discovery algorithms: if we cannot distinguish the language of two process trees via their directly follows graphs, then no directly follows based algorithm can be expected to discover the correct model of these two.

Therefore, we first introduce a basic class of process trees (C_B) that, as we will prove later, can be uniquely identified by directly follows graphs. We characterise this class by limiting certain nestings of operators, and illustrate the need for these limitations using some counterexamples: trees with different languages but equivalent directly follows graphs. In this section, we limit ourselves to process tree structures that correspond to free choice structures, as these are supported by many discovery algorithms, such as α and its derivatives, Heuristic Miner (HM) and Fodina (FO). In later sections, we will weaken some requirements further.

The class of process trees is described in Section 5.2.1. Second, the footprints of process tree operators in directly follows graphs are discussed in Section 5.2.2. Finally, in Section 5.2.3 we prove that these footprints are distinctive enough to distinguish all language-different trees of C_B.

5.2.1 A Class of Trees: C_B

The class of process trees for which we will prove language uniqueness is given below. After the class, we illustrate the necessity of the requirements with some counterexamples.

Definition 5.2 (Class C_B) Let M be a process tree. Then M belongs to C_B if for each reduced (sub)tree M' of M, it holds that

$C_B.1$ The subtree is not a silent activity:
$$M' \neq \tau$$
$C_B.2$ No activity appears in two children of the subtree:
$$M' = \oplus(M'_1, \ldots M'_n) : \forall_{1 \leq i < j \leq n} \Sigma(M'_i) \cap \Sigma(M'_j) = \emptyset$$
with $\oplus \in \{\times, \rightarrow, \wedge, \vee, \circlearrowleft, \leftrightarrow\}$
$C_B.3$ The body of a loop has disjoint start and end activities:
$$M' = \circlearrowleft(M'_1, \ldots M'_n) : \text{Start}(M'_1) \cap \text{End}(M'_1) = \emptyset$$
$C_B.4$ The subtree is not interleaved:
$$M' \neq \leftrightarrow(\ldots)$$
$C_B.5$ The subtree is not an inclusive choice:
$$M' \neq \vee(\ldots)$$

Next, we illustrate each requirement by giving some weaker constraints and counterexamples why those weaker constraints do not suffice. Notice that in later sections, we will show that some requirements can be relaxed, but in this section, we limit ourselves to the process tree operators that correspond to free-choice structures, which form the boundary of many discovery algorithms.

- Requirement $C_B.1$: no silent activities. Silent activities, i.e. τ leafs, are invisible to execution: they bring the system in a new state without directly visible effects. A counterexample are the trees $\wedge$ and $\wedge$, which have the same directly

  $a \; b$ $a \; \times$

  $\tau \; b$

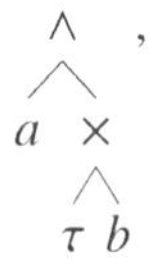

  follows graph but not the same language: $\gtrsim a \leftrightarrows b \lesssim$. We will study the influence of τ on directly follows graphs in more detail and we will increase the class of trees that can be identified in Section 5.6. We added this requirement here as most directly follows based algorithms do not support the τ.

- Requirement $C_B.2$: no duplicate activities. This restriction follows directly from the use of directly follows graphs: each activity is represented by one node in the graph. For instance, Figure 5.2 shows two trees that do not have the same language, but for which the directly follows graphs are identical.

$$M_{22} = \circlearrowleft(a, a)$$
$$M_{23} = \rightarrow(a, a)$$

(a) Language-different process trees.   (b) $\twoheadrightarrow(M_{22}) = \twoheadrightarrow(M_{23})$

Fig. 5.2: A counterexample for duplicate activities.

- Requirement $C_B.3$: disjoint start and end activities for loops. Figure 5.3 shows a counterexample, i.e. the two trees do not have the same language, both violate Requirement $C_B.3$ and have the same directly follows graph. Therefore, directly follows-based algorithms will not be able to distinguish these trees. This overlap is caused by the presence of loops of which the start and end activities overlap: $Start(M_{24}) = End(M_{24}) = Start(M_{25}) = End(M_{25}) = \{a, c\}$. This requirement has some similarity with the so-called short loops of the α algorithm [2].

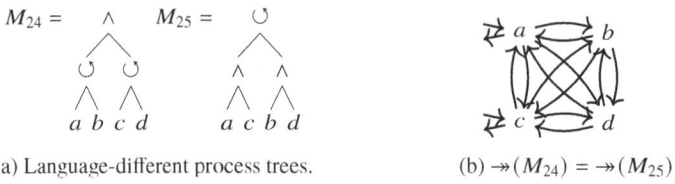

(a) Language-different process trees.   (b) $\twoheadrightarrow(M_{24}) = \twoheadrightarrow(M_{25})$

Fig. 5.3: A counterexample for $\circlearrowright$ having overlapping start and end activities.

A weaker constraint one could consider is that at least one child of the loop should have disjoint start and end activities: $M' = \circlearrowright(M'_1, \ldots M'_n) : \exists_{1 \leq i \leq n} \ Start(M'_i) \cap End(M'_i) = \emptyset$. This constraint would allow both M_{26} and M_{27} (see Figure 5.4), i.e. c can be in the redo of the inner or outer $\circlearrowright$-node. These trees have the same directly follows graph and both adhere to the weaker constraint. However, they do not have the same language (e.g. $\langle a, b, c, a, b, d \rangle$ is in $\mathcal{L}(M_{26})$ but not in $\mathcal{L}(M_{27})$), thus the weaker constraint is not strong enough.

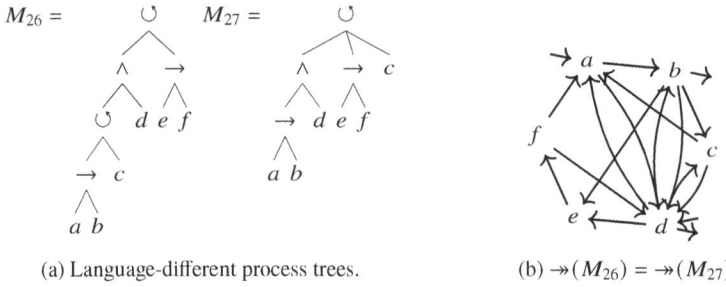

(a) Language-different process trees.   (b) $\twoheadrightarrow(M_{26}) = \twoheadrightarrow(M_{27})$

Fig. 5.4: A counterexample that a weaker constraint could not replace Requirement $C_B.3$.

- Requirement $C_B.4$: no interleaved operators. Figure 5.5 contains a counterexample: the two trees in normal form do not have the same language, e.g. $\langle b, a, c, d \rangle$ is in $\mathcal{L}(M_{28})$ but not in $\mathcal{L}(M_{29})$, but have the same directly follows graph.

  A property of the interleaved operator is that its children can be executed only once. Furthermore, it describes that the execution of its children cannot overlap. Both of these properties cannot be verified using a directly follows graph: executing a child a second time will not add edges to the graph, neither as executing a nested child in between (as in Figure 5.5).

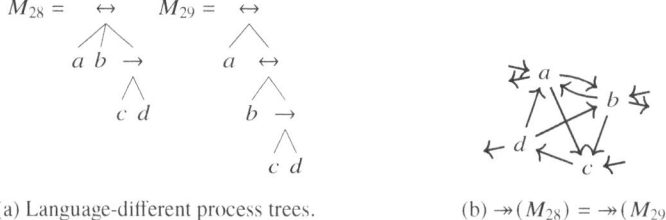

(a) Language-different process trees.   (b) $\twoheadrightarrow(M_{28}) = \twoheadrightarrow(M_{29})$

Fig. 5.5: A counterexample for interleaved operators.

In Section 5.4, we study this operator in more detail, weaken this requirement and prove that the directly follows graph can distinguish some interleaved operators. However, the interleaved operator brings process trees outside the class of free choice Petri nets, thus in the remainder of this section, we do not consider it.

- Requirement $C_B.5$: no inclusive choice operators. Figure 5.6 shows a directly follows graph of $\vee(a, b, c)$. However, the tree $\wedge(a, b, c)$ has the same directly follows graph, just as $\vee(a, \wedge(b, c))$, $\wedge(a, \vee(a, b))$ and so on. In Section 5.6 we discuss this in more detail.

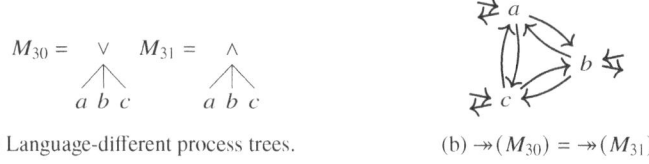

(a) Language-different process trees.   (b) $\twoheadrightarrow(M_{30}) = \twoheadrightarrow(M_{31})$

Fig. 5.6: Counterexample for inclusive choice operators.

Even though the requirements might seem rather restrictive, they describe a lower bound, i.e. there might be more process trees which normal form can be uniquely identified by its directly follows graph. For instance, the process tree $\circlearrowleft(a, b)$ has a directly follows graph that uniquely identifies it. However, we chose to limit ourselves to process trees that can be arbitrarily nested: the tree $\circlearrowleft(a, b)$ in isolation has a directly follows graph that is unique to it, but this does not hold if this tree is nested, e.g. $\wedge(\circlearrowleft(a, b), c)$ has the same directly follows graph as $\wedge(\circlearrowleft(c, b), a)$ but not the same language. In future research, we intend to study such non-arbitrarily nestable trees and extend the classes of process trees for which the directly follows graph is distinguishable to include them.

Future work 5.2: Extend C_B with non-arbitrarily nestable trees.

In the next sections, we prove that the directly follows graph uniquely determines the normal form for process trees of C_B.

5.2.2 Footprints

In the previous sections, we introduced a set of reduction rules and proved that these lead to a single normal form, and we introduced a class of process trees. We will prove that the directly follows graph can distinguish all languages that can be represented by this class of process trees. This proof will be given in Section 5.2.3 and will use a set of characteristics (*footprints*) for each process tree operator. In this section, we introduce these footprints.

Earlier, we defined the language of a process tree as the language defined by its root operator applied to the languages of its children. Here, we follow the same structure to investigate the influence of the root operator on the directly follows graph. The footprints introduced here are also a key part of the cut detection in the discovery algorithms that will be presented in Chapter 6. That is, these algorithms will search for these footprints in the directly follows graph of an event log.

A tuple $(\oplus, \Sigma_1, \ldots \Sigma_m)$ that adheres to the $\oplus$-footprint is an $\oplus$-cut, for instance $(\times, \Sigma(M_1), \ldots \Sigma(M_m))$ is an exclusive choice cut. In a *non-trivial cut*, $m > 1$ and no Σ_i is empty. We first define the footprints as patterns in directly follows relations, after which we establish their relationship with the process tree operators.

Definition 5.3 (directly follows footprints) Let $\twoheadrightarrow$ be a directly follows relation, let Start be the start activities of $\twoheadrightarrow$, let End be the end activities of $\twoheadrightarrow$ and let $c = (\oplus, \Sigma_1, \ldots \Sigma_n)$ be a cut, consisting of a process tree operator $\oplus \in \{\times, \rightarrow, \wedge, \circlearrowleft\}$ and a partition of activities with parts $\Sigma_1 \ldots \Sigma_n$ such that $\Sigma(\twoheadrightarrow) = \bigcup_{1 \leq i \leq n} \Sigma_i$ and $\forall_{1 \leq i < j \leq n} \Sigma_i \cap \Sigma_j = \emptyset$.

- Exclusive choice: c is an *exclusive choice cut* if $\oplus = \times$ and

    ×.1 No part is connected to any other part:
    $$\forall_{1 \leq i \leq n, 1 \leq j \leq n, i \neq j} \forall_{a \in \Sigma_i, b \in \Sigma_j} \ a \not\twoheadrightarrow b \wedge b \not\twoheadrightarrow a$$

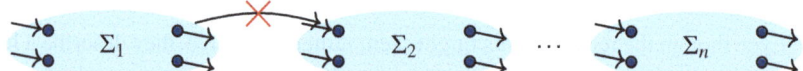

- Sequential. c is a *sequence cut* if $\oplus = \rightarrow$ and

    →.1 Each node in a part is indirectly and only connected to all nodes in the parts "after" it:
    $$\forall_{1 \leq i < j \leq n} \forall_{a \in \Sigma_i, b \in \Sigma_j} \ a \twoheadrightarrow^+ b \wedge b \not\twoheadrightarrow^+ a$$

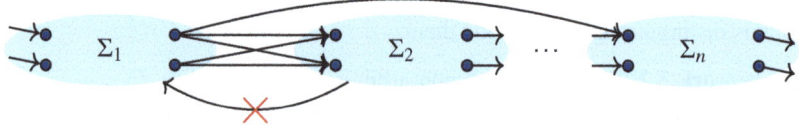

- Concurrent. c is a *concurrent cut* if $\oplus = \wedge$ and

∧.1 Each part contains a start and an end activity:

$\forall_{1 \leq i \leq n} \ \text{Start} \cap \Sigma_i \neq \emptyset \wedge \text{End} \cap \Sigma_i \neq \emptyset$

∧.2 All parts are fully interconnected:

$\forall_{1 \leq i < n, 1 \leq j \leq n, i \neq j} \ \forall_{a \in \Sigma_i, b \in \Sigma_j} \ a \twoheadrightarrow b \wedge b \twoheadrightarrow a$

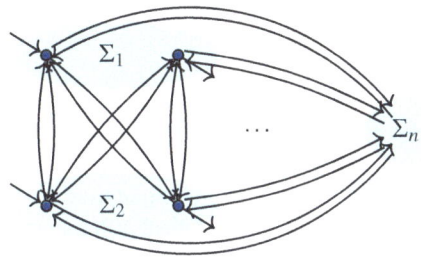

• Loop. c is a *loop cut* if $\oplus = \circlearrowright$ and

↻.1 All start and end activities are in the body (i.e. the first) part:

Start $\cup$ End $\subseteq \Sigma_1$

↻.2 Only start/end activities in the body part have connections from/to other parts:

$\forall_{2 \leq j \leq n} \ \forall_{a \in \Sigma_1, b \in \Sigma_j} \ a \twoheadrightarrow b \Rightarrow a \in \text{End}$

$\forall_{2 \leq j \leq n} \ \forall_{a \in \Sigma_1, b \in \Sigma_j} \ b \twoheadrightarrow a \Rightarrow a \in \text{Start}$

↻.3 Redo parts have no connections to other redo parts:

$\forall_{2 \leq i \leq n, 2 \leq j \leq n, i \neq j} \ \forall_{a \in \Sigma_i, b \in \Sigma_j} \ a \not\twoheadrightarrow b \wedge b \not\twoheadrightarrow a$

↻.4 If an activity from a redo part has a connection to/from the body part, then it has connections to/from all start/end activities:

$\forall_{2 \leq i \leq n} \ \forall_{a \in \text{Start}, b \in \Sigma_i} \ b \twoheadrightarrow a \Leftrightarrow \forall_{c \in \text{Start}} \ b \twoheadrightarrow c$

$\forall_{2 \leq i \leq n} \ \forall_{a \in \text{End}, b \in \Sigma_i} \ a \twoheadrightarrow b \Leftrightarrow \forall_{c \in \text{End}} \ c \twoheadrightarrow b$

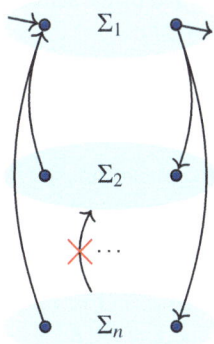

An inspection of the semantics of the process tree operators (Definition 2.6) reveals that these footprints are indeed present in directly follows graphs of process trees.

Lemma 5.3 (Directly follows footprints) *Let $M = \oplus(M_1, \ldots M_m)$ be a process tree without duplicate activities (Requirement $C_B.2$), with $\oplus \in \{\times, \rightarrow, \wedge, \circlearrowleft\}$. Then, the footprints of Definition 5.3 hold, i.e. $\twoheadrightarrow(M)$ contains the footprint of the cut $(\oplus, \Sigma(M_1), \ldots \Sigma(M_n))$.*

5.2.3 Language Uniqueness

Using the reduction rules and the restrictions mentioned earlier, we prove that no two reduced trees have equal languages, unless they are syntactically equal (notice that we do not consider the order of children of commutative operators for equivalence). We do this by proving that if there is a syntactical difference somewhere in the trees, then their directly follows graphs must differ, and consequently their language. For readability, the proof is split in three parts: we first prove this for operators, then for *activity partitions* (i.e. the division of activities over children), and finally prove the main result.

Lemma 5.4 (Operators are mutually exclusive) *Take two reduced process trees of C_B $K = \oplus(K_1, \ldots K_n)$ and $M = \otimes(M_1, \ldots M_m)$ such that $\oplus \neq \otimes$. Then $\mathcal{L}(K) \neq \mathcal{L}(M)$.*

Proof Towards contradiction, assume that $\mathcal{L}(K) = \mathcal{L}(M)$. Then, $\twoheadrightarrow(K) = \twoheadrightarrow(M)$. By Corollary 5.1, $n \geq 2$ and $m \geq 2$. Perform case distinction on $\oplus$ to prove that $\twoheadrightarrow(K) \neq \twoheadrightarrow(M)$ or $\mathcal{L}(K) \neq \mathcal{L}(M)$.

$\oplus = \times$ By semantics of the $\times$ operator and the reduction rules, there exist at least n unconnected parts in $\twoheadrightarrow(K)$ (see Lemma 5.3). As $\otimes \neq \times$ and by the semantics of the other operators, $\twoheadrightarrow(M)$ is connected, so $\twoheadrightarrow(K) \neq \twoheadrightarrow(M)$.

$\oplus = \rightarrow$ By semantics of the the $\rightarrow$ operator, $\twoheadrightarrow(K)$ is a chain of at least n clusters (see Lemma 5.3). As $\otimes \neq \rightarrow$ and by the semantics of the other operators, $\twoheadrightarrow(M)$ is not a chain, so $\twoheadrightarrow(K) \neq \twoheadrightarrow(M)$.

$\oplus = \wedge$ By semantics of the $\wedge$ operator, $\twoheadrightarrow(K)$ consists of at least n fully interconnected clusters (see Lemma 5.3). Perform case distinction on the (due to symmetry) remaining cases of $\otimes$:

$\otimes = \circlearrowleft$ We try to construct a concurrent cut $(\wedge, \Sigma_1, \ldots \Sigma_p)$ for M. Take an activity $a \in Start(M_1)$. By Requirement $C_B.3$, $a \notin End(M_1)$. Take an activity $b \in \Sigma(M) \setminus \Sigma(M_1)$. Then, by semantics of $\circlearrowleft$, $a \not\twoheadrightarrow b$ and by Requirement $\wedge.2$, a and b are part of the same Σ in the cut we are constructing, e.g. Σ_1. This holds for all a and b, thus $\Sigma(M) = \Sigma_1$. Hence, there is no non-trivial concurrent cut, and $\twoheadrightarrow(K) \neq \twoheadrightarrow(M)$.

Obviously, these arguments are symmetric in $\oplus$ and $\otimes$, so we conclude that $\twoheadrightarrow(K) \neq \twoheadrightarrow(M)$. As directly follows graphs are defined based on languages, one language obviously has one directly follows graph. Hence, $\mathcal{L}(K) \neq \mathcal{L}(M)$, which is a contradiction. $\square$

To provide some intuition about the interplay of the reduction rules and the restrictions imposed by C_B, consider the process trees denoted in Figure 5.7. Trees M_{33} and M_{34} have the same language, and indeed, M_{34} can be reduced to M_{33} using the reduction rules of Definition 5.1. However, even though the language of M_{33} and M_{34} differs from the language of M_{32}, all three process trees have the same directly follows graph (Figure 5.7). Hence, no directly follows based algorithm can distinguish them: for the IM framework, it is not decidable whether the root operator should be $\wedge$ or $\circlearrowleft$. This issue is (formally) solved by C_B, which puts these trees outside the scope being considered. In Section 5.5, we will show what information could be used to discriminate these trees.

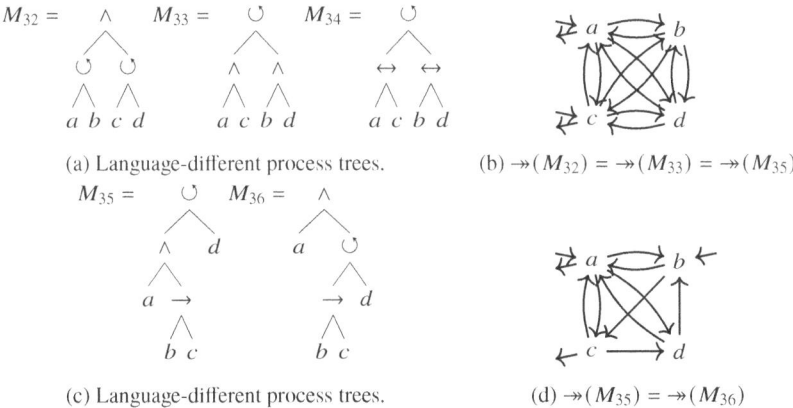

(a) Language-different process trees. (b) $\twoheadrightarrow(M_{32}) = \twoheadrightarrow(M_{33}) = \twoheadrightarrow(M_{35})$

(c) Language-different process trees. (d) $\twoheadrightarrow(M_{35}) = \twoheadrightarrow(M_{36})$

Fig. 5.7: Counterexamples for Lemma 5.4 on trees not in C_B.

Lemma 5.5 (Partitions are mutually exclusive) *Take two reduced process trees of C_B $K = \oplus(K_1 \ldots K_n)$ and $M = \oplus(M_1 \ldots M_m)$ such that their activity partition is different, i.e. there is a w such that $1 \le w \le min(n, m)$ and $\Sigma(K_w) \ne \Sigma(M_w)$. Then, $\mathcal{L}(K) \ne \mathcal{L}(M)$.*

Proof Without loss of generality, we assume that children of the commutative operators $(\rightarrow, \circlearrowleft)$ have a fixed order. Towards contradiction, assume that $\twoheadrightarrow(K) = \twoheadrightarrow(M)$. Perform case distinction on $\oplus$ (the case for K and M swapped is symmetric):

$\oplus = \times$ Take a pair of activities a, b such that $a \in \Sigma(K_x)$, $a \in \Sigma(M_y)$, $b \in \Sigma(K_x)$ and $b \notin \Sigma(M_y)$ (choose x and y as desired). Obviously, if the activity partitions of K and M are different such a pair exists. By Corollary 5.1, no child $K_1 \ldots K_n$ is an exclusive-choice subtree itself, and by semantics of the other operators there is an undirected path in $\twoheadrightarrow(K)$, i.e. $a \leftrightsquigarrow b$ in $\twoheadrightarrow(K)$. However, as $a \in \Sigma(M_y) \wedge b \notin \Sigma(M_y)$, $a \not\leftrightsquigarrow b$ in $\twoheadrightarrow(M)$. Hence, $\twoheadrightarrow(K) \ne \twoheadrightarrow(M)$.

$\oplus = \rightarrow$ Take $a \in \Sigma(K_i)$ and $b \in \Sigma(K_j)$ such that $i < j$. Then by the $\rightarrow$-cut, $a \twoheadrightarrow^+ b \wedge b \not\twoheadrightarrow^+ a$. By Corollary 5.1, all children of K and M are not $\rightarrow$-nodes themselves, thus, by the semantics of the other operators ($\times$ is unconnected, $\wedge$ and $\circlearrowleft$ are strongly connected), either $a \not\rightarrow b$ or $b \rightarrow^+ a$. Then, $a \in \Sigma(M_x) \wedge b \in \Sigma(M_y)$ with $x < y$. This holds for all such a and b, hence $\forall_{1 \leq i \leq n = m} \Sigma(K_i) = \Sigma(M_i)$, which contradicts the initial assumption.

$\oplus = \wedge$ To prove the equality of the activity partitions, we consider two symmetrical directions: a) if two activities are in the same Σ_i in K, then they are in the same Σ_i in M. b) if two activities are in the same Σ_i in M, then they are in the same Σ_i in K.

Consider a child M_x. Perform case distinction on the structure of M_x:

$M_x = a$ A single activity cannot be split. Therefore, $\Sigma(K_x) \subseteq \Sigma(M_x)$.

$M_x = \times(M_{x_1}, \ldots M_{x_p})$ Take two activities $a \in \Sigma(M_{x_1})$ and $b \in \Sigma(M_{x_2})$. By semantics of $\times$, $a \not\rightarrow b$. Thus, in a concurrent cut, a and b should be part of the same Σ. This holds for all such activities of all children of M_x, thus $\Sigma(K_x) \subseteq \Sigma(M_x)$.

$M_x = \rightarrow(M_{x_1}, \ldots M_{x_p})$ Similar, using that either $a \not\rightarrow b$ or $b \not\rightarrow a$.

$M_x = \wedge(M_{x_1}, \ldots M_{x_p})$ Excluded by the reduction rules.

$M_x = \circlearrowleft(M_{x_1}, \ldots M_{x_p})$ By C_B, there is at least one child M_{x_i} such that $\text{Start}(M_{x_i}) \cap \text{End}(M_{x_i}) = \emptyset$. Take such a M_{x_i} and an a from $\Sigma(M_{x_i})$. Furthermore, take b from any other child. There are three cases for a: $a \notin \text{Start}(M_{x_i})$, $a \notin \text{End}(M_{x_i})$ or both. For all these three cases, $a \not\rightarrow b \vee b \not\rightarrow a$. Thus, by argumentation similar to the $\times$ case, $\Sigma(K_x) \subseteq \Sigma(M_x)$.

Hence, $\Sigma(K_x) \subseteq \Sigma(M_x)$. This holds for all $\Sigma(M_x)$ and by symmetry for all $\Sigma(K_x)$. Hence, $\forall_{1 \leq i \leq n} \Sigma(K_i) = \Sigma(M_i)$, which contradicts the initial assumption.

$\oplus = \circlearrowleft$ Consider $\Sigma(K_i)$ for some $2 \leq i \leq n$. By Corollary 5.1, K_i is of the form $\times(\ldots)$. By semantics of the other operators, for all $a, b \in \Sigma(K_i)$, there exists an undirected path $a \leftrightsquigarrow b$ in $\rightarrow(K)$, such that all activities on this undirected path are in K_i. Between all the activities on this path, there exists a connection in $\rightarrow(K_i)$, and none of the activities on this path is in $\text{Start}(K)$ or $\text{End}(K)$. By Lemma 5.3, in a non-trivial loop cut, (without loss of generality) $\Sigma(K_i) \subseteq \Sigma(M_i)$. Let $K_1 = \otimes(K_{1_1}, \ldots K_{1_p})$. Perform case distinction on $\otimes$:

$\otimes = \times$ Take a child K_{1_i}. By the reduction rules, this child is not an $\times$. For all activities $a \in \text{Start}(K_{1_i})$, $b \in \text{End}(K_{1_i})$, there exist a directed path $a \rightarrow^+ b$, such that this path is completely in $\Sigma(K_{1_i})$. Furthermore, take an activity $c \in \text{End}(K_{1_{j \neq i}})$. By semantics of $\times$, c has no directly follows connection to any node on the path. Towards contradiction, assume there's a first node d on the path $\notin \Sigma(M_1)$. Then, by semantics of $\circlearrowleft$, there should be a connection $c \rightarrow d$. This holds for all activities d and children i, so $\Sigma(K_1) \subseteq \Sigma(M_1)$.

$\otimes = \rightarrow$ Similar to the $\times$-case.

$\otimes = \wedge$ $\text{Start}(K) \cup \text{End}(K) \subseteq \Sigma(M_1)$, thus we only need to consider non-start non-end activities. Take such an activity a in child K_{1_i}, and take an activity $b \in \text{End}(K_{1_{j \neq i}})$. By semantics of $\wedge$, $a \rightarrow b$; by C_B, $b \notin \text{Start}(K_1)$; thus by Lemma 5.3, $a \in \Sigma(M_1)$. This holds for all a, so $\Sigma(K_1) \subseteq \Sigma(M_1)$.

$\otimes = \circlearrowleft$ Excluded by the reduction rules.

By contradiction, we conclude $\mathcal{L}(K) \neq \mathcal{L}(M)$.   □

Additional examples showing the necessity of the restrictions of C_B are given in Figure 5.8, e.g. trees M_{37} and M_{38} do not have the same language, but share their directly follows graph. In this example, IM cannot decide between the cuts $(\wedge, \{a, b\}, \{c\})$ (blue dotted line) and $(\wedge, \{b, c\}, \{a\})$ (red dashed line). In Section 5.5, we will show what information could be used to discriminate these trees.

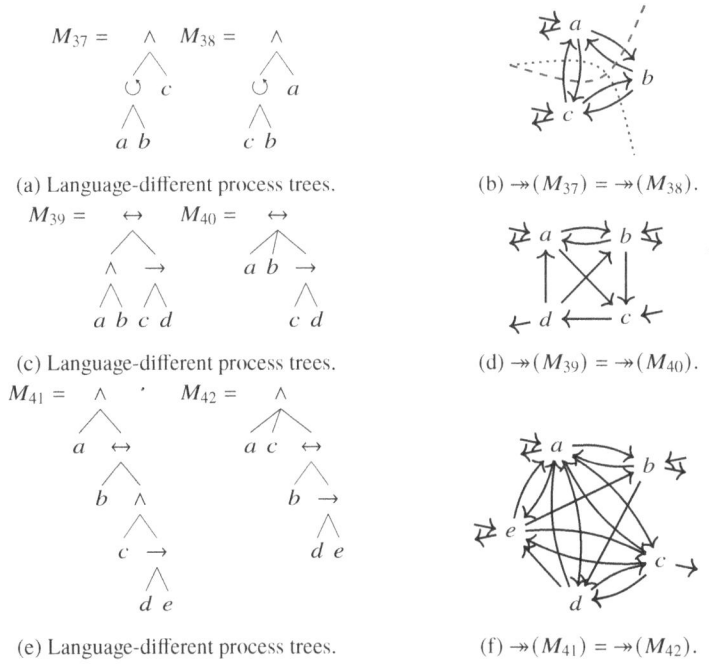

(a) Language-different process trees.

(b) $\twoheadrightarrow(M_{37}) = \twoheadrightarrow(M_{38})$.

(c) Language-different process trees.

(d) $\twoheadrightarrow(M_{39}) = \twoheadrightarrow(M_{40})$.

(e) Language-different process trees.

(f) $\twoheadrightarrow(M_{41}) = \twoheadrightarrow(M_{42})$.

Fig. 5.8: Examples showing that Lemma 5.5 might not hold for trees not in C_B: these trees have different languages but the same directly follows graph.

Lemma 5.6 (Language uniqueness for C_B) *For trees of class C_B, the normal form of Definition 5.1 is language unique.*

Proof Towards contradiction, assume that there exist two reduced process trees K and M, both of C_B, such that $\mathcal{L}(K) = \mathcal{L}(M)$, but $K \neq M$. Then there exist topmost subtrees K' in K and M' in M such that $\mathcal{L}(K') = \mathcal{L}(M')$ and such that K', M' are structurally different in their activity, operator or activity partition, i.e. either

- K' or M' is a τ while the other is not. Then obviously their language cannot be equivalent.

- K' or M' is a single activity while the other is not. Then, by the restrictions of C_B, their language cannot be equivalent.
- $K' = \otimes(K_1' \ldots K_n')$ and $M' = \oplus(M_1' \ldots M_n')$ such that $\oplus \neq \otimes$. By Lemma 5.4, $\mathcal{L}(K') \neq \mathcal{L}(M')$.
- $K' = \oplus(K_1' \ldots K_n')$ and $M' = \oplus(M_1' \ldots M_n')$ such that the activity partition is different, i.e. there is an i such that $\Sigma(K_i') \neq \Sigma(M_i')$. By Lemma 5.5, $\mathcal{L}(K') \neq \mathcal{L}(M')$.

Hence, there cannot exists such K and M and therefore the reduction rules yield a language unique normal form. $\qquad\qquad\qquad\qquad\qquad\qquad\qquad\qquad\qquad\qquad\qquad$ □

Lemmas 5.4, 5.5 and 5.6 were all proven using directly follows graphs as intermediate steps, i.e. we proved that if there is a structural difference between two process trees, then their directly follows graphs are different as well. Consequently, we concluded that their languages must be different as well. However, from this intermediate result, we derive:

Corollary 5.3 (Directly follows graph uniqueness) *There are no two different reduced process trees of C_B with equal directly follows graphs.*

This result can easily be extended to other process tree operators, as witnessed by the addition of the $\leftrightarrow$ operator, which was not present in [8] and which will be added in Section 5.4.

5.3 Language Uniqueness with Activity Relations

The previous section studied the effect of process tree operators on the directly follows graph globally. Besides global influence, the process tree operators have local influence on pairs of activities as well. In this section, we study this influence and use it to define relations between pairs of activities, and we also prove that this local information on its own suffices to distinguish the same class of models (C_B) as the global footprints. Later, we will exploit this fact when the information in an event log is not complete and hence global information is incomplete (Section 6.3). There, we will see that we can use some statistics on partially complete local information about behaviour to infer the missing local information about the behaviour of a system. This will allow us to discover complete models from event logs with very few traces.

The local information in this section is derived from the directly follows relation $\twoheadrightarrow$. Using this relation and its transitive closure $\twoheadrightarrow^+$, we first identify 9 cases (*activity relations*) that correspond to the 4 basic process tree operators used in Section 5.2. The combination of the activity relations between all pairs of activities in a language is an abstraction. Second, we show that this abstraction is language unique for the class of process trees C_B, i.e. two different reduced trees in C_B have different activity relations.

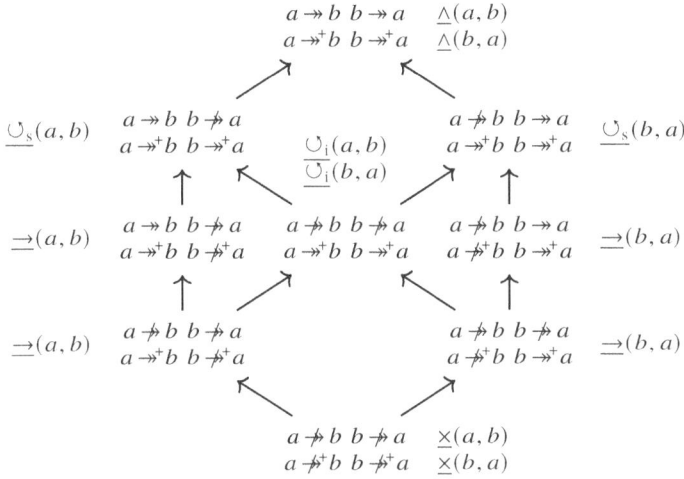

Fig. 5.9: Activity relations; the arrows define a lattice.

5.3.1 Activity Relations

The most basic behavioural information between two activities is given by the directly follows relation $\twoheadrightarrow$ and its transitive closure $\twoheadrightarrow^+$. $\twoheadrightarrow$ and $\twoheadrightarrow^+$ can be combined in nine cases. Notice that $a \twoheadrightarrow b$ implies $a \twoheadrightarrow^+ b$ and therefore, three possibilities exist: $(a \twoheadrightarrow b \wedge a \twoheadrightarrow^+ b)$, $(a \not\twoheadrightarrow b \wedge a \twoheadrightarrow^+ b)$ and $(a \not\twoheadrightarrow b \wedge a \not\twoheadrightarrow^+ b)$. Figure 5.9 identifies these nine cases for two activities a and b, and organises these cases in a lattice. The structure of the lattice follows from $\twoheadrightarrow$ and $\twoheadrightarrow^+$: an edge in the lattice corresponds to an extension of the $\twoheadrightarrow$ or $\twoheadrightarrow^+$-relation with one pair of activities.

We group the nine cases of the lattice into five distinct *activity relations*: $\underline{\times}$, $\underline{\wedge}$, $\underline{\circlearrowleft}_i$, $\underline{\rightarrow}$ and $\underline{\circlearrowleft}_s$. For instance, if $b \twoheadrightarrow a$ and $a \not\twoheadrightarrow^+ b$, then $\underline{\rightarrow}(a, b)$, and if $a \twoheadrightarrow^+ b$, $b \twoheadrightarrow^+ a$, $a \not\twoheadrightarrow b$ and $b \not\twoheadrightarrow a$, then $\underline{\circlearrowleft}_i(a, b)$. Informally, $\underline{\times}(a, b)$ denotes that a and b are in an exclusive choice relation, $\underline{\rightarrow}(a, b)$ denotes that a and b are in a sequence relation, and $\underline{\wedge}(a, b)$ denotes that a and b are in a concurrent relation. These are similar to the α-relations $\#_W$, $\rightarrow_W$ and $\|_W$ [1]. Furthermore, both $\underline{\circlearrowleft}_i(a, b)$ (*loop indirect*) and $\underline{\circlearrowleft}_s(a, b)$ (*loop single*) denote that a and b are in a loop relation. We do not group these cases, as we will need them to distinguish the loop body and redo parts. Using these five relations, all process trees of C_B can be distinguished.

The activity relations in the centre of the lattice ($\underline{\times}$, $\underline{\circlearrowleft}_i$ and $\underline{\wedge}$) are associative. Therefore, we consider associative cases, for instance $\underline{\wedge}(a, b)$ and $\underline{\wedge}(b, a)$, to be equivalent.

These five relations correspond to local footprints of process tree operators. Say that the process tree under consideration is $M = \oplus(M_1, \ldots M_n)$, and a and b are activities from different children M_i and M_j. Then, the relation between a and b is determined by the root operator $\oplus$. For readability, we denote the local activity relation between a and b as $\oplus(a, b)$. This correspondence between activity relations and process tree operators only holds for root operators. For instance, in $\rightarrow(a, b)$, it holds that $\underrightarrow{\quad}(a, b)$. However, in $\circlearrowleft(\rightarrow(a, b), c)$, $\circlearrowleft_s(a, b)$ holds.

Lemma 5.7 (Loop activity relations) *Let $M = \oplus(M_1, \dots M_m)$ be a process tree from C_B, and let a, b be activities from different $\Sigma(M_i)$ and $\Sigma(M_j)$. Then, $a \oplus b$ if $\oplus \in \{\times, \rightarrow, \wedge\}$. If $\oplus = \circlearrowleft$, then either $\underline{\circlearrowleft}_s(a, b)$ or $\underline{\circlearrowleft}_i(a, b)$.*

Proof The cases $\times$, $\rightarrow$ and $\wedge$ follow from the semantics of these operators. For $\circlearrowleft$, obviously for all pairs of activities, $a \twoheadrightarrow^+ b$ and $b \twoheadrightarrow^+ a$, therefore either $\underline{\circlearrowleft}_i(a, b)$, $\underline{\circlearrowleft}_s(a, b)$, $\underline{\circlearrowleft}_s(b, a)$ or $\underline{\wedge}(a, b)$. Consider a pair of activities (u, v) such that u and v are not in the same $\Sigma(M_i)$. By Requirement $C_B.3$, $\text{Start}(M_1) \cap \text{End}(M_1) = \emptyset$. Then, by semantics of $\circlearrowleft$, $u \not\twoheadrightarrow v$ or $v \not\twoheadrightarrow u$ and hence $\underline{\not\wedge}(a, b)$. Therefore, either $\underline{\circlearrowleft}_i(a, b)$, $\underline{\circlearrowleft}_s(a, b)$, $\underline{\circlearrowleft}_s(b, a)$.   □

The $\circlearrowleft_s$ and $\circlearrowleft_i$ both correspond to the $\circlearrowleft$ operator. If we combined them into a single relation, this single relation would not give sufficient information to partition the activities. The two relations $\circlearrowleft_s$ and $\circlearrowleft_i$, as given by the lattice, give enough information as will be proven in Section 5.3.3.

Let $\underrightarrow{\quad}_L$ combine all the activity relations for a language L, i.e. for each pair of activities, $\underrightarrow{\quad}_L$ denotes their relation. For instance, if $\underline{\times}_L(a, b)$ then $\underrightarrow{\quad}(a, b)_L = \underline{\times}$. If L is clear from the context, we will omit this subscript.

5.3.2 Binary Trees

The normal form of Definition 5.1 contains associativity rules for all operators of C_B: $A_\times$, $A_\rightarrow$, $A_\wedge$, $A_{\circlearrowleft B}$ and $A_{\circlearrowleft R}$. Therefore, even though process trees can be n-ary trees, all process trees in C_B have language equivalent binary trees. Such binary trees are not in normal form, i.e. one might have to apply reduction rules backwards to obtain binary trees, but this nevertheless allows us to limit our analysis to binary trees.

A binary tree *conforms* to an n-ary tree if reducing the binary tree using the reduction rules $A_\times$, $A_\rightarrow$, $A_\wedge$, $A_{\circlearrowleft B}$ and $A_{\circlearrowleft R}$ would yield the n-ary tree. Similarly, a cut conforms to an n-ary tree if the partition of the cut corresponds to the n-ary tree:

Definition 5.4 (cut conformance) Let $c = (\oplus, \Sigma_1, \dots \Sigma_n)$ be a non-trivial cut and let $M = \oplus(M_1 \dots M_m)$ be a process tree in normal form. Then c *conforms* to M if no $\Sigma(M_i)$ is partitioned: $\forall_{1 \leq i \leq m} \exists_{1 \leq j \leq n} \Sigma(M_i) \subseteq \Sigma_j$. Furthermore,

- if $\oplus = \rightarrow$, then the order of subtrees is maintained, i.e. let $f: \{1 \dots m\} \rightarrow \{1 \dots n\}$ such that $\forall_{1 \leq i < m} f(i) \leq f(i + 1)$, then $\forall_{1 \leq i \leq n} \Sigma_i = \bigcup_{f(j) = i} \Sigma(M_j)$;
- if $\oplus = \circlearrowleft$, the body is preserved: $\Sigma(M_1) \subseteq \Sigma_1$.

Discovery algorithms of the IM framework can discover the language of any tree by searching for conforming binary cuts, i.e. a cut c conforms to M if selecting c

does not disable discovery of a process tree that is language equivalent to M. For example, if $M = \rightarrow(A, B, C)$, it is perfectly fine to discover either $\rightarrow(A, \rightarrow(B, C))$ or $\rightarrow(\rightarrow(A, B), C)$.

5.3.3 Language Uniqueness

In the previous parts of this section, we have introduced an abstraction of languages: the activity relations. A desirable property of abstractions is whether they are unique for a large class of languages, i.e. there are no two different languages of the class with the same abstraction. If this language uniqueness property holds for an abstraction, process discovery algorithms can use it to distinguish the languages in the class for which language uniqueness was proven, and will it be guaranteed that no confusion can arise from the abstraction. In this section, we prove this for the class of languages represented by process trees of C_B.

In the main lemma of this section, we will use a very general property of partitions (i.e. a proper division of activities over two or more sets): any two partitions share at least one crossing pair of activities: a pair of activities will *cross a cut* if both activities are not in the same Σ.

Lemma 5.8 (Two cuts share a crossing edge) *Take two binary partitions Σ_1, Σ_2 and Σ_1', Σ_2', both of the same Σ. Then there is a pair of activities (a, b) that is partitioned by both partitions:* $\exists_{1 \leq i \leq 2, 1 \leq j \leq 2, i \neq j}\ a \in \Sigma_i, b \in \Sigma_j$, $\exists_{1 \leq i \leq 2, 1 \leq j \leq 2, i \neq j}\ a \in \Sigma_i', b \in \Sigma_j'$.

Proof Perform case distinction on whether $|\Sigma_1 \cap \Sigma_2| = 2$. If both Σ_1 and Σ_2 consist of a single activity, there is one pair of activities that crosses Σ_1, Σ_2 and this pair also crosses Σ_1', Σ_2'.

Otherwise, assume without loss of generality that $|\Sigma_1| \geq 2$. Towards contradiction, assume there is no pair that is partitioned by both Σ_1, Σ_2 and Σ_1', Σ_2'. Then, take $a_1, a_1' \in \Sigma_1$, $a_2 \in \Sigma_2$. Pairs (a_1, a_2) and (a_1', a_2) are partitioned by Σ_1, Σ_2, so by assumption they are not partitioned by Σ_1', Σ_2'. Thus, there is an $1 \leq i \leq 2$ such that $a_1, a_1', a_2 \in \Sigma_i'$. As we posed no restrictions on a_1 and a_1', for some $1 \leq i \leq 2$, $\Sigma_1 \subseteq \Sigma_i'$. By symmetry, $\Sigma_2 \subseteq \Sigma_i'$, so $\Sigma_1 \cup \Sigma_2 \subseteq \Sigma_i'$. Therefore, $\Sigma_i' = \Sigma$ and hence Σ_1', Σ_2' is not a partition. $\qquad\square$

In Lemma 5.6, we established the language uniqueness of the reduction rules of Definition 5.1 and C_B. Therefore, we only need to prove that each normal form of the class C_B has a unique set of activity relations.

Lemma 5.9 (Language uniqueness with activity relations) *Take two reduced process trees $K = \oplus(K_1, \ldots K_n)$ and $M = \otimes(M_1, \ldots M_m)$ of class C_B. Then, $K = M$ if and only if $\twoheadrightarrow_{\mathcal{L}(K)} = \twoheadrightarrow_{\mathcal{L}(M)}$.*

Proof If $K = M$, then as $\twoheadrightarrow$ is a language based relation, $\twoheadrightarrow_{\mathcal{L}(K)} = \twoheadrightarrow_{\mathcal{L}(M)}$. If $K \neq M$, then without loss of generality, assume that either $\oplus \neq \otimes$ or there is a child i such that $\Sigma(K_i) \neq \Sigma(M_i)$.

We first prove the $\oplus \neq \otimes$ case. Consider a cut $(\oplus, \Sigma_1, \Sigma_2)$ conforming to K, and consider a cut $(\otimes, \Sigma_1', \Sigma_2')$ conforming to M. By Lemma 5.8, a pair of activities (a, b) exists that crosses both cuts. Then, by Lemma 5.7, $\twoheadrightarrow_{\mathcal{L}(K)}(a, b) = \underline{\oplus} \neq \underline{\otimes} = \twoheadrightarrow_{\mathcal{L}(M)}(a, b)$ (abusing notation a bit by combining $\underline{\circlearrowleft}_s$ and $\underline{\circlearrowleft}_i$). Hence, $\twoheadrightarrow_{\mathcal{L}(K)} \neq \twoheadrightarrow_{\mathcal{L}(M)}$.

Second, we prove the $\Sigma(K_i) \neq \Sigma(M_i)$ case by proving that there is a "misclassified" activity relation. Perform case distinction on whether $\oplus = \circlearrowleft$:

- $\oplus \neq \circlearrowleft$ As K and M are reduced and structurally different, a cut $c = (\oplus, \Sigma_1, \Sigma_2)$ exists such that c conforms to K but not to M. As c does not conform to M, there is a $\Sigma(M_j)$ that is partitioned by c: $\Sigma_1 \cap \Sigma(M_j) \neq \emptyset$ and $\Sigma_2 \cap \Sigma(M_j) \neq \emptyset$. Consider this $M_j = \ominus(\ldots)$, then $c' = (\ominus, \Sigma(M_j) \cap \Sigma_1, \Sigma(M_j) \cap \Sigma_2)$ is a cut of M_j. Take an arbitrary cut c'' that conforms to M_j, then by Lemma 5.8 a pair of activities (u, v) exists that crosses both c'' and c'. As c'' conforms to M_j and $\oplus \neq \circlearrowleft$, $\ominus(u, v)$ holds. However, in c', $\oplus(u, v)$ holds, and as K and M are reduced, $\ominus \neq \oplus$.

- $\oplus = \circlearrowleft$ By Requirement $\circlearrowleft$.1, $\mathrm{Start}(K) \cup \mathrm{End}(K) = \subseteq \Sigma_1$ and $\mathrm{Start}(M) \cup \mathrm{End}(M) \subseteq \Sigma_1$. Take an activity $a \in \Sigma(K_i)$ but $a \notin \Sigma(M_i)$ and $a \notin \mathrm{Start}(M_i) \cup \mathrm{End}(M_i)$. Without loss of generality, assume that $i = 1$ (the case $i > 1$ is similar). Then, there exists a start activity s and an activity b such that there is a $\twoheadrightarrow$-path $s \leadsto a$ such that s is the only start or end activity on the path, and the entire path is in $\Sigma(M_1)$. Let b the activity on this path just before a, i.e. $b \twoheadrightarrow a$, and therefore $\underline{\circlearrowleft}_s(b, a)$ or $\underline{\wedge}(a, b)$. If $a \notin \Sigma(K_1)$, then by semantics of process trees $b \not\twoheadrightarrow a$ and $a \not\twoheadrightarrow b$, thus $\underline{\circlearrowleft}_s(a, b)$ or $\underline{\circlearrowleft}_i(a, b)$. Hence, $\Sigma(K_1) = \Sigma(M_1)$.

We conclude that $\twoheadrightarrow(K) \neq \twoheadrightarrow(M)$ and hence, there cannot exist such K and M. $\square$

The influence of the silent activity τ and the operators $\leftrightarrow$ and $\vee$ on the activity relations has not been studied in detail yet.

If one would introduce the $\underline{\leftrightarrow}$ activity relation, corresponding to the $\leftrightarrow$ operator, this relation would overlap with $\underline{\circlearrowleft}_i$, $\underline{\circlearrowleft}_s$ and $\underline{\wedge}$:

$$\underline{\leftrightarrow}(a, b) \equiv \underline{\circlearrowleft}_s(a, b) \vee \underline{\circlearrowleft}_i(a, b) \vee \underline{\circlearrowleft}_s(b, a) \vee \underline{\wedge}(a, b)$$

Furthermore, the $\leftrightarrow$-operator is not associative, thus its activity relations are inherently ambiguous:

$\leftrightarrow(a, b, c)$	$\underline{\leftrightarrow}(a, b)$ $\underline{\leftrightarrow}(b, c)$ $\underline{\leftrightarrow}(a, c)$
$\leftrightarrow(a, \leftrightarrow(b, c))$	$\underline{\leftrightarrow}(a, b)$ $\underline{\leftrightarrow}(b, c)$ $\underline{\leftrightarrow}(a, c)$
$\leftrightarrow(\leftrightarrow(a, b), c)$	$\underline{\leftrightarrow}(a, b)$ $\underline{\leftrightarrow}(b, c)$ $\underline{\leftrightarrow}(a, c)$
$\leftrightarrow(b, \leftrightarrow(a, c))$	$\underline{\leftrightarrow}(a, b)$ $\underline{\leftrightarrow}(b, c)$ $\underline{\leftrightarrow}(a, c)$

Hence, the current activity relations cannot be used to discover interleaved behaviour. Whether this is possible remains subject of further study.

Future work 5.3: Study the influence of τ, $\leftrightarrow$ and $\vee$ on activity relations.

5.4 Language Uniqueness with Interleaving

In the previous sections, we studied the four basic process tree operators $\times$, $\rightarrow$, $\wedge$ and $\circlearrowright$. When a process tree of these four operators is translated to a Petri net, that net is free-choice, unlabelled (i.e. without duplicate activities/transitions), but might contain silent transitions. Therefore, trees of C_B have some similarities with models that can be discovered by algorithms such as α and HM.

In this section, we add the interleaved operator $\leftrightarrow$ to the considerations. The interleaved operator is similar to the concurrent operator, i.e. all its children need to be executed. However, the interleaved operator specifies that the executions of its children cannot overlap. For instance, consider the tree $\leftrightarrow(\rightarrow(a, b), c)$, then once execution of the $\rightarrow(a, b)$ child begins, the c child cannot begin execution until b has finished. That is, $\langle a, c, b \rangle$ is not part of the language of this model.

In a Petri net, two ways to model interleaved behaviour are to model the interleaved children concurrently and limiting execution to one branch at the same time using a so-called critical section place, or to model the different possible sequences explicitly (thereby duplicating activities). Figure 5.10 shows these two strategies applied to our example tree. Both of these strategies result in either non-free choice (critical section place) Petri nets or nets with unlabelled transitions(duplicate activities)[1], which are outside the class of trees that can be discovered by some process discovery algorithms.

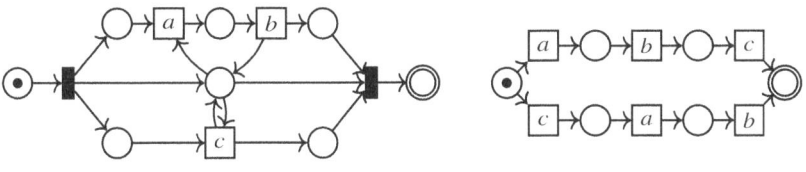

(a) Using a critical section place (in the middle).   (b) Using duplicate activities.

Fig. 5.10: A tree with interleaved behaviour $\leftrightarrow(\rightarrow(a, b), c)$ translated to a Petri net.

In this section, we show that interleaved behaviour can however be identified using a directly follows graph, and that language uniqueness holds for interleaved behaviour as well. We start with the footprint of interleaved behaviour in directly follows graphs, after which we introduce a new class of process trees, for which we prove language uniqueness, thereby establishing the one to one mapping of semantics and syntax of process trees with interleaved operators.

---

[1] i.e. if l in Definition 2.2 is not bijective

5.4.1 Footprint

Next, we characterise the footprint that the $\leftrightarrow$-operator leaves in a directly follows graphs, and by which this operator can be identified.

Definition 5.5 (directly follows footprint ($\leftrightarrow$)) Let $\twoheadrightarrow$ be a directly follows relation, let Start be the start activities of $\twoheadrightarrow$, let End be the end activities of $\twoheadrightarrow$ and let $c = (\leftrightarrow, \Sigma_1 \ldots \Sigma_n)$ be a cut, consisting of an interleaved operator and a partition of activities with parts $\Sigma_1 \ldots \Sigma_n$ such that $\Sigma(\twoheadrightarrow) = \bigcup_{1 \le i \le n} \Sigma_i$ and $\forall_{1 \le i < j \le n} \Sigma_i \cap \Sigma_j = \emptyset$.
  c is an *interleaved cut* if

$\leftrightarrow$.1 Between parts, all and only connections exist from an end to a start activity:
  $$\forall_{1 \le i \le n, 1 \le j \le n, i \ne j} \forall_{a \in \Sigma_i, b \in \Sigma_j} a \twoheadrightarrow b \Leftrightarrow (a \in \text{End} \land b \in \text{Start})$$

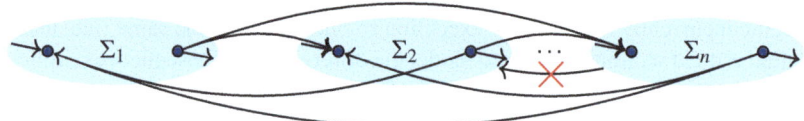

An inspection of the semantics of the interleaved operator (Definition 5.5) reveals that this footprint is present in directly follows graphs of process trees:

Lemma 5.10 (Directly follows footprint ($\leftrightarrow$)) *Let $M = \oplus(M_1, \ldots M_m)$ be a process tree without duplicate activities (Requirement $C_B.2$) with $\oplus \in \{\times, \rightarrow, \leftrightarrow, \wedge, \circlearrowright\}$. Then, the footprints of definitions 5.3 and 5.5 hold, i.e. $\twoheadrightarrow(M)$ contains the footprint of the cut $(\oplus, \Sigma(M_1), \ldots \Sigma(M_n))$.*

5.4.2 A Class of Trees: C_I

We introduce the class of process trees for which we will later prove language uniqueness. That is, for the following class of models, we will prove that no two different trees reduced by the rules of Definition 5.1 have equal languages. Moreover, we will prove that no two such trees have equal directly follows graphs. In this class, interleaved operators are allowed, if they have at least one child with disjoint start and end activities. Certain nestings of interleaved and concurrency are disallowed.

Definition 5.6 (C_I) Let M be a process tree, and let $\bigoplus = \{\times, \rightarrow, \wedge, \circlearrowright, \leftrightarrow\}$. Then M belongs to C_I if for each reduced (sub)tree M' at any position in M, it holds that

$C_I.1$ The subtree adheres to all restrictions of C_B, however $\leftrightarrow$-operators are allowed (Requirement $C_B.4$ is dropped).
$C_I.2$ An interleaving has at least one child with disjoint start and end activities:
  $$M' = \leftrightarrow(M'_1, \ldots M'_n) : \exists_{1 \le i \le n} \text{Start}(M'_i) \cap \text{End}(M'_i) = \emptyset$$

$C_1.3$ An interleaving has no interleaved child:
$$M' = \leftrightarrow(M'_1, \ldots M'_n) : \forall_{1 \leq i \leq n} M'_i \neq \leftrightarrow(\ldots)$$
$C_1.4$ A concurrent child of an interleaving has at least one child with disjoint start and end activities:
$$M' = \leftrightarrow(M'_1, \ldots \wedge(M'_{m_1}, \ldots M'_{m_x}), \ldots M'_n) : \exists_{1 \leq i \leq x} \text{Start}(M'_{m_i}) \cap \text{End}(M'_{m_i}) = \emptyset$$

We explain each requirement that differs from C_B:

- Requirement $C_1.2$: disjoint start and end activities for interleaving. Figure 5.11 shows a counterexample: the trees $M_{43} = $ 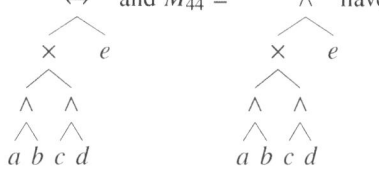 and $M_{44} = $ have a different root operator and language, but have the same directly follows graph. Specifically, activity e can be both interleaved and concurrent to the other activities.

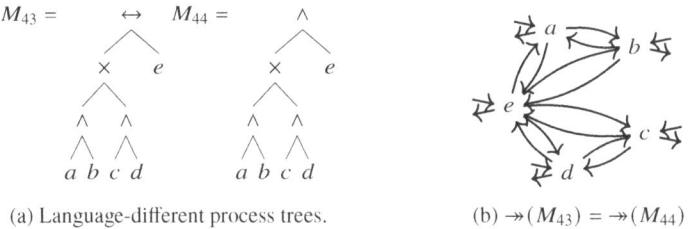

(a) Language-different process trees.   (b) $\twoheadrightarrow(M_{43}) = \twoheadrightarrow(M_{44})$

Fig. 5.11: A counterexample for $\leftrightarrow$ without disjoint start and end activities: trees having different languages but the same directly follows graph.

- Requirement $C_1.3$: no nested interleavings. Figure 5.12 shows a counterexample, i.e. the four trees (in which Q, R and S can be any subtrees) do not have the same activity partition and not the same language, but share their directly follows graph. The difference between these trees are "semi-long-dependencies", e.g. in M_{45}, S cannot be executed between Q and R, and such dependencies cannot be captured by a directly follows relation. In contrast to the $\wedge$-operator, the $\leftrightarrow$-operator is not associative.
- Requirement $C_1.4$: $\wedge$ nested under $\leftrightarrow$ has at least one child with disjoint start and end activities. Figure 5.13 shows a counterexample in which activity e witnesses ambiguity: e can be concurrent to $\times(c, d)$ (M_{49}) or interleaved to the rest of the tree (M_{50}). That is, the directly follows graph does not give enough information to determine the location of e in the tree. We consider this restriction rather inelegant as it concerns three layers of process tree operators. However, we

$$M_{45} = \leftrightarrow(S, \leftrightarrow(Q, R))$$
$$M_{46} = \leftrightarrow(Q, \leftrightarrow(R, S))$$
$$M_{47} = \leftrightarrow(R, \leftrightarrow(Q, S))$$
$$M_{48} = \leftrightarrow(Q, R, S)$$

(a) Language-different process trees.   (b) $\twoheadrightarrow(M_{45}) = \twoheadrightarrow(M_{46}) = \twoheadrightarrow(M_{47}) = \twoheadrightarrow(M_{48})$

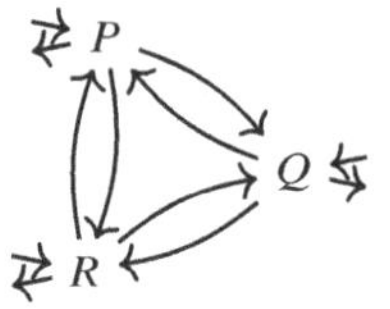

Fig. 5.12: A counterexample for nested $\leftrightarrow$: trees having different languages but the same directly follows graph.

chose this constraint over the stronger constraint stating that all children should have disjoint start and end activities, as that would be more restricting.

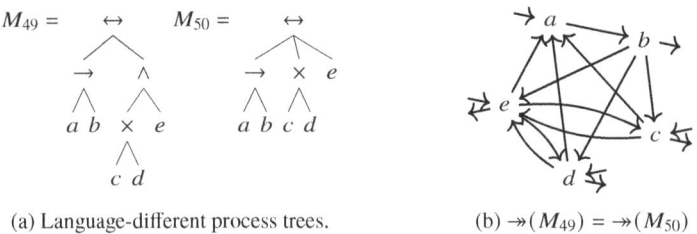

(a) Language-different process trees.   (b) $\twoheadrightarrow(M_{49}) = \twoheadrightarrow(M_{50})$

Fig. 5.13: A counterexample for nested $\wedge$ under $\leftrightarrow$: trees having different languages but the same directly follows graph.

5.4.3 Language Uniqueness

Using the footprint of Lemma 5.10, in this section we establish the link between syntax and semantics of reduced process trees of C_l. That is, we prove that there are no two different reduced process trees having the same directly follows graph, and hence no two such trees having the same language. The proof strategy resembles that of Section 5.2.3: we first prove that if the root operator of two trees differs, their language differs as well (Lemma 5.11). Second, we prove that if the *activity partitions*, i.e. the division of activities over children, of two trees differs, their language differs as well (Lemma 5.12). Finally, language uniqueness follows directly from these two lemmas.

Lemma 5.11 (Operators are mutually exclusive (with $\leftrightarrow$)) *Take two reduced process trees of C_l $K = \oplus(K_1, \ldots K_n)$ and $M = \otimes(M_1, \ldots M_m)$ such that $\oplus \neq \otimes$. Then* $\mathcal{L}(K) \neq \mathcal{L}(M)$.

We prove the lemma by showing for each operator that the footprint of the operator in the directly follows graph is different from the footprints of all other operators.

Proof Towards contradiction, assume that $\mathcal{L}(K) = \mathcal{L}(M)$. Then, $\twoheadrightarrow(K) = \twoheadrightarrow(M)$. By Corollary 5.1, $n \geq 2$ and $m \geq 2$. Perform case distinction on $\oplus$ to prove that $\twoheadrightarrow(K) \neq \twoheadrightarrow(M)$ or $\mathcal{L}(K) \neq \mathcal{L}(M)$, thereby leaving out cases already discussed in the proof of Lemma 5.4.

$\oplus = \times$ The graph $\twoheadrightarrow(M)$ is a connected component, thus this case in Lemma 5.4 holds.

$\oplus = \rightarrow$ The graph $\twoheadrightarrow(M)$ is not a chain, thus this case in Lemma 5.4 holds.

$\oplus = \wedge$ Perform case distinction on the remaining cases of $\otimes$:

> $\otimes = \leftrightarrow$ By C_B, $\exists_{1 \leq i \leq n} \exists_{a \in \Sigma(M_i)} a \notin \text{Start}(M_i) \vee a \notin \text{End}(M_i)$. Take such an M_i and a. As either $a \notin \text{Start}(M_i)$ or $a \notin \text{End}(M_i)$, there's no connection to/from a to any other subtree, i.e. $\forall_{1 \leq j \leq n, j \neq i} \forall_{b \in \Sigma(M_j)} b \not\twoheadrightarrow a \vee a \not\twoheadrightarrow b$. If we would construct a concurrent cut $(\wedge, \Sigma_1 \ldots \Sigma_p)$, then both a and all such b's would be in the same Σ, e.g. $\{a\} \cup (\Sigma(K) \setminus \Sigma(M_j)\}) \subseteq \Sigma_1$. This holds for all activities of $\text{Start}(M_i)$ and $\text{End}(M_i)$. Hence, if we would construct a concurrent cut, all $\text{Start}(K)$ and $\text{End}(K)$ activities would be part of the same Σ. Therefore, there cannot be a non-trivial concurrent cut for K and hence, $\twoheadrightarrow(K) \neq \twoheadrightarrow(M)$.

$\oplus = \circlearrowright$ By semantics of the $\wedge$ operator, $\twoheadrightarrow(K)$ is a single strongly connected component (see Lemma 5.3). Perform case distinction on the remaining case of $\otimes$:

> $\otimes = \leftrightarrow$ We try to construct a loop cut $(\circlearrowright, \Sigma_1, \ldots \Sigma_n)$. Consider a child M_i, and an activity s from the start activities of another child. Moreover, consider a path $a_1 \twoheadrightarrow a_2 \twoheadrightarrow \ldots a_p$ such that all activities on the path are in $\Sigma(M_i)$, and $a_1 \in \text{Start}(M_i)$ and $a_p \in \text{End}(M_i)$. By Lemma 5.3, $a_1 \in \Sigma_1 \wedge a_p \in \Sigma_1$. Consider activity a_2. If $a_2 \in \text{Start}(M_i)$, then $a_2 \in \Sigma_1$. If $a_2 \in \text{End}(M_i)$, then $a_2 \in \Sigma_1$. If $a_2 \notin \text{Start}(M_i) \wedge a_2 \notin \text{End}(M_i)$, then by the semantics of $\leftrightarrow$, $s \not\twoheadrightarrow a_2$. If a_2 would be in Σ_2, as it has a connection $a_1 \twoheadrightarrow a_2$, by the semantics of $\circlearrowright$ there should be a connection $s \twoheadrightarrow a_2$. Thus, $a_2 \in \Sigma_1$. This argument holds for the entire path, and by construction of $\twoheadrightarrow(M)$ each activity is on such a path, thus $\Sigma(M_i) \subseteq \Sigma_1$. This holds for all children M_i, so there cannot be a non-trivial loop cut. Hence, $\twoheadrightarrow(K) \neq \twoheadrightarrow(M)$.

Obviously, these arguments are symmetric in $\oplus$ and $\otimes$, so we conclude that $\twoheadrightarrow(K) \neq \twoheadrightarrow(M)$, which contradicts that $\mathcal{L}(K) = \mathcal{L}(M)$. $\qquad\qquad\square$

Lemma 5.12 (Partitions are mutually exclusive (with $\leftrightarrow$)) *Take two reduced process trees of C, $K = \oplus(K_1 \ldots K_n)$ and $M = \oplus(M_1 \ldots M_m)$ such that their activity partition is different, i.e. there is a w such that $1 \leq w \leq min(n, m)$ and $\Sigma(K_w) \neq \Sigma(M_w)$. Then, $\mathcal{L}(K) \neq \mathcal{L}(M)$.*

Proof Without loss of generality, we assume that children of the commutative operators $(\rightarrow, \circlearrowright)$ have a fixed order. Towards contradiction, assume that $\twoheadrightarrow(K) = \twoheadrightarrow(M)$. Perform case distinction on $\oplus$ (the case for K and M swapped is symmetric), thereby leaving out cases already discussed in the proof of Lemma 5.5.

$\oplus = \times$ For a subtree K_i with $K_i = \leftrightarrow(\ldots)$, $\twoheadrightarrow(K_i)$ is a connected component, so this case in Lemma 5.5 holds.

$\oplus = \rightarrow$ All children of K and M that are $\leftrightarrow$ are strongly connected components, so this case in Lemma 5.5 holds.

$\oplus = \wedge$ Consider a child M_x. Perform case distinction on the structure of M_x:

$M_x = \leftrightarrow(M_{x_1}, \ldots M_{x_p})$ Similar to the $\circlearrowright$ case.

$\oplus = \circlearrowright$ Let $K_1 = \otimes(K_{1_1}, \ldots K_{1_p})$. Perform case distinction on $\otimes$:

$\otimes = \leftrightarrow$ Similar to the $\times$-case.

$\oplus = \leftrightarrow$ Take a w such that $\Sigma(K_w) \neq \Sigma(M_w)$ and let $K_w = \otimes(K_{w_1} \ldots K_{w_p})$. Perform case distinction on $\otimes$:

$\otimes = \times$ By semantics of $\times$, no end activity of K_{w_1} has a connection to any start activity of any other K_{w_j}. Thus, as M contains an interleaved activity partition, $\Sigma(K_w) \subseteq \Sigma(M_w)$.

$\otimes = \rightarrow$ Similar to the $\times$ case.

$\otimes = \wedge$ By C_1, at least one child of K_w has disjoint start and end activities. Take such a child K_{w_y}, and consider two activities: $a \notin \text{Start}(K_{w_y})$ and $b \in \Sigma(K_w) \setminus K_{w_y}$. By semantics of $\wedge$, $b \twoheadrightarrow a$. Then, by Lemma 5.3, $a \in \Sigma(M_w)$ and $b \in \Sigma(M_w)$. This holds for all b and by symmetry for $\text{Start}(K_{w_y}) \cup \text{End}(K_{w_y})$. By semantics of $\leftrightarrow$, non-start non-end activities only have connections with start/end activities of K_w. Therefore, $\Sigma(K_w) \setminus (\text{Start}(K_w) \cup \text{End}(K_w)) \subseteq \Sigma(M_w)$. Hence, $\Sigma(K_w) \subseteq \Sigma(M_w)$.

$\otimes = \circlearrowright$ By semantics of $\leftrightarrow$, non-start non-end activities only have connections with start/end activities of K_w. Therefore, $\Sigma(K_w) \setminus (\text{Start}(K_w) \cup \text{End}(K_w)) \subseteq \Sigma(M_w)$. All activities $\in \text{Start}(K_w) \cup \text{End}(K_w)$ have connections from/to $\text{End}(K_{w_2}) \cup \text{Start}(K_{w_2})$, thus $\text{Start}(K_w) \cup \text{End}(K_w) \subseteq \Sigma(M_w)$. Hence, $\Sigma(K_w) \subseteq \Sigma(M_w)$.

$\otimes = \leftrightarrow$ Excluded by C_1.

By contradiction, we conclude $\mathcal{L}(K) \neq \mathcal{L}(M)$. $\square$

Lemma 5.13 (Language uniqueness for C_1) *Take two different trees of class C_1 in normal form (of Definition 5.1). Then, the languages of these two trees are different.*

The proof for this lemma is similar to the proof of Lemma 5.6, using lemmas 5.11 and 5.12.

In these proofs, showed that if there is a structural difference between two process trees, then their directly follows graphs are different as well. Consequently, we concluded that their languages must be different as well. However, from this intermediate result, we derive:

Corollary 5.4 (Directly follows graph uniqueness with $\leftrightarrow$) *There are no two different reduced process trees of C_1 with equal directly follows graphs.*

5.5 Language Uniqueness with Minimum Self-Distance

In previous sections, several examples were given of process trees with equivalent directly follows graphs, e.g. in Figure 5.3, the following process trees were shown to have the same directly follows graphs:

$$M_{24} = \quad \wedge \qquad\qquad\qquad M_{25} = \quad \circlearrowleft$$

$$\circlearrowleft \quad \circlearrowleft \qquad\qquad\qquad\qquad \wedge \quad \wedge$$

$$\wedge \quad \wedge \qquad\qquad\qquad\qquad \wedge \quad \wedge$$

$$a\ b\ c\ d \qquad\qquad\qquad\qquad a\ c\ b\ d$$

In this section, we will introduce an abstraction that is able to distinguish such trees: the minimum self-distance relation. Specifically, we attempt to drop Requirement $C_B.3$, i.e. that a loop body should have disjoint start and end activities.

As in sections 5.2 and 5.3, the aim is to establish language uniqueness for the new abstraction, such that discovery algorithms can use it to enhance discovery. We proceed as follows: first we introduce the notion of *minimum self-distance* as an additional abstraction of process behaviour, and we characterise a class of process trees (C_M) that is larger than C_B. Second, we consider which operators produce which characteristics in this minimum self-distance abstraction, and finally show that in the larger class C_M, any two trees in normal form can be distinguished based on their combination of directly follows graph and minimum-self-distance abstraction.

5.5.1 Minimum Self-Distance

Two activities a and b are in a *minimum self-distance* relation if in order to execute a twice with a minimum number of events in between, b might be executed in between these two executions of a. Formally, we define the minimum self-distance relation $\circledcirc$ on a language L as follows:

Definition 5.7 (Minimum self-distance) Let L be a language, and let $a, b \in \Sigma(L)$. Then, the minimum self-distance of a is the minimum number of events in between two executions of a:

$$m(a) = \begin{cases} \min_{\langle \ldots, a, \ldots_1, a, \ldots \rangle \in L} |\ldots_1| & \text{if } \exists_{\langle \ldots, a, \ldots_1, a, \ldots \rangle \in L} \\ \infty & \text{otherwise} \end{cases}$$

Then, b is a *witness* of this minimum self-distance of a, denoted by $a \circledcirc b$, if and only if it can appear in between two minimum-distant executions of a:

$$a \circledcirc b \equiv \exists_{\langle \ldots, a, \ldots_1, a, \ldots \rangle \in L}\ b \in \ldots_1 \wedge |\ldots_1| = m(a)$$

For instance, figure 5.14d and 5.14e show two minimum self-distance graphs. In the first graph, the trace $\langle c, a, b, a \rangle$ in the language of M_{51} witnesses that b can be executed between two a's ($a \otimes b$). Further inspection reveals that there is no way to reduce the number of events in between the two as: at least one activity (a b) should be executed between them. Therefore, $a \otimes b$ holds, which is denoted with a double-bordered edge from a to b in Figure 5.14d.

In the tree of the second graph, such a trace would be $\langle c, a, b, d, a, c \rangle$, i.e. $a \otimes b$ and $a \otimes d$. Figure 5.14f shows another example ($\circlearrowleft(a, b, c)$): there is at least one event between two executions of a and this one event can be either a b or a c, thus $a \otimes b$ and $a \otimes c$.

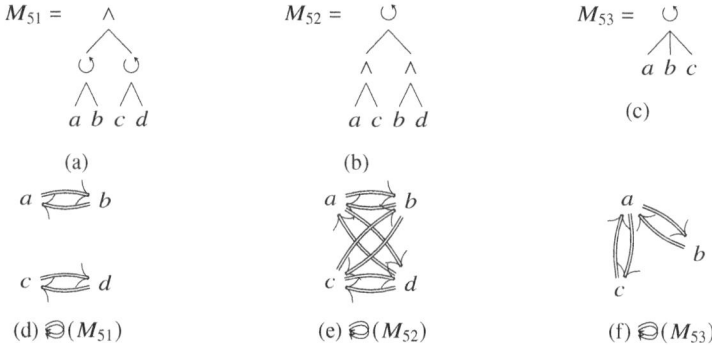

Fig. 5.14: Examples of minimum self-distance graphs.

5.5.2 A Class of Trees: C_M

Previously, we identified that concurrency nested under loop in some cases has the same directly follows footprint as loop under concurrency. In this section, we drop this restriction by introducing a new class of process trees C_M. In the remainder of this section, we will introduce footprints and prove language uniqueness for trees of C_M.

Using Definition 5.7, we extend the class of process trees for which the normal forms are language unique:

Definition 5.8 (Class C_M) Let M be a process tree. Then, M belongs to C_M if all requirements of C_B hold for all reduced subtrees of M, except Requirement $C_B.3$, i.e. the loop body is not required to have disjoint start and end activities.

5.5.3 Footprints

In the previous sections, we showed that a directly follows graph does not identify process trees with concurrency nested under loop uniquely, i.e. there might be process trees with different languages but equal directly follows graphs. Therefore, we introduced the new minimum self-distance abstraction. In this section, we first illustrate when minimum self-distance edges appear, after which we give footprints of process tree operators in the minimum self-distance graphs. In the next section, we prove language uniqueness.

For the loop operator, let b be an activity in the redo of its lowest loop ancestor X, i.e. $X = \circlearrowleft(X_1, \dots X_n)$ such that for an i, $b \in \Sigma(X_i)$ such that b has no loop ancestor in X_i. Then, because the tree is of C_M, Requirement $C_B.1$ holds, so X_1 cannot produce the empty trace. Therefore, b should have a minimum self-distance connection with at least one activity $a \in \Sigma(X_1)$. Moreover, b cannot have any minimum self-distance connection to any activity outside $\Sigma(X)$. For instance, Figure 5.15 shows the $\ominus$-graph of $M_{54} = \circlearrowleft(\wedge(\circlearrowleft(a, b), c), d)$. In this $\ominus$-graph, b is in the redo of its lowest loop ancestor $\circlearrowleft(a, b)$. Thus, $b \ominus a$ but $b \not\ominus c$, $b \not\ominus d$.

$M_{55} =$ (a) (b) $\ominus(M_{55})$

Fig. 5.15: A process tree and its minimum self-distance graph.

Using these observations, we extend the definition of concurrent and interleaved footprints to use both the directly follows graph and the minimum self-distance graph. Notice that we do not need to extend the footprints of $\times$ and $\rightarrow$, as these are not influenced by the removal of Requirement $C_B.3$.

Definition 5.9 (minimum self-distance footprints) Let $\ominus$ be a minimum self-distance relation and let $c = (\oplus, \Sigma_1, \dots \Sigma_n)$ be a cut, consisting of a process tree operator $\oplus \in \{\times, \rightarrow, \leftrightarrow, \wedge, \circlearrowleft\}$ and a partition of activities with parts $\Sigma_1 \dots \Sigma_n$ such that $\Sigma(\ominus) = \bigcup_{1 \leq i \leq n} \Sigma_i$ and $\forall_{1 \leq i < j \leq n} \Sigma_i \cap \Sigma_j = \emptyset$.

• Concurrent and interleaved. If $\oplus = \wedge$ or $\oplus = \leftrightarrow$, then in $\ominus$:

$\wedge \leftrightarrow.1$ There are no $\ominus$ connections between parts:
$$\forall_{1 \leq i \leq n, 1 \leq j \leq n, i \neq j} \forall_{a \in \Sigma_i, b \in \Sigma_j} a \not\ominus b$$

- Loop. If $\oplus = \circlearrowleft$ then in ⊝:

$\circlearrowleft$.1 Each activity has an outgoing edge:

$$\forall_{a\in\Sigma(\circlearrowleft)}\ \exists_{b\in\Sigma(\circlearrowleft),b\neq a}\ a \mathbin{⊝} b$$

$\circlearrowleft$.2 All redo activities that have a connection to a body activity, have connections to the same body activities:

$$\forall_{2\leq i\leq n,2\leq j\leq n}\ \forall_{a\in\Sigma_i,b\in\Sigma_j}\ \{c \mid a \mathbin{⊝} c\} \cap \Sigma_1 = \emptyset \ \vee$$
$$\{c \mid b \mathbin{⊝} c\} \cap \Sigma_1 = \emptyset \ \vee$$
$$\{c \mid a \mathbin{⊝} c\} \cap \Sigma_1 = \{c \mid b \mathbin{⊝} c\} \cap \Sigma_1$$

$\circlearrowleft$.3 All body activities that have a connection to a redo activity, have connections to the same redo activities:

$$\forall_{a,b\in\Sigma_1}\ \{c \mid a \mathbin{⊝} c\} \cap \bigcup_{2\leq i\leq n}\Sigma_i = \emptyset \ \vee$$
$$\{c \mid b \mathbin{⊝} c\} \cap \bigcup_{2\leq i\leq n}\Sigma_i = \emptyset \ \vee$$
$$\{c \mid a \mathbin{⊝} c\} \cap \bigcup_{2\leq i\leq n}\Sigma_i = \{c \mid b \mathbin{⊝} c\} \cap \bigcup_{2\leq i\leq n}\Sigma_i$$

$\circlearrowleft$.4 No two activities from different redo children have an ⊝-connection:

$$\forall_{2\leq i<j\leq n}\ \forall_{a\in\Sigma_i,b\in\Sigma_j}\ a \mathbin{⊘} b \wedge b \mathbin{⊘} a$$

We illustrate these properties:

$\wedge\leftrightarrow$.1   For instance, in our example tree $M_{51} = \wedge(\circlearrowleft(a,b), \circlearrowleft(c,d))$, consider a and c. Notice that in order to execute a twice, it is not necessary to enter $\circlearrowleft(c,d)$ in between these executions of a, as the $\wedge$ operator does not enforce execution of the $\circlearrowleft(c,d)$ child at any particular moment. As a shorter trace is available that avoids c, $a \mathbin{⊘} c$. This holds for all operators that do not enforce a particular sequence of executions, i.e. $\wedge$ and $\leftrightarrow$. Therefore, any two activities with a lowest common ancestor being $\wedge$ or $\leftrightarrow$ cannot be in a minimum self-distance relation.

$\circlearrowleft$.1   Each activity that has a $\circlearrowleft$ ancestor can be executed multiple times in a trace. Therefore, there must be at least one trace in which the activity occurs twice with a minimal number of events in between. Hence, the activity must have an outgoing ⊝ edge.

$\circlearrowleft$.2   Consider an activity r in a redo child $\Sigma(M_{i>1})$, such that r has an outgoing ⊝-connection to activities $b_1 \ldots b_x$ in the body $\Sigma(M_1)$, and there is a trace $t = \langle \ldots r \ldots b_1 \ldots b_x \ldots r \ldots \rangle$ such that the number of events between the two r's is minimal. Then, the subtrace $\langle b_1 \ldots b_x \rangle$ is a shortest path through M_1. Obviously, any other activity r' in any redo child $\Sigma(M_{j>1})$ that has an outgoing ⊝-connection to any body activity has ⊝-connections to all activities $b_1 \ldots b_x$ on the shortest path.

$\circlearrowleft$.3   Similar to Requirement $\circlearrowleft$.2.

$\circlearrowleft$.4 Take two activities from different redo children a and b. If $a \otimes b$, then the shortest path between a and a would pass through the body of the loop twice. Therefore, this cannot be a shortest path and thus $a \not\otimes b$.

Finally, using the semantics of the process tree operators (Definition 2.6), we derive that these footprints are present in minimum self-distance graphs of process trees.

Lemma 5.14 (Minimum self-distance footprints) *Let* $M = \oplus(M_1, \ldots M_m)$ *be a process tree without duplicate activities (Requirement C_n.2), with $\oplus$ being a process tree operator* $\in \{\times, \rightarrow, \leftrightarrow, \wedge, \circlearrowleft\}$. *Then, the footprints of Definition 5.9 hold, i.e. $\otimes(M)$ contains the footprint of the cut* $(\oplus, \Sigma(M_1), \ldots \Sigma(M_n))$.

5.5.4 LC-Property

For some classes of process trees, the footprints of Lemma 5.14 do not suffice to conclude language uniqueness. That is, there are process trees of C_M that have a different normal form, and have different languages and $\otimes$ relations, but cannot be distinguished by these footprints.

For instance, $M_{56} =$ and $M_{57} =$ are such trees: they are both in

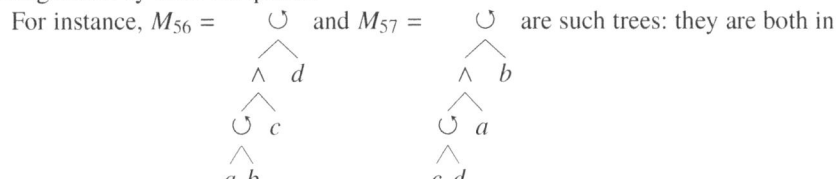

C_M and have different languages. These trees have an equivalent $\twoheadrightarrow$-graph, which is shown in Figure 5.16b. Figures 5.16c and 5.16d show their $\otimes$-graphs.

These $\otimes$-graphs are clearly different, so these trees are not a counterexample to language uniqueness, i.e. one could still distinguish them by their $\otimes$-graphs. However, the footprint described in Definition 5.9 applies to both graphs, i.e. the requirements hold for $\otimes(M_{56})$ using $\Sigma_1 = \{a, c, d\}$, $\Sigma_2 = \{b\}$, which corresponds to $\otimes(M_{57})$. This implies that a discovery algorithm using the footprints cannot distinguish these two trees.

This problem occurs in certain nestings of loops and concurrent operators. The current footprints (Definition 5.9) are not strong enough to distinguish these trees: we did not find an $\otimes$-footprint property to distinguish such trees, but we also did not find a counterexample that disproves language uniqueness. Therefore, the proof of language uniqueness further on in this section (Lemma 5.15) will contain a gap. Therefore, we first characterise the process trees of C_M for which we did not identify a footprint by introducing the *loop-concurrent-property* (LC-property), which is a footprint of the identified trees in the $\otimes$-graph. If such a property exists, then language uniqueness holds. Second, we conjecture that such a property exists.

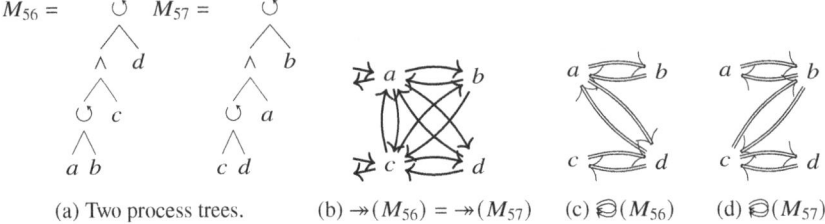

(a) Two process trees.   (b) $\twoheadrightarrow(M_{56}) = \twoheadrightarrow(M_{57})$ (c) $\ominus(M_{56})$ (d) $\ominus(M_{57})$

Fig. 5.16: A counterexample for language uniqueness using $\ominus$ footprints: M_{56} and M_{57} have a different language and $\ominus$-relations, but this doesn't become apparent in the $\ominus$-footprint.

Definition 5.10 (LC-property) Let K and M be process trees in normal form such that $K = $ ⟳ , $M = $ ⟳ , $M, K \in C_M$,

$K_1 = \wedge \quad\quad K_2 \ldots K_n \quad\quad M_1 = \wedge \quad\quad M_2 \ldots M_n$

$K_{1,1} \ldots K_{1,p} \quad\quad\quad M_{1,1} \ldots M_{1,q}$

and $\twoheadrightarrow(K) = \twoheadrightarrow(M)$. Then, an *LC-property LC* is a function that distinguishes the cuts of K and M in their minimum self-distance graphs, i.e. $LC(\ominus(K)) \wedge LC(\ominus(M))$ if and only if the cut $(⟳, \Sigma(K_1), \ldots \Sigma(K_n))$ conforms to both K and M.

We believe that there exists such a property that is able to distinguish process trees of class C_M. However, finding one remains future work.

Conjecture 5.1 (LC-property) There exists an LC-property (Definition 5.10).

5.5.5 Language Uniqueness

In this chapter, we study abstractions from languages to establish a one to one link between syntax and semantics of process trees. We do this by introducing a set of reduction rules, such that footprints in abstractions can distinguish all languages of particular classes of process trees. These footprints will be used by process discovery algorithms in Chapter 6. In this section (5.5), we have introduced the minimum self-distance abstraction, footprints using this abstraction, and a larger class of process trees (C_M). Furthermore, we described the unknown LC-property: a footprint property that is currently missing. In this section, we first prove language uniqueness, i.e. that there are no two process trees of C_M with the same abstraction but a different language. This proof will assume that the LC-property exists. Second, we discuss this property and its implications further.

Lemma 5.15 (Language uniqueness for C_M) *Assume that there exists an LC-property. Take two reduced process trees of class C_M: $K = \oplus(K_1, \ldots K_n)$ and $M = \otimes(M_1, \ldots M_m)$. Then, $K = M$ if and only if $\twoheadrightarrow(K) = \twoheadrightarrow(M)$ and $\ominus(K) = \ominus(M)$.*

Proof The proof strategy is to first show that if either the operators $\oplus$ and $\otimes$ differ, or the activity partitions differ, then $\twoheadrightarrow(K) \neq \twoheadrightarrow(M)$ or $\oslash(K) \neq \oslash(M)$. The lemma follows from these properties, similarly to Lemma 5.6. Both properties are proven by contradiction, i.e. two structurally different but language equivalent process trees are assumed and a contradiction is shown. We do not replicate the entire proof of Lemma 5.6, but limit ourselves to the cases in its proof in which the dropped restriction is involved.

The first property corresponds to Lemma 5.4, i.e. towards contradiction, assume $\oplus \neq \otimes$ and $\mathcal{L}(K) = \mathcal{L}(M)$. Then, $\twoheadrightarrow(K) = \twoheadrightarrow(M)$ and $\oslash(K) = \oslash(M)$.

$\oplus = \wedge$ and $\otimes = \circlearrowleft$. We try to construct a concurrent cut $\Sigma_1 \ldots \Sigma_q$ for M. By Requirement $\wedge.1$, every such Σ_i must have a start and an end activity. Thus, we only need to prove that $\mathrm{Start}(M_1) \cup \mathrm{End}(M_1) \subseteq \Sigma_1$. Perform case distinction on M_1:

$M_1 = \times(M_{1_1}, \ldots M_{1_p})$ Each $a \in \Sigma(M_{1_i})$ has no $\twoheadrightarrow$-connection to any activity in $\Sigma(M_{j \neq i})$. Therefore, $\mathrm{Start}(M_1) \cup \mathrm{End}(M_1) \subseteq \Sigma_1$.

$M_1 = \rightarrow(M_{1_1}, \ldots M_{1_p})$ Each $a \in \Sigma(M_{1_i})$ has no $\twoheadrightarrow$-connection to any activity in $\Sigma(M_{j < i})$. Therefore, $\mathrm{Start}(M_1) \cup \mathrm{End}(M_1) \subseteq \Sigma_1$.

$M_1 = \wedge(\ldots)$ Consider three cases:

- If any of the $M_{2 \leq i \leq p}$ contains a $\circlearrowleft$, consider an activity a in the redo of that $\circlearrowleft$. By semantics of $\circlearrowleft$, there is no $\twoheadrightarrow$-connection between a and any activity in $\Sigma(M_1)$. Therefore, $\mathrm{Start}(M_1) \cup \mathrm{End}(M_1) \subseteq \Sigma_1$.
- If none of the $M_{2 \leq i \leq p}$ contains a $\circlearrowleft$ and M_1 does not contain a $\circlearrowleft$, then the $\oslash$-graph is connected and therefore by Requirement $\wedge \leftrightarrow.1$, $\Sigma(M) \subseteq \Sigma_1$.
- If none of the $M_{2 \leq i \leq p}$ contains a $\circlearrowleft$ and M_1 contains a $\circlearrowleft$, then consider an activity a under a redo of any such $\circlearrowleft$, and any activity $b \in \Sigma(M_{2 \leq i \leq m})$. By semantics of $\circlearrowleft$, $a \not\twoheadrightarrow b$ and $b \not\twoheadrightarrow a$, thus a and b must be in the same Σ_1. All activities $\mathrm{Start}(M_1) \cup \mathrm{End}(M_1)$ have at least an $\oslash$-connection with at least some activity in the redo of a $\circlearrowleft$. Thus, by Requirement $\wedge \leftrightarrow.1$, $\mathrm{Start}(M_1) \cup \mathrm{End}(M_1) \subseteq \Sigma_1$.

$M_1 = \circlearrowleft(\ldots)$ Excluded by C_M.

$M_1 = \leftrightarrow(\ldots)$ By C_M, there exists a child M_{1_i} such that $\mathrm{Start}(M_{1_i}) \cap \mathrm{End}(M_{1_i}) = \emptyset$. Thus, all activities in $\mathrm{End}(M_{1_{j \neq i}})$ have no $\twoheadrightarrow$-connection to $\mathrm{End}(M_{1_i})$, and similarly for the activities of $\mathrm{Start}(M_{1_j})$. Therefore, $\mathrm{Start}(M_1) \cup \mathrm{End}(M_1) \subseteq \Sigma_1$.

Hence, there is no concurrent cut in M and therefore $\twoheadrightarrow(K) \neq \twoheadrightarrow(M)$.

$\oplus = \circlearrowleft$ and $\otimes = \leftrightarrow$. No change necessary.

The second property corresponds to Lemma 5.5, i.e. towards contradiction, assume that $\oplus = \otimes$ and that there is a w such that $\Sigma(K_w) \neq \Sigma(M_w)$ and $\mathcal{L}(K) = \mathcal{L}(M)$. Then, $\twoheadrightarrow(K) = \twoheadrightarrow(M)$ and $\oslash(K) = \oslash(M)$.

$\oplus = \wedge$ and $M_x = \circlearrowleft(M_{x_1}, \ldots M_{x_p})$. Try to construct a $\wedge$-cut and prove that $\Sigma(M_x) \subseteq \Sigma_x$. Consider three cases:

- If any of the $M_{x_{2 \le i \le p}}$ contains a $\circlearrowleft$, consider an activity a in the redo of that $\circlearrowleft$. By semantics of $\circlearrowleft$, there is no $\twoheadrightarrow$-connection between a and any activity in $\Sigma(M_{x_1})$. Therefore, $\Sigma(M_{x_1}) \subseteq \Sigma_x$. This holds for all such a, thus all such redo-activities are in Σ_x. Consider all remaining activities, i.e. $b \in \Sigma(M_{x_{j \ne i}})$ such that b is in no other $\circlearrowleft$-redo than M_x. For each of these activities b, there is a $\oslash$-relation with an activity in Σ_{x_1} or an activity such as a. Thus, $\Sigma(M_x) \subseteq \Sigma_x$.
- If none of the $M_{x_{2 \le i \le p}}$ contains a $\circlearrowleft$ and M_{x_1} does not contain a $\circlearrowleft$, then the $\oslash$-graph is connected and therefore $\Sigma(M_x) \subseteq \Sigma_x$.
- If none of the $M_{x_{2 \le i \le p}}$ contains a $\circlearrowleft$ and M_{x_1} contains a $\circlearrowleft$, then consider an activity a under a redo of any such $\circlearrowleft$, and any activity $b \in \Sigma(M_{x_{2 \le i \le m}})$. By semantics of $\circlearrowleft$, $a \not\twoheadrightarrow b$ and $b \not\twoheadrightarrow a$, thus a and b must be in the same Σ_x. All activities in $\Sigma(M_{x_1})$ have at least an $\oslash$-connection with at least some activity in the redo of a $\circlearrowleft$, Thus, $\Sigma(M_x) \subseteq \Sigma_x$.

$\oplus = \circlearrowleft$ and $K_1 = \wedge(K_{1,1}, \ldots K_{1,p})$. Try to construct a $\circlearrowleft$-cut and prove that $\Sigma(K_1) \subseteq \Sigma_1$. By semantics of $\circlearrowleft$, $\text{Start}(K_1) \cup \text{End}(K_1) \subseteq \Sigma_1$. Take an activity $a \in \Sigma(K_1)$, such that $a \notin \text{Start}(K_1) \cup \text{End}(K_1)$, and take another $b \in \bigcup_{1 \le i \le n} \Sigma(K_i)$ such that $b \in \text{Start}(K_1) \cup \text{End}(K_1)$. Then, $b \in \Sigma_1$. Perform case distinction on b:

$b \notin \text{End}(K_1)$	Then, $b \twoheadrightarrow a$ and thus $a \in \Sigma_1$.
$b \notin \text{Start}(K_1)$	Then, $a \twoheadrightarrow b$ and thus $a \in \Sigma_1$.
$b \in \text{Start}(K_1) \cap \text{End}(K_1)$	Then, as there exists an LC-property (Conjecture 5.1), $a \in \Sigma_1$. $\square$

Corollary 5.5 (Minimum self-distance uniqueness) *If an LC-property exists, then for all reduced process trees $K \ne M$ of C_M, $\twoheadrightarrow(K) \ne \twoheadrightarrow(M)$ or $\oslash(K) \ne \oslash(M)$.*

From the proof of this lemma it follows that the footprint of Lemma 5.14 suffices to distinguish concurrency from loop behaviour. However, the unknown LC-property is used in the proof of one particular case: a loop with concurrency as its body, in which there are activities that are both start and end, i.e. $M = \circlearrowleft(\wedge(K_{1,1}, \ldots K_{1,p}), \ldots)$. This characterises the class LC, and an example of such trees was given in Figure 5.16.

Notice that this class LC resembles Requirement $\circlearrowleft$.1 of C_B i.e. that the start and end activities of a loop are disjoint. This requirement served two purposes: (1) distinguish concurrent and loop behaviour, and (2) distinguish loop body from loop redo behaviour. We showed that concurrent and loop behaviour can be distinguished using the $\oslash$-relation, what remains to be shown is that loop body and loop redo behaviour can be distinguished as well. We did not find an extension of the footprint properties, however we also did not find a counterexample. Therefore, it remains unknown whether such an LC-property exists.

Future work 5.4: Find or disprove a footprint LC-property of $\oslash$-graphs to distinguish all trees of C_M.

5.6 Language Uniqueness with Optionality & Inclusive Choice

In the previous sections, several abstractions and footprints of process tree operators in these abstractions were identified. Using these footprints, we proved that a set of reduction rules (Definition 5.1) provides language uniqueness, i.e. there are no two process trees in normal form that have an equal language. From these sections, two process tree constructs were left out: the inclusive choice $\vee$ and the invisible activity τ. In this section, we address these constructs: we focus on $\vee$ nodes and $\times(\tau, .)$ constructs.

We first introduce the $\times(\tau, .)$ construct and introduce *optionality*, which denotes that a process tree is not necessarily executed. Second, we study the influence of optionality on directly follows graphs. Third, we study the influence of inclusive choice on directly follows graphs, and note that many reduced process trees have the same directly follows graph. In the previous sections, in case two reduced process trees had the same abstraction, we chose to limit the class of process trees for which we proved language uniqueness. For the inclusive choice however, we do not pose such requirements, but instead investigate which abstraction suffices to distinguish all process trees. Fourth, we introduce this abstraction, the *concurrent-optional-or relations*. The abstraction will be used by our discovery algorithms to discover inclusive choice behaviour, as we will show in Chapter 6. In this section, we lay the formal foundation by proving that no two process trees of a certain class (which we introduce in Section 5.6.3) have the same combination of coo-relations and directly follows graph, and hence no two such trees have the same language (they are language unique).

5.6.1 Optionality

A tau (τ) denotes the silent activity, i.e. executing τ will change the state of a system, but will not generate an event. From the reduction rules (Corollary 5.1), it follows that τ leafs are only relevant in two constructs: $\circlearrowleft$ and $\times$, i.e. as a redo part

$$\overset{\curvearrowleft}{\ldots \tau \ldots} \qquad \overset{\wedge}{\tau \ldots}$$

(non-first child) of $\circlearrowleft$ or as a child of $\times$. In this chapter, we focus on τ leafs that are children of $\times$ nodes; τ's as children of $\circlearrowleft$ nodes will be discussed at the end of this chapter.

The $\times(\tau, .)$ construct explicitly adds the empty trace to the language of nodes. The empty trace might propagate to nodes higher up in the tree, e.g. the language of $\wedge(\times(\tau, a), \times(\tau, b))$ contains the empty trace (semantically) without the root being a (structural) $\times(\tau, .)$ construct. We refer to a process tree whose language contains the empty trace as a tree with *optionality*.

Definition 5.11 (optionality) A process tree is *optional* ($\overline{?}$) if its language contains the empty trace:

$$\overline{?}(M) \equiv \epsilon \in \mathcal{L}(M)$$

5.6.2 Optionality in the Directly Follows Graph

Optionality is difficult or impossible to discover for directly follows based algorithms, as it might leave footprints that are indistinguishable from other process tree constructs. For instance, Figure 5.17 shows two process trees having a different language. These trees differ in the subtrees that can be skipped: in M_{58}, b and c can only be skipped together, while in M_{59}, they can be skipped independently. Their directly follows graphs are the same, thus no directly follows based algorithm can distinguish these trees.

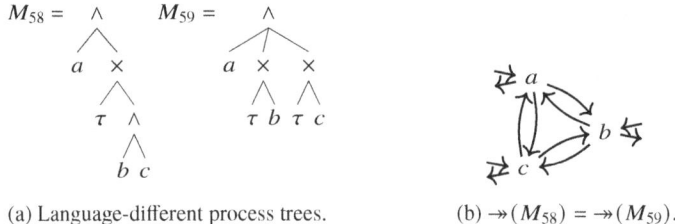

(a) Language-different process trees.   (b) $\twoheadrightarrow(M_{58}) = \twoheadrightarrow(M_{59})$.

Fig. 5.17: The $\times(\tau, .)$ construct might not be captured by $\twoheadrightarrow$-footprints.

In this section, we study the influence of optionality on directly follows graphs when abstracting behaviour and recognising operators in abstractions. First, we introduce the footprint of the $\times(\tau, .)$ construct. Second, we observe that this construct might have influence on the footprints of nodes higher in the tree, and that the footprint is not a sufficient condition to conclude the $\times(\tau, .)$ construct and analyse this influence in more detail. We perform this analysis in two steps: the influence of the operator, and the influence of the activity partition. From the analysis, it follows that two types of constructs, i.e. nested sequences and nested concurrency/inclusive choice, need a stricter footprint and a new abstraction, which will be introduced in the next sections. In the remaining part of this section, we introduce a class of process trees for which we prove that the stricter footprint and the new abstraction provide language uniqueness.

5.6.2.1 Footprint of Optionality

We first define the footprint of the $\times(\tau, .)$ construct: the $\times(\tau, .)$ construct will manifest in the directly follows relation as an edge from start to end:

Lemma 5.16 ($\times(\tau, .)$ **footprint**) *Let* $M = \times(\tau, M_1)$ *be a process tree. Then, the following property holds in* $\twoheadrightarrow(M)$:

$\times(\tau).1$ *There is a connection from start straight to end:*

$$\top \twoheadrightarrow \bot$$

This footprint does not suffice to conclude the $\times(\tau,.)$ construct however: for instance, the reduced tree $\wedge(\times(\tau,a),\times(\tau,b))$ has the empty trace in its language and hence $\top \twoheadrightarrow \bot$ in its directly follows graph, however its root is not the $\times(\tau,.)$ construct. Although the footprint is not sufficient to conclude the $\times(\tau,.)$ construct, we will nevertheless use it later on to discover an over-approximation of structural $\times(\tau,.)$ constructs, which are in a second step removed by applying Reduction Rule $T_\times$ of Definition 5.1.

5.6.2.2 The Influence of Optionality on Operators

The process tree $\wedge(\times(\tau,a),\times(\tau,b))$ illustrates that the influence of $\times(\tau,.)$ constructs is not limited to their level in the process tree: even though the root is a concurrent operator, it can still produce the empty trace. Therefore, we analyse the influence of $\times(\tau,.)$ constructs on several levels below the root. Consider a process tree $P = \oplus(Q_1,\dots,Q_l)$, $Q_i = \otimes(R_1,\dots R_m)$ and $R_j = \ominus(\dots)$ (Figure 5.18), with arbitrary operators $\oplus$, $\otimes$ and $\ominus$. For this tree, we discuss what happens to the root level behaviour and directly follows graph if an optional operator is added at several levels:

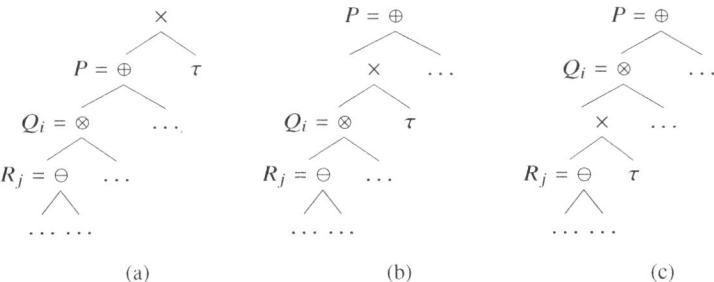

Fig. 5.18: Process trees with optionality at different levels.

- If we make P optional by adding the $\times(\tau,.)$ construct above the root (Figure 5.18a), then we add an explicit empty trace to the language of the model, and the $\top \twoheadrightarrow \bot$ edge manifests itself in the directly follows relation. Obviously, this addition has no further influence on the footprint of $\oplus$ as defined in definitions 5.3 and 5.5. Therefore, in this section, we may limit ourselves to discussing non-optional root nodes.
- How a subtree $\times(\tau,Q_i)$ manifests itself in the footprint of the directly follows graph of P (Figure 5.18b) depends on the root operator $\oplus$:
  - If $\oplus = \times$, the entire tree is optional if at least one child of the root is optional. In the directly follows relation, the edge $\top \twoheadrightarrow \bot$ appears, i.e. the footprint of

the $\times(\tau, .)$ construct. No further edges are added or removed, so the footprint of Lemma 5.3 is preserved.

- If $\oplus = \rightarrow$, then the entire tree is optional if all children of the root are optional. In that case, $\top \twoheadrightarrow \bot$ is part of the directly follows graph.
  Furthermore, in the directly follows graph, edges are introduced that bypass the optional child. For instance, the directly follows graph of $\rightarrow(a, b, c)$ is $\overset{\rightsquigarrow}{} a \longrightarrow b \longrightarrow c \rightsquigarrow$, and the directly follows graph of $\rightarrow(a, \times(\tau, b), \times(\tau, c))$ is $\rightsquigarrow a \rightleftharpoons b \rightleftharpoons c \rightsquigarrow$. All bypassing edges go in the same direction as edges due to visible steps. Therefore, Requirement $\rightarrow$.1, i.e. that there are no "backwards" edges, remains satisfied.

- If $\oplus \in \{\vee, \leftrightarrow, \wedge\}$, then the entire tree is optional if all children of the root are optional. In that case, $\top \twoheadrightarrow \bot$ is part of the directly follows relation. However, there are no further changes in the directly follows relation of the tree caused by the addition of the $\times(\tau, .)$ construct. For instance, the directly follows graphs of $\wedge(a, b)$ and of $\wedge(a, \times(\tau, b))$ are equivalent: $\rightleftharpoons a \rightleftharpoons b \rightleftharpoons$. Hence, the footprints of lemmas 5.3 and 5.10 remain distinguishable (we will discuss $\vee$ in more detail in Section 5.6.5).

- If $\oplus = \circlearrowleft$, then the entire tree is optional if and only if the body child (i.e. the first child) of the root is optional (notice that by Corollary 5.1, in a reduced tree the body of a loop cannot be τ). We give an example to illustrate the influence of optionality on the loop: consider the trees 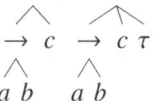 and

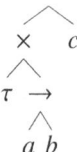 . Their directly follows graphs are shown in Figure 5.19. In the

second tree, τ has been added as a redo child (or, equivalently, c has been made optional). In the directly follows graph of this tree, the edge $b \twoheadrightarrow a$ has been added and the footprint of the $\circlearrowleft$-operator is still present. In the third tree, the body child has been made optional using a $\times(\tau, .)$ construct. In the directly follows graph, this adds the empty trace ($\top \twoheadrightarrow \bot$), and the addition of c as a start and end activity. This violates the footprint: Requirement $\circlearrowleft$.1 states that all start and end activities should be in the loop body, and c is both.

In the remainder of this section, we will not consider loop nodes of which any child is optional, and assume that such nodes are not present. Thus, we do not consider the trees shown in figures 5.19b and 5.19c. We discuss $\circlearrowleft$ nodes with optional children at the end of the section.

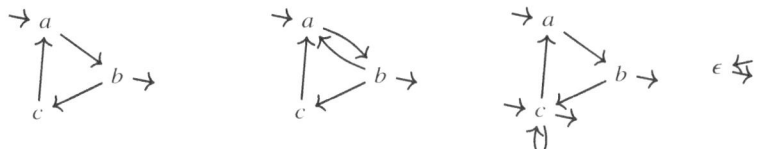

(a) $\rightarrow$ of $\circlearrowleft(\rightarrow(a,b),c)$.   (b) $\rightarrow$ of $\circlearrowleft(\rightarrow(a,b),c,\tau)$.   (c) $\rightarrow$ of $\circlearrowleft(\times(\tau,\rightarrow(a,b)),c)$.

Fig. 5.19: The influence of τ on the directly follows graph of $\circlearrowleft$.

- Assume that Q_i is not optional (otherwise, see the previous case). If we add an $\times(\tau,.)$ construct at the bottom level (just above R_j) (Figure 5.18c), then the influence on the $\twoheadrightarrow$ graph of P depends on the operator $\otimes$:
  - If $\otimes = \times$, then Q_i must be optional, which was excluded by the assumption of the Q_i-case.
  - If $\otimes = \rightarrow$, then not all children of R_j are optional. As added edges only strengthen the footprint of the $\rightarrow$ of $\twoheadrightarrow(Q_i)$, the footprint of $\oplus$ will be present in $\twoheadrightarrow(P)$.
  - If $\otimes \in \{\vee,\leftrightarrow,\wedge\}$, then, as argued before, the directly follows graph does not change. Therefore, the footprint of $\oplus$ will be present in $\twoheadrightarrow(P)$.
  - The $\otimes = \circlearrowleft$ case is excluded.

5.6.2.3 The Influence of Optionality on Activity Partitions

In the previous section, we addressed the influence of $\times(\tau,.)$ constructs at various levels in the process tree, and found that for all operators except $\circlearrowleft$, the footprint of the operator was preserved in the directly follows relation. In this section, we consider the influence of the $\times(\tau,.)$ construct on activity partitions.

For most operators, Definition 5.1 provides a reduction rule that uses associativity, e.g.  would be reduced to $\underset{a\ b\ c}{\rightarrow}$. Therefore, in previous parts of this chapter,

we could assume that an operator had no child of the same operator, and therefore its children had a distinct footprint. The $\times(\tau,.)$ construct disables this property: now, children of an operator can have the same footprint as the operator. For instance,

$$\begin{array}{c}\rightarrow\\ \diagup\diagdown\\ \times\quad c\\ \diagup\diagdown\\ \tau\quad\rightarrow\\ \diagup\diagdown\\ a\ b\end{array}$$

is in normal form and cannot be reduced further, but still exhibits the

footprint of the cut $(\rightarrow,\{a\},\{b\},\{c\})$: the $\times$-subtree has a sequential footprint but not a sequential root operator. Therefore, the abstractions used by process discovery algorithms need to be able to distinguish such trees from similar "flattened" trees.

We study the influence of this property using a process tree $P = \oplus(P_1, \ldots P_m)$, in which one such P_i is $\times(\tau, \oplus(Q_1, \ldots Q_l))$. We perform this analysis for each operator $\oplus$:

$\oplus = \times$ By Corollary 5.1, no child P_i can be a $\times(\tau, .)$ construct, as this would imply that three $\times$-operators are directly nested, and these would vanish in reduction.

$\oplus = \rightarrow$ Optionality poses a challenge in combination with sequential trees, as the sequential footprint (Lemma 5.3) applies to nested sequential subtrees as well. For instance, consider the trees M_{60} and M_{61}. Both trees are in normal form and have a different language. Figure 5.20 shows their directly follows graphs, which are different. However, for both, the footprint of Lemma 5.3 yields a cut $(\rightarrow, \{a\}, \{b\}, \{c\}, \{d\}, \{e\})$. This cut does not conform to M_{61}, as $\{b, c, d\}$ (the activities of the middle child) are partitioned. Therefore, the sequence footprint does not suffice to distinguish process trees in normal form when τ is involved.

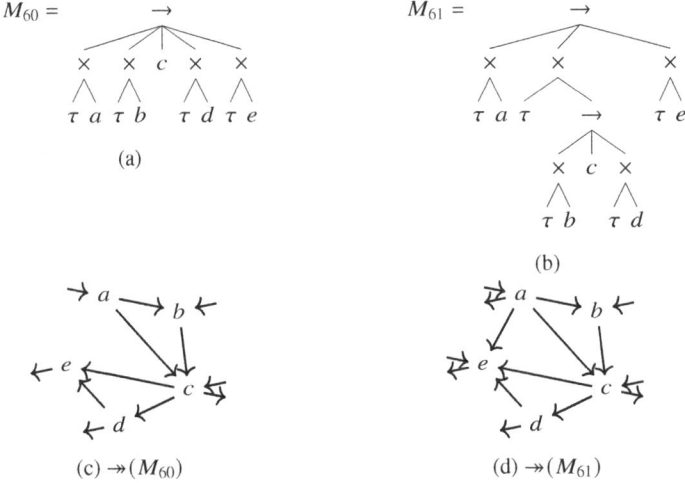

Fig. 5.20: Two process trees with optionality in different places and their directly follows graphs.

In Section 5.6.4, we study this in more detail, and introduce a stricter footprint of the directly follows graph to distinguish such trees.

$\oplus = \leftrightarrow$ As discussed with Requirement $C_1.3$, nested interleaved operators cannot be distinguished by directly follows graphs.

For instance, the trees $\leftrightarrow$ and $\leftrightarrow$ do not have the same language, but have the same directly follows graph (a clique), hence a discovery algorithm using the directly follows abstraction cannot distinguish these trees. As the addition of a $\times(\tau, .)$ construct does not change the directly follows graph, this argument holds for the $\leftrightarrow(\times(\tau, \leftrightarrow(\ldots)))$ case as well. For instance,

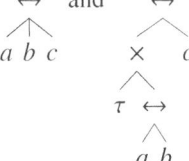

have a different language, but the same directly follows graph.

$\oplus = \wedge$ Concurrency suffers from the same issue as $\leftrightarrow$: the directly follows graph of concurrency nested with optionality and concurrency. For instance,

 have a different language but equivalent directly follows graph.

In Section 5.6.5, we introduce a new abstraction that solves this problem.

$\oplus = \vee$ The $\vee$ operator suffers from the same issue as $\wedge$ and $\leftrightarrow$ case. Moreover, the directly follows footprint of $\vee$ is equivalent to the footprint of $\wedge$. For instance, $\wedge$ and $\vee$ have the same directly follows graph: $\supset$ a $\rightleftharpoons$ b $\circlearrowleft$.
$a\ b$ $a\ b$

In Section 5.6.5, we introduce a new abstraction that solves this problem.

$\oplus = \circlearrowright$ As discussed in the previous section, we do not consider loop nodes with optional children in this section.

Using these observations, we introduce the class of process trees for this section.

5.6.3 A Class of Trees: C_{coo}

In the previous section, we studied the influence of $\times(\tau, .)$ constructs on directly follows graphs. Furthermore, we showed that $\wedge$ and $\vee$ have equal footprints in directly follows graphs. Therefore, later on in this section, we will introduce a new abstraction (the *concurrent-optional-or* (coo) abstraction) that is able to distinguish $\wedge$ and $\vee$. However, even with this new abstraction, not all process trees can be distinguished, so we introduce a new class of process trees. For this class, we will prove that the directly follows graph and the new coo abstraction yield language uniqueness. The class introduced here (C_{coo}) is the largest class of process trees that we will consider: it allows for inclusive choice and τ nodes. However, $\times(\tau, .)$

constructs necessitate new (but smaller) requirements for $\circlearrowleft$ and $\leftrightarrow$. We first give the class definition, after which we explain its requirements.

Definition 5.12 ($\mathbf{C_{coo}}$) Let M be a reduced process tree. Then M belongs to C_{coo} if for each reduced (sub)tree M' at any position in M, it holds that

C_{coo}.1 The subtree adheres to all restrictions of C_B and C_I, however $\leftrightarrow$-operators are allowed (Requirement C_B.4 is dropped), $\vee$-operators are allowed (Requirement C_B.5 is dropped), and τ leafs are allowed (Requirement C_B.1 is dropped).
C_{coo}.2 No redo child of a loop can produce the empty trace:
$$M' = \circlearrowleft(M'_1, \ldots M'_n) : \forall_{2 \leq i \leq n} \, \epsilon \notin \mathcal{L}(M'_i)$$
C_{coo}.3 Interleaving cannot be nested using optionality:
$$M' = \leftrightarrow(M'_1, \ldots M'_n) : \forall_{1 \leq i \leq n} \, M'_i \neq \times(\tau, \leftrightarrow(\ldots))$$
C_{coo}.4 An inclusive choice child of an interleaving has at least one child with disjoint start and end activities:
$$M' = \leftrightarrow(M'_1, \ldots \vee(M'_{m_1}, \ldots M'_{m_x}), \ldots M'_n) :$$
$$\exists_{1 \leq i \leq x} \, \mathrm{Start}(M'_{m_i}) \cap \mathrm{End}(M'_{m_i}) = \emptyset$$

We illustrate the relaxed or newly added requirements:

- Requirement C_{coo}.2: no silent activities under a loop-redo. By Corollary 5.1, in reduced process trees, silent activities will only appear as a child of a $\times$-node or as a redo-child of a $\circlearrowleft$-node. As shown in Section 5.1, the reduction rules of Definition 5.1 are not strong enough. For instance, the process trees

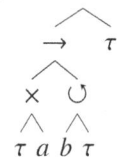

and 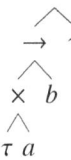 are both in normal form but have the same language. Therefore,

language uniqueness of the current reduction rules does not hold for such trees.
- Requirement C_{coo}.3: no optional nested interleaving. Such a nested optional interleaving has the same directly follows graph as a nested interleaving, which is ambiguous (see Requirement C_I.3) and the previous section. As discovery algorithms require models to be distinguishable by their abstraction, at least one of these models could not be discovered by discovery algorithms.
- Requirement C_{coo}.4. This requirement corresponds to Requirement C_I.4.

From the observations in Section 5.6.2, it follows that one can arbitrarily add $\times(\tau, .)$ constructs to trees of C_I: the footprints will remain visible in the directly follows graph.

Corollary 5.6 (optionality preserves cuts) *Take two reduced process trees* $M \in C_l$, *and* $M' \in C_{coo}$, *such that* $M = \oplus(M_1, \ldots M_m)$, $M' = \oplus(M'_1, \ldots M'_n)$ *and each* M'_i *is equal to either* M_i *or* $\times(\tau, M_i)$. *Then,* $\twoheadrightarrow(M')$ *contains a cut* $(\oplus, \Sigma(M_1) \ldots \Sigma(M_m))$, *i.e. a footprint according to Lemma 5.3.*

5.6.4 Optionality under Sequence

In the previous sections, we analysed the influence of $\times(\tau, .)$ constructs on directly follows graphs. Two challenges were identified: the directly follows graph does not distinguish nested $\wedge$ and $\vee$ operators, and the footprint of Lemma 5.3 does not distinguish nested $\rightarrow$ operators. Both challenges prevent discovery algorithms from identifying these constructs. In Section 5.6.5, we introduce a new abstraction for $\wedge$ and $\vee$. In this section, we introduce a stricter footprint for $\rightarrow$.

The challenge occurs in all process trees in which $\rightarrow$ and $\times(\tau, .)$ constructs appear nested at the top of the tree. We refer to the top part of such a tree as a *sequence-optional stem* (so stem), and to the subtrees that are not part of the stem as *non-so subtrees*. Figure 5.21 shows an example, in which the non-so subtrees have been highlighted in blue.

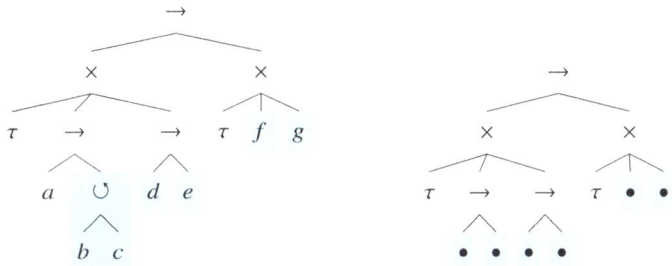

(a) A process tree M_{62}. The non-so-subtrees are denoted in blue.

(b) The so stem of M_{62}; the dots denote non-so subtrees.

Fig. 5.21: An example of a process tree showing its so stem and non-so subtrees.

Formally, we define the so stem of a process tree as follows:

Definition 5.13 (sequence-optional stem) Let Σ be an alphabet of activities such that $\tau \notin \Sigma$ and $\bullet \notin \Sigma$, then

- activity $a \in \Sigma$ is a non-so subtree;
- τ is a non-so subtree;
- let $M_1 \ldots M_n$ with $n > 0$ be process trees (Definition 2.6) and let $\oplus \in \{\leftrightarrow, \wedge, \vee, \circlearrowleft\}$, then $\oplus(M_1, \ldots M_n)$ is a non-so subtree;

– let $M_1 \ldots M_n$ with $n > 0$ be process trees (Definition 2.6) such that no M_i is τ, then $\times(M_1, \ldots M_n)$ is a non-so subtree.

Furthermore,

– • is a so stem;
– let $M_1 \ldots M_n$ with $n > 0$ be so stems, then $\rightarrow(M_1, \ldots M_n)$ is a so stem;
– let $M_1 \ldots M_n$ with $n > 0$ be so stems, $\times(\tau, M_1, \ldots M_n)$ is a so stem.

A process tree M *has* an so stem S if and only if $S \neq •$ and S can be transformed into M by replacing each • in S with a non-so subtree.

Observe that if all children of a $\times(\tau, \rightarrow(\ldots))$ construct would be optional, then the $\rightarrow$-node itself would be optional, and then the reduction rules (Definition 5.1) would remove the $\times(\tau, .)$ construct. (Consequently, the nested $\rightarrow$ would be removed as well.)

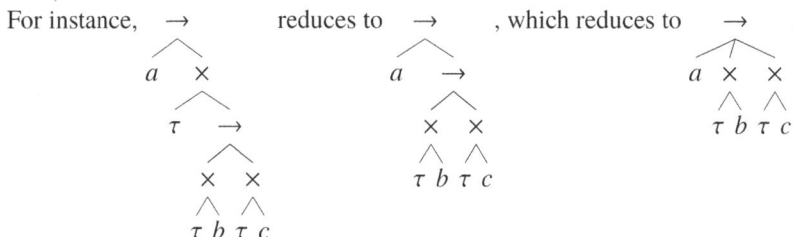

Corollary 5.7 (pivot) *In a reduced process tree, each $\times(\tau, \rightarrow(\ldots))$ construct has at least one subtree (a* pivot*) that is not optional. By Corollary 5.1, a pivot cannot be a sequential node itself.*

Secondly, observe that while non-root pivots are not necessarily executed in a trace (due to their $\times(\tau, \rightarrow())$ parent), execution of the pivot is implied by the execution of any sibling. For instance, in our example tree M_{61} (Figure 5.20d), not every trace contains the pivot c, but c is implied by b and d (siblings of c). We use this observation to introduce a new, stricter, footprint for sequential behaviour in directly follows graphs. This footprint also contains all the non-root siblings (the *scope* of the pivot).

Lemma 5.17 (Nested $\times(\tau, \rightarrow(\ldots))$ footprint) *Let $M = \rightarrow(\ldots)$, having a subtree $M' = \times(\tau, \rightarrow(M'_1, \ldots M'_n))$:*

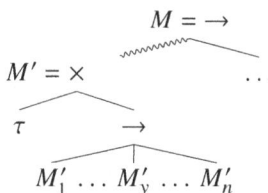

Then there is a non-so subtree pivot M'_y, such that in $\twoheadrightarrow(M)$:

$\times(\tau \rightarrow (\ldots)).1$ *It is possible to not execute the pivot:*

$$\exists_{1 \leq i' \leq i} \exists_{a_{i'} \in I_{i'}} a_{i'} \in \text{End} \vee$$

$$\exists_{1 \leq j' \leq j} \exists_{a_{j'} \in J_{j'}} a_{j'} \in \text{Start} \vee$$

$$\exists_{1 \leq i' \leq i, 1 \leq j' \leq j} \exists_{a_{i'} \in I_{i'}, a_{j'} \in J_{j'}} a_{i'} \twoheadrightarrow a_{j'}$$

In which $(\rightarrow, I_1 \ldots I_i, \Sigma(M'_y), J_1 \ldots J_j)$ is the maximal sequence cut of $\twoheadrightarrow(M)$.

$\times(\tau \rightarrow (\ldots)).2$ *Children before the pivot are not end activities:*

$$\forall_{1 \leq x < y} \forall_{a_x \in \Sigma(M'_x)} a_x \notin \text{End}$$

$\times(\tau \rightarrow (\ldots)).3$ *Children before the pivot only have outgoing connections to children before the pivot, or to the pivot itself:*

$$\forall_{1 \leq x < y} \forall_{a_x \in \Sigma(M'_x)} \forall_{a_x \twoheadrightarrow b} \exists_{x < x' \leq y} b \in \Sigma(M'_{x'})$$

$\times(\tau \rightarrow (\ldots)).4$ *Children after the pivot are not start activities:*

$$\forall_{y < z \leq n} \forall_{a_z \in \Sigma(M'_z)} a_z \notin \text{Start}$$

$\times(\tau \rightarrow (\ldots)).5$ *Children after the pivot only have incoming connections from children after the pivot, or from the pivot itself:*

$$\forall_{y < z \leq n} \forall_{a_z \in \Sigma(M'_z)} \forall_{b \twoheadrightarrow a_z} \exists_{y \leq z' < z} b \in \Sigma(M'_{z'})$$

We refer to the set $\Sigma(M_x) \cup \Sigma(M_y) \cup \Sigma(M_z)$ as the scope of a pivot, *we refer to a scope with at least one other subtree besides the pivot as a* nontrivial *scope of a pivot, and we refer to a scope to which no other activities can be added as a* maximal scope. *Let the children before the pivot (M_x) be the* pre-scope, *and let the children after the pivot (M_z) be the* post-scope.

In the following illustration, the scope of the pivot (M_y) is denoted by a blue coloured region, M_x is the pre-scope, M_z is the post-scope and I and J are not part of the scope of M_y.

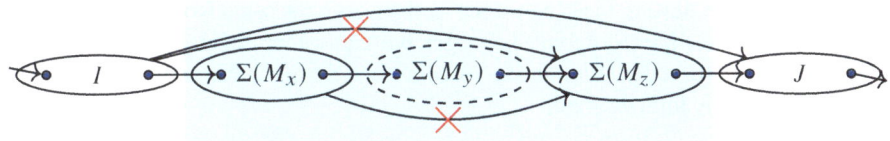

In Section 5.6.6, we prove that this footprint suffices to distinguish all so stems in process trees of C_{coo}. However, we first address the remaining challenge: distinguishing nested $\vee$, $\wedge$ and $\times(\tau, .)$.

5.6.5 Optionality under Inclusive Choice & Concurrency

Process discovery algorithms often use footprints of behaviour in an abstraction to identify language constructs. Therefore, as many languages as possible should be distinguishable by the abstraction and the footprint, i.e. for a class of process models, there should not be two different models with the same abstraction or the same footprint in the abstraction. In this chapter, we studied several abstractions and classes of process trees. In this section (5.6), we address the class of process trees that includes the $\times(\tau, .)$, i.e. optional, construct.

In the previous sections, we introduced the $\times(\tau, .)$ construct, which makes a process tree optional. Furthermore, we analysed the influence of this construct in combination with the process tree operators on directly follows graphs: under $\times$, the construct disappears in reduction, under $\rightarrow$, a stricter footprint was necessary that was introduced in Section 5.6.4, under $\leftrightarrow$, the construct doesn't have any influence on the directly follows graph, and $\circlearrowright$ has some inherent challenges and will not be considered here. In this section, we focus on $\vee$ and $\wedge$: we show that they have equivalent directly follows graphs, and hence that these graphs are not sufficient as an abstraction. We first study the properties of these *concurrent-optional-or* constructs, and introduce a new abstraction. This abstraction differs from the abstractions used before: it does not consist of a single graph or relation, but it is a hierarchical relation.

5.6.5.1 Coo Stem

The concurrent, inclusive choice and optionality constructs have equivalent directly follows graphs, hence discovery by directly follows based algorithms is impossible. For instance, the trees $\wedge$, $\vee$ and $\wedge$ all have different languages, but their

$$\begin{matrix} \wedge & \vee & \wedge \\ a\ b & a\ b & \times\ b \\ & & \wedge \\ & & \tau\ a \end{matrix}$$

directly follows graphs are equivalent: $\rightleftarrows a \rightleftharpoons b \leftrightarrows$.

Furthermore, nesting these constructs suffers from the same issue: $\wedge(a, \vee(b, c))$ and $\vee(a, \wedge(b, c))$ have different languages but equivalent directly follows graphs. For instance, Figure 5.22 shows two process trees having a different language, but the same directly follows graph.

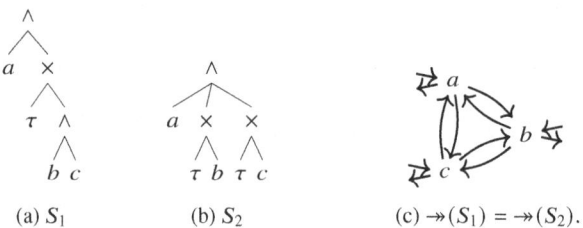

$$\text{(a) } S_1 \qquad\qquad \text{(b) } S_2 \qquad\qquad \text{(c) } \twoheadrightarrow(S_1) = \twoheadrightarrow(S_2).$$

Fig. 5.22: The $\times(\tau, .)$ construct might interfere with $\twoheadrightarrow$-footprints (Figure 5.17 revisited).

This problem occurs for any nesting of $\wedge$, $\vee$ and $\times(\tau, .)$ constructs (the *concurrent-optional-or operators*): for any such nesting, the directly follows graph contains a concurrent footprint of Lemma 5.3 of all children of the nesting. We refer to such a nesting as the *concurrent-optional-or stem*(coo-stem) of a process tree, and to the children that are not coo operator themselves as the *non-coo subtrees* of the coo stem. Intuitively, the coo stem of a process tree is the topmost part of the tree consisting of only $\wedge$, $\vee$ and $\times(\tau, .)$ constructs (similar to the so stem). For instance, Figure 5.23 shows a process tree and its coo stem; the non-coo subtrees, which are denoted in blue, are $\rightarrow(a, b)$, c, d and e. The coo stem of this tree is $\wedge(\vee(\bullet, \wedge(\bullet, \bullet)), \times(\tau, \bullet))$.

(a) A process tree M_{63}. The non-coo-subtrees are denoted in blue. (b) The coo stem of M_{63}; the dots denote non-coo subtrees.

Fig. 5.23: An example of a process tree showing its coo stem and non-coo subtrees.

Formally, we define the non-coo subtrees and the coo stem as follows:

Definition 5.14 (concurrent-optional-or stem) Let Σ be an alphabet of activities such that $\tau \notin \Sigma$ and $\bullet \notin \Sigma$, then

– activity $a \in \Sigma$ is a non-coo subtree;
– τ is a non-coo subtree;
– let $M_1 \ldots M_n$ with $n > 0$ be process trees (Definition 2.6) and let $\oplus \in \{\rightarrow, \leftrightarrow, \circlearrowleft\}$, then $\oplus(M_1, \ldots M_n)$ is a non-coo subtree;
– let $M_1 \ldots M_n$ with $n > 0$ be process trees (Definition 2.6) such that no M_i is τ, then $\times(M_1, \ldots M_n)$ is a non-coo subtree.

Furthermore,

– $\bullet$ is a coo stem;
– let $M_1 \ldots M_n$ with $n > 0$ be coo stems and let $\oplus \in \{\wedge, \vee\}$, then $\oplus(M_1, \ldots M_n)$ is a coo stem;
– let $M_1 \ldots M_n$ with $n > 1$ be coo stems, then $\times(\tau, M_1, \ldots M_n)$ is a coo stem.

A process tree M *has* a coo stem S if and only if $S \neq \bullet$ and S can be transformed into M by replacing each $\bullet$ in S with a non-coo subtree.

5.6.5.2 Preliminaries

Before we introduce the coo relations, we introduce some terminology and concepts that aid reasoning over process trees with coo stems. First, we introduce a function that gives the activities that are directly below the coo stems, grouped by their subtree (*activity sets of non-coo subtrees*). Second, we define the language of the process tree, in terms of such activity sets, i.e. if these sets were to be the activities of the process tree, what the language of that process tree would be. This language can obviously only be expressed over sets of activities that "correspond" to a process tree, which is formalised by the third function (*merge superset of coo subtrees*).

We denote the sets of activities of the non-coo subtrees with $\Sigma^{\oslash}$. Formally:

Definition 5.15 (activity sets of non-coo subtrees) Let M be a process tree in normal form of C_{coo}. Then, $\Sigma^{\oslash}(M)$ returns the activity sets of the non-coo subtrees of M:

$$\Sigma^{\oslash} : \mathbb{T} \to 2^{\Sigma}$$
$$\Sigma^{\oslash}(a) = \{\{a\}\}$$
$$\Sigma^{\oslash}(\times(\tau, \ldots)) = \Sigma^{\oslash}(\times(\ldots))$$
$$\Sigma^{\oslash}(\times(M_1, \ldots M_m)) = \{\Sigma(\oplus(M_1, \ldots M_m))\} \qquad \text{with } \forall_i M_i \neq \tau$$
$$\Sigma^{\oslash}(\oplus(M_1, \ldots M_m)) = \{\Sigma(\oplus(M_1, \ldots M_m))\} \qquad \text{with } \oplus \in \{\to, \leftrightarrow, \circlearrowleft\}$$
$$\Sigma^{\oslash}(\oplus(M_1, \ldots M_m)) = \bigcup_{1 \leq i \leq m} \Sigma^{\oslash}(M_i) \qquad \text{with } \oplus \in \{\vee, \wedge\}$$

For instance, in our example tree $M_{64} =$

$$\Sigma^{\oslash}(M_{64}) = \{\{a, b\}, \{c\}, \{d\}, \{e\}\}$$

A coo stem does not express order over its non-coo subtrees: if a subtree is executed, its execution is concurrent with all other non-coo subtrees. Furthermore, due to the absence of loops in the coo stem, each non-coo subtree is executed at most once. Thus, a coo stem only expresses which combinations of its subtrees can be executed.

To enable reasoning about the combinations of subtrees that can be executed, we introduce a language abstraction that abstracts from the subtrees. We refer to the set of such possible combinations that can be produced by a process tree as an *activity set language* ($\mathcal{L}^{\Sigma}$). Notice that this language depends on the set of sets of activities under consideration. Formally:

Definition 5.16 (activity set language ($\mathcal{L}^{\Sigma}$)) For a process tree M and a set of sets of activities S, $\mathcal{L}^{\Sigma}(M, S)$ expresses the language of M over the sets of activities in S.

Let $q(n)$ be the set of all combinations of the numbers $\{1 \ldots n\}$ without the empty combination.

$$\mathcal{L}^{\Sigma}(M, S) = \{\Sigma(M)\} \text{ if } \Sigma(M) \in S \text{ and } M \text{ not optional}$$

$$\mathcal{L}^{\Sigma}(\times(\tau, \ldots), S) = \mathcal{L}^{\Sigma}(\times(\ldots), S) \cup \{\emptyset\}$$

$$\mathcal{L}^{\Sigma}(\wedge(M_1, \ldots M_m), S) = \{X \mid \forall_{1 \leq i \leq m} A_i \in \mathcal{L}^{\Sigma}(M_i, S) \wedge X = \bigcup_{1 \leq i \leq m} A_i\}$$

$$\text{if } \bigcup_{1 \leq i \leq m} \Sigma(M_i) \notin S$$

$$\mathcal{L}^{\Sigma}(\vee(M_1, \ldots M_m), S) = \bigcup_{(i_1 \ldots i_n) \in q(m)} \mathcal{L}^{\Sigma}(\wedge(M_{i_1}, \ldots, M_{i_n}), S)$$

$$\text{if } \bigcup_{1 \leq i \leq m} \Sigma(M_i) \notin S$$

For instance, in our example tree $M_{64} =$

$\mathcal{L}^{\Sigma}(M_{64}, \{\{a, b\}, \{c\}, \{d\}, \{e\}\}) = \{\langle\{a, b\}\rangle, \qquad \langle\{c\}, \{d\}\rangle,$
$\langle\{a, b\}, \{c\}, \{d\}\rangle, \quad \langle\{a, b\}, \{e\}\rangle,$
$\langle\{c\}, \{d\}, \{e\}\rangle, \qquad \langle\{a, b\}, \{c\}, \{d\}, \{e\}\rangle\}$

A trace from an activity set language is an *activity set trace*, e.g. $\langle\{a, b\}, \{c\}, \{d\}\rangle$. We denote the removal of a set of activities A from an activity set trace with $\setminus$, e.g. $\langle\{a, b\}, \{c\}, \{d\}\rangle \setminus \{c\} = \langle\{a, b\}, \{d\}\rangle$.

Notice that the activity set language definition is a partial definition: an activity set language is only defined on sets of activities that are actually "in the tree". This final concept we formalise using the *merge superset of coo subtrees*, i.e. $\mathcal{L}^{\Sigma}$ is only defined for S if $S \in \mathbb{M}^{\Sigma}(M)$. The merge superset is obtained from $\Sigma^{\circledcirc}$ by recursively merging all activity subsets according to the process tree. Formally:

Definition 5.17 (merge superset of coo subtrees)

$$\mathbb{M}^{\Sigma}(a) = \{\{a\}\}$$

$$\mathbb{M}^{\Sigma}(\times(\tau, \ldots)) = \mathbb{M}^{\Sigma}(\times(\ldots))$$

$$\mathbb{M}^{\Sigma}(\times(M_1, \ldots M_m)) = \{\Sigma(\times(M_1, \ldots M_m))\} \text{ with } \forall_i M_i \neq \tau$$

$$\mathbb{M}^{\Sigma}(\oplus(M_1, \ldots M_m)) = \{\Sigma(\oplus(M_1, \ldots M_m))\} \text{ with } \oplus \in \{\rightarrow, \leftrightarrow, \circlearrowright\}$$

$$\mathbb{M}^{\Sigma}(\oplus(M_1, \ldots M_m)) = \{ \bigcup_{1 \leq i \leq m} \Sigma(M_i)\} \cup \{\{T_1 \ldots T_m\} | \forall_{1 \leq i \leq m} T_i \in \mathbb{M}^{\Sigma}(M_i)\}$$

$$\text{with } \oplus \in \{\vee, \wedge\}$$

For instance, in our example tree $M_{64} =$

$$\mathbb{M}^{\Sigma}(M_{64}) = \{\{\{a, b\}, \{c\}, \{d\}, \{e\}\}, \qquad \{\{a, b\}, \{c, d\}, \{e\}\},$$
$$\{\{a, b, c, d\}, \{e\}\}, \qquad \{\{a, b, c, d, e\}\}\}$$

Each activity set in the superset corresponds to a certain "view" of the process tree in which some nodes are "collapsed", e.g. in our example, the set $\{\{a, b\}, \{c, d\}, \{e\}\}$ corresponds to

5.6.5.3 Coo Relations & Abstraction

Using the concepts defined in the previous section, we define the the *coo abstraction*. This abstraction is hierarchical and contains relations for each view in the process tree, i.e. for each set of sets of activities in $\mathbb{M}^{\Sigma}$. For each such set (e.g. $\{\{a, b\}, \{c\}, \{d\}, \{e\}\}$, we define *coo relations* that relate the sets of activities. In reasoning (and algorithms in Chapter 6), we will use the relations bottom-up in the abstraction, i.e. we first reason about $\Sigma^{\otimes}$ and gradually, by merging activity sets, we obtain the singleton set Σ. Notice that the coo relations have a different nature than the activity relations introduced in Section 5.3: activity relations are defined on single activities and are applied in a top-down fashion, while the coo relations are defined on sets of activities are are applied in a bottom-up fashion.

In the remainder of this section, we first introduce these coo relations. Second, we give footprints that connect these relations to the process tree operators $\vee$ and $\wedge$. Third, we use that property to prove that the reduction rules yield a language unique normal form for the class of process trees C_{coo}. That is, we prove that for every two different reduced trees of C_{coo}, their coo abstractions are different.

We identified three coo relations on sets of activities: optionality $\overline{?}$, implication $\overline{\Rightarrow}$ and interchangeability $\overline{\vee}$. The unary optionality relation expresses that there cannot be an obligation to execute an activity of the set. Second, the binary directed implication expresses that if at least one activity of the first set is executed, then an activity of the second set should be executed as well. Finally, the binary undirected interchangeability expresses that if at least one activity of one set is executed, then that execution can be replaced by (or added to) an execution of the other set. Formally:

Definition 5.18 (coo relations) Let L be an activity set language, and let $A, B \subseteq \Sigma$ denote non-overlapping sets of activities. Then,

$$\overline{?}\, A \equiv \forall_{t \in L}\ A \in t \Rightarrow t \setminus \{A\} \in L$$

$$A \overline{\Rightarrow} B \equiv \forall_{t \in L}\ A \in t \Rightarrow B \in t$$

$$A \overline{\vee} B \equiv \forall_{t \in L}\ A \in t \vee B \in t \Rightarrow$$
$$t \cup \{A, B\} \in L \wedge (t \cup \{A\}) \setminus \{B\} \in L \wedge (t \cup \{B\}) \setminus \{A\} \in L$$

$$A \overline{\wedge} B \equiv A \overline{\Rightarrow} B \wedge B \overline{\Rightarrow} A$$

$$A \overline{\wedge?} B \equiv \overline{?}\, A \wedge A \overline{\Rightarrow} B \wedge$$
$$\neg \exists_{C \subseteq \Sigma \setminus (A \cup B)}\ (B \overline{\not\Rightarrow} C \wedge C \overline{\Rightarrow} B \wedge \forall_{a \in A \cup B, c \in C}\ a \twoheadrightarrow c \wedge c \twoheadrightarrow a)$$

Notice that we overload optionality here: before, it was defined on process trees, now on sets of activities; if and only if a process tree M is optional, then $\overline{?}(\Sigma(M))$ holds.

We illustrate the relations using our example tree $T = $

$$\Sigma^{\circledcirc}(T) = \{\{a, b\}, \{c\}, \{d\}, \{e\}\}$$
$$\mathcal{L}^{\Sigma}(T, \Sigma^{\circledcirc}(T)) = \{\{\{a, b\}\},$$
$$\{\{c\}, \{d\}\},$$
$$\{\{a, b\}, \{c\}, \{d\}\},$$
$$\{\{a, b\}, \{e\}\},$$
$$\{\{c\}, \{d\}, \{e\}\},$$
$$\{\{a, b\}, \{c\}, \{d\}, \{e\}\}\}$$

Here, $\overline{?}\{e\}$ holds, as each trace with $\{e\}$ also appears without $\{e\}$. Furthermore, $\{c\} \overline{\Rightarrow} \{d\}$ holds, as in each trace where $\{c\}$ occurs, $\{d\}$ occurs as well. Similarly, $\{d\} \overline{\Rightarrow} \{c\}$ holds, and consequently $\{c\} \overline{\wedge} \{d\}$ holds. Notice that this relation corresponds to the bottom-most $\wedge$ node in T.

As another example, merge $\{c\}$ and $\{d\}$ to obtain S_2:

$$S_2 = \{\{a,b\},\{c,d\},\{e\}\}$$
$$\mathcal{L}^{\Sigma}(T,S_2) = \{\{\{a,b\}\},$$
$$\{\{c,d\}\},$$
$$\{\{a,b\},\{c,d\}\},$$
$$\{\{a,b\},\{e\}\},$$
$$\{\{c,d\},\{e\}\},$$
$$\{\{a,b\},\{c,d\},\{e\}\}\}$$

(We don't report on $\Rightarrow$ from now on.) Here, the relation $\overline{?}\{e\}$ holds, as well as $\{a,b\}\,\overline{\vee}\,\{c,d\}$, as for each trace in which either $\{a,b\}$, $\{c,d\}$ or both occur, traces with the other two of these three options also occur. For instance, there is a trace $\{\{a,b\},\{e\}\}$, which implies that there should also be traces $\{\{c,d\},\{e\}\}$ and $\{\{a,b\},\{c,d\},\{e\}\}$. This relation corresponds to the $\vee$ node in T.

As a last example, merge $\{a,b\}$ and $\{c,d\}$ to obtain S_3:

$$S_2 = \{\{a,b,c,d\},\{e\}\}$$
$$\mathcal{L}^{\Sigma}(T,S_2) = \{\{\{a,b,c,d\}\},$$
$$\{\{a,b,c,d\},\{e\}\}\}$$

Here, the relation $\{e\}\,\overline{\wedge?}\,\{a,b,c,d\}$ holds, as $\overline{?}\{e\}$ holds and whenever $\{e\}$ appears, $\{a,b,c,d\}$ appears as well. This relation corresponds to the root $\wedge$ node in T.

5.6.5.4 Footprint

The coo relations have a one-to-one or one-to-two relationship with the process tree operators $\wedge$ and $\vee$. Therefore, the footprint of these operators is simply the presence of a coo relation: for $\vee$, this is $\overline{\vee}$, for $\wedge$ this is $\overline{\wedge}$ or $\overline{\wedge?}$.

Finally, in this section, we prove that the relations are necessary and sufficient for children of coo stem operators. We do this for both operators separately in two lemmas. In the next section, we will use these lemmas prove language uniqueness, i.e. that two different reduced process trees of C_{coo} have different directly follows graphs and/or different coo abstractions.

First, we show the correspondence for $\overline{\vee}$ and $\vee$:

Lemma 5.18 ($\overline{\vee}$ **corresponds to** $\vee$) *Let M be a reduced process tree of class C_{coo} with a coo stem, and let $M' = \vee(M'_1, \ldots M'_m)$ be one of its coo stem nodes. Take any child M'_i, and let $S \in \mathbb{M}^{\Sigma}(M)$ such that $\Sigma(M'_i) \in S$. Then for any $A \in S$ such that $A \neq \Sigma(M'_i)$, it holds that $\Sigma(M'_i)\,\overline{\vee}\,A$ if and only if $\exists_{1 \leq j \leq m}\, A = \Sigma(M'_j)$.*

Proof Prove both directions separately:

$\Leftarrow$ Take such an M'_j. By semantics of $\vee$, $\Sigma(M'_i)\,\overline{\vee}\,\Sigma(M'_j)$. Hence, children of M' are in $\vee$-relations with each other.

$\Rightarrow$ Towards contradiction, assume that there exists such a set of activities A, such that $A \overline{\vee} \Sigma(M_i')$. As $A \in S$ and $S \in \mathbb{M}^\Sigma$, A corresponds to a node in M. Let M'' be this node. Then, of M'' and M', the lowest common parent is either M' itself or a parent of M'.

- The lowest common parent is M'. Then by the assumptions made, M'' is not a direct child of M'. Furthermore, by Corollary 5.1, the direct child of M' must be a $\wedge$, and there must be a node X such that (the wiggled edge denotes M'' might be an indirect child):

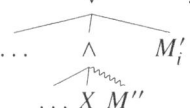

In this case, $\Sigma(M'') \overline{\Rightarrow} \Sigma(X)$, thus $\Sigma(M'')$ and $\Sigma(M_i')$ cannot be interchangeable, and therefore $A \overline{\vee} \Sigma(M_i')$.

- The lowest common parent is a parent of M'. Then, by Corollary 5.1, the parent of M' must be a $\wedge$, and there must be a node X such that

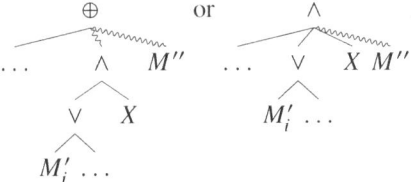

In both cases, $\Sigma(M_i') \overline{\Rightarrow} \Sigma(X)$, thus $\Sigma(M_i')$ and $\Sigma(M'')$ cannot be interchangeable, and therefore $A \overline{\vee} \Sigma(M_i')$.

Hence, such a set A cannot exist, and thus children of M' are in $\vee$-relations with each other only.

Thus, the presence of $\overline{\vee}$ is a necessary and sufficient condition for two subtrees to have a $\vee$-parent. $\square$

Second, we prove the correspondence for $\overline{\wedge}$ and $\overline{\wedge?}$, and $\wedge$.

Lemma 5.19 ($\overline{\wedge}$ corresponds to $\overline{\wedge?}$ and $\wedge$) *Let M be a reduced process tree of C_{coo} with a coo stem, and let $M' = \wedge(M_1', \dots M_m')$ be one of its coo stem nodes. Take any child M_i', and let $S \in \mathbb{M}^\Sigma(M)$ such that $\Sigma(M_i') \in S$. Then for any $A \in S$ such that $A \neq \Sigma(M_i')$, it holds that $\Sigma(M_i') \overline{\wedge} A \vee \Sigma(M_i') \overline{\wedge?} A \vee A \overline{\wedge?} \Sigma(M_i')$ if and only if $\exists_{1 \leq j \leq m} A = \Sigma(M_j')$.*

Proof Prove both directions separately:

$\Leftarrow$ Take such an M_j'. By Corollary 5.1, at most one of M_i' and M_j' is optional. If neither M_i' nor M_j' is optional, by semantics of $\wedge$, $\Sigma(M_i') \overline{\wedge} \Sigma(M_j')$. If M_i' is optional, by semantics of $\wedge$, $\Sigma(M_i') \overline{\wedge?} \Sigma(M_j')$. If M_j' is optional, by semantics of $\wedge$, $\Sigma(M_j') \overline{\wedge?} \Sigma(M_i')$. Hence, all children of M' are in either $\overline{\wedge}$ or $\overline{\wedge?}$ with each other.

⇒ Towards contradiction, assume that there exists such a set of activities A. As $A \in S$ and $S \in \mathbb{M}^{\Sigma}$, A corresponds to a node in M. Let M'' be this node. Then, of M'' and M', the lowest common parent is either M' itself or a parent of M'.

- The lowest common parent is M'. By the assumptions made, M'' is not a direct child of M'. Furthermore, by Corollary 5.1, the direct child of M' must be either $\vee$ or $\times(\tau, \wedge(\ldots))$, and there must be a node X:

$$\ldots \vee / \times(\tau, \wedge()) \ M_i' \quad \overset{\wedge}{\underset{}{\mid}}$$
$$\ldots M'' \ X$$

  In both cases, $\Sigma(M_i') \not\Rightarrow \Sigma(M'')$, therefore $\Sigma(M_i') \; \overline{\wedge} \; \Sigma(M'')$ and $\Sigma(M_i') \; \overline{\wedge?} \; \Sigma(M'')$. For every $Y \in \mathbb{M}^{\Sigma}(X)$, $Y \Rightarrow \Sigma(M_i')$, and one such Y is in S. Therefore, $\Sigma(M'') \; \overline{\wedge?} \; \Sigma(M_i')$.

- The lowest common parent is a parent of M'. Using Corollary 5.1, four cases apply as shown in Figure 5.24. In all these cases, $\Sigma(M'') \not\Rightarrow \Sigma(M_i')$. Thus,

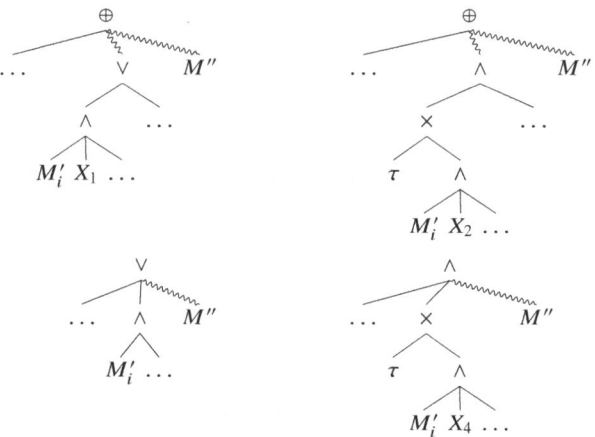

Fig. 5.24: Cases for Lemma 5.19.

$\Sigma(M_i') \; \overline{\wedge} \; \Sigma(M'')$ and $\Sigma(M'') \; \overline{\wedge?} \; \Sigma(M_i')$. Left to prove: $\Sigma(M_i') \; \overline{\wedge?} \; \Sigma(M'')$.
- In the first case, if $\Sigma(M_i') \Rightarrow \Sigma(M'')$ then for every $Y \in \mathbb{M}^{\Sigma}(X_1)$ it holds that $Y \Rightarrow \Sigma(M'')$, and at least one such Y is in S, thus $\Sigma(M_i') \; \overline{\wedge?} \; \Sigma(M'')$.
- The second case is similar to the first.
- In the third case, $\Sigma(M_i') \not\Rightarrow \Sigma(M'')$, thus $\Sigma(M_i') \; \overline{\wedge?} \; \Sigma(M'')$.
- In the fourth case, for every $Y \in \mathbb{M}^{\Sigma}(X_4)$, $Y \Rightarrow \Sigma(M'')$, and one such Y is in S, thus $\Sigma(M_i') \; \overline{\wedge?} \; \Sigma(M'')$.

Thus, the presence of $\overline{\wedge}$ or $\overline{\wedge?}$ is a necessary and sufficient condition for two subtrees to have a $\wedge$-parent. □

From these two lemmas, it straightforwardly follows that the language of non-coo subtrees is unique. That is, there exists only one reduced coo stem with the same language.

Lemma 5.20 (Coo-stem uniqueness) *There are no two reduced (Definition 5.1) coo stems $A \neq B \in C_{coo}$ such that $\mathcal{L}^{\Sigma}(A, \Sigma^{\oslash}(A)) = \mathcal{L}^{\Sigma}(B, \Sigma^{\oslash}(B))$.*

5.6.6 Language Uniqueness

In this section, we have analysed the influence of $\times(\tau, .)$ and addressed the $\vee$ operator. Two cases required special attention: $\rightarrow$ and $\vee/\wedge$. For the other operators, no changes were necessary.

As a final step, we prove language uniqueness, i.e. the one-to-one correspondence between syntax and semantics of reduced process trees, languages, and the combination of coo abstractions and directly follows graphs: each two reduced process trees from C_{COO} have a different directly follows graph or a different coo abstraction, and a different language. We prove language uniqueness by proving that for reduced trees with the same language, the root operators cannot be different (Lemma 5.21), that the root activity partitions cannot be different (Lemma 5.22), and finally that the entire trees need to be equivalent (Lemma 5.23).

Lemma 5.21 (Operators are mutually exclusive for C_{coo}) *Take two reduced process trees of C_{coo} $K = \oplus(K_1, \ldots K_n)$ and $M = \otimes(M_1, \ldots M_m)$ such that $\oplus \neq \otimes$. Then $\mathcal{L}(K) \neq \mathcal{L}(M)$.*

Proof Towards contradiction, assume that $\mathcal{L}(K) = \mathcal{L}(M)$. Then, $\rightarrow\!\!\!\rightarrow(K) = \rightarrow\!\!\!\rightarrow(M)$ and $\overline{\wedge}(K) = \overline{\wedge}(M)$. Perform case distinction on $\oplus$:

$\oplus = \times$ and one child K_i is a τ. As described before, the footprint of $\times(\tau, \ldots)$ applies whenever the root is optional. Thus, we need to consider the case in which M is optional, but does not have the $\times(\tau, .)$ construct as root. Let $K = \times(\tau, (\oplus'(K_1', \ldots K_k')))$ (or $K = \times(K_1', \ldots, K_k')$ if $\oplus = \times$) and perform case distinction on $\otimes$:

$\otimes = \times$ By semantics of $\times$, $\rightarrow\!\!\!\rightarrow(M)$ consists of unconnected clusters. As $\rightarrow\!\!\!\rightarrow(M) = \rightarrow\!\!\!\rightarrow(K)$, and by semantics of the operators, $\oplus' = \times$. At least one child (say M_j) is optional, but does not have the $\times(\tau, .)$ construct as root. Let K_i' be the corresponding child in K. Then, $\mathcal{L}(K_i') \cup \{\epsilon\} = M_j$. M_j cannot be a single activity (cannot be optional without the $\times(\tau, .)$ construct), or $\times$ (by Rule $A_\times$). For the other operators, see the other cases (termination of the argument guaranteed as K_i and M_j are strictly smaller than M).

$\otimes = \rightarrow$ By semantics of $\rightarrow$, $\rightarrow\!\!\!\rightarrow(M)$ consists of a chain of clusters. As $\rightarrow\!\!\!\rightarrow(M) = \rightarrow\!\!\!\rightarrow(K)$, and by semantics of the operators, $\oplus' = \rightarrow$. By semantics of $\rightarrow$, all children M_j are optional. By rule $T_\times$, at least one child (say K_i) is not optional. Therefore, there is a non-empty trace in $\mathcal{L}(K)$ in which no activity of $\Sigma(K_i)$ occurs. There is no such M_j, thus $\mathcal{L}(K) \neq \mathcal{L}(M)$.

$\otimes = \wedge$ By semantics of $\wedge$, all children M_j must be optional. However, by rule $C_\vee$, this situation cannot occur.

$\otimes = \vee$ By semantics of $\vee$, at least one child M_j is optional. Consider the options for M_j exhaustively: $\times(\tau, \ldots)$ (would be reduced by Rule $T_{\vee,\times}$), $\vee(\ldots)$ (would be reduced by Rule $A_\vee$), $\wedge(\ldots)$ with all children optional (would be reduced by Rule $C_\vee$), a (cannot be optional without $\times(\tau, .)$ construct), or, hence, an optional non coo subtree without $\times(\tau, .)$ as root. For the other operators, see the other cases (termination of the argument guaranteed as K_i and M_j are strictly smaller than M).

$\otimes = \leftrightarrow$ By semantics of the process tree operators, $\oplus' = \leftrightarrow$. By reduction rule $T_\times$, at least one child K_i' is not optional. By Requirement $C_{coo}.3$, all children M_i must be optional.

Take a child $K_{j \neq i}'$. Then, execution of some activity in K_j' implies execution of some activity in K_i', while there can be no child $M_{j \neq i}$ with such a dependency can exist in M_i, as $\leftrightarrow$ cannot be nested by Requirement $C_{coo}.3$. Hence, $\mathcal{L}(K) \neq \mathcal{L}(M)$.

$\otimes = \circlearrowleft$ In this case, $\circlearrowleft$ is optional and this is excluded by Requirement $C_{coo}.2$.

Hence, $\mathcal{L}(K) \neq \mathcal{L}(M)$.

$\oplus = \times$ and no child is a τ. The graph $\twoheadrightarrow(M)$ consists of several unconnected components, while as $\otimes$ is either $(\rightarrow, \wedge, \vee, \leftrightarrow, \circlearrowleft)$, $\twoheadrightarrow(M)$ is connected. Thus, $\twoheadrightarrow(K) \neq \twoheadrightarrow(M)$.

$\oplus = \rightarrow$ The graph $\twoheadrightarrow(M)$ is a chain, while as $\otimes$ is either $(\times, \wedge, \vee, \leftrightarrow, \circlearrowleft)$, $\twoheadrightarrow(M)$ is either unconnected or strongly connected. Thus, $\twoheadrightarrow(K) \neq \twoheadrightarrow(M)$.

$\oplus = \wedge$ We consider the remaining cases of $\otimes$:

  $\otimes = \vee$ The children of K have $\overline{\wedge}$ or $\overline{\wedge?}$ relations, while the children of M have $\overline{\vee}$ relations (lemmas 5.18 and 5.19). Hence, $\mathcal{L}(K) \neq \mathcal{L}(M)$.

  $\otimes = \leftrightarrow$ As shown in Section 5.6.2, optionality does not influence the footprint of $\wedge$ or $\leftrightarrow$. Therefore, Lemma 5.13 applies. Hence, $\mathcal{L}(K) \neq \mathcal{L}(M)$.

  $\otimes = \circlearrowleft$ By Requirement $C_{coo}.2$, children of $\circlearrowleft$ are not allowed to be optional. Therefore, Lemma 5.13 applies. Hence, $\mathcal{L}(K) \neq \mathcal{L}(M)$.

$\oplus = \vee$ $\vee$ has the same directly follows footprint as $\wedge$. Therefore, the arguments given at $\oplus = \wedge, \otimes = \leftrightarrow$ and $\otimes = \circlearrowleft$ apply.

$\oplus = \leftrightarrow$ We consider the remaining case of $\otimes$, being $\otimes = \circlearrowleft$.

By Requirement $C_{coo}.2$, children of $\circlearrowleft$ are not allowed to be optional. Therefore, Lemma 5.13 applies.

As $\mathcal{L}(K) \neq \mathcal{L}(M)$, two reduced process trees of C_{coo} with the same language cannot have different root operators. □

Lemma 5.22 (Partitions are mutually exclusive for C_{coo}) *Take two reduced process trees of C_{coo} $K = \oplus(K_1 \ldots K_n)$ and $M = \oplus(M_1 \ldots M_m)$ such that their activity partition is different, i.e. there is a $1 \leq w \leq n$ such that $\Sigma(K_w) \neq \Sigma(M_w)$. Then, $\mathcal{L}(K) \neq \mathcal{L}(M)$.*

Proof Without loss of generality, we assume a fixed order of subtrees for all operators. Towards contradiction, assume that $\twoheadrightarrow(K) = \twoheadrightarrow(M)$. Perform case distinction on $\oplus$ (the case for K and M swapped is symmetric).

$\oplus = \times$ If a child K_i is τ, see the proof of Lemma 5.21.
 As K is reduced, $\twoheadrightarrow(K)$ contains n unconnected clusters, corresponding to $\Sigma(K_i)$'s. These clusters themselves are connected (by Rule $A_\times$ and semantics of the other operators), hence $\twoheadrightarrow(K)$ contains a maximal $\times$ cut. The same holds for $\twoheadrightarrow(M)$, hence $\Sigma(K_w) = \Sigma(M_w)$.

$\oplus = \rightarrow$ In case no child K_i has the $\times(\tau, \rightarrow(\dots))$ structure, $\twoheadrightarrow(K)$ is a chain of strongly connected or unconnected clusters, which correspond to $\Sigma(K_i)$'s. Notice that $\twoheadrightarrow$-edges can skip clusters, hence $\twoheadrightarrow(K)$ contains a maximal $\rightarrow$ cut. The same holds for $\twoheadrightarrow(M)$.
 In case at least one child K_i has the $\times(\tau, \rightarrow(\dots))$ structure, the corresponding cluster $\Sigma(K_i)$ is a chain itself. By Rule $T_\times$, at least one child of K_i (say K_{i_p}) is a pivot according to Lemma 5.17. By semantics of $\rightarrow$, $\Sigma(K_i)$ is a pivot scope. This holds for all such $\Sigma(K_i)$, hence $\Sigma(K_w) = \Sigma(M_w)$.

$\oplus = \wedge$ In K, $\Sigma(K_w) \overline{\wedge} \Sigma(K_{v \neq w})$ or $\Sigma(K_w) \overline{\wedge?} \Sigma(K_v)$. By Lemma 5.19 and as $\mathcal{L}(K) = \mathcal{L}(M)$, $\Sigma(M_w) \overline{\wedge} \Sigma(M_{v \neq w})$ or $\Sigma(M_w) \overline{\wedge?} \Sigma(M_v)$. Hence, $\Sigma(K_w) = \Sigma(M_w)$.

$\oplus = \leftrightarrow$ Let $K_w = \otimes(K_{w_1} \dots K_{w_p})$. Perform case distinction on $\otimes$:

  $\otimes = \times$ and a child M_i is τ. The $\leftrightarrow$ operator has a distinct directly follows graph footprint, on which $\times(\tau, .)$ has no influence. Therefore, refer to the other cases as if $\otimes$ is the child of $\times(\tau, .)$, using the requirements of C_{coo}.

  $\otimes = \times$ and no child M_i is τ. By semantics of $\times$, no end activity of K_{w_1} has a connection to any start activity of any other K_{w_j}. Thus, as M contains an interleaved activity partition, $\Sigma(K_w) \subseteq \Sigma(M_w)$.

  $\otimes = \rightarrow$ Similar to the $\times$ case.

  $\otimes = \wedge$ and $\otimes = \vee$. By requirements $C_1.2$ and $C_{coo}.4$, at least one child of K_w has disjoint start and end activities. Take such a child K_{w_y}, and consider two activities: $a \notin \text{Start}(K_{w_y})$ and $b \in \Sigma(K_w) \setminus K_{w_y}$. By semantics of $\wedge$ and $\vee$, $b \twoheadrightarrow a$. Then, by Lemma 5.3, $a \in \Sigma(M_w)$ and $b \in \Sigma(M_w)$. This holds for all b and by symmetry for $\text{Start}(K_{w_y}) \cup \text{End}(K_{w_y})$. By semantics of $\leftrightarrow$, non-start non-end activities only have connections with start/end activities of K_w. Therefore, $\Sigma(K_w) \setminus (\text{Start}(K_w) \cup \text{End}(K_w)) \subseteq \Sigma(M_w)$. Hence, $\Sigma(K_w) \subseteq \Sigma(M_w)$.

  $\otimes = \leftrightarrow$ Excluded by Requirement $C_1.3$.

  $\otimes = \circlearrowleft$ By semantics of $\leftrightarrow$, non-start non-end activities only have connections with start/end activities of K_w. Therefore, $\Sigma(K_w) \setminus (\text{Start}(K_w) \cup \text{End}(K_w)) \subseteq \Sigma(M_w)$. All activities $\in \text{Start}(K_w) \cup \text{End}(K_w)$ have connections from/to $\text{End}(K_{w_2}) \cup \text{Start}(K_{w_2})$, thus $\text{Start}(K_w) \cup \text{End}(K_w) \subseteq \Sigma(M_w)$. Hence, $\Sigma(K_w) \subseteq \Sigma(M_w)$.

 By symmetry, $\Sigma(K_w) = \Sigma(M_w)$.

$\oplus = \vee$ In K, $\Sigma(K_w) \overline{\vee} \Sigma(K_{v \neq w})$. By Lemma 5.18 and as $\mathcal{L}(K) = \mathcal{L}(M)$, it holds that $\Sigma(M_w) \overline{\vee} \Sigma(M_{v \neq w})$. Hence, $\Sigma(K_w) = \Sigma(M_w)$.

$\oplus = \circlearrowleft$ By Requirement $C_{coo}.2$, children of $\circlearrowleft$ are not allowed to be optional. Therefore, Lemma 5.13 applies.

By contradiction, we conclude $\mathcal{L}(K) \neq \mathcal{L}(M)$. $\square$

Lemma 5.23 (Language uniqueness for C_{coo}) *For trees of class C_{coo}, the normal form of Definition 5.1 is language unique: for any two reduced process trees $K \neq M$ of C_{coo}, $\mathcal{L}(K) \neq \mathcal{L}(M)$.*

The proof for this lemma is similar to the proof of Lemma 5.6, using lemmas 5.21 and 5.22.

As a side effect of the proofs of lemmas 5.21, 5.22 and 5.23, we conclude that the directly follows graphs and coo relations are unique:

Corollary 5.8 (Directly follows graph and coo relations uniqueness) *There are no two different reduced process trees of C_{coo} with equal directly follows graphs and equal coo relations: for all reduced $K \neq M$ of C_{coo}, $\twoheadrightarrow(K) \neq \twoheadrightarrow(M)$ or $coo(K) \neq coo(M)$.*

In this section, we have analysed the influence of the $\times(\tau, .)$ construct and addressed the $\vee$ operator: the $\times(\tau, .)$ construct explicitly introduces the empty trace to the language of a process tree, but its influence can be larger: we introduced a stricter footprint to address nested sequences (Lemma 5.17), and we introduced the new coo abstraction and coo relations to be able to distinguish inclusive choice and concurrency, such that these can be arbitrarily nested as well. Due to Corollary 5.8, discovery algorithms can use the stricter footprint and the new abstraction to identify these constructs.

Several combinations of constructs were excluded in this section: for nested $\vee$ and $\wedge$ operators, we introduced a new abstraction, but for $\leftrightarrow$, we limited the class of process trees C_{coo}. An interesting subject of further study would be to reverse these, i.e. to identify a new abstraction to distinguish more nested $\leftrightarrow$ operators and find requirements such that the new abstraction is not necessary for $\vee$ and $\wedge$.

Future work 5.5: Identify requirements such that nested $\vee$ and $\wedge$ can be handled without coo abstractions, and identify an abstraction to identify nested $\leftrightarrow$.

Furthermore, as shown in Section 5.1, the reduction rules of Definition 5.1 are not strong enough to reduce all language equivalent trees to the same normal form, e.g. we have not yet identified rules to reduce the following trees to a common normal form:

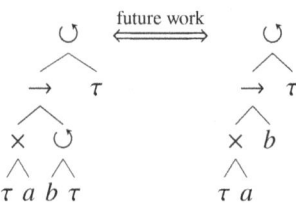

Therefore, language uniqueness of the current reduction rules does not hold for such trees.

5.7 Language Uniqueness with non-Atomic Process Models

In this chapter, we analyse abstractions of languages, which are used by discovery algorithms to not have to consider a full event log, as event logs are typically assumed to be incomplete. A desirable property of such abstractions is that they represent a large class of models, such that no two different models or languages of the class have the same abstraction. Furthermore, we studied process trees and established a one-to-one mapping between the semantics and syntax of process trees, by introducing a normal form and showing that for a large class of trees, the abstraction is unique to the normal form, i.e. there are no two trees with different normal forms and the same abstraction. Using the one-to-one mapping, discovery algorithms can focus on finding a tree for an abstraction, disregarding unimportant syntactical variations in models having the same language.

The abstractions addressed before include directly follows graphs, minimum self-distance graphs and concurrent-optional-or relations. For each of these abstractions, we proved language uniqueness for a certain class of process trees. In this section, we address a different variant of process models: *non-atomic process models*, i.e. in the following, we consider how to describe non-atomic activity executions: the execution of an activity starts at some point, and at a later point it completes, represented by two different but related steps in the model/symbols in the language.

In the event logs used in the previous parts of this chapter, each execution of an activity is *atomic*, i.e. instantaneous. However, as discussed in Chapter 2, event logs might contain information about the start and completion of executions of activities, i.e. they provide a duration and make these executions *non-atomic*. In this section, we study abstractions for non-atomic process models, i.e. process models in which the steps consist of a distinguishable start and completion step. For instance, in Petri nets, steps are atomic, i.e. the occurrence of a Petri net transition labeled with some activity a denotes an atomic and instantaneous execution of a. In this section, we extend the notions of process trees and Petri nets to include non-atomic activities, and we explore the limitations of these formalisms. Non-atomic Petri nets and process trees require adapted abstractions: we introduce an adapted directly follows graph and study its limitations. Furthermore, in non-atomic process trees, single activities in a concurrent or interleaved relation have different languages, a difference that cannot be noticed using the directly follows abstraction. Therefore, we introduce a new abstraction in which these types of behaviour have different footprints. We introduce these footprints and a class of process trees that drops a restriction of C_l, and we prove for this class that two trees having different languages also have different abstractions.

We introduce the non-atomic process model formalisms in Section 5.7.1. In Section 5.7.2, we explore the boundaries of non-atomic formalisms. We adapt the directly follows abstraction in Section 5.7.3, and introduce the new abstraction in Section 5.7.4. In Section 5.7.5, we introduce the class of trees, for which we prove language uniqueness in Section 5.7.6.

5.7.1 Non-Atomic Process Models

As shown in the previous sections, process discovery formalisms typically assume
that their execution steps are atomic. However, sometimes event logs contain more
information, i.e. in event logs adhering to the XES standard [5], events can be
annotated as being start or completion events (for more information, please refer
to Section 2.3.2). A simple strategy for a discovery algorithm would be to treat
start and completion events as different activities, i.e. for an an activity a, consider
a *start event* a_s and a *completion event* a_c as the "activities", such that likely no
further change in the algorithm is necessary. However, this poses a new challenge
for discovery algorithms: these a_s and a_c need to be related and kept in sync in
the discovered model, e.g. a model like $\rightarrow(a_c, a_s)$ would make little sense. Thus,
discovery algorithms face the challenge of keeping the start and completion events
together, such that each trace of the model is *consistent*. That is, for each activity
a in the model there should be a one-to-one mapping between a_s and a_c events,
such that each start event in the mapping appears before its mapped end event (see
Definition 2.13).

We apply a different strategy: we support non-atomic steps by using higher-level
building blocks that denote the execution of activities. These higher-level blocks
consist of low-level start and completion steps, but the discovery algorithm can
safely ignore these. If the discovery algorithm discovers these higher-level building
blocks, consistency is guaranteed as long as the translation from higher-level to
low-level constructs is consistent. A downside of this choice is that constraints such
as 'a can start as soon as b has started' cannot be expressed.

For Petri nets we use a *non-atomic transition*, as shown in Figure 5.25a. A non-
atomic transition can be directly translated to a normal Petri net construct, as shown
in Figure 5.25b. We refer to this translated net as an *expanded* net. For Petri nets, a
similar technique has been proposed in [3].

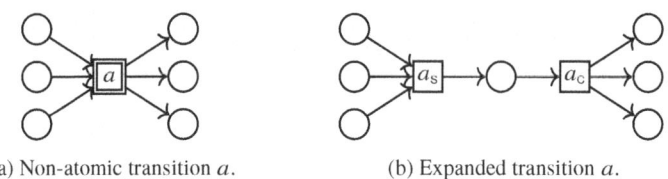

(a) Non-atomic transition a.   (b) Expanded transition a.

Fig. 5.25: Non-atomic Petri nets.

For process trees, we introduce the *non-atomic leaf* or *non-atomic activity*, which
we denote with $\tilde{a}$ for an activity a. A non-atomic leaf can be expanded into a normal
process tree construct by putting a_s a_c in sequence:

Definition 5.19 (non-atomic process tree) Let a be an activity, then

$$\tilde{a} \equiv \quad \overset{\longrightarrow}{\underset{a_{\mathrm{s}} \; a_{\mathrm{c}}}{\wedge}}$$

A process tree with non-atomic leaves is a *non-atomic process tree*.

A non-atomic process trees is inherently consistent, and denotes a process tree in which execution of activities takes time. Notice that in this section, we consider process trees without any atomic leaf. However, there is no reason why a process tree could have both atomic and non-atomic leaves.

5.7.2 Representational Bias of Non-Atomic Models

The concept of non-atomic process models brings some limitations. First, the notion of regular languages suddenly becomes very limiting: as all regular languages can be constructed using $|$, $\cdot$ and $*$, there is no concurrency. Luckily, both non-atomic Petri nets and non-atomic process trees can express concurrency and therefore are able to express more languages than non-atomic regular languages.

Second, non-atomic trees and workflow nets cannot express all expanded regular languages. As a counterexample, consider Figure 5.26. In this example, a_{s} is first executed, after which b can start concurrently. The restriction that b can start after a started is not expressible in non-atomic process models, as it inherently involves targeting the 'hidden' start in a non-atomic activity, regardless of the formalism used.

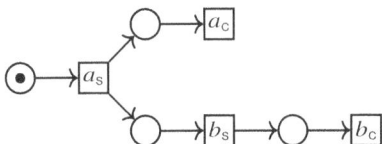

Fig. 5.26: A counterexample showing that non-atomic Petri nets are not restricted to regular languages.

Third, neither non-atomic nor expanded non-atomic process trees nor sound workflow nets can express unbounded concurrency. For instance, consider the infinite set of traces $\mathcal{L} = \{\langle a_{\mathrm{s}}, a_{\mathrm{s}}, \ldots a_{\mathrm{c}}, a_{\mathrm{c}} \rangle\}$, i.e. a is concurrent with itself arbitrarily often. The YAWL [7] language supports unbounded concurrency by means of 'multiple-instance activities'. However, correctly handling multi-instance activities requires support from the BPM system, as keeping track of the *number* of started multiple-instance activities cannot be modelled in a regular language, and hence in neither process trees nor sound workflow nets [9]. This choice for absence of unbounded concurrency also implies that a non-atomic flower model cannot exist, i.e. there is no non-atomic process tree or sound workflow net of which the language contains only and all consistent traces over an alphabet.

Fourth, the language of any non-atomic process model can obviously only contain consistent traces. Therefore, a restriction applies to fitness: on logs with inconsistent traces, perfect traditional fitness is unachievable, e.g. there is no non-atomic process model on which the trace $\langle a_s, a_s \rangle$ is fitting. Therefore, we assume that all non-atomic traces are consistent.

5.7.3 Non-Atomic Directly Follows Graphs & Footprints

Atomic directly follows graphs were not defined over start and completion events. Therefore, in this section we define directly follows graphs on non-atomic languages, such that the footprints and formal results of the previous sections apply to non-atomic languages as well.

Similar to an atomic directly follows relation, a *non-atomic directly follows relation* ($\tilde{\rightarrow}$) is a relation of activities and the special elements $\top$ denoting start of traces and $\bot$ denoting completion of traces. The part of the relation in which only activities are involved is the *non-atomic directly follows graph*. Let $\tilde{L}$ be a non-atomic language.

In a trace, an activity instance f follows an activity instance e directly if there is no full activity instance between e_c and f_s: a full activity instance is an unmatched completion event or a combination of a start and completion event of the same activity. Furthermore, an activity instance e that is not preceded by a full activity instance is a *start activity instance*. Then, a *start activity* has a start activity instance in at least one trace in $\tilde{L}$. Similarly, an activity instance after which no full activity instance of another activity occurs in a trace is an *end activity instance*. Its corresponding activity is an *end activity*.

We give an example, which is shown in Figure 5.27. Let $\tilde{L_{66}}$ be a non-atomic language consisting of the single trace $t = \langle a_s, a_s, a_c, b_s, b_c, c_s, a_c, c_c \rangle$ be a non-atomic trace. The start activities of $\tilde{L_{66}}$ are a and b, but not c as b_c precedes c_s in the only trace t. Similarly, the end activities are a and c.

Formally,

Definition 5.20 (Non-Atomic Directly follows Relation) Let $\tilde{L}$ be a consistent non-atomic language. Then,

$$a \tilde{\rightarrow} b \Leftrightarrow \exists_{t \in \tilde{L}} \ t = \langle \ldots, a_c, \ldots_1, b_s, \ldots \rangle$$
$$\text{such that } \neg\exists_d \langle \ldots d_s \ldots d_c \ldots \rangle = \ldots_1$$
$$\top \tilde{\rightarrow} a \Leftrightarrow \exists_{t \in \tilde{L}} \ t = \langle \ldots_2, a_s, \ldots \rangle$$
$$\text{such that } \neg\exists_d \langle \ldots, d_c, \ldots \rangle = \ldots_2$$
$$a \tilde{\rightarrow} \bot \Leftrightarrow \exists_{t \in \tilde{L}} \ t = \langle \ldots, a_c, \ldots_3 \rangle$$
$$\text{such that } \neg\exists_d \langle \ldots, d_s, \ldots \rangle = \ldots_3$$
$$\top \tilde{\rightarrow} \bot \Leftrightarrow \exists_{t \in \tilde{L}} \ t = \epsilon$$

Notice that it is not necessary to know which start event belongs to which completion event. We chose not to make this assumption to increase the class of event logs the techniques will be able to handle.

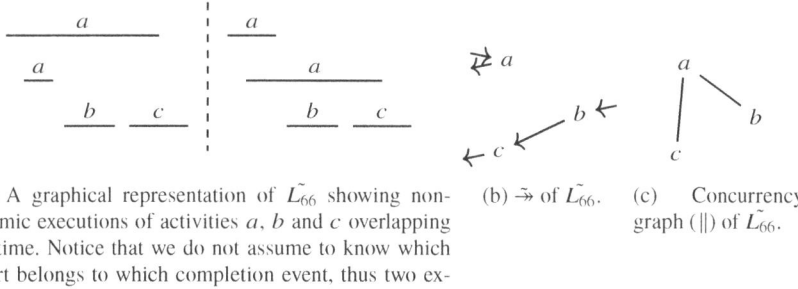

(a) A graphical representation of $\tilde{L}_{66}$ showing non-atomic executions of activities a, b and c overlapping in time. Notice that we do not assume to know which start belongs to which completion event, thus two explanations are possible.

(b) $\tilde{\rightarrow}$ of $\tilde{L}_{66}$.

(c)   Concurrency graph ($\tilde{\|}$) of $\tilde{L}_{66}$.

Fig. 5.27: Graphs of a trace $\tilde{L}_{66} = [\langle a_s, a_s, a_c, b_s, b_c, c_s, a_c, c_c \rangle]$.

The non-atomic directly follows relation resembles the atomic relation: using the mapping of non-atomic events (a) to non-atomic events (a_s followed by their a_c), the graphs of definitions 5.20 and 2.14 are equivalent. As atomic process trees only differ in leaves from non-atomic trees (and these leaves have a straightforward translation to atomic trees), all directly follows footprints for tree operators introduced earlier in this chapter (e.g. Lemma 5.10) apply as well.

A difference in semantics between atomic and non-atomic trees can be found in the $\leftrightarrow$ and $\wedge$ operators with leaves as children: in non-atomic semantics, these represent the same behaviour, while in non-atomic semantics these children can now be executed concurrently. For instance, the language of the atomic trees $\leftrightarrow(a, b)$ and $\wedge(a, b)$ is $\{\langle a, b \rangle, \langle b, a \rangle\}$, while the languages of their non-atomic counterparts are:

$$\mathcal{L}(\leftrightarrow(\tilde{a}, \tilde{b})) = \{\langle a_s, a_c, b_s, b_c \rangle, \qquad \mathcal{L}(\wedge(\tilde{a}, \tilde{b})) = \{\langle a_s, a_c, b_s, b_c \rangle,$$
$$\langle b_s, b_c, a_s, a_c \rangle\} \qquad\qquad\qquad \langle a_s, b_s, a_c, b_c \rangle,$$
$$\langle a_s, b_s, b_c, a_c \rangle,$$
$$\langle b_s, b_c, a_s, a_c \rangle,$$
$$\langle b_s, a_s, b_c, a_c \rangle,$$
$$\langle b_s, a_s, a_c, b_c \rangle\}$$

As shown before, these process trees have the same directly follows graph, and therefore no directly follows footprint can distinguish them. However, non-atomic languages provide information about concurrency as well. Next, we introduce an abstraction and footprints to benefit from this information, after which we prove that this abstraction is able to distinguish a larger class of process models than considered before.

5.7.4 Concurrency Graphs & Footprints

To distinguish interleaved and concurrent behaviour, in this section we introduce a new abstraction that describes concurrency explicitly: the *concurrency graph*. In the remainder of this section, we introduce footprints of process tree operators in the concurrency graph abstraction, and we introduce a class of process trees for which we later prove that no two different reduced trees of this class have the same directly follows graph and concurrency graph.

Definition 5.21 (Concurrency Graph) Let $\tilde{L}$ be a consistent non-atomic language. Activities a and b are *concurrent* if somewhere in $\tilde{L}$, activity instances of a and b are executed and these instances overlap in time:

$$a \parallel b \Leftrightarrow \exists_{t \in \tilde{L}} \ t = \langle \ldots a_s \ldots_1 b_s \ldots a_c \ldots \rangle$$
$$\text{such that } |[a_s \in \ldots_1]| \geq |[a_c \in \ldots_1]| \ \lor$$
$$t = \langle \ldots a_s \ldots_1 b_c \ldots a_c \ldots \rangle$$
$$\text{such that } |[a_s \in \ldots_1]| \geq |[a_c \in \ldots_1]| \ \lor$$
$$t = \langle \ldots b_s \ldots_1 a_s \ldots_2 a_c \ldots b_c \rangle$$
$$\text{such that } |[a_s \in \ldots_1 \cdot \ldots_2]| \geq |[a_c \in \ldots_1 \cdot \ldots_2]|$$

An example is given in Figure 5.27c: as a overlaps in time with b, this witnesses that a is concurrent with b ($a \parallel b$). This also illustrates that we do not need to assume knowledge of which a_s belongs to which a_c, as for the concurrency relation, it is only important that there is *an* activity instance of a being executed when b starts, not *which* activity instance of a. Similarly, $a \parallel c$, but $b \nparallel c$ as these do not overlap in time.

Notice that as we assume all traces to be consistent, $\parallel$ is commutative, i.e. $a \parallel b \Leftrightarrow b \parallel a$. We use $\parallel(\tilde{L})$ and $\parallel(M)$ for the complete concurrency relation over an non-atomic language $\tilde{L}$ or a non-atomic process tree M.

5.7.4.1 Concurrency Footprints

We first give the relevant characteristic footprints, after which we introduce a new, larger, class of process trees and prove language uniqueness for this class.

Using this concurrency graph abstraction, we introduce concurrent, interleaved and loop footprints:

Definition 5.22 (Concurrency footprints) Let $\parallel$ be a concurrency graph and let $c = (\oplus, \Sigma_1, \ldots \Sigma_n)$ be a cut, consisting of a process tree operator $\oplus \in \{\leftrightarrow, \land, \circlearrowleft\}$ and a partition of activities with parts $\Sigma_1 \ldots \Sigma_n$ such that $\Sigma(\parallel) = \bigcup_{1 \leq i \leq n} \Sigma_i$ and $\forall_{1 \leq i < j \leq n} \Sigma_i \cap \Sigma_j = \emptyset$.

- c is an *concurrent cut* if $\oplus = \land$ and

∧.1 All activities in all parts are connected to all activities in all other parts in the concurrency graph:

$$\forall_{1 \leq i < n, 1 \leq j \leq n, i \neq j} \; \forall_{a \in \Sigma(M_i), b \in \Sigma(M_j)} \; a \parallel b$$

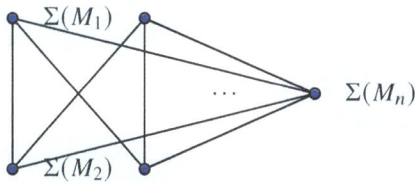

• c is an *interleaved* or *loop* cut if $\oplus = \leftrightarrow$ or $\oplus = \circlearrowright$ and

↔ ○.1 No activities have connections to other parts in the concurrency graph:

$$\forall_{1 \leq i \leq n, 1 \leq j \leq n, i \neq j} \; \forall_{a \in \Sigma(M_i), b \in \Sigma(M_j)} \; a \not\parallel b$$

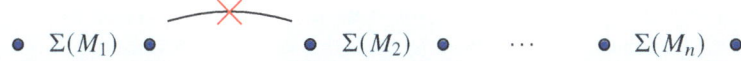

Lemma 5.24 (Concurrency footprints) *Let* $M = \oplus(M_1, \ldots M_m)$ *in which* $\oplus$ *is a process tree operator* $\in \{\times, \rightarrow, \wedge, \circlearrowright, \leftrightarrow\}$ *be a process tree for which Requirement* $C_B.2$ *holds, i.e.* M *does not contain duplicate activities. Then, the footprints of Definition 5.24 hold, i.e.* $\parallel(M)$ *contains the footprint of the cut* $(\oplus, \Sigma(M_1), \ldots \Sigma(M_n))$.

5.7.5 A Class of Trees: C_{LC}

Previously in this section we showed that directly follows graphs cannot distinguish concurrent and interleaved activities, even if these are not atomic. We introduced non-atomic process trees and non-atomic directly follows graphs, and argued that the normal, atomic, footprints are valid for these new graphs. Furthermore, we introduced the concurrency graph abstraction. In this section, we extend the class of process trees C_l for which we will prove language uniqueness, i.e. that two different reduced trees of this new class C_{lc} have different non-atomic directly follows graphs or different concurrency graphs.

Definition 5.23 (Class C_{LC}) Let M be a non-atomic process tree. Then, M belongs to C_{lc} if all requirements except Requirement $C_l.4$ hold for all reduced and expanded subtrees of M as if they were normal subtrees.

Compared to C_l, Requirement $C_l.4$ is dropped: the concurrency graph allows for the distinction of concurrent subtrees of interleaved trees.

Requirement $C_R.3$ cannot be dropped. Even though the concurrency footprints can distinguish $\circlearrowleft$ and $\wedge$ in all cases, it does not aid in distinguishing the activity partition of some $\circlearrowleft$ trees (see the discussion of the LC-property in Section 5.5.4).

5.7.6 Language Uniqueness

In this section, we have analysed languages with non-atomic activities, i.e. activities that take time and have an explicit start and completion moment. We identified the limitations of non-atomic directly follows graphs and introduced a new abstraction and footprints. In the remainder of this section, we first introduce a new normal form for these non-atomic trees. Second, we prove that two trees of C_{LC} in this new normal form have a different language, using the new concurrency footprints.

As we can now distinguish concurrency and interleaving for single activities, we need to adjust the set of reduction rules that lead to a normal form.

Definition 5.24 (Reduction rules for non-atomic process trees) The reduction rules for non-atomic trees coincide with the rules of Definition 5.1, excluding Rule $C_{\leftrightarrow}$, i.e. $\leftrightarrow(\tilde{a}, \tilde{b})$ cannot be reduced to $\wedge(\tilde{a}, \tilde{b})$.

Correctness of these rules, i.e. applying a rule will not change the language of the tree, follows from Definition 5.1: all rules except Rule $C_{\leftrightarrow}$ pose no restrictions on their subtrees, and non-atomic trees have a direct translation into atomic trees. The exception is Rule $C_{\leftrightarrow}$, which requires its subtrees to be leaves, thus this rule is not applicable to non-atomic process trees. Obviously, this set of reduction rules is still terminating, locally confluent, confluent and thus canonical, as of Corollary 5.2.

Finally, we prove that for trees of class C_{LC}, the normal form of Definition 5.24 is language unique.

Lemma 5.25 (Language uniqueness for C_{LC}) *Take two process trees of C_{LC}: $K = \oplus(K_1, \ldots K_n)$ and $M = \otimes(M_1, \ldots M_m)$ such that $K \neq M$. Then, $\mathcal{L}(K) \neq \mathcal{L}(M)$.*

Proof We prove this lemma in three steps: first, we show that if the operators $\oplus$ and $\otimes$ are different, then the abstractions of the trees are different. Second, we prove that if their activity partitions are different, then their abstractions are different. Third, we reuse Lemma 5.6 that says that for any two structurally different trees, one of the preceding cases holds.

Given the close resemblance with lemmas 5.11 and 5.12, we only need to address some cases of these lemmas:

$\oplus = \wedge$ We need to consider the cases $M_x = \circlearrowleft$ and $M_x = \leftrightarrow$. For both, $\Sigma(K_x) \subseteq \Sigma(M_x)$ follows from argumentation similar to the $\times$-case, however using the $\parallel$-relation instead of the $\twoheadrightarrow$-relation.

$\oplus = \circlearrowleft$ We only need to consider the case $\otimes = \wedge$, which trivially holds using the $\parallel$-relation.

$\oplus = \leftrightarrow$ We only need to consider the case $\otimes = \wedge$, which trivially holds using the $\parallel$-relation. $\square$

Corollary 5.9 (concurrency uniqueness) *All reduced process trees of $C_{I,C}$ have unique combinations of directly follows graphs and concurrency graphs.*

Using this final corollary, discovery algorithms can distinguish process trees of $C_{I,C}$ using the directly follows and concurrency relations. In the next section, we explore some more boundaries of non-atomic process models. The non-atomicity concept can easily be combined with the minimum self-distance concept, by only considering completion events in the computation of minimum self-distances.

Another way to derive a concurrency graph from an event log is to use the intermediate concept of partial orders, i.e. instead of considering a trace to be a fully ordered list of events, the events have predecessors and successors. Partial orders can be derived from timestamps, e.g. to resolve timestamp accuracy issues, resources, e.g. assuming resources are busy all the time, or other data fields. For more information about partial orders and their use in process discovery, please refer to [10]. To show that the analysis performed in this chapter, e.g. Corollary 5.9, holds for partially ordered traces, one could verify that concurrency graphs derived from partially ordered traces are semantically equivalent to the concurrency graphs introduced in this section.

5.8 Classes of Process Trees: Revisited

In process discovery, it is typically assumed that an event log does not contain all possible behaviour. The model should generalise and not just show the observed behaviour. Moreover, one cannot assume to have seen all possible traces. To infer a loop one need to see infinitely many traces. Moreover, concurrency constructs may generate much more traces than the number of observed traces. For instance, there are 10! = 3.628.800 possibilities to execute 10 activities concurrently, hence an event log with all behaviour would need to contain at least 3.628.800 traces. Notice that even if the log would have more traces the likelihood to have observed all traces is close to zero. Therefore, instead of considering an event log directly, many process discovery algorithms first construct an abstraction of the behaviour in the event log, and use that abstraction to discover a model. As a consequence, typically only all behaviour of the abstraction needs to be present in the event log. For instance, a directly follows graph abstraction of 10 concurrent activities has only $10 \cdot 9 = 90$ edges, so an event log with all behaviour according to this abstraction could be smaller than 90 traces instead of at least 3.628.800.

Using an abstraction allows discovery algorithms to generalise behaviour, however there might be multiple languages with the same abstraction, which therefore cannot be distinguished, and hence might reduce precision. In this chapter, we analysed what classes of languages have equivalent abstractions, and therefore cannot be discovered reliably by algorithms that use these abstractions. For several abstractions, we introduced a class of languages such that no two languages of the class have the same abstractions.

Furthermore, in this chapter we addressed the many-to-many relation between semantics and syntax of process trees: there can be many process trees with the same language, and we are not interested in the difference between two process models if they have the same language. Therefore, we introduced a set of reduction rules for process trees, such that applying these rules exhaustively yields a normal form. Ideally, these normal forms have one-to-one mapping to languages, i.e. for each language in our class of process tree languages there is precisely one process tree in normal form and vice versa. We proved this property for several classes of process trees using abstractions, i.e. we proved that two process trees in normal form have different abstractions. This establishes the close relation between the syntax of process trees in normal form, the abstraction under consideration and the semantics (i.e. the language) of process trees (as the abstractions are language based, obviously two different abstractions represent different languages).

In this section, we summarise the classes of process trees that were addressed in this chapter, illustrate the hierarchy between these classes and show a theoretical boundary to these abstractions (Corollary 5.10).

Figure 5.28 shows the classes of process trees addressed in this chapter, and shows their hierarchy. Furthermore, it shows which abstractions are sufficient such that no two process trees of the class have the same combination of abstractions, and a reference to the proof of this property.

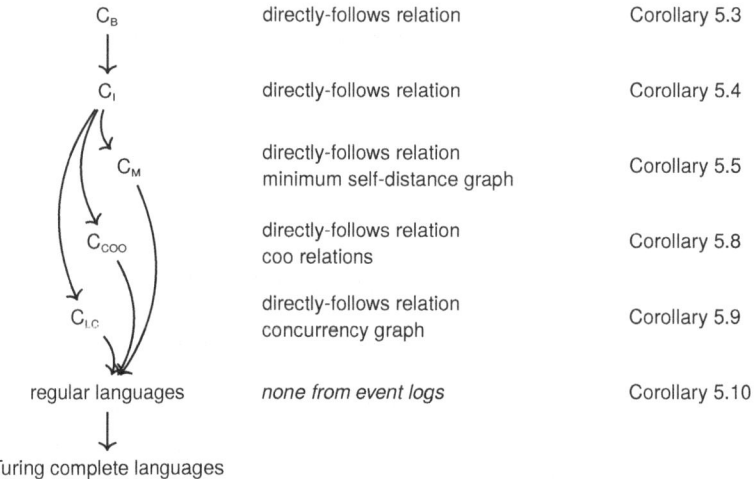

C_B	directly-follows relation	Corollary 5.3
C_I	directly-follows relation	Corollary 5.4
C_M	directly-follows relation minimum self-distance graph	Corollary 5.5
C_{coo}	directly-follows relation coo relations	Corollary 5.8
C_{LC}	directly-follows relation concurrency graph	Corollary 5.9
regular languages	*none from event logs*	Corollary 5.10

Turing complete languages

Fig. 5.28: A hierarchy of the classes of process trees used in this chapter, the abstractions that distinguish their trees and a reference to where this was proven.

The edges in Figure 5.28 denote the inclusion of classes, and follow directly from the definitions of these classes.

The class C_B consists of process trees with the four basic operators $\times$, $\rightarrow$, $\wedge$ and $\circlearrowleft$, which can be arbitrarily nested, except for concurrency and activities under loops. We included this class as it resembles the class of unlabeled free-choice Petri nets, which is used by many process discovery algorithms (e.g. α, HM). However, the class of unlabeled free-choice Petri nets is incomparable with C_B, as witnessed by Figure 5.29. Figure 5.29 illustrates that the process tree $\wedge(a, b)$, which is of C_B cannot be translated to a unlabeled free-choice Petri net: either the thread of control is split at the start of the process, which requires a silent (thus, labeled with an explicit "no label") transition, or all possible traces of the concurrency are explicitly denoted, which inherently introduces duplicated (thus labeled) transitions. Hence, the class of languages that can be expressed using process trees of C_B is not a subset of unlabeled free-choice Petri nets.

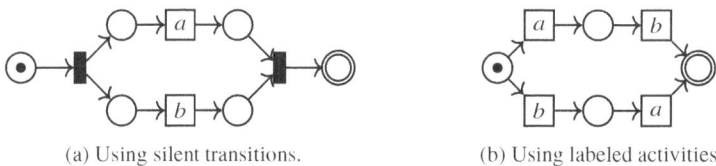

(a) Using silent transitions.   (b) Using labeled activities.

Fig. 5.29: Two Petri nets with the same language as $\wedge(a, b)$.

The other way around, Figure 5.30 shows an unlabeled free-choice Petri net. There is no process tree of C_B with the same language, as nesting of sequential and exclusive choice construct is inherently non-block-structured. Therefore, the only way to represent this Petri net in a process tree is by duplication of activities, e.g. . Hence, the class of unlabeled free-choice Petri nets is not a subset of

```
        ×
      ⌒⌒
    →     →
   ⌒    ⌒
 d c a    ×
          ⌒
        →   e
        ⌒
       b c
```

the class of languages that can be expressed using process trees of C_B, and hence, these classes are incomparable.

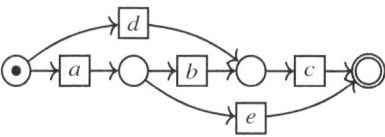

Fig. 5.30: An unlabeled free-choice Petri net, which cannot be translated to a process tree without duplicate activities.

The class C_I adds the interleaved operator to the basic four operators of C_B. As shown in Section 5.4, this operator necessitates either duplication of activities (to produce a language-equivalent Petri net) or a (non-free choice) critical section place. The class C_I does not require a larger abstraction: as for C_B, the directly follows relation suffices. However, the interleaved operator cannot be nested with itself or with concurrency, as the difference between such trees is not observable from a directly follows graph.

The following three classes of process trees (C_M, C_{COO} and C_{LC}) all extend C_I by dropping restrictions, in particular restrictions related to concurrency, thereby all requiring new abstractions. With C_M we attempted to drop the restriction of C_B that concurrency cannot be nested under loop operators. This was partially achieved: concurrent behaviour can be distinguished from loop behaviour using the minimum self-distance graph, but we did not identify a footprint to distinguish loops from loops with a different activity partition yet. We believe that such a footprint exists, but we have not found a counterexample or proof yet (Conjecture 5.1). Second, the class C_{COO} adds both the inclusive choice operator and a skip construct to C_I. The inclusive choice operator has the same directly follows footprint as the concurrency operator, thus a new abstraction is necessary. This abstraction uses the occurrences of subtrees to distinguish skipping constructs, concurrency and inclusive choices. Finally, the class C_{LC} adds non-atomic activities to C_I, i.e. activities that take time. As a consequence, concurrency becomes explicit: non-atomic activities can overlap in time. This explicit concurrency notion is captured by the concurrency graph abstraction, and used to allow concurrency nested under interleaving.

As a final step in this chapter, we show a limit to what can be achieved using event logs and language abstractions. In [4], it was proven that a regular language cannot be rediscovered from an event log by any process discovery technique, as an event log contains only traces from the system. In order to rediscover all regular languages, event logs should contain "negative" traces, i.e. traces that do not adhere to the system, and these negative traces need to be clearly marked. The unrestricted class of process trees coincides with the class of regular languages, as each regular language can be expressed using $|$, $\cdot$ and $*$ (see Section 2.2.1), which correspond to the process tree constructs $\times$, $\rightarrow$ and $\circlearrowleft(\tau, \ldots)$). Therefore, the unrestricted class of process trees cannot be rediscovered from event logs without the use of negative traces. Consequently:

Corollary 5.10 (language-uniqueness for regular languages) *If there is a language abstraction such that no two different regular languages have the same abstraction, then this abstraction cannot be derived from an event log.*

In the next chapter, we introduce several process discovery algorithms that use the abstractions presented in this chapter. Using the language uniqueness properties, we will prove that if the event log contains the entire abstraction that an algorithm uses, then the discovery algorithm rediscovers a process tree with the same abstraction.

References

1. van der Aalst, W., Weijters, A., Maruster, L.: Workflow mining: Discovering process models from event logs. IEEE Trans. Knowl. Data Eng. **16**(9), 1128–1142 (2004)
2. van der Aalst, W.M.P.: Process mining - discovery, conformance and enhancement of business processes. Springer (2011). DOI 10.1007/978-3-642-19345-3. URL http://dx.doi.org/10.1007/978-3-642-19345-3
3. Adriansyah, A.: Aligning Observed and Modeled Behavior. Ph.D. thesis, Eindhoven University of Technology (2014)
4. Gold, E.M.: Language identification in the limit. Information and Control **10**(5), 447–474 (1967). DOI 10.1016/S0019-9958(67)91165-5. URL http://dx.doi.org/10.1016/S0019-9958(67)91165-5
5. Günther, C., Verbeek, H.: XES v2.0 (2014). URL http://www.xes-standard.org/
6. Günther, C.W., Rozinat, A.: Disco: Discover your processes. In: N. Lohmann, S. Moser (eds.) Proceedings of the Demonstration Track of the 10th International Conference on Business Process Management (BPM 2012), Tallinn, Estonia, September 4, 2012, *CEUR Workshop Proceedings*, vol. 940, pp. 40–44. CEUR-WS.org (2012). URL http://ceur-ws.org/Vol-940/paper8.pdf
7. ter Hofstede, A.H.M., van der Aalst, W.M.P., Adams, M., Russell, N. (eds.): Modern Business Process Automation - YAWL and its Support Environment. Springer (2010). URL http://www.yawlbook.com/home/
8. Leemans, S.J.J., Fahland, D., van der Aalst, W.M.P.: Discovering block-structured process models from event logs - a constructive approach. In: J.M. Colom, J. Desel (eds.) Application and Theory of Petri Nets and Concurrency - 34th International Conference, PETRI NETS 2013, Milan, Italy, June 24-28, 2013. Proceedings, *Lecture Notes in Computer Science*, vol. 7927, pp. 311–329. Springer (2013). DOI 10.1007/978-3-642-38697-8_17. URL http://dx.doi.org/10.1007/978-3-642-38697-8_17
9. Linz, P.: An introduction to formal languages and automata (4. ed.). Jones and Bartlett Publishers (2006)
10. Lu, X., Fahland, D., van der Aalst, W.M.P.: Conformance checking based on partially ordered event data. In: F. Fournier, J. Mendling (eds.) Business Process Management Workshops - BPM 2014 International Workshops, Eindhoven, The Netherlands, September 7-8, 2014, Revised Papers, *Lecture Notes in Business Information Processing*, vol. 202, pp. 75–88. Springer (2014). DOI 10.1007/978-3-319-15895-2_7. URL http://dx.doi.org/10.1007/978-3-319-15895-2_7
11. Newman, M.H.A.: On theories with a combinatorial definition of "equivalence". Annals of mathematics **43**(2), 223–243 (1942)
12. Redlich, D., Molka, T., Gilani, W., Blair, G.S., Rashid, A.: Constructs competition miner: Process control-flow discovery of bp-domain constructs. In: S.W. Sadiq, P. Soffer, H. Völzer (eds.) Business Process Management - 12th International Conference, BPM 2014, Haifa, Israel, September 7-11, 2014. Proceedings, *Lecture Notes in Computer Science*, vol. 8659, pp. 134–150. Springer (2014). DOI 10.1007/978-3-319-10172-9_9. URL http://dx.doi.org/10.1007/978-3-319-10172-9_9
13. Weijters, A.J.M.M., Ribeiro, J.T.S.: Flexible heuristics miner (FHM). In: Proceedings of the IEEE Symposium on Computational Intelligence and Data Mining, CIDM 2011, part of the IEEE Symposium Series on Computational Intelligence 2011, April 11-15, 2011, Paris, France, pp. 310–317. IEEE (2011). DOI 10.1109/CIDM.2011.5949453. URL http://dx.doi.org/10.1109/CIDM.2011.5949453

Chapter 6
Discovery Algorithms

Sander J. J. Leemans: Robust Process Mining with Guarantees, LNBIP 440, pp. 215–325, 2022
https://doi.org/10.1007/978-3-030-96655-3_6

Abstract In Chapter 4, we introduced the IM framework, which recursively discovers process trees from event logs. In Chapter 5, we identified footprints of behaviour in abstractions and corresponding classes of process trees that can be uniquely identified using these footprints. In this chapter, we introduce actual process discovery algorithms by defining the four functions of the IM framework, and prove the guarantees they provide. That is, we introduce a basic algorithm (IM), an algorithm to handle deviating and infrequent behaviour (IM$_F$), an algorithm to handle incomplete behaviour (IM$_C$), an algorithm that handles silent activities, inclusive choices and interleaving operators (IM$_A$), an algorithm that handles these constructs and infrequent behaviour (IM$_{FA}$), an algorithm that handles non-atomic behaviour (IM$_{LC}$), an algorithm that handles non-atomic and infrequent behaviour (IM$_{FLC}$), and an algorithm that handles non-atomic and incomplete behaviour (IM$_{CLC}$). Furthermore, we introduce an adapted framework that performs recursion on a directly follows graph rather than a log (the Inductive Miner - directly follows framework), with several algorithms implementing it. Finally, we describe the implementations of these algorithms and provide a choose-your-miner flow diagram.

Every algorithm that implements the IM framework is characterised by four functions: one to detect cuts (i.e. process tree operators) in the event log, one to split the event log, one to handle base cases (e.g. individual events), and one as a fall through (i.e. to handle exceptional cases). In Chapter 4, we discussed several guarantees that the IM framework supports, such as perfect fitness, perfect log precision and rediscoverability. As many process discovery algorithms use an abstraction, we studied rediscoverability in terms of these abstractions. We showed that a desirable property of these abstractions is that no two models with different languages have the same abstraction (for a certain class of systems). In Chapter 5, we studied this property, *language uniqueness* (Definition 4.3), for combinations of abstractions and process trees, thereby showing the expressive power and limits of these abstractions.

In this chapter, we introduce actual process discovery algorithms. As argued in Chapter 3, a process discovery algorithm should ideally adhere to certain requirements (see sections 3.2.4 and 3.3.3):

- all discovered models should be sound (DR1),
- balance log-conformance measures (DR4),
- distinguish deviating, infrequent and incomplete behaviour (DR3),
- be fast (DR5),
- and provide rediscoverability on systems with several challenging constructs (DR2, DR6, DR7, DR8 and DR9).

Furthermore, we argued that no single discovery algorithm can satisfy all use cases and all these requirements.

Therefore, in this chapter we introduce several discovery algorithms. For each algorithm, we describe how it implements the IM framework, i.e. we first give an example, after which we describe their base case, cut detection, log splitting and fall-through functions. For each of these functions, we show whether they preserve fitness and log precision locally. We finish each algorithm with a discussion of guarantees provided by the algorithms, using the rediscoverability framework of Section 4.2.2.

We start with a basic discovery algorithm (Section 6.1), that guarantees perfect fitness, maximises log precision and guarantees rediscoverability. Rediscoverability is guaranteed for systems consisting of the four basic operators $\times$, $\rightarrow$, $\wedge$ and $\circlearrowleft$ adhering to some restrictions (i.e. from C_B), and assuming that the event log is fitting with respect to and has the same directly follows graph as the system. Even though this basic algorithm has rather strong assumptions on the input log for rediscoverability, it illustrates the principles of the IM framework, and we extend it in subsequent algorithms to handle event logs with more challenges. One such challenge is deviating or infrequent behaviour, i.e. behaviour that is not in the system model but ends up in the event log anyway (deviating behaviour) or behaviour of the system that occurs so infrequently that the user may want it to be excluded from the model, e.g. to obtain a "80% model" (infrequent behaviour). In Section 6.2, we show that deviating or infrequent behaviour might prevent rediscovery of the original system, and we show a variant of the basic algorithm that can filter out infrequent behaviour and thereby becomes suitable to more practical use cases.

Another challenge is incompleteness of information, i.e. the event log not containing enough behaviour of the system model to rediscover that system model. In Section 6.3, we show how missing information influences the basic discovery algorithm, and we show how this can be solved in the IM framework: we introduce an algorithm to deal with less information than the basic algorithm.

In Section 6.4, we show how to extend the algorithm to discover the remaining process tree operators we consider. The basic algorithm is unable to discover interleaved $\leftrightarrow$, inclusive choice $\vee$ constructs and is not proven to handle silent activity τ constructs, and we discuss extensions to discover these constructs. Furthermore, we show how the IM framework can handle non-atomic event logs (Section 6.5).

In Chapter 8, we will show that algorithms of the IM framework are efficient and capable of handling event logs with millions of events and hundreds of activities on common (anno 2016) hardware. However, the IM framework requires the event log to reside in main memory, and therefore has difficulties handling event logs of billions of events and thousands of activities. Therefore, in Section 6.6, we show that ideas and parts of the IM framework can be applied in algorithms that do not implement the full framework: we introduce an algorithm that applies a divide-and-conquer strategy on directly follows graphs, instead of on event logs. A directly follows graph can be computed by a single pass over the event log, and therefore the algorithms introduced in Section 6.6 are able to handle much larger event logs, at the cost of using less information of the event log. In our evaluation (Chapter 8), we will show that using less information manifests positively when dealing with infrequent and deviating behaviour (as using less information may decrease the influence of such behaviour), while having a negative effect on incomplete behaviour handling.

In Section 6.7, we describe the implementation of these algorithms in the ProM framework, after which the chapter is concluded in Section 6.8, in which we summarise the discussed process discovery algorithms, provide an overview when to choose which miner, and discuss the guarantees provided by each algorithm.

6.1 Inductive Miner (IM)

In this section, we introduce a basic algorithm that uses the IM framework: the *Inductive Miner* (IM). This algorithm guarantees to preserve fitness and aims to maximise precision, i.e. for any input event log, IM discovers a model that has at least the behaviour of the log. Furthermore, IM guarantees rediscoverability, i.e. is able to rediscover the language of a system, if the event log fits the system and the system and the event log have the same directly follows graph, and if the system is representable by a process tree using the four operators $\times$, $\rightarrow$, $\wedge$ and $\circlearrowright$ and adhering to some restrictions, i.e. of class C_B, which was defined in Section 5.2.1.

Rather than being a practical algorithm, IM illustrates the IM framework by providing a straightforward implementation, for which we prove local fitness preservation and rediscoverability, and that illustrates the boundaries of local log-precision preservation.

We start with an example, after which we explain how IM uses the parameters functions of the IM framework: cut detection, log splitting, base cases and fall throughs in Section 6.1.2. The section is finished with a summary of these four functions (Section 6.1.2.5) and a discussion of the guarantees provided by IM (Section 6.1.3).

6.1.1 Example

We revisit the example given in Section 4.1.3, considering the event log

$$L_{67} = [\langle a, b, c, d, e \rangle, \langle a, d, b, e \rangle, \langle a, e, b \rangle, \langle a, c, b \rangle, \langle a, b, d, e, c \rangle]$$

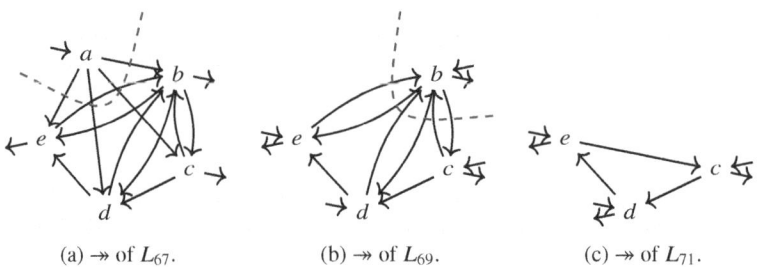

(a) $\twoheadrightarrow$ of L_{67}.   (b) $\twoheadrightarrow$ of L_{69}.   (c) $\twoheadrightarrow$ of L_{71}.

Fig. 6.1: Directly follows graphs of logs used in the recursion. The dashed red curves denote cuts.

Following the steps of the IM framework, IM first considers a base case, however as L_{67} contains multiple activities, the BASECASE$_{IM}$ function does not detect a base case in L_{67}. Second, cut detection is attempted, for which IM considers the directly follows graph abstraction, which is shown in Figure 6.1a. In the cut detection, IM looks for the footprints as defined in Definition 5.3, e.g. for L_{67}, the sequence cut $c_1 = (\rightarrow, \{a\}, \{b, c, d, e\})$ is present, as all edges cross this line in one direction (hence, the sequence). Third, using this cut, the log is split accordingly, e.g. SPLITLOG$_{IM}(L_{67}, c_1)$ splits the log in sublogs L_{68} and L_{69} as follows:

$$L_{68} = [\langle a \rangle^5]$$
$$L_{69} = [\langle b, c, d, e \rangle, \langle d, b, e \rangle, \langle e, b \rangle, \langle c, b \rangle, \langle b, d, e, c \rangle]$$

Fourth, IM records the choice and recurses, i.e. IM$(L_{67}) = \rightarrow(\text{IM}(L_{68}), \text{IM}(L_{69}))$.

We first consider the recursive step on L_{68}. As L_{68} consists of a single activity (a), BASECASE$_{IM}(L_{68})$ returns a base case, being the process tree a:

$$\text{BASECASE}_{IM}(L_{68}) = a$$

On L_{69}, no base case applies as it contains multiple activities, and thus cut detection is applied to its directly follows graph, which is shown in Figure 6.1b. In this graph, the concurrent cut $(\wedge, \{b\}, \{c, d, e\})$ is present, as all edges that could cross the dashed red line in Figure 6.1b are present. Using this cut, the log is split: SPLITLOG$_{IM}(L_{69}, c_3) = L_{70}, L_{71}$ with

$$L_{70} = [\langle b \rangle^5]$$
$$L_{71} = [\langle c, d, e \rangle, \langle d, e \rangle, \langle e \rangle, \langle c \rangle, \langle d, e, c \rangle]$$

The choice recorded is IM$(L_{69}) = \wedge(\text{IM}(L_{70}), \text{IM}(L_{71}))$.

Log L_{70} contains a single activity and is again a base case, i.e. IM$(L_{70}) = b$.

Log L_{71} does not contain a base case, and a cut cannot be found as its directly follows graph, shown in Figure 6.1c, does not contain a footprint of any of the four operators $\times$, $\rightarrow$, $\wedge$ or $\circlearrowright$. Therefore, IM applies a fall through. The fall-through function *must* return a process tree, and FALLTHROUGH$_{IM}$ has several options; IM aims to get a perfect fitness and an as high as possible log precision, thus all fall throughs preserve fitness, but the most precise one is chosen: we will discuss this in more detail in Section 6.1.2.4. For L_{71}, FALLTHROUGH$_{IM}$ takes activity d out, puts it concurrent and splits the event log, i.e. IM$(L_{71}) = \wedge(\text{IM}(L_{72}), \text{IM}(L_{73}))$, with

$$L_{72} = [\langle d^3 \rangle, \epsilon^2]$$
$$L_{73} = [\langle c, e \rangle, \langle e \rangle^2, \langle c \rangle, \langle e, c \rangle]$$

On L_{73}, no base case applies, however the cut $(\wedge, \{c\}, \{e\})$ is detected, and the log is split into L_{74} and L_{75}:

$$L_{74} = [\langle c \rangle^3, \epsilon^2]$$
$$L_{75} = [\langle e \rangle^4, \epsilon]$$

The log L_{74} contains only a single activity. However, it contains an empty trace ϵ as well, which cannot be ignored as IM aims to preserve fitness. Therefore, no base case or cut detection applies, and a fall through is chosen. This fall through denotes the possibility of skipping explicitly by discovering a model $\times(\tau, \text{IM}(L_{76}))$ and continuing the recursion on a log L_{76} from which the empty traces have been removed:

$$L_{76} = [\langle c \rangle^3]$$

Log L_{76} contains a base case, thus $\text{IM}(L_{76}) = c$. Similarly, for log L_{72} the model $\times(\tau, d)$, and for L_{75} the model $\times(\tau, e)$ is discovered. Combining all intermediate steps, IM will discover the process tree $T = \quad \rightarrow \quad$, which is optionally

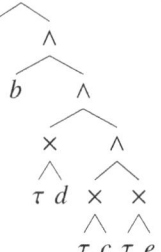

reduced to $\quad \rightarrow \quad$ using the reduction rules of Definition 5.1.

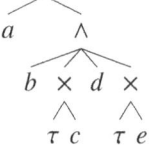

In the remainder of this section, we first introduce the algorithm formally, after which we discuss its guarantees.

6.1.2 Inductive Miner (IM)

In this section, we formally introduce IM: for each of the four parameter functions of the IM framework, we describe how they are implemented by IM. Furthermore, for each of these parameter functions, we show whether local fitness and log-precision preservation holds. We start with cut detection, after which we discuss log splitting, base cases and fall throughs.

6.1.2.1 Cut Detection

The IM searches for several cuts using the cut footprints discussed in Section 5.2.2: it attempts to find cuts in the order $\times$, $\rightarrow$, $\wedge$ and $\circlearrowright$. As soon as a non-trivial cut is encountered, that cut is returned by the FINDCUT$_{\text{IM}}$ function. We first give the pseudo code of these cuts, after which we prove their correctness and discuss their guarantees. We will prove correctness for event logs without empty traces, this assumption will be satisfied by the fall through EMPTYTRACES.

In Definition 4.1, we defined local fitness and log-precision preservation on combinations of cut finders and log splitters. Therefore, we will discuss local fitness and log-precision preservation after introducing the log splitters. However, some cut finders already disable local log-precision preservation, irrespective of the used log splitter, so these are discussed here.

The cut detection algorithms construct a partition of the alphabet of the event log: they start with the largest partition, i.e. each activity has its own set, and the algorithms repeatedly merge sets until the requirements of the particular operator footprint are met. In case the event log does not contain a footprint, then all activities will be merged and the partition will consist of a single set. The final FINDCUT$_{\text{IM}}$ function applies the following footprint detection functions, until one finds a cut with a partition consisting of multiple sets.

Exclusive Choice.

To detect an exclusive choice cut, perform the following steps:

function XORCUT($\twoheadrightarrow$)
    $\Sigma_1 \ldots \Sigma_k \leftarrow$ nodes of connected components of $\twoheadrightarrow$
    return $(\times, \Sigma_1 \ldots \Sigma_k)$
end function

We prove that XORCUT coincides with Definition 5.3:

Lemma 6.1 (XORCUT returns $\times$-cuts) *For any log L such that $\epsilon \notin L$, XORCUT($\twoheadrightarrow(L)$) returning a cut corresponds to L containing a maximal $\times$-cut according to Definition 5.3.*

Proof By construction of connected components, no connections exist between the parts corresponding to the cut's activity partition, which coincides with Requirement $\times$.1. $\qquad\square$

Sequence.

To detect a sequence cut, perform the following steps:

function SEQUENCECUT($\twoheadrightarrow$)
    $P \leftarrow \{\{a\} | a \in \Sigma(\twoheadrightarrow)\}$
    for all $a, b \in \Sigma(\twoheadrightarrow)$ **do**
        if $a \twoheadrightarrow^+ b \wedge b \twoheadrightarrow^+ a$ **then**   ▷ merge pairwise reachable nodes
            let $a \in P_x$ and $b \in P_y$, then $P \leftarrow P \setminus \{P_x, P_y\} \cup \{P_x \cup P_y\}$
        end if
        if $a \not\twoheadrightarrow^+ b \wedge b \not\twoheadrightarrow^+ a$ **then**   ▷ merge pairwise unreachable nodes
            let $a \in P_x$ and $b \in P_y$, then $P \leftarrow P \setminus \{P_x, P_y\} \cup \{P_x \cup P_y\}$

end if
end for
sort $P_1 \dots P_k$ on reachability, i.e. $P_i < P_j \Leftrightarrow \forall_{a \in P_i, b \in P_j} \, a \twoheadrightarrow^+ b$
return $(\to, P_1 \dots P_k)$
end function

We prove that SEQUENCECUT coincides with Definition 5.3.

Lemma 6.2 (SEQUENCECUT returns $\to$-cuts) *For any log L such that $\epsilon \notin L$, it holds that if SEQUENCECUT$(\twoheadrightarrow(L))$ returns a cut, then this cut corresponds a maximal $\to$- cut in L according to Definition 5.3.*

Proof For each pair of activities a and b from different P_i and P_j, Requirement $\to$.1 states that $a \twoheadrightarrow^+ b \not\Leftrightarrow b \twoheadrightarrow^+ a$. This corresponds to the two commented checks in SEQUENCECUT, i.e. the two cases of merging sets of activities when this condition does not hold.   □

The SEQUENCECUT function can be locally fitness preserving (with a proper log splitting function), but not locally log-precision preserving, as for some event logs, extra behaviour might be introduced. For instance, consider the event log in Figure 6.2. This log contains a so-called *long-distance dependency*, i.e. the choice between d and e depends on the choice between a and b. For this log L_{77}, the function SEQUENCECUT$(\twoheadrightarrow(L))$ returns the cut $c = (\to, \{a, b\}, \{c\}, \{d, e\})$. (Eventually, the process tree $\to(\times(a, b), c, \times(d, e))$ will be discovered.) However, the long-distance dependency is not captured by this cut, thus precision is not preserved locally, e.g. $\langle a, c, e \rangle$ will be part of the discovered model, while it is not present in L_{77}. A solution to this problem would be to only report a $\to$-cut after verifying that choosing this cut will not introduce new behaviour. For efficiency considerations, we did not include such a step in the IM algorithm.

$L_{77} = [\langle a, c, d \rangle, \langle b, c, e \rangle]$

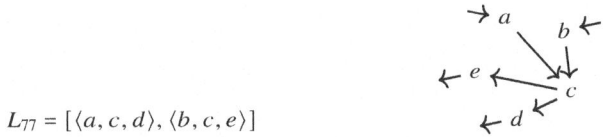

Fig. 6.2: A log and its directly follows graph.

Concurrency.

To detect a concurrent cut, perform the following steps:
function CONCURRENTCUT$(\twoheadrightarrow, \ominus)$
    $P \leftarrow \{\{a\} | a \in \Sigma(\twoheadrightarrow)\}$
                                                            ▷ merge not-fully connected sets
    for all $a, b \in \Sigma(\twoheadrightarrow), a \neq b$ **do**

        if $a \not\rightarrow b \vee b \not\rightarrow a$ **then**

              let $a \in P_x$ and $b \in P_y$, then $P \leftarrow P \setminus \{P_x, P_y\} \cup \{P_x \cup P_y\}$

        else if $a \ominus b \vee b \ominus a$ **then**

              let $a \in P_x$ and $b \in P_y$, then $P \leftarrow P \setminus \{P_x, P_y\} \cup \{P_x \cup P_y\}$

        end if

      end for

                                       ▷ merge sets without start or end activities

      for all $C \in P$ **do**

        if $C \cap \mathrm{Start}(\twoheadrightarrow) = \emptyset \vee C \cap \mathrm{End}(\twoheadrightarrow) = \emptyset$ **then**

              merge C with an arbitrary other set in P

        end if

      end for

      return $(\wedge, P)$

    end function

We prove that concurrentCut coincides with definitions 5.3 and 5.9.

Lemma 6.3 (concurrentCut returns ∧-cuts) *For any log L such that $\epsilon \notin L$, concurrentCut$(\twoheadrightarrow(L), \ominus(L))$ returns a cut $(\wedge, \Sigma_1, \dots \Sigma_n)$ according to definitions 5.3 and 5.9.*

Proof The second for-loop of concurrentCut coincides with Requirement ∧.1; the first if in the first for-loop coincides with Requirement ∧.2; and the second if coincides with Requirement $\wedge \leftrightarrow$.1.   □

For local log-precision preservation, an argument similar to sequenceCut holds, i.e. as the directly follows graph cannot capture the full behaviour of loops, an extension is necessary to preserve log precision locally.

Loop.

To detect a loop cut, perform the following steps:

    function loopCut$(\twoheadrightarrow)$

      $P_1 \leftarrow \mathrm{Start}(\twoheadrightarrow) \cup \mathrm{End}(\twoheadrightarrow)$ ▷ Requirement ↻.1

                                        ▷ Requirement ↻.3

      $P_2 \dots P_n \leftarrow$ maximal partition of $\Sigma(\twoheadrightarrow) \setminus P_1$ such that $\forall_{2 \le i < j \le n, a \in P_i, b \in P_j}\ a \not\rightarrow b$

      $P \leftarrow P_1 \dots P_n$

                        ▷ exclude sets that are connected from a start activity

      for all $a \in \mathrm{Start}(\twoheadrightarrow) \setminus \mathrm{End}(\twoheadrightarrow)$ **do**

        for all b such that $a \twoheadrightarrow b$ **do**

              let $b \in P_y$, then $P \leftarrow P \setminus \{P_1, P_y\} \cup \{P_1 \cup P_y\}$

        end for

      end for

                                     ▷ exclude sets that are connected to an end activity

for all $b \in \mathrm{End}(\twoheadrightarrow) \setminus \mathrm{Start}(\twoheadrightarrow)$ **do**

    for all a such that $a \twoheadrightarrow b$ **do**

        let $a \in P_x$, then $P \leftarrow P \setminus \{P_x, P_1\} \cup \{P_x \cup P_1\}$

    end for

end for

                     ▷ sets should have all connections (Requirement $\circlearrowleft$.4)

for all $2 \leq i \leq n, a \in P_i$ **do**

    if $\exists_{b \in \mathrm{Start}(\twoheadrightarrow)}\ a \twoheadrightarrow b \land \neg\forall_{b \in \mathrm{Start}(\twoheadrightarrow)}\ a \twoheadrightarrow b$ **then**

        let $a \in P_x$, then $P \leftarrow P \setminus \{P_x, P_1\} \cup \{P_x \cup P_1\}$

    end if

    if $\exists_{b \in \mathrm{End}(\twoheadrightarrow)}\ b \twoheadrightarrow a \land \neg\forall_{b \in \mathrm{End}(\twoheadrightarrow)}\ b \twoheadrightarrow a$ **then**

        let $a \in P_x$, then $P \leftarrow P \setminus \{P_x, P_1\} \cup \{P_x \cup P_1\}$

    end if

end for

return $(\circlearrowleft, P_1, \ldots P_n)$

end function

We prove that LOOPCUT coincides with Definition 5.3.

Lemma 6.4 (LOOPCUT returns $\circlearrowleft$-cuts) *For any log L such that $\epsilon \notin L$, LOOPCUT($\twoheadrightarrow(L)$) returning a cut corresponds to L containing a maximal $\circlearrowleft$-cut according to Definition 5.3.*

Proof Requirements $\circlearrowleft$.1, $\circlearrowleft$.3 and $\circlearrowleft$.4 coincide with the parts denoted in the pseudocode. Requirement $\circlearrowleft$.2 coincides with the two remaining for-loops. Thus, the remaining non-first sets $P_2 \ldots P_n$ are the redo parts.   □

Local fitness preservation will be discussed in combination with the log-splitting functions in the next section. Local log-precision preservation is not possible for loops, as discussed in Section 4.1.4.2.

Local Log-Precision Preservation.

The log-precision discussions of the functions SEQUENCECUT and CONCURRENTCUT show the limitations of using an abstraction: a directly follows graph does not contain as much information as an event log, thus the directly follows based cut detection techniques cannot guarantee to not introduce extra behaviour. However, the IM framework is not limited to directly follows based techniques: one could easily define cut detection techniques that use the entire log and, as discussed before, preserve log precision locally. The only exception to this is the $\circlearrowleft$-operator that, as discussed in Section 4.1.4.2, cannot guarantee to preserve log precision, as it describes unbounded behaviour while the event log is always bounded.

6.1.2.2 Log Splitting

After finding a cut, the IM framework splits the log into several sub-logs, on which recursion continues. The IM algorithm uses several log splitting functions. For each of these log splitting functions, we give pseudocode, an example and we prove their local guarantees.

Exclusive Choice.

To split a log L according to an exclusive choice cut, IM puts each trace in its respective sublog:

function xorSplit($L, (\times, \Sigma_1, \ldots, \Sigma_n)$)
    $\forall i : L_i \leftarrow [t \in L | \forall e \in t : e \in \Sigma_i]$
    return $L_1, \ldots, L_n$
end function

For instance, the log $L = [\langle a, b \rangle, \langle c, c, c \rangle]$ would be split using the cut $(\times, \{a, b\}, \{c\})$ into $[\langle a, b \rangle], [\langle c, c, c \rangle]$. Due to the cut detection of xorCut, all cuts to which xorSplit is applied adhere to Definition 5.3 and consequently, each trace contains events of at most one $\Sigma_{1 \le i \le n}$.

Lemma 6.5 (xorCut & xorSplit are locally fitness preserving) *Let L be a log and $c = (\oplus, \Sigma_1, \ldots \Sigma_n)$ be a non-trivial exclusive choice cut (Definition 5.3). Then $set(L) \subseteq \times_{\mathcal{L}}(xorSplit(L, c))$ (see Definition 2.7).*

Proof Let $L_1, \ldots, L_n$ be the result of xorSplit(L, c), i.e. the split logs. By construction of xorSplit and the fact that $\bigcup_i \Sigma_i = \Sigma(L)$, every $t \in L$ is in at least one L_i. Hence, $set(L) \subseteq \bigcup_i L_i$ and thus by semantics of $\times$, $set(L) \subseteq \times_{\mathcal{L}}(xorSplit(L, c))$.□

Lemma 6.6 (xorSplit is locally log-precision preserving) *Let L be a log. Then $\times_{\mathcal{L}}(xorSplit(L, c)) \subseteq set(L)$ for any cut c.*

Proof Let $L_1, \ldots, L_n$ be the result of xorSplit(L, c), i.e. the split logs. Pick a trace t in any L_i. By construction of xorSplit, $t \in L$. Hence, $\times_{\mathcal{L}}(xorSplit(L, c)) \subseteq set(L)$.   □

Sequence.

To split a log L according to a sequence cut, IM searches for the split points in each trace:

function sequenceSplit($L, (\rightarrow, \Sigma_1, \ldots, \Sigma_n)$)
    $\forall j : L_j \leftarrow [t_j | t_1 \cdot t_2 \cdots t_n \in L \land \forall i \le n \land \Sigma([t_i]) \subseteq \Sigma_i]$
    return $L_1, \ldots, L_n$
end function

For instance, the log $L = [\langle a,b,c\rangle, \langle b,a,c\rangle]$ would be split using the cut $(\rightarrow, \{a,b\}, \{c\})$ into $[\langle a,b\rangle, \langle b,a\rangle], [\langle c\rangle^2]$. Similar to xorCut, all cuts to which sequenceSplit is applied adhere to Definition 5.3.

Lemma 6.7 (sequenceCut & sequenceSplit are locally fitness preserving) *Let L be a log and $c = (\rightarrow, \Sigma_1, \dots \Sigma_n)$ be a non-trivial sequence cut (Definition 5.3). Then $set(L) \subseteq \rightarrow_L(\textsc{sequenceSplit}(L, c))$ (see Definition 2.8).*

Proof Let $L_1, \dots, L_n$ be the result of sequenceSplit(L, c), i.e. the split logs. Pick a trace $t \in L$. Divide $t = z_1 \cdot t_1 \cdot t_2 \cdots t_n \cdot z_2$ such that $\forall i : \Sigma([t_i]) \subseteq \Sigma_i$ and both z_1 and z_2 are as small as possible. For $|t| = 0$ and $|t| = 1$, z_1 and z_2 are trivially empty. Towards contradiction, assume $|t| > 1$ and $z_1 \neq \epsilon \vee z_2 \neq \epsilon$. Then there must be two activities a_i and a_{i+1} somewhere in t with $a_i \in \Sigma_k, a_{i+1} \in \Sigma_l$ and $k > l$. By definition of $G(L)$, $a_i \twoheadrightarrow^+ a_{i+1}$ in L and therefore, by Definition 5.3, $l \leq k$. Hence, both z_1 and z_2 must be empty and t can be written as $t_1 \cdot t_2 \cdots t_n$ such that $\forall i : \Sigma(\{t_i\}) = \Sigma_i$. By construction of sequenceSplit and semantics of $\rightarrow$, $t \in \rightarrow_L(L_1, \dots, L_n)$ and hence $set(L) \subseteq \rightarrow_L(\textsc{sequenceSplit}(L, c))$. $\qquad\square$

In Section 6.1.2.1, we showed that the sequenceCut is not locally log-precision preserving. Therefore, neither the combination of sequenceCut and sequenceSplit is locally log-precision preserving.

Concurrency.

To split a log L according to a concurrent cut, IM divides events over their corresponding subtraces:

    function concurrentSplit$(L, (\wedge, \Sigma_1, \dots, \Sigma_n))$
        $\forall i : L_i \leftarrow [t|_{\Sigma_j} | t \in L]$
        return $L_1, \dots, L_n$
    end function

For instance, the log $L = [\langle a,b,c\rangle, \langle a,c,b\rangle, \langle c,a,b\rangle]$ would be split using the cut $(\wedge, \{a,b\}, \{c\})$ into $[\langle a,b\rangle^3], [\langle c\rangle^3]$.

Lemma 6.8 (concurrentSplit is locally fitness preserving) *Let L be a log. Then $set(L) \subseteq \wedge_L(\textsc{concurrentSplit}(L, c))$ for any cut c.*

Proof Pick a trace $t \in L$. By definition of concurrentSplit, each L_i contains a t_i, being the projection of t to Σ_i. Obviously, for each t there is a corresponding trace in $t_1 \sqcup\!\sqcup t_2 \dots t_n$. Hence, $set(L) \subseteq \wedge_L(L_1, \dots, L_n)$ and thus $set(L) \subseteq \wedge_L(\textsc{concurrentSplit}(L, c))$. $\qquad\square$

Loop

To split a log L according to a loop cut, IM starts a new trace whenever it detects that execution left a Σ_i:

function LOOPSPLIT$(L, (\circlearrowleft, \Sigma_1, \ldots, \Sigma_n))$
$\quad \forall i : L_i \leftarrow [t_2 | t_1 \cdot t_2 \cdot t_3 \in L \wedge$
$\qquad\qquad \Sigma([t_2]) \subseteq \Sigma_i \wedge$
$\qquad\qquad (t_1 = \epsilon \vee (t_1 = \langle \cdots, a_1 \rangle \wedge a_1 \notin \Sigma_i)) \wedge$
$\qquad\qquad (t_3 = \epsilon \vee (t_3 = \langle a_3, \cdots \rangle \wedge a_3 \notin \Sigma_i))]$
$\quad$**return** $L_1, \ldots, L_n$
end function

For instance, the log $L_{78} = [\langle a, b \rangle, \langle a, b, c, a, b \rangle, \langle a, b, c, a, b, c, a, b \rangle]$ would be split using the cut $(\circlearrowleft, \{a, b\}, \{c\})$ into $L_{79} = [\langle a, b \rangle^6]$ and $L_{80} = [\langle c \rangle^3]$, as illustrated in Figure 6.3.

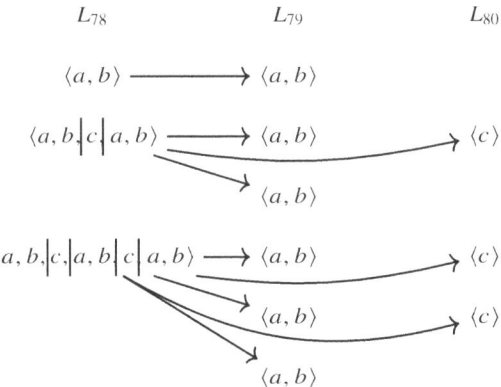

Fig. 6.3: Example of log splitting for a $\circlearrowleft$-cut.

Lemma 6.9 (LOOPSPLIT is locally fitness preserving) *Let L be a log. Then* $set(L) \subseteq \circlearrowleft_{\mathcal{L}}(\text{LOOPSPLIT}(L, c))$ *for any cut c.*

Proof Pick a trace $t \in L$. Apply case distinction on whether t consists exclusively of activities in Σ_1:

$\Sigma([t]) = \Sigma_1$ By construction of LOOPSPLIT, L_1 contains t.
$\Sigma([t]) \neq \Sigma_1$ By Definition 5.3, $Start(L) \cup End(L) \subseteq \Sigma_1$ and therefore there exist
$\quad$ t_i such that $t = t_1 \cdot t_2 \cdots t_{2m+1}$, such that $\forall j : \Sigma(\{t_{2j+1}\}) = \Sigma_1$. Definition 5.3
$\quad$ guarantees that no $t_{2m'}$ contains activities from two different Σ_i. Then, LOOPSPLIT
$\quad$ puts all t_{2m} in some $L_{i \neq 1}$ intact and all L_{2m+1} in L_1 intact.

By semantics of $\circlearrowleft, t \in \circlearrowleft_{\mathcal{L}}(L_1, \ldots, L_n)$ and hence $set(L) \subseteq \circlearrowleft_{\mathcal{L}}(L_1, \ldots, L_n)$.$\square$

Local Guarantees.

Fitness and log-precision preservation are defined on combinations of cut finders and log splitters (Definition 4.1). Table 6.1 summarises the local guarantee lemmas and the descriptions in Section 6.1.2.1.

Table 6.1: Local guarantees provided by the cut detection and log splitting functions of IM.

	locally fitness preserving	locally log precision preserving
xorCut & Split	yes (Lemma 6.5)	yes (Lemma 6.6)
sequenceCut & Split	yes (Lemma 6.7)	when extended
concurrentCut & Split	yes (Lemma 6.8)	when extended
loopCut & Split	yes (Lemma 6.9)	no (see Section 4.1.4.2)

6.1.2.3 Base Cases

The base cases for the IM algorithm provide an end to the recursion. IM implements the baseCase$_{IM}$-function of the IM framework using several steps: it tries several base cases and returns the first matching one. As the base cases are mutually exclusive, so their order is irrelevant. We distinguish several cases:

- emptyLog applies when the log contains no traces, i.e. $set(L) = \emptyset$. The only thing any discovery algorithm could do is return τ, i.e. the model consisting of an empty step.
- singleActivity applies when the event log contains only traces with a single activity, i.e. $\Sigma = \{a\} \wedge \forall_{t \in L} |t| = 1$ for some activity a. This activity a is returned as a leaf.

All of these base cases obviously preserve both fitness and precision (Definition 4.1):

	locally fitness preserving	locally log precision preserving
emptyLog	yes	yes
singleActivity	yes	yes

6.1.2.4 Fall Throughs

For some input event logs, no base case applies (if the log contains multiple activities) and no cut applies (because e.g. the directly follows graph is not complete, the system is not from C_B, the log contains empty traces, or the log contains deviating behaviour,

...). However, IM should return a process tree under all circumstances, hence a *fall through* needs to be selected. As a last resort a *flower model* can be returned, i.e. a model that allows for any behaviour over a given set of activities. However, such a flower model would have a bad precision. Therefore, we identified some patterns that could improve precision over a flower model. The fall-through function of IM, FALLTHROUGH$_{IM}$, applies these patterns in order until one matches.

As this is the last resort for the IM framework, the final one, i.e. FLOWERMODEL, always applies. We introduce each fall through and illustrate each of them using the same log, to which no cut applies:

$$L_{81} = [\langle a, b, c, d \rangle, \langle d, a, b \rangle, \langle a, d, c \rangle, \langle b, c, d \rangle]$$

- EMPTYTRACES applies when the event log contains empty traces, i.e. if $\epsilon \in L$. To preserve fitness, the empty traces are accounted for using $\times(\tau, \ldots)$, and recursion continues on a log without the empty traces, i.e. the model $\times(\tau, \text{IM}(L \setminus \{\epsilon\}))$ is returned. (notice that EMPTYTRACES does not apply to our example log L_{81}).
- ACTIVITYONCEPERTRACE applies when an activity a appears precisely once in every trace of the log L. Then, this fall through discovers that activity a can be put concurrent to L', which is obtained by filtering a from L, i.e. $\wedge(a, \text{IM}(L'))$. In case this applies to multiple activities a, an arbitrary one is chosen.
  For instance, in L_{81}, d appears once in every trace. Thus, d is filtered out of L_{81} and a new event log L_{82} is obtained:

$$L_{81} = [\langle a, b, c, \cancel{d} \rangle, \langle \cancel{d}, a, b \rangle, \langle a, \cancel{d}, c \rangle, \langle b, c, \cancel{d} \rangle]$$
$$L_{82} = [\langle a, b, c \rangle, \langle a, b \rangle, \langle a, c \rangle, \langle b, c \rangle]$$

Then recursion continues on L_{82} and the tree $\wedge(d, \text{IM}(L_{82}))$ is discovered.
- ACTIVITYCONCURRENT leaves out an activity a from the log L and tries to find a cut. If this succeeds, a is put concurrently to L', which is obtained by filtering a from L, and recursion continues on L'. In case this applies to multiple activities, an arbitrary one is chosen.
  For instance, in L_{81}, activity d is filtered out and logs L_{83} and L_{84} are obtained:

$$L_{81} = [\langle a, b, c, \underline{d} \rangle, \langle \underline{d}, a, b \rangle, \langle a, \underline{d}, c \rangle, \langle b, c, \underline{d} \rangle]$$
$$L_{85} = [\langle d \rangle^4]$$
$$L_{86} = [\langle a, b, c \rangle, \langle a, b \rangle, \langle a, c \rangle, \langle b, c \rangle]$$

This log contains the non-trivial cut $(\rightarrow, \{a\}, \{b\}, \{c\})$. Thus, the tree $\wedge(\text{IM}(L_{85}), \text{IM}(L_{88}))$ is discovered and recursion continues on L_{85} and L_{88}.
The fall through ACTIVITYCONCURRENT potentially leads to a lower log precision than ACTIVITYONCEPERTRACE: if ACTIVITYCONCURRENT selects an activity (e.g. a) that is executed multiple times in one of the traces (e.g. $\langle a, b, a \rangle$), recursive calls will put this activity in a loop (e.g. $\circlearrowleft(a, \tau)$) and thereby lower log precision. This is a rather expensive fall through, as the event log is split repeatedly and after each split cut finding is applied. Even though our implementation executes these

calls in parallel, we observed that in large event logs with lots of activities, this fall through is the most time-consuming task of the IM algorithm. However, this fall through is the last fall through that does not introduce unbounded behaviour (i.e. loops), which would cause a lower log precision.

- STRICTTAULOOP applies when looping behaviour is present. To verify this, each trace of the log L is split on each occurrence of an end activity followed by a start activity, and the result is stored in a log L'. If L' has more traces than L, i.e. at least one trace was split, a tau loop, i.e. $\circlearrowleft(\text{IM}(L'), \tau)$, is discovered and the recursion continues.

  For instance, the start activities of L_{81} are $\{a, b, d\}$, the end activities $\{b, c, d\}$. Therefore, L_{81} is split into L_{87}:

  $$L_{81} = [\langle a, b, c \mid d \rangle, \langle d \mid a, b \rangle, \langle a, d, c \rangle, \langle b, c, \mid d \rangle]$$
  $$L_{88} = [\langle a, b, c \rangle, \langle a, b \rangle, \langle a, d, c \rangle, \langle b, c \rangle, \langle d \rangle^3]$$

  and the model $\circlearrowleft(\text{IM}(L_{88}), \tau)$ is discovered and recursion continues.

- TAULOOP applies when looping behaviour is present. To verify this, each trace of the log L is split on every occurrence of a start activity, and the result is stored in a log L'. If L' has more traces than L, i.e. at least one trace was split, a tau loop, i.e. $\circlearrowleft(\text{IM}(L'), \tau)$, is discovered and the recursion continues.

  For instance, the start activities of L_{81} are $\{a, b, d\}$. Therefore, L_{81} is split into L_{89} as follows:

  $$L_{81} = [\langle a \mid b, c \mid d \rangle, \langle d \mid a \mid b \rangle, \langle a \mid d, c \rangle, \langle b, c, \mid d \rangle]$$
  $$L_{89} = [\langle a \rangle^3, \langle b, c \rangle, \langle d \rangle^3, \langle b \rangle, \langle d, c \rangle, \langle b, c \rangle]$$

  the model $\circlearrowleft(\text{IM}(L_{89}), \tau)$ is discovered and recursion continues. Notice the difference between TAULOOP and STRICTTAULOOP: STRICTTAULOOP leaves longer subtraces in the log. This might preserve information in the event log and thus potentially increase log precision.

- FLOWERMODEL applies to an event log without empty traces, i.e. $\epsilon \notin L$. Given the activities of the event log $\Sigma(L)$, it returns the model that allows for any behaviour without ϵ: $\circlearrowleft(\times(a_1, \ldots a_n), \tau)$ with $a_1 \ldots a_n = \Sigma(L)$.

  For instance, given log L_{81} this fall through discovers the model $\circlearrowleft(\times(a, b, c, d), \tau)$.

We illustrated these fall throughs using a single example event log to illustrate the loss of log precision that these fall throughs imply. The partial models that would be returned for L_{81} are:

$$\text{EMPTYTRACES} \quad \times(\tau, \ldots)^1$$
$$\text{ACTIVITYONCEPERTRACE} \quad \wedge(d, \ldots)$$
$$\text{ACTIVITYCONCURRENT} \quad \wedge(\ldots, \ldots)$$
$$\text{STRICTTAULOOP} \quad \circlearrowleft(\ldots, \tau)$$
$$\text{TAULOOP} \quad \circlearrowleft(\ldots, \tau)$$
$$\text{FLOWERMODEL} \quad \circlearrowleft(\times(a, b, c, d), \tau)$$

The order in which the fall throughs are applied was chosen to preserve log precision as much as possible: the last resort (FLOWERMODEL) allows for any behaviour except the empty trace and therefore has almost the lowest precision possible, TAULOOP and STRICTTAULOOP introduce a loop and remove a lot of information from the event log but at least continue the recursion, ACTIVITYCONCURRENT sacrifices log precision of one activity (in our example: d) to continue the recursion normally on the other activities, and ACTIVITYONCEPERTRACE applies this to the special case that an activity appears once in every trace.

For completeness, we describe two more fall throughs: the first is the TRACEMODEL that locally preserves log precision: TRACEMODEL applies to any event log L, and returns a *trace model*, i.e. the choice between sequences corresponding to all traces. For instance, if $L = [\langle a, b \rangle, \langle a, c, b \rangle]$, then TRACEMODEL$(L) = \times(\rightarrow(a, b), \rightarrow(a, c, b))$. We chose to not include this fall through in IM as a trace model has a poor generalisation and simplicity. However, in case log precision should be preserved, a TRACEMODEL can be used to guarantee this. Second is the FLOWERMODELWITHEP-SILON, which applies to any event log, even if it contains empty traces: given the activities of the event log $\Sigma(L)$, it returns the model that allows for any behaviour, i.e. $\circlearrowleft(\tau, a_1, \ldots a_n)$ with $a_1 \ldots a_n = \Sigma(L)$. For instance, given log L this fall through discovers the model $\circlearrowleft(\tau, a, b, c, d)$.

Local Guarantees.

The following table shows the preservation guarantees of these fall throughs. Notice that ACTIVITYONCEPERTRACE and ACTIVITYCONCURRENT could be extended to preserve log precision locally, similar to SEQUENCECUT. However, this would limit their applicability to cases in which all behaviour is present in the event log, which would render them useless as in such cases, a cut will be detected and the fall through will never be reached.

---

[1] EMPTYTRACES does not apply to L_{81}, but has been included for the sake of completeness.

	locally fitness preserving	locally log precision preserving
EMPTYTRACES	yes	yes
ACTIVITYONCEPERTRACE	yes	when extended
ACTIVITYCONCURRENT	yes	when extended
STRICTTAULOOP	yes	no
TAULOOP	yes	no
FLOWERMODEL	yes	no
FLOWERMODELWITHEPSILON	yes	no
TRACEMODEL	yes	yes

6.1.2.5 Summary

To summarise, the *Inductive Miner* (IM) implements the functions of the IM framework as follows. In these functions, strategies (i.e. base cases, cut detections, fall throughs) are tried until one matches (if a strategy does not apply, it returns nothing ($\square$)). For instance, in BASECASE$_{\text{IM}}$, the variable bc holds the result of the base cases.

function BASECASE$_{\text{IM}}(L)$
    if $\epsilon \notin L$ **then**
        $bc \leftarrow$ EMPTYLOG(L)
        if $bc = \square$ **then** $bc \leftarrow$ SINGLEACTIVITY(L) **end if**
        if $bc \neq \square$ **then return** bc **end if**
    end if
    return $\square$
end function
function FINDCUT$_{\text{IM}}(L)$
    if $\epsilon \notin L$ **then**
        $(\oplus, \Sigma_1 \ldots \Sigma_k) \leftarrow$ XORCUT$(\twoheadrightarrow(L))$
        if $k \leq 1$ **then** $(\oplus, \Sigma_1 \ldots \Sigma_k) \leftarrow$ SEQUENCECUT$(\twoheadrightarrow(L))$ **end if**
        if $k \leq 1$ **then** $(\oplus, \Sigma_1 \ldots \Sigma_k) \leftarrow$ CONCURRENTCUT$(\twoheadrightarrow(L), \bigcirc\!\!\!\!\!\; (L))$ **end if**
        if $k \leq 1$ **then** $(\oplus, \Sigma_1 \ldots \Sigma_k) \leftarrow$ LOOPCUT$(\twoheadrightarrow(L))$ **end if**
        if $k \geq 2$ **then return** $(\oplus, \Sigma_1 \ldots \Sigma_k)$ **end if**
    end if
    return $\square$
end function
function SPLITLOG$_{\text{IM}}(L, (\oplus, \Sigma_1, \ldots, \Sigma_n))$
    if $\oplus = \times$ **then return** XORSPLIT$(L, (\oplus, \Sigma_1, \ldots, \Sigma_n))$
    else if $\oplus = \rightarrow$ **then return** SEQUENCESPLIT$(L, (\oplus, \Sigma_1, \ldots, \Sigma_n))$
    else if $\oplus = \wedge$ **then return** CONCURRENTSPLIT$(L, (\oplus, \Sigma_1, \ldots, \Sigma_n))$
    else if $\oplus = \circlearrowleft$ **then return** LOOPSPLIT$(L, (\oplus, \Sigma_1, \ldots, \Sigma_n))$
    end if
end function
function FALLTHROUGH$_{\text{IM}}(L)$

$ft \leftarrow$ EMPTYTRACES(L)
if $ft = \square$ **then** $ft \leftarrow$ ACTIVITYONCEPERTRACE(L) **end if**
if $ft = \square$ **then** $ft \leftarrow$ ACTIVITYCONCURRENT(L) **end if**
if $ft = \square$ **then** $ft \leftarrow$ STRICTTAULOOP(L) **end if**
if $ft = \square$ **then** $ft \leftarrow$ TAULOOP(L) **end if**
if $ft \neq \square$ **then return** ft
else return FLOWERMODEL(L)
end if
end function

The run time of IM depends on the size of the event log L and on the size of the alphabet $\Sigma(L)$. Three recursive paths are relevant for run time:

- BASECASE$_{IM}$ stops recursion, and has $O(|L|)$ run time.
- FINDCUT$_{IM}$ has a polynomial run time: $O(|L|)$ to construct a directly follows graph, and at most $O(|\Sigma(L)|^3)$ to compute reachability in SEQUENCECUT. Second, SPLITLOG$_{IM}$ takes $O(|L|)$. Notice that this step decreases the size of the alphabet by at least one.
- most fall throughs of FALLTHROUGH$_{IM}$ take at most $O(|L|)$, however ACTIVITY-CONCURRENT takes $|\Sigma(L)| \cdot |\Sigma(L)|^3)$, i.e. alphabet size times cut finding time. Furthermore, four functions recurse: ACTIVITYONCEPERTRACE and ACTIVITYCON-CURRENT, which reduce the size of the alphabet by one, and STRICTTAULOOP and TAULOOP, which can never be applied twice consecutively.

Hence, the run time of IM is $O(|L| \cdot |\Sigma(L)|^5)$.

Besides the fall throughs mentioned in this section, other techniques that could be included as fall throughs are: (1) trace clustering [2], i.e. clustering similar traces and discovering an exclusive choice between these clusters (and continuing the recursion), (2) hybrid approaches, for instance [8] that would mine a Declare model and treat this as a process tree operator, and (3) other process tree discovery algorithms, for instance the Evolutionary Tree Miner [3]. In the future, we would like to explore such options.

Future work 6.1: Explore other techniques as fall throughs.

6.1.3 Guarantees

In the previous sections, we introduced the basic IM algorithm, and showed how it implements the IM framework. As a final step, we discuss the guarantees provided by IM: soundness, perfect fitness and rediscoverability.

Soundness is guaranteed as IM implements the IM framework that returns process trees, which are sound by construction.

In the previous section, we proved that all steps of IM are locally fitness preserving, so by Corollary 4.1 we conclude that IM always returns a fitting model:

Corollary 6.1 (IM guarantees fitness) *As all steps of IM are locally fitness preserving, by Corollary 4.1 for any log L it holds that set(L) $\subseteq \mathcal{L}(IM(L))$.*

Next, we show rediscoverability for IM, i.e. we show that if a system model S is of class C_B, and a log L is given to IM that is fitting to S and has the same directly follows graph, then IM will return a model that is language equivalent to S. Let $LA_{IM}(S)$ be the log assumption function of IM, i.e. $L \in LA_{IM}(S) \equiv (set(L) \subseteq \mathcal{L}(S) \wedge \twoheadrightarrow(S) = \twoheadrightarrow(L))$. In order to prove this, we perform three steps: we first show that the directly follows graph survives log splitting, second we prove that IM is abstraction preserving, and third we prove that IM only discovers trees of C_B. These three arguments provide rediscoverability directly.

Lemma 6.10 (IM: log splitting preserves log assumptions) *Let $S = \oplus(S_1, \ldots S_n)$ with $S \in C_B$, let $c = (\oplus, \Sigma_1, \ldots \Sigma_m)$ be a cut conforming to S, let $L_1 \ldots L_m = $ SPLITLOG(L, c) and let $L \in LA_{IM}(S)$. Then, there exist subtrees $M_1 \ldots M_m$ such that $\twoheadrightarrow(\oplus(M_1, \ldots M_m)) = \twoheadrightarrow(S)$ and $\forall_{1 \leq i \leq m} L_i \in LA_{IM}(M_i)$.*

Proof We prove this lemma by constructing trees $M_1 \ldots M_m$ corresponding to $S_1 \ldots S_n$ and showing that the log assumptions hold for these $M_1 \ldots M_m$, i.e. that the sublogs returned by SPLITLOG$_{IM}$ are fitting to their respective M_i and have the same directly follows graph.

As c is conforming, each $\Sigma_1 \ldots \Sigma_m$ is the conjunction of one or more $\Sigma(S_i)$. Let each $M_1 \ldots M_m$ be the trees corresponding to the subtrees S_i, combined with $\oplus$ if necessary. (for instance, if $S = \rightarrow(a, b, c)$ and $c = (\rightarrow, \{a, b\}, \{c\})$, then $M_1 = \rightarrow(a, b)$ and $M_2 = c$).

We prove the log assumptions LA_{IM} for these sublogs, i.e. $\forall_{1 \leq i \leq m} (set(L_i) \subseteq \mathcal{L}(M_i) \wedge \twoheadrightarrow(M_i) = \twoheadrightarrow(L_i)$ by case distinction on $\oplus$:

$\oplus = \times$ Let $i \leq n$ and $t \in L_i$. By EXCLUSIVECHOICESPLIT, $t \in L$. Then, $t \in \mathcal{L}(S)$ and by semantics of $\times$, $t \in \mathcal{L}(M_i)$. Hence, $set(L_i) \subseteq \mathcal{L}(M_i)$ and $\Sigma(L_i) = \Sigma(M_i)$.

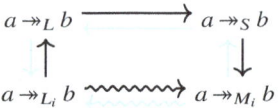

Fig. 6.4: Proof strategy to prove that $\twoheadrightarrow(L_i) = \twoheadrightarrow(M_i)$ (Lemma 6.10).

Left to prove: $\twoheadrightarrow(L_i) = \twoheadrightarrow(M_i)$, for which we follow a strategy shown in Figure 6.4. As $set(L_i) \subseteq \mathcal{L}(M_i)$, $a \twoheadrightarrow_{L_i} b \Rightarrow a \twoheadrightarrow_{M_i} b$. Reversely, assume $a \twoheadrightarrow_{M_i} b$. By Definition 2.14, there is a $t = \langle \ldots a, b, \ldots \rangle \in L$. By EXCLUS-IVECHOICESPLIT, $t \in L_i$, hence $a \twoheadrightarrow_{L_i} b \Leftrightarrow a \twoheadrightarrow_{M_i} b$. By similar arguments, $\forall_{a \in \Sigma(M_i)} \top \twoheadrightarrow_{M_i} a \Leftrightarrow \top \twoheadrightarrow_{L_i} a$ and $\forall_{a \in \Sigma(M_i)} a \twoheadrightarrow_{M_i} \bot \Leftrightarrow a \twoheadrightarrow_{L_i} \bot$. As $S \in C_B$, $\epsilon \notin \mathcal{L}(S) \supseteq L$. Thus, neither $\top \twoheadrightarrow_{M_i} \bot$ nor $\top \twoheadrightarrow_{L_i} \bot$. Hence, $\twoheadrightarrow(L_i) = \twoheadrightarrow(M_i)$.

$\oplus = \rightarrow$ Let $i \leq n$ and $t \in L_i$. By SEQUENCESPLIT, there must be a trace $t' \cdot t \cdot t'' \in L$, such that $\Sigma(\{t'\}) \cap \Sigma(M_i) = \emptyset = \Sigma(\{t''\}) \cap \Sigma(M_i)$. As $L \in LA_{IM}(S)$, $t' \cdot t \cdot t'' \in \mathcal{L}(S)$. Then by semantics of $\rightarrow$, t must have been produced by M_i. Hence, $set(L_i) \subseteq \mathcal{L}(M_i)$.

Left to prove: $\twoheadrightarrow(L_i) = \twoheadrightarrow(M_i)$. As $set(L_i) \subseteq \mathcal{L}(M_i)$, $a \twoheadrightarrow_{L_i} b \Rightarrow a \twoheadrightarrow_{M_i} b$. Reversely, assume $a \twoheadrightarrow_{M_i} b$. Each $\Sigma(M_j)$ can be recognised as a cluster of of nodes in $\twoheadrightarrow(S)$. Consider an internal edge $a \twoheadrightarrow_{M_j} b$ in this cluster. As $L \in LA_{\mathrm{IM}}(S)$, there exists a trace $t = \langle \ldots a, b, \rangle \in L$. As SEQUENCESPLIT only splits t on the boundaries of the cluster, $\exists \langle \ldots a, b, \ldots \rangle \in L_i$, so $a \twoheadrightarrow_{L_i} b \Leftrightarrow a \twoheadrightarrow_{M_i} b$. By arguments similar to the $\times$ case, $df(L_i) = \twoheadrightarrow(M_i)$.

$\oplus = \wedge$ Let $i \le n$ and $t \in L_i$. By construction of CONCURRENTSPLIT, there must be a trace $t' \in L$ such that t is a projection of t'. As $L \in LA_{\mathrm{IM}}(S)$, $t' \in \mathcal{L}(S)$. By Requirement $C_B.2$, the activities of t' in t can only be produced by M_i. Therefore, M_i must have produced t' and hence $set(L_i) \subseteq \mathcal{L}(M_i)$.

Left to prove: $\twoheadrightarrow(L_i) = \twoheadrightarrow(M_i)$, which holds by an argument similar to the $\oplus = \rightarrow$ case.

$\oplus = \circlearrowleft$ Let $i \le n$ and $t \in L_i$. Apply case distinction on whether $i = 1$:

i = 1 By LOOPSPLIT, there exists a trace $t' \cdot t \cdot t'' \in L$, such that t' is either empty or ends with an activity $\notin \Sigma(M_1)$, and t'' is either empty or starts with an activity $\notin \Sigma(M_i)$.

$i \ne 1$ By LOOPSPLIT, there exists a trace $t' \cdot \langle a' \rangle \cdot t \cdot \langle a'' \rangle \cdot t'' \in L$, such that $a', a'' \notin \Sigma(M_i)$.

By Requirement $C_B.2$, the semantics of $\circlearrowleft$ and $LA_{\mathrm{IM}}(S)$, t must have been produced by M_i. Hence, $set(L_i) \subseteq \mathcal{L}(M_i)$.

Left to prove: $\twoheadrightarrow(L_i) = \twoheadrightarrow(M_i)$, which holds by an argument similar to the $\oplus = \rightarrow$ case.

Hence, subtrees $M_1 \ldots M_m$ exist such that $\twoheadrightarrow(\oplus(M_1, \ldots M_m)) = \twoheadrightarrow(S)$ and $\forall_{1 \le i \le m}$ $L_i \in LA_{\mathrm{IM}}(M_i)$.   □

Next, we prove that IM is abstraction preserving.

Lemma 6.11 (IM is abstraction preserving) *IM is abstraction preserving, i.e. the combination of the class of process trees C_B, the directly follows abstraction $\twoheadrightarrow$, the log assumptions function $L \in LA_{\mathrm{IM}}(S) \equiv (set(L) \subseteq \mathcal{L}(S) \wedge \twoheadrightarrow(S) = \twoheadrightarrow(L))$, and the algorithm IM implementing the IM framework with BASECASE$_{\mathrm{IM}}$, FINDCUT$_{\mathrm{IM}}$, SPLITLOG$_{\mathrm{IM}}$ and FALLTHROUGH$_{\mathrm{IM}}$, is abstraction preserving.*

Proof We discuss the requirements of Definition 4.5:

AP.1 An activity base case preserves the abstraction.
    As $L \in LA_{\mathrm{IM}}(S)$ holds, $set(L) = \{\langle a \rangle\}$. By code inspection, the base case SINGLEACTIVITY applies, which returns a. Hence, $\twoheadrightarrow(\text{BASECASE}_{\mathrm{IM}}(L)) = \twoheadrightarrow(a)$.

AP.2 A τ base case preserves the abstraction.
    As $\tau \notin C_B$, this case cannot occur and the requirement holds.

AP.3 The base case parameter function preserves the abstraction.
    If $S = \oplus(S_1, \ldots S_n)$, with $S \in C_B$, then $\Sigma(S) \ge 2$. As $L \in LA_{\mathrm{IM}}(S)$ holds, by code inspection, no base case in BASECASE$_{\mathrm{IM}}$ applies. Thus, if BASECASE$_{\mathrm{IM}}$ applies, then $\twoheadrightarrow(\text{BASECASE}_{\mathrm{IM}}(L)) = \twoheadrightarrow(S)$.

AP.4 Every cut that is detected conforms to S.

As $L \in LA_{IM}(S)$, $\twoheadrightarrow(S) = \twoheadrightarrow(L)$. By Lemma 5.3, $\twoheadrightarrow(L)$ contains a cut $c = \oplus(\Sigma(S_1), \dots \Sigma(S_n))$. By Corollary 5.3, no other footprint is present in $\twoheadrightarrow(L)$. By code inspection of FINDCUT$_{IM}$, this cut c is returned, hence FINDCUT$_{IM}(L)$ conforms to S (Definition 5.4).

AP.5 Log splitting preserves the log assumptions.

This requirement follows from Lemma 6.10.

AP.6 Fall throughs preserve the abstraction.

By the previous requirements and Lemma 5.3, for all systems $S \in C_B$, either BASECASE$_{IM}$ or FINDCUT$_{IM}$ applies, i.e. FALLTHROUGH$_{IM}$ is never reached for $S \in C_B$. Therefore, this case cannot occur and the requirement holds. $\square$

We show that IM is language-class preserving (Definition 4.6), i.e. that the discovered model is of C_B:

Lemma 6.12 (IM is language-class preserving) *For all systems $S \in C_B$ and logs L such that $L \in LA_{IM}(S)$, it holds that $IM(L) \in C_B$.*

Proof As discussed in the previous requirements, FALLTHROUGH$_{IM}$ is never executed. We consider the requirements of C_B separately:

$C_B.1$ As shown for Requirement AP.2, no τ is returned in a base case for $S \in C_B$. Similarly, FALLTHROUGH$_{IM}$ is not reached, thus IM does not return τ leafs.

$C_B.2$ This requirement is guaranteed by the cuts discovered by FINDCUT$_{IM}$, which guarantee that all Σ_i are disjoint, and SPLITLOG$_{IM}$ being fitting (Requirement AP.5.

$C_B.3$ As the FALLTHROUGH$_{IM}$ function is never reached if $S \in C_B$, we limit ourselves to the case in which a cut is detected and the log is split into sublogs $L_1 \dots L_n$. By the previous requirements, FINDCUT$_{IM}$ only selects a $\circlearrowleft$ if $S = \oplus(S_1, \dots S_n)$. As $S \in C_B$, $\text{Start}(S_1) \cap \text{End}(S_1) = \emptyset$. By Requirement AP.5 and the log assumptions LA_{IM}, $\text{Start}(L_1) \cap \text{End}(L_1) = \emptyset$. By lemmas 6.11 and 4.1, $\twoheadrightarrow(IM(L_1)) = \twoheadrightarrow(IM(S_1))$, hence $\text{Start}(IM(L)) \cap \text{End}(IM(L)) = \emptyset$.

$C_B.4$ By code inspection, IM never returns $\leftrightarrow$.

$C_B.5$ By code inspection, IM never returns $\vee$.

Hence, $IM(L) \in C_B$ and thus IM is language-class preserving (Definition 4.6).$\square$

Then, by Theorem 4.2, IM guarantees rediscoverability for C_B:

Theorem 6.2 (IM rediscoverability) *Let L be a log and $S \in C_B$ be a system such that $set(L) \subseteq \mathcal{L}(S) \wedge \twoheadrightarrow(L) = \twoheadrightarrow(S)$. Then, $\mathcal{L}(IM(L)) = \mathcal{L}(S)$.*

To summarise, IM guarantees to return a fitting and sound model for all event logs. Furthermore, if a system S is from C_B and a log L has a perfect fitness with respect to S and has the same directly follows graph, then IM applies to L returns a model that is language equivalent to S.

In Section 5.5, we showed that the minimum self-distance relation suffices to distinguish the larger class of process trees C_M (Corollary 5.5), if an LC-property exists

(Conjecture 5.1) that distinguishes of a particular class of process trees containing nested $\circlearrowleft$ and $\wedge$ operators. As shown in Section 5.5, the directly follows relation does not suffice to distinguish all process trees of C_M, hence if an LC-property will be discovered in the future, this property will have to be incorporated into the cut detection functions of IM in order to guarantee rediscoverability for C_M. Notice that to incorporate this property, the algorithm would need to be changed in a limited scope: only the LOOPCUT function would need to be changed, which shows the flexibility of the IM framework.

Using the functions described in this section, a log-precision-guaranteeing algorithm can be constructed (using EMPTYLOG, EMPTYTRACES, SINGLEACTIVITY, XORCUT, XORSPLIT and TRACEMODEL, and when extended SEQUENCECUT, SEQUENCESPLIT, CONCURRENTCUT, CONCURRENTSPLIT, ACTIVITYONCEPERTRACE and ACTIVITYCONCURRENT). The current functions would result in an algorithm with a limited scope, i.e. only supporting the ×-operator, but it nevertheless shows that the IM framework is flexible enough to support such guarantees. In future research, the mentioned extensions could be based on other abstractions.

Future work 6.3: Research more elegant locally log-precision preserving IM framework functions.

6.2 Handling Deviating & Infrequent Behaviour

In the previous section, we introduced the basic IM algorithm that returns a fitting model, and returns a model that is language equivalent to the system, if the event log contains no deviations and is directly follows complete to the system. Deviating behaviour is behaviour that appears in the event log but is not part of the system. Infrequent behaviour is behaviour that is part of the system but does not happen or happens in just a few cases in the event log. As described in Section 3.2.2, both deviating and infrequent behaviour challenge discovery algorithms. In this section, we study the influence of deviating and infrequent behaviour on directly follows graphs, process discovery and the IM algorithm in Chapter 8. Next, we introduce a variant of IM to exclude such behaviour from event logs in Section 6.2.1. Finally, we give an example in Section 6.2.3, and we finish with a discussion of the guarantees provided by the new variant (Section 6.2.4).

6.2.1 Deviating & Infrequent Behaviour

Figure 6.5 contains an example of an event log in which each trace occurs 25 times, except the last trace, which occurs once. Given this event log, IM returns

, which is fitting (Corollary 6.1), but is neither log precise nor

simple. In case the use case does not require perfect fitness, an obviously better
model is achievable: → . This model would fit 100 of the 101 traces, and be

perfectly precise, as each trace of the model appears in the event log as well.

$$L_{90} = [\langle a, b \rangle^{25}, \langle a, c \rangle^{25}, \langle d, b \rangle^{25},$$
$$\langle d, c \rangle^{25}, \langle a, b, a, c \rangle]$$

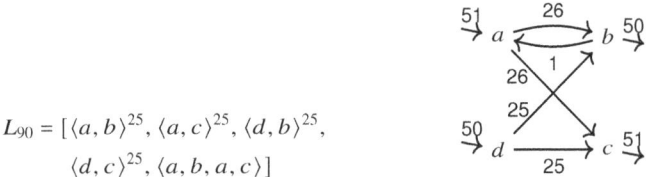

Fig. 6.5: A log and its directly follows graph.

IM fails to discover the more log-precise second model as the first cut should be
$(\rightarrow, \{a, d\}, \{b, c\})$, which is not a valid sequence cut in the directly follows graph of
L_{90} (edge $b \twoheadrightarrow a$ prevents it). In general, if the behaviour of the event log does not fit
the representational bias of the discovery algorithm well, then leaving out behaviour
might help to improve the balance between fitness and log precision.

As the system is unknown to the algorithm, deviating behaviour is not always
distinguishable from infrequent behaviour, algorithms enjoy the freedom to classify
behaviour as deviating (and filter it) if it does not fit nicely in the model class
the algorithm considers, or if it does not lead to "nice" models. It might even be
necessary to include deviating or infrequent behaviour to discover an elegant model:
due to concurrency, trace might appear few times in the log as well. For instance,
consider the process tree   ∧   , consisting of two concurrent branches, and

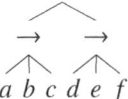

having a language of 20 different traces. Furthermore, consider the following event
log of 50 correct traces (randomly generated) and two deviating traces (the trace
$\langle a, b, a, b, a, b, c \rangle$ that occurs twice):

$$L_{91} = [\langle d, a, b, c, e, f\rangle^5, \quad \langle a, b, d, e, f, c\rangle^4, \quad \langle d, a, e, b, c, f\rangle^4, \quad \langle a, d, b, c, e, f\rangle^3,$$
$$\langle a, d, b, e, f, c\rangle^3, \quad \langle a, d, b, e, c, f\rangle^3, \quad \langle d, a, e, b, f, c\rangle^3, \quad \langle d, a, e, f, b, c\rangle^3,$$
$$\langle d, a, b, e, f, c\rangle^3, \quad \langle d, e, f, a, b, c\rangle^3, \quad \langle d, e, a, f, b, c\rangle^3, \quad \langle a, b, c, d, e, f\rangle^2,$$
$$\langle d, e, a, b, f, c\rangle^2, \quad \langle a, b, d, c, e, f\rangle^2, \quad \langle d, a, b, e, c, f\rangle^2, \quad \langle a, d, e, f, b, c\rangle^2,$$
$$\langle a, d, e, b, f, c\rangle^2, \quad \langle d, e, a, b, c, f\rangle, \quad \langle a, b, a, b, a, b, c\rangle^2]$$

Figure 6.6 shows the corresponding directly follows graph (for readability reasons, frequencies have been denoted in a separate table). The deviating edge $b \twoheadrightarrow a$ occurs twice, which is not that often, as the correct edges, i.e. the edges corresponding to the process tree, between a and f and between c and d occur only 2 or 3 times. Thus, deviating behaviour might have a big impact, as it requires only a couple of deviating traces to introduce an edge that is stronger than some 'real' edges.

$\twoheadrightarrow$	a	b	c	d	e	f
a		28		13	10	3
b	2		27	6	11	7
c				2	10	10
d	20	9	2		19	
e	6	9	5			30
f	3	8	17			

(a) $\twoheadrightarrow$ of L_{91}.   (b) Weights of $\twoheadrightarrow(L_{91})$.

Fig. 6.6: The directly follows graph and edge weights of log L_{91}.

Hence, deviation filtering involves a trade-off between removing unwanted behaviour while keeping less-occurring wanted behaviour. We discuss several strategies, after which we discuss our choice for IM$_F$:

- Decide for each pair of activities what their most likely relation is, i.e. which of the two possible edges between them is really present. This technique, presented in [12], compares the frequencies of both edges and decides the relation between activities based on a heuristic. This would work on both example logs L_{90} and L_{91}, but is vulnerable to parameter settings: if the event log is not balanced, this method might remove correct edges as well.
- Filter the infrequent traces from the log before discovery, i.e. choose a threshold f and remove all traces that occur less than f times. On L_1, it would obviously be easy to choose f: any value between 2 and 24 will remove the faulty trace. However, on L_2, there is no suitable f, as choosing it to be 1 or 2 will already remove valid behaviour. This is due to the concurrency: on a model of n concurrent activities, there are $O(n!)$ different traces, which makes the expected frequency of edge cases low.
- Filter the least-occurring edges from the directly follows graph, i.e. choose a threshold f and remove all edges from the directly follows graph that do not occur more than f times. On L_{90}, this would easily remove the faulty directly

follows edge. On L_{91}, this would remove the faulty edge, but it is clear that in larger examples, deviations would overshadow concurrent behaviour. This method would perform better than the previous method, as in a model of n concurrent activities, there are at most $O(n^2)$ directly follows edges.

- Filter the locally least-occurring edges from the directly follows graph, i.e. choose a threshold f such that $0 \leq f \leq 1$, consider the outgoing edges of each activity, and remove all outgoing edges that occur less than f times the occurrences of the most-occurring outgoing edge. On L_{90}, this would easily remove the faulty directly follows edge. On L_{91}, this would remove the faulty edge, but the method would fail on larger examples. Nevertheless, this method would perform better than the previous method, as with a model of n concurrent activities, there are at most $O(n)$ outgoing directly follows edges for each activity.

Many more deviation-filtering techniques could be applied, and different real-life event logs might require different techniques. In the next section, we illustrate how IM can be adapted by using the last of these deviation-filtering techniques. The modularity of the IM framework allows for easy implementation of other techniques.

Future work 6.4: Consider other deviation-filtering techniques to distinguish concurrency and deviating/infrequent behaviour.

6.2.2 Inductive Miner - infrequent (IMF)

To handle deviating and infrequent behaviour, we introduce a second algorithm: *Inductive Miner - infrequent* (IMF). IMF applies filtering to all four steps of the IM framework, using a deviation-threshold parameter f. In this section, we discuss how IMF implements the IM framework by giving its cut detection, log splitting, base cases and fall through parameter functions.

6.2.2.1 Cut Detection

To guarantee rediscoverability, in each recursion, IMF first applies the cut-detection functions of IM. If these do not succeed, the directly follows graph is filtered, after which cut detection is applied again. We first formalise the deviations filtering step, after which we give an example of a filtering cut detection function. The other filtering cut detection functions are similar. A user-chosen deviation threshold parameter f is assumed to be available.

function FILTER($\twoheadrightarrow$)
    copy $\twoheadrightarrow$ into $\twoheadrightarrow'$
    for $a \in \Sigma(L) \cup \{\top\}, b \in \Sigma(L) \cup \{\bot\}$ such that $a \twoheadrightarrow b$ **do**
        if $|a \twoheadrightarrow b| < f \cdot \max_{c \in \Sigma(\twoheadrightarrow) \cup \bot} |a \twoheadrightarrow c|$ in L **then**
            $\twoheadrightarrow' \leftarrow \twoheadrightarrow' \setminus$ all (a, b)
        end if

      end for
      return $\twoheadrightarrow'$
  end function
  function XORCUTFILTERING($\twoheadrightarrow$)
      return XORCUT(FILTER($\twoheadrightarrow$))
  end function
  function SEQUENCECUTFILTERING($\twoheadrightarrow$)
      return SEQUENCECUT(FILTER($\twoheadrightarrow$))
  end function
  function CONCURRENTCUTFILTERING($\twoheadrightarrow$, $\ominus$)
      return CONCURRENTCUT(FILTER($\twoheadrightarrow$), $\ominus$)
  end function
  function LOOPCUTFILTERING($\twoheadrightarrow$)
      return LOOPCUT(FILTER($\twoheadrightarrow$))
  end function

6.2.2.2 Log Splitting

As the cut detection functions of IMF might return cuts that do not adhere to the footprints of Lemma 5.3, log splitting must be robust to deviating behaviour. A strategy could be to simply ignore the deviating behaviour and making sure that log splitting is not influenced. However, deviations might accumulate over recursions and influence discovery in later recursions of the IM framework. Therefore, the log splitting functions defined below filter deviating events whenever they are detected. As a consequence, in absence of deviating events, these log splitting functions perform the same split as the log splitting functions of IM, i.e. the filtering log splitting functions of IMF could be used for IM as well. We describe the log splitting functions of IMF.

Exclusive Choice.

To split a log L according to an exclusive choice cut, IMF needs to put each trace in one sublog; all events not from the Σ_i belonging to that sublog are decided to be deviating. To choose a sublog for a trace, IMF selects the Σ_i with the most events in the trace, which minimises the number of deviating events.

  function XORSPLITFILTERING($L, (\times, \Sigma_1, \ldots, \Sigma_n)$)
      $L_1 \ldots L_n \leftarrow [\,] \ldots [\,]$
      for $t \in L$ **do**
          $i \leftarrow$ the Σ_i with the most events in t
          $t' \leftarrow t|_{\Sigma_i}$
          $L_i \leftarrow L_i \uplus [t']$
      end for
      return $L_1, \ldots, L_n$

end function

For instance, the log $L = [\langle a, b \rangle, \langle c, c, c \rangle, \langle a, b, c \rangle]$ would be split using the cut $(\times, \{a, b\}, \{c\})$ into $[\langle a, b \rangle^2]$ and $[\langle c, c, c \rangle]$.

Notice that as a side effect, XORSPLITFILTERING might return empty sublogs. These sublogs will be recursed on and they will be handled by the EMPTYLOG base case. For instance, when the log $[\langle a, b, b \rangle]$ is split using the cut $(\times, \{a\}, \{b\})$, the sublog for $\{a\}$ will be empty, while the sublog for $\{b\}$ contains the trace $\langle b, b \rangle$.

Obviously, in case of a cut according the footprint of Lemma 5.3, no events are filtered. Therefore, Lemma 6.5 holds as well, assuming that a valid cut is provided.

Sequence.

To split a log L according to a sequence cut, IMF aims to minimise the number of events that are classified as deviations. Let a *split point (sequence split)* be the point in a trace where execution changed branches, e.g. for the cut $(\rightarrow, \{a\}, \{b\})$ and trace $\langle a, b \rangle$, the split point would be in between a and b, denoted with $\langle a|b \rangle$. All events that are on the wrong side of the split point according to the cut are classified as deviations. For instance, given the cut $(\rightarrow, \{a\}, \{b\})$, the trace $\langle a, b, a, a, b \rangle$ could have several split points (the deviations have been striked out):

$$\langle|\ \not{a}, b, \not{a}, \not{a}, b \rangle$$
$$\langle a\ |\ b, \not{a}, \not{a}, b \rangle$$
$$\langle a, \not{b}|\ \not{a}, \not{a}, b \rangle$$
$$\langle a, \not{b}, a\ |\ \not{a}, b \rangle$$
$$\langle a, \not{b}, a, a\ |\ b \rangle$$
$$\langle a, \not{b}, a, a, \not{b}|\ \rangle$$

The second-last split point is optimal, as it introduces the least number of deviations. The function SEQUENCESPLITFILTERING first decides the split points in each trace, which it does iteratively. Second, all deviating events are removed.

function SEQUENCESPLITFILTERING$(L, (\rightarrow, \Sigma_1, \ldots, \Sigma_n))$
    $L_1 \ldots L_n \leftarrow [\ ] \ldots [\ ]$
    for $1 \le i \le n$ **do**
        $splitPoint \leftarrow 0$
        for $t \in L$ **do**
            $newSplitPoint \leftarrow$ FINDSPLITPOINT$(t, \Sigma_i, splitPoint, \cup_{1 \le j < i} \Sigma_j)$
            $t' \leftarrow t[splitPoint, newSplitPoint)|_{\Sigma_i}$
            $L_i \leftarrow L_i \uplus [t']$
            $splitPoint \leftarrow newSplitPoint$
        end for
    end for
    return $L_1, \ldots, L_n$
end function

in which $t[a, b)$ denotes the subtrace of t starting at position a up to (exclusive) position b.

The main problem of the log splitting is to find the split point in the trace, such that splitting the trace on that point introduces the least number of deviations. This algorithm searches for the optimal split point in trace t where the set of activities Σ begins its subtrace. The algorithm walks over t once and keeps track of the cost (or gain) that is involved with including each event in the final subtrace. To limit the search space, the algorithm ignores the part of the trace before position *start* and all activities in *ignore*.

> **function** FINDSPLITPOINT($t, \Sigma, start, ignore$)
> $leastCost \leftarrow start$
> $positionWithLeastCost \leftarrow start$
> $cost \leftarrow 0$
>   **for** $i \leftarrow start \ldots |t|$ **do**
>   **if** $t[i] \in \Sigma$ **then** $cost \leftarrow cost - 1$
>   **else if** $t[i] \notin ignore$ **then** $cost \leftarrow cost + 1$
>   **end if**
>   **if** $cost < leastCost$ **then**
> $leastCost \leftarrow cost$
> $positionWithLeastCost \leftarrow i$
>   **end if**
>   **end for**
>   **return** $positionWithLeastCost$
> **end function**

For instance, the log $[\langle a, b, c\rangle, \langle b, a, c\rangle, \langle c, a, b, c\rangle]$ would be split using the cut $(\rightarrow, \{a, b\}, \{c\})$ into $[\langle a, b\rangle^2, \langle b, a\rangle], [\langle c\rangle^3]$, i.e. the first c in the third trace is classified as deviating and removed.

Notice that as a side effect, SEQUENCESPLITFILTERING might return sublogs with empty traces. For instance, when the trace $\langle b, b, a\rangle$ is split using the cut $(\rightarrow, \{a\}, \{b\})$, the sublog for $\{a\}$ contains the empty trace, while the sublog for $\{b\}$ contains the trace $\langle b, b\rangle$.

In case the given cut corresponds to a footprint of Lemma 5.3, the log splitting points returned by FINDSPLITPOINT correspond to the switch from Σ_i to Σ_{i+1} and the log split corresponds to the regular non-filtering sequence split. Therefore, if such a cut is provided, Lemma 6.7 holds as well.

Concurrency.

In the IM framework, a log splitting function has no knowledge at all about the subtrees that will be discovered later, i.e. the cut and the log are the only information available. A concurrent operator combines the languages of its subtrees in a nonrestrictive way, i.e. all behaviour of all subtrees can be present in any interleaved way. As the concurrent operator does not restrict behaviour, at log splitting using a concurrent cut no behaviour could be considered deviating. Therefore, IMF uses the CONCURRENTSPLIT function of IM and no changes are necessary.

Loop.

In the IM framework, a log splitting function is assumed to have no knowledge about the subtrees. Therefore, few deviations can be detected while splitting a log using a loop cut. For instance, consider the cut $(\circlearrowleft, \{a\}, \{b\})$, the trace $t = \langle a, b, a, a, b, a \rangle$ and the model $\underset{a \; b}{\overset{\wedge}{\circlearrowleft}}$. Trace t does not fit the model: the two consecutive a's are not

supported by the model. In the IM framework, the log splitting function is assumed to have no knowledge about subtrees, thus cannot determine that the two consecutive a's are deviating.

The only deviations that can be detected using log splitting with a loop cut is if a trace starts or ends with an activity from the body of the loop. For instance, consider the system $\underset{a \; b}{\overset{\wedge}{\circlearrowleft}}$, the cut $(\circlearrowleft, \{a\}, \{b\})$ and the trace $\langle a, b \rangle$, which does not

fit the model. In this case, log splitting can detect the deviation, as the trace ends with the non-body activity b. To handle these deviations, we decided to "repair" the log by inserting an empty trace in the sublog for a, i.e. in our example we discover the sublogs $L_1 = [\langle a \rangle, \epsilon]$ and $L_2 = [\langle b \rangle]$. In a subsequent recursion on L_1, the fall through EMPTYTRACES, which will be introduced later on in this section, will decide whether this empty trace occurs frequent enough to be included in the model.

We introduce the filtering log splitting function for loop cuts LOOPSPLITFILTERING:

function LOOPSPLITFILTERING$(L, (\circlearrowleft, \Sigma_1, \ldots, \Sigma_n))$
    $\forall_{1 \leq i \leq n} L_i \leftarrow [\,]$
    for $t \in L$ **do**
        $S \leftarrow \Sigma_1$
        $st \leftarrow \epsilon$
        for $a \in t$ **do**
            if $a \in S$ **then**
                $st \leftarrow \langle a \rangle \cdot st$
            else
                $L_j \leftarrow L_j \uplus [st]$ with $\Sigma_j = S$
                $st \leftarrow \epsilon$
                $S \leftarrow \Sigma_i$ such that $a \in \Sigma_i$
            end if
        end for
        $L_j \leftarrow L_j \uplus [st]$ with $\Sigma_j = S$
        if $S \neq \Sigma_1$ **then** $L_1 \leftarrow L_1 \uplus [\epsilon]$ **end if**
    end for
    end function

6.2.2.3 Base Cases

In the base cases, infrequent behaviour can be detected and accounted for as follows:

- in EMPTYLOG, no filtering can be applied, thus we reuse this base case of IM.
- SINGLEACTIVITYFILTERING applies when the event log contains a single activity, i.e. $\Sigma(L) = \{a\}$ for some activity a. Then, the event log might contain empty traces, traces with a single a or traces with multiple a's. If the log contains "enough" traces with a single a, we consider the base case a appropriate, which only produces traces with a single a. Therefore, we assume a geometric distribution with parameter p, which we estimate as $\widehat{p} = |L|/(||L|| + |L|)$, in which $|L|$ is the number of traces in log L and $||L||$ is the number of events in L. If the log contains only traces with a single a, then $\widehat{p} = 0.5$. If this $\widehat{p}$ is 'close enough' to 0.5, i.e. $|\widehat{p} - 0.5| \le f$, the activity a is returned as a leaf.

The new base case EMPTYTRACESFILTERING obviously does not preserve fitness, and does not preserve log precision, as it applies to logs in which there is no trace with a single event, thus it might introduce new behaviour.

	locally fitness preserving	locally log precision preserving
EMPTYLOG	yes	yes
SINGLEACTIVITYFILTERING	no	no

6.2.2.4 Fall Throughs

All fall throughs of IM preserve all behaviour of the log, i.e. are locally fitness preserving, thus also work in case of deviating behaviour, i.e. they preserve all deviating behaviour. Thus, IMF mostly uses the same fall throughs as IM. However, the EMPTYTRACES fall through is sensitive to deviating and infrequent behaviour: a single empty trace in the event log triggers the discovery of optionality, even if all other traces occur thousands of times. Therefore, the EMPTYTRACESFILTERING fall through applies when the event log contains empty traces, i.e. $\epsilon \in L$. If the event log contains "enough" empty traces, i.e. $|\epsilon \in L| \ge |L| \times f$, the model $\times(\tau, \text{IMF}(L \setminus \mathbb{M}(\{\epsilon\})))$ is returned and recursion continues on a log without the empty traces. Otherwise, the empty traces are filtered out and recursion continues, i.e. $\text{IMF}(L \setminus \mathbb{M}(\{\epsilon\}))$.

The new fall through does not preserve fitness, as in the second case the empty traces are simply removed.

	locally fitness preserving	locally log precision preserving
EMPTYTRACESFILTERING	no	yes

6.2.2.5 Summary

To summarise, the *Inductive Miner - infrequent* (IMF) implements the functions of the IM framework as follows, using a user-chooseable deviation-threshold-filtering parameter f:

> **function** BASECASE$_{\text{IMF}}(L)$
>   **if** $\epsilon \notin L$ **then**
> $bc \leftarrow$ EMPTYLOG(L)
>   **if** $bc \neq \square$ **then** $bc \leftarrow$ SINGLEACTIVITYFILTERING(L) **end if**
>   **return** bc
>   **end if**
>   **return** $\square$
> **end function**

Notice that FINDCUT$_{\text{IMF}}$ first calls FINDCUT$_{\text{IM}}$ to, as we will show in Theorem 6.5.

> **function** FINDCUT$_{\text{IMF}}(L)$
>   **if** $\epsilon \notin L$ **then**
> $(\oplus, \Sigma_1 \ldots \Sigma_k) \leftarrow$ FINDCUT$_{\text{IM}}(\twoheadrightarrow(L))$
>   **if** $k \leq 1$ **then** $(\oplus, \Sigma_1 \ldots \Sigma_k) \leftarrow$ XORCUTFILTERING$(\twoheadrightarrow(L))$ **end if**
>   **if** $k \leq 1$ **then** $(\oplus, \Sigma_1 \ldots \Sigma_k) \leftarrow$ SEQUENCECUTFILTERING$(\twoheadrightarrow(L))$ **end if**
>   **if** $k \leq 1$ **then** $(\oplus, \Sigma_1 \ldots \Sigma_k) \leftarrow$ CONCURRENTCUTFILTERING$(\twoheadrightarrow(L))$ **end if**
>   **if** $k \leq 1$ **then**
>   **return** LOOPCUTFILTERING$(\twoheadrightarrow(L))$
>   **else**
>   **return** $(\oplus, \Sigma_1 \ldots \Sigma_k)$
>   **end if**
>   **end if**
>   **return** $\square$
> **end function**

> **function** SPLITLOG$_{\text{IMF}}(L, (\oplus, \Sigma_1, \ldots, \Sigma_n))$
>   **if** $\oplus = \times$ **then return** XORSPLITFILTERING$(L, (\oplus, \Sigma_1, \ldots, \Sigma_n))$
>   **else if** $\oplus = \rightarrow$ **then return**
>   SEQUENCESPLITFILTERING$(L, (\oplus, \Sigma_1, \ldots, \Sigma_n))$
>   **else if** $\oplus = \wedge$ **then return** CONCURRENTSPLIT$(L, (\oplus, \Sigma_1, \ldots, \Sigma_n))$
>   **else if** $\oplus = \circlearrowleft$ **then return** LOOPSPLITFILTERING$(L, (\oplus, \Sigma_1, \ldots, \Sigma_n))$
>   **end if**
> **end function**

> **function** FALLTHROUGH$_{\text{IMF}}(L)$
> $bc \leftarrow$ EMPTYTRACESFILTERING(L)
>   **if** $bc = \square$ **then return** bc
>   **else return** FALLTHROUGH$_{\text{IM}}(L)$
>   **end if**
> **end function**

The run time of IMF equals the run time of IM: $O(|L| \cdot |\Sigma(L)|^4)$.

6.2.3 Example

We revisit the example given in Section 4.1.3, using an event log in which all traces happen 10 times, except for the last trace, which occurs once:

$$L_{92} = [\langle a, b, c, d, e\rangle^{10}, \langle a, d, b, e\rangle^{10}, \langle a, e, b\rangle^{10}, \langle a, c, b\rangle^{10}, \langle a, b, d, e, c\rangle^{10}, \langle c, a, b\rangle]$$

Figure 6.7a shows the directly follows graph of L_{92}. As a first step, IMF applies FINDCUT$_{IM}$, which returns nothing. Second, the directly follows graph is filtered, here using a threshold of 0.15. The result is shown in Figure 6.7b. In this filtered directly follows graph, the sequence cut $(\rightarrow, \{a\}, \{b, c, d, e\})$ is present.

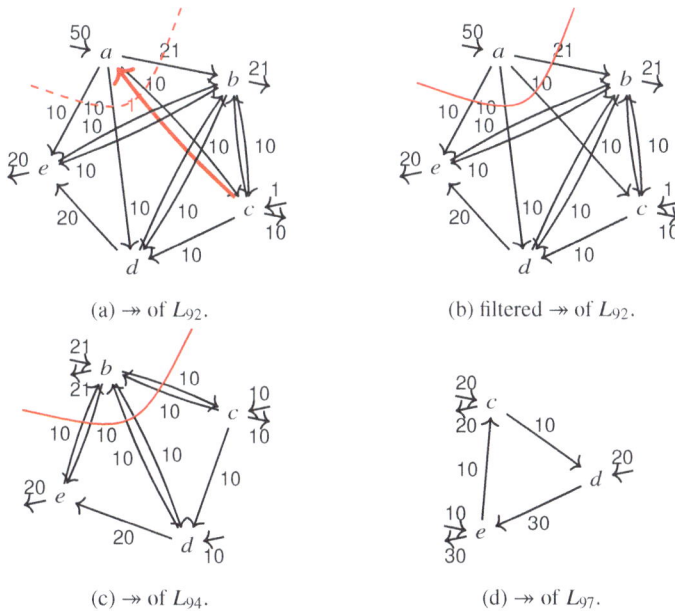

(a) $\twoheadrightarrow$ of L_{92}.

(b) filtered $\twoheadrightarrow$ of L_{92}.

(c) $\twoheadrightarrow$ of L_{94}.

(d) $\twoheadrightarrow$ of L_{97}.

Fig. 6.7: Directly follows graphs of logs used in the recursion. The dashed red curve does not denote a cut as of the red thick edge $c \twoheadrightarrow a$. The non-dashed red curves denote cuts.

Then, SEQUENCESPLITFILTERING splits the log in sublogs L_{93} and L_{94} (notice that in the last trace, c is considered a deviation and is removed):

$$L_{93} = [\langle a \rangle^{51}]$$
$$L_{94} = [\langle b, c, d, e\rangle^{10}, \langle d, b, e\rangle^{10}, \langle e, b\rangle^{10}, \langle c, b\rangle^{10}, \langle b, d, e, c\rangle^{10}, \langle b\rangle]$$

IMF records the choice and recurses, i.e. IMF$(L_{92}) = \rightarrow(\text{IMF}(L_{93}), \text{IMF}(L_{94}))$. We first consider the recursive step on L_{93}, for which BASECASE$_{IMF}(L_{93})$ returns a base case, being the process tree a:

$$\textsc{baseCase}_{\textsc{IM}}(L_{93}) = a$$

Next, we give the computation steps taken and the results of the recursive calls:

$$\textsc{IM}_{\textsc{F}}(L_{93}) = a$$
$$\textsc{baseCase}_{\textsc{IM}_{\textsc{F}}}(L_{94}) = \square$$
$$\textsc{findCut}_{\textsc{IM}_{\textsc{F}}}(L_{94}) = c_3 = (\wedge, \{b\}, \{c, d, e\}) \text{ (see Figure 6.7c)}$$
$$\textsc{splitLog}_{\textsc{IM}_{\textsc{F}}}(L_{94}, c_3) = L_{95}, L_{96}$$
$$L_{95} = [\langle b \rangle^{51}]$$
$$L_{96} = [\langle c, d, e \rangle^{10}, \langle d, e \rangle^{10}, \langle e \rangle^{10}, \langle c \rangle^{10}, \langle d, e, c \rangle^{10}, \epsilon]$$
$$\textsc{IM}_{\textsc{F}}(L_{94}) = \wedge(\textsc{IM}_{\textsc{F}}(L_{95}), \textsc{IM}_{\textsc{F}}(L_{96}))$$
$$\textsc{baseCase}_{\textsc{IM}_{\textsc{F}}}(L_{95}) = b$$
$$\textsc{IM}_{\textsc{F}}(L_{95}) = b$$
$$\textsc{baseCase}_{\textsc{IM}_{\textsc{F}}}(L_{96}) = \textsc{IM}_{\textsc{F}}(L_{97}) \text{ (remove } \epsilon)$$
$$L_{97} = [\langle c, d, e \rangle^{10}, \langle d, e \rangle^{10}, \langle e \rangle^{10}, \langle c \rangle^{10}, \langle d, e, c \rangle^{10}]$$
$$\textsc{IM}_{\textsc{F}}(L_{96}) = \textsc{IM}_{\textsc{F}}(L_{97})$$
$$\textsc{findCut}_{\textsc{IM}_{\textsc{F}}}(L_{97}) = \square \text{ (see Figure 6.7d)}$$
$$\textsc{fallThrough}_{\textsc{IM}_{\textsc{F}}}(L_{97}) = \circlearrowright(\times(c, d, e), \tau)$$
$$\textsc{IM}_{\textsc{F}}(L_{97}) = \circlearrowright(\times(c, d, e), \tau)$$

Combining all these intermediate steps, $\textsc{IM}_{\textsc{F}}$ will discover the process tree $T =$

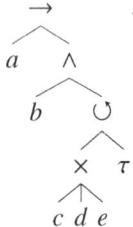

6.2.4 Guarantees

Filtering deviating behaviour excludes behaviour of the event log from the discovered model, hence $\textsc{IM}_{\textsc{F}}$ does not guarantee fitness: Corollary 6.1 does not hold for $\textsc{IM}_{\textsc{F}}$.

To preserve rediscoverability, $\textsc{IM}_{\textsc{F}}$ extends $\textsc{IM}$, i.e. before any filtering is applied to cut detection, the cut detection functions of $\textsc{IM}$ are applied (i.e. the second line in $\textsc{findCut}_{\textsc{IM}_{\textsc{F}}}$ calls $\textsc{findCut}_{\textsc{IM}}$). Thus, in case the system model is of class $C_{\textsc{B}}$, the deviation filtering is not reached, and therefore, lemmas 6.11 and 6.12 hold for IM$_F$, and hence by Theorem 4.2, IM$_F$ guarantees rediscoverability for C_B. Using the flexibility of the IM framework, we inserted deviation filtering in several steps, without losing rediscoverability.

Theorem 6.5 (IM$_F$ rediscoverability) *Let L be a log and $S \in C_B$ be a system such that $set(L) \subseteq \mathcal{L}(S) \wedge \twoheadrightarrow(L) = \twoheadrightarrow(S)$. Then, $\mathcal{L}(IM_F(L)) = \mathcal{L}(S)$.*

Notice that result only holds in case the event log contains no infrequent or deviating behaviour ($set(L) \subseteq \mathcal{L}(S)$). We introduce the rediscoverability framework of Theorem 4.2, which enables rediscoverability proofs using formal characterisations and quantifications of deviating and infrequent behaviour, which remain future work:

Future work 6.6: Prove rediscoverability of IM$_F$ for logs with deviating and infrequent behaviour.

In this section, we described several techniques to detect deviating and infrequent behaviour, and used one of these in IM$_F$. However, many more deviation-filtering techniques could be applied, and different real-life event logs might require different techniques.

Future work 6.7: Consider other deviation-filtering techniques to distinguish concurrency and deviating/infrequent behaviour.

IM$_F$ uses a user-specified threshold f to filter deviating and infrequent behaviour, which is fixed for the entire application of IM$_F$ to a log. We chose to use a single f for simplicity and to limit the number of interactions a user has with the algorithm, however one could choose different thresholds for different cut detections, base cases, fall throughs and recursions. Future work might reveal whether different thresholds may be used sensibly.

In Chapter 8, we investigate the influence of deviating and infrequent behaviour on rediscoverability, and evaluate how IM$_F$ handles such behaviour. In the next section, we introduce an algorithm to handle the opposite of deviating and infrequent behaviour: incomplete behaviour.

6.3 Handling Incomplete Behaviour

In Section 3.2, we identified three types of behaviour that challenge discovery techniques: deviating behaviour, infrequent behaviour and incompleteness. In the previous section, we showed how the IM framework can be used to handle deviating and infrequent behaviour. In this section, we address the last challenge: incompleteness, i.e. behaviour that is part of the system but which is not present in the event log. We first discuss incompleteness and its influence on directly follows graphs and process discovery in Section 6.3.1. In Section 6.3.2, we introduce an algorithm that handles incompleteness, the *Inductive Miner - incompleteness* (IM$_C$). We give an example in Section 6.3.3, discuss the guarantees it provides in Section 6.3.4 and discuss some of its implementation details in Section 6.3.5.

6.3.1 Incomplete Behaviour

As discussed in Chapter 3, in process discovery it is assumed that an event log does not contain all behaviour of the system, as it is infeasible or impossible that this assumption holds for complex systems with, for instance, a high degree of concurrency or looping behaviour. Furthermore, if one would assume that all behaviour is present in the event log, then the event log could serve as a model and it is not necessary to discover a model. Therefore, in process discovery, it is typically assumed that not all behaviour of the system is present in the event log, i.e. that the event log is merely a sample of the possible behaviour. In the previous sections, we used this assumption already. However, in this section, we push the lower boundary of behaviour that needs to be present in the event log to enable rediscovery.

Instead of assuming that the event log contains all behaviour, many discovery algorithms assume a weaker notion of completeness to provide rediscoverability. For instance, in the previous sections, we showed that IM and IMF provide rediscoverability, assuming that the directly follows relation of the event log contains all relations of the directly follows relation of the system. We refer to this property of log and system as the log being *directly follows complete* to the system. In this section, we explore what happens if the information in the event log does not suffice to obtain an equivalent directly follows graph: in the relation of the log, edges, start activities or end activities are missing compared to the relation of the system.

For instance, consider the tree $M_{98} = $

$$
\begin{array}{c}
\circlearrowleft \\
\wedge \qquad \rightarrow \\
\wedge \qquad \wedge \\
\rightarrow \quad \rightarrow \; e \; f \\
\wedge \; \wedge \\
a \; b \; c \; d
\end{array}
$$

and the following event log:

$$
\begin{aligned}
L_{99} = [&\langle a, b, c, d, e, f, a, c, b, d \rangle, \\
&\langle a, c, d, b, e, f, a, b, c, d \rangle, \\
&\langle c, a, b, d \rangle, \\
&\langle c, a, d, b \rangle, \\
&\langle c, d, a, b \rangle]
\end{aligned}
$$

Figure 6.8 shows the directly follows graph of L_{99}, in which the edge $f \twoheadrightarrow c$ of M_{98} is missing in L_{99}. Due to this missing edge, the cut $(\circlearrowleft, \{a, b, c, d\}, \{e, f\})$ is not a cut according to Lemma 5.2.2 and therefore IM will not rediscover M_{98}, but will select a fall through (ACTIVITYCONCURRENT in this case).

In the next section, we introduce an algorithm to handle such incomplete behaviour.

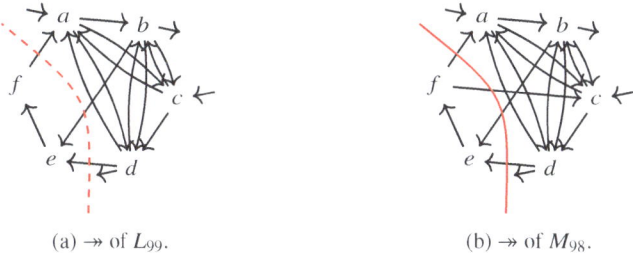

(a) → of L_{99}.   (b) → of M_{98}.

Fig. 6.8: Directly follows graphs of L_{99} and M_{98}. The dashed red line denotes that the loop cut $(\circlearrowleft, \{a, b, c, d\}, \{e, f\})$ is not present.

6.3.2 Inductive Miner - incompleteness (IMc)

In the following, we explore a way to use information from incomplete logs that could help to rediscover the original model. That is, introduce an algorithm that instead of selecting a cut that perfectly matches a footprint of Lemma 5.2.2, searches for the cut that comes closest, using the activity relations of Section 5.3. In the remainder of this section, we describe how this algorithm, *Inductive Miner - incompleteness* (IMc), implements the IM framework: cut detection, log splitting, base cases and fall throughs.

6.3.2.1 Cut Detection

Cut detection in IMc changed completely compared to IM and IMF: instead of searching for a cut that perfectly matches the footprints of Lemma 5.3, it selects the cut that comes closest to a footprint, i.e. the most probable cut. To explain the steps necessary to select the most probable cut, we reuse the example process tree M_{98} and log L_{99} of the previous section, i.e $M_{98} =$ 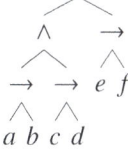, from which the event

log L_{99} was derived. The directly follows graph of L_{99} was shown in Figure 6.8, and the edge $f \twoheadrightarrow c$ of M_{98} is missing in L_{99}. Without this edge, the footprint of cut $(\circlearrowleft, \{a, b, c, d\}, \{e, f\})$, which conforms to M_{98}, is not present as Requirement $\circlearrowleft.4$ does not hold for it. Furthermore, no other cut is present.

To determine the probability that a cut would show up when more behaviour would be added to the log, IMc uses the activity relations introduced in Section 5.3, which express the type of relation of two activities: we identified the five relations $\times$, $\rightarrow$, $\wedge$, $\circlearrowleft_i$ and $\circlearrowleft_s$. Information in the log may allow us to conclude that a particular relation between two activities cannot hold. For instance, in L_{99}, $\underline{\circlearrowleft_i}(f, c)$

has been observed, i.e. a trace $\langle \dots f \dots c \dots \rangle$ and a trace $\langle \dots c \dots f \dots \rangle$, hence observing more behaviour will not allow us to conclude that e.g. $\rightarrow(f, c)$ holds, as $\langle \dots c \dots f \dots \rangle$ has already been observed, which violates $\rightarrow(f, c)$ (we assume that the event log contains neither deviating nor infrequent behaviour). These violations follow from the lattice recalled in Figure 6.9: if the log contains information that a relation $\oplus$ holds, then any weaker relation, i.e., not reachable from $\oplus$ in the lattice, cannot hold after seeing more traces: one can only move up in the lattice when more behaviour is observed.

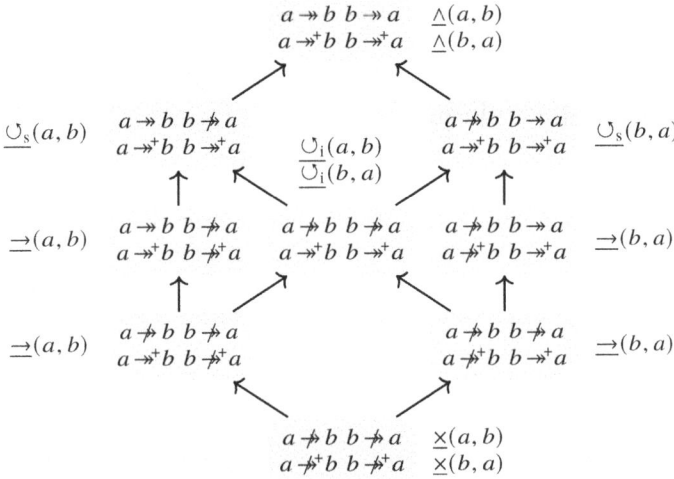

Fig. 6.9: Activity relations; the arrows define a lattice.

However, even when knowing that a weaker relation does not hold, stronger relations than $\otimes$ might still hold. As we do not have precise information about which relations hold, IMc uses an estimated probability that a relation holds instead of a binary choice: for each of the activity relations $\oplus$, we introduce a probabilistic version $p_\oplus$: for activities a and b, $p_\oplus(a, b)$ denotes an estimated probability that (a, b) are in a $\oplus$-relation. These probabilistic versions make it easier for techniques to handle incompleteness, e.g. instead of a binary choice whether $\circlearrowright_i(c, f)$ or $\circlearrowright_s(f, c)$ hold, we can compare the probabilities $p_{\circlearrowright_i}(a, b)$ and $p_{\circlearrowright_s}(a, b)$ to make this choice (notice that for $\circlearrowright_s$, there should be a connection $f \twoheadrightarrow c$). The actual probabilities could be chosen in several ways; in our example, we chose $p_{\circlearrowright_s}(c, f)$ to be 0.045.

These estimated probabilities are accumulated into a probabilities for cuts, and IMc searches for and selects the cut with the highest accumulated probability. In L_{99}, the cut with the highest probability is $(\circlearrowright, \{a, b, c, d\}, \{e, f\})$, and its probability is 0.72.

We first discuss how probabilities are estimated from the activity relations, second we show how they are accumulated into probabilities for cuts, and third we show

how IMc performs the cut selection. We finish the section with an example, after which, in the next section, we discuss log splitting, base cases and fall throughs.

Probabilistic Activity Relations.

In this part, we describe how we estimate a probability $p_\oplus(a, b)$ for two activities a and b and an activity relation $\oplus$. Our choice for these $p_\oplus$ is shown in Table 6.2, next we explain our rationale, after which we characterise the conditions for other choices that lead to rediscoverability. In this table, if the relations given in the first column hold, the table denotes the probability that each $\oplus$ is the 'true' relation. Let M be a model and L a log of M. Then, using Figure 6.9, we distinguish three cases and choose $p_\oplus(a, b)$ as follows:

- if $\oplus(a, b)$ holds in L, it makes sense to choose $p_\oplus(a, b)$ as the highest of all relations for the pair (a, b). The more frequent activities a and b occur in L, the more confident we are that $\oplus(a, b)$ holds for M, and not some stronger relation. We choose $p_\oplus(a, b)$ as follows: let $z(a, b) = \frac{|a|+|b|}{2}$ denote the average number of occurrences of a and b, then we define $p_\oplus(a, b) = 1 - \frac{1}{z(a,b)+1}$, yielding a number between $\frac{1}{2}$ and 1.
- if some relation $\otimes(a, b)$, holds in L from which $\oplus(a, b)$ is unreachable, then L contains a violation to $p_\oplus(a, b)$, as we assumed L to be deviation-free and the behavioural relations cannot cease to hold by adding observations. Therefore, we choose $p_\oplus(a, b) = 0$.
- if some relation $\otimes'(a, b)$ holds in L from which $\oplus(a, b)$ can be reached, i.e. $\oplus(a, b)$ could hold by adding more traces to L, we choose to divide the remaining $\frac{1}{z(a,b)+1}$ evenly over all remaining entries, such that the probabilities for each pair (a, b) sum up to 1.

In our example, in case of L_{99}, we obtain $p_{\cup_1}(a, e) = 0.82$ and $p_{\cup_\times}(c, f) = 0.045$.

One could define Table 6.2 differently, as long as for each pair of activities (a, b) and each relation $\oplus$, a probability $p_\oplus(a, b)$ is available, and as long as $p_\oplus(a, b) \geq 0.5$ if $\oplus(a, b)$ has been observed in the event log, and $p_\oplus(a, b) < 0.5$ if $\oplus(a, b)$ has not been observed in the event log yet. In Section 6.3.4, we will show that such a choice for $p_\oplus$ leads to an algorithm that provides rediscoverability.

Accumulated Probabilities.

Given such activity relation probabilities, we compute an accumulated probability for a cut. We first explain $\times$, $\rightarrow$ and $\wedge$, after which we explain $\circlearrowleft$. Informally, for $\oplus \in \{\times, \rightarrow, \wedge\}$, the accumulated probability $p_\oplus$ is the average $p_\oplus$ over all pairs of activities that are partitioned:

Definition 6.1 (accumulated probability for $\times$, $\rightarrow$ and $\wedge$) Let $c = (\oplus, \Sigma_1, \Sigma_2)$ be a cut, with $\oplus \in \{\times, \rightarrow, \wedge\}$. Then $p_\oplus(\Sigma_1, \Sigma_2)$ denotes the accumulated probability of

Table 6.2: Our proposal for probabilistic activity relations for activities a and b, with $z(a,b) = (|a| + |b|)/2$. Negations of relations are omitted from the first column.

	$p_\times(a,b)$	$p_\to(a,b)$	$p_\to(b,a)$	$p_{\cup_i}(a,b)$	$p_{\cup_s}(a,b)$	$p_{\cup_s}(b,a)$	$p_\wedge(a,b)$
(nothing)	$1-\frac{1}{z+1}$	$\frac{1}{6}\cdot\frac{1}{z+1}$	$\frac{1}{6}\cdot\frac{1}{z+1}$	$\frac{1}{6}\cdot\frac{1}{z+1}$	$\frac{1}{6}\cdot\frac{1}{z+1}$	$\frac{1}{6}\cdot\frac{1}{z+1}$	$\frac{1}{6}\cdot\frac{1}{z+1}$
$a \to^+ b$	0	$1-\frac{1}{z+1}$	0	$\frac{1}{4}\cdot\frac{1}{z+1}$	$\frac{1}{4}\cdot\frac{1}{z+1}$	$\frac{1}{4}\cdot\frac{1}{z+1}$	$\frac{1}{4}\cdot\frac{1}{z+1}$
$b \to^+ a$	0	0	$1-\frac{1}{z+1}$	$\frac{1}{4}\cdot\frac{1}{z+1}$	$\frac{1}{4}\cdot\frac{1}{z+1}$	$\frac{1}{4}\cdot\frac{1}{z+1}$	$\frac{1}{4}\cdot\frac{1}{z+1}$
$a \to^+ b \wedge b \to^+ a$	0	0	0	$1-\frac{1}{z+1}$	$\frac{1}{3}\cdot\frac{1}{z+1}$	$\frac{1}{3}\cdot\frac{1}{z+1}$	$\frac{1}{3}\cdot\frac{1}{z+1}$
$a \twoheadrightarrow b$	0	$1-\frac{1}{z+1}$	0	0	$\frac{1}{2}\cdot\frac{1}{z+1}$	0	$\frac{1}{2}\cdot\frac{1}{z+1}$
$a \twoheadrightarrow b \wedge b \to^+ a$	0	0	0	0	$1-\frac{1}{z+1}$	0	$\frac{1}{z+1}$
$b \twoheadrightarrow a$	0	0	$1-\frac{1}{z+1}$	0	0	$\frac{1}{2}\cdot\frac{1}{z+1}$	$\frac{1}{2}\cdot\frac{1}{z+1}$
$b \twoheadrightarrow a \wedge a \to^+ b$	0	0	0	0	0	$1-\frac{1}{z+1}$	$\frac{1}{z+1}$
$a \twoheadrightarrow b \wedge b \twoheadrightarrow a$	0	0	0	0	0	0	1

c:

$$p_\oplus(\Sigma_1, \Sigma_2) = \frac{\sum_{a\in\Sigma_1, b\in\Sigma_2} p_\oplus(a,b)}{|\Sigma_1| \cdot |\Sigma_2|}$$

For instance, in L_{99}, the accumulated probability of the cut $(\wedge, \{a\}, \{b, c, d, e, f\})$ is the average over $p_\wedge(a,b)$, $p_\wedge(a,c)$, $p_\wedge(a,d)$, $p_\wedge(a,e)$ and $p_\wedge(a,f)$.

By Definition 6.1, a $\to$, $\times$, or $\wedge$ cut requires all pairs of activities to be in the same relation sufficiently often. For a loop cut, this is not sufficient, as all pairs with an activity on both sides of the partition (all *crossing* pairs) in a loop are in a loop relation, i.e. $\cup_s$ or $\cup_i$. The combination of both loop relations suffices to describe the probability whether all activities are indeed in a loop, but on its own cannot distinguish the body of a loop from its redo parts. For this, we have to explicitly pick the start and end activities of the redo parts, such that a *redo start activity* follows a *body end activity*, and a redo end activity is followed by a body start activity. This direct succession in a loop is expressed in $\cup_s$. Next, we define the probability that $c = (\cup, \Sigma_1, \Sigma_2)$ is a loop cut, given a set of redo start activities S_2 and a set of redo end activities E_2. In the next section, we show how S_2 and E_2 could be chosen (one can always try all possibilities to find the possibility with the highest probability).

Definition 6.2 (accumulated probability for $\cup$) Let $c = (\cup, \Sigma_1, \Sigma_2)$ be a cut, L be a log, and $S_2 E_2 \subseteq \Sigma_2$ be sets of activities. We aggregate over three parts: start of a redo part, end of a redo part and everything else:

$$redo_{start} = \sum_{(a,b)\in End(L)\times S_2} \mathrm{p}_{\underline{\cup_s}}(a,b)$$

$$redo_{end} = \sum_{(a,b)\in E_2\times Start(L)} \mathrm{p}_{\underline{\cup_s}}(a,b)$$

$$indirect = \sum_{\substack{a\in \Sigma_1, b\in \Sigma_2 \\ (a,b)\notin (End(L)\times S_2)\cup(E_2\times Start(L))}} \mathrm{p}_{\underline{\cup_l}}(a,b)$$

Then, $\mathrm{p}_\cup(\Sigma_1, \Sigma_2, S_2, E_2)$ denotes the accumulated probability of c:

$$\mathrm{p}_\cup(\Sigma_1, \Sigma_2, S_2, E_2) = \frac{redo_{start} + redo_{end} + indirect}{|\Sigma_1| \cdot |\Sigma_2|}$$

In this definition, $redo_{start}$ and $redo_{end}$ capture the strength of S_2 and E_2 really being the start and end of the redo parts; *indirect* captures the strength that all other pairs of activities that cross Σ_1, Σ_2 are in a loop relation. For readability reasons, in the following, we will omit the parameters S_2 and E_2.

For instance, in the example log L_{99} for the cut $(\circlearrowright, \{a,b,c,d\}, \{e,f\})$, we assume knowledge that $S_2 = \{e\}$ and that $E_2 = \{f\}$:

$$\mathrm{Start}(L) = \{a, c\}$$
$$\mathrm{End}(L) = \{b, d\}$$
$$S_2 = \{e\}$$
$$E_2 = \{f\}$$
$$redo_{start} = \mathrm{p}_{\underline{\cup_s}}(b, e) + \mathrm{p}_{\underline{\cup_s}}(d, e)$$
$$= 0.818 + 0.818$$
$$= 1.636$$
$$redo_{end} = \mathrm{p}_{\underline{\cup_s}}(f, a) + \mathrm{p}_{\underline{\cup_s}}(f, c)$$
$$= 0.818 + 0.045$$
$$= 0.864$$
$$indirect = \mathrm{p}_{\underline{\cup_l}}(a, e) + \mathrm{p}_{\underline{\cup_l}}(b, f) + \mathrm{p}_{\underline{\cup_l}}(c, e) + \mathrm{p}_{\underline{\cup_l}}(d, f)$$
$$= 0.818 + 0.818 + 0.818 + 0.818$$
$$= 3.273$$
$$\mathrm{p}_\cup(\{a, b, c, d\}, \{e, f\}, S_2, E_2) = (1.636 + 0.864 + 3.273)/8$$
$$= 0.722$$

Performing the Cut Detection.

To detect a cut, IMc uses the accumulated estimates of definitions 6.1 and 6.2 to select the cut with the highest accumulated probability over all cuts, for the operators

$\times$, $\rightarrow$, $\wedge$ and $\circlearrowleft$. To select a cut with highest $p_\oplus$, our implementation uses an SMT solver: in Section 6.3.5, we will discuss the translation to an SMT problem in more detail. In case $p_\oplus = \circlearrowleft$, the SMT solver will choose S_2 and E_2 as well.

As solving the SMT problem might take a long time, and to guarantee rediscoverability, in each recursion, IMc first applies the cut-detection functions of IM, before attempting to find a cut with the method described.

Applied on our example log L_{99}, IMc starts with the probabilistic activity relations:

$\underline{\times}$	a	b	c	d	e	f
a	.	0.00	0.00	0.00	0.00	0.00
b	0.00	.	0.00	0.00	0.00	0.00
c	0.00	0.00	.	0.00	0.00	0.00
d	0.00	0.00	0.00	.	0.00	0.00
e	0.00	0.00	0.00	0.00	.	0.00
f	0.00	0.00	0.00	0.00	0.00	.

$\underline{\rightarrow}$	a	b	c	d	e	f
a	.	0.00	0.00	0.00	0.00	0.00
b	0.00	.	0.00	0.00	0.00	0.00
c	0.00	0.00	.	0.00	0.00	0.00
d	0.00	0.00	0.00	.	0.00	0.00
e	0.00	0.00	0.00	0.00	.	0.00
f	0.00	0.00	0.00	0.00	0.00	.

$\underline{\wedge}$	a	b	c	d	e	f
a	.	0.13	1.00	1.00	0.05	0.18
b	0.13	.	1.00	1.00	0.18	0.05
c	1.00	1.00	.	0.13	0.05	0.05
d	1.00	1.00	0.13	.	0.18	0.05
e	0.05	0.18	0.05	0.18	.	0.33
f	0.18	0.05	0.05	0.05	0.33	.

$\circlearrowright_s$	a	b	c	d	e	f
a	.	0.88	0.00	0.00	0.05	0.00
b	0.00	.	0.00	0.00	0.82	0.05
c	0.00	0.00	.	0.88	0.05	0.05
d	0.00	0.00	0.00	.	0.82	0.05
e	0.05	0.00	0.05	0.00	.	0.67
f	0.82	0.05	0.05	0.05	0.00	.

$\circlearrowright_i$	a	b	c	d	e	f
a	.	0.00	0.00	0.00	0.82	0.00
b	0.00	.	0.00	0.00	0.00	0.82
c	0.00	0.00	.	0.00	0.82	0.82
d	0.00	0.00	0.00	.	0.00	0.82
e	0.82	0.00	0.82	0.00	.	0.00
f	0.00	0.82	0.82	0.82	0.00	.

Second, IMc considers all possible cuts that can be made using the activities a, b, c, d, e and f. We do not list all of them here, but instead provide some examples:

$$p_{\circlearrowright}(\{a, b, c, d\}, \{e, f\}) = 0.722$$
$$p_{\times}(\{a, b\}, \{c, d, e, f\}) = 0$$
$$p_{\wedge}(\{a, b, c\}, \{d, e, f\}) = 0.299$$
$$p_{\rightarrow}(\{a, b\}, \{c, d, e, f\}) = 0$$

$$\cdots$$

From all these cuts, the cut with the highest accumulated probability is selected, i.e. here $(\circlearrowright, \{a, b, c, d\}, \{e, f\})$. Thus, even though the directly follows graph (shown in Figure 6.8) is not complete and does not contain any cut, IMc manages to select a cut according to tree M_{98} that underlies the log L_{99} by choosing the most probable cut.

Local Guarantees.

The aim of IMc is to handle event logs with incomplete information, i.e. to rediscover a system even though the event log is not directly follows complete with respect to that system. Inherently, log precision is not preserved locally: by the introduction of extra $\twoheadrightarrow$-edges in the behaviour of the model, the model contains more behaviour than the event log. Furthermore, IMc selects a cut by maximising the accumulated cut probability, thus individual activity relations might be violated in the final cut, hence local fitness cannot be guaranteed neither. Nevertheless, our evaluation (Chapter 8) shows that IMc is more robust to incomplete behaviour than IM and other algorithms.

6.3.2.2 Log Splitting

The cuts chosen by IMc are not guaranteed to be locally fitness preserving, i.e. in the process choosing cuts and splitting logs, deviating behaviour might surface. Therefore, IMc uses the filtering log splitting functions of IMF.

6.3.2.3 Base Cases

As the aim of IMc is to handle incompleteness, we choose to use the non-filtering base cases of IM.

6.3.2.4 Fall Throughs

As IMc always discovers a cut, a fall through is only necessary if the event log consists of a single activity, or if the event log contains empty traces ϵ. That is, EMPTYTRACES and FLOWERMODEL.

6.3.2.5 Summary

To summarise, the *Inductive Miner - incompleteness* (IMc) implements the functions of the IM framework as follows:

function BASECASE$_{\text{IMc}}(L)$
    return BASECASE$_{\text{IM}}$
end function

function FINDCUT$_{\text{IMc}}(L)$
    if $\epsilon \notin L$ **then**
        return cut $(\oplus, \Sigma_1, \Sigma_2)$ of $\Sigma(L)$ with highest $p_\oplus(\Sigma_1, \Sigma_2); \oplus \in \{\times, \rightarrow, \wedge, \circlearrowleft\}$
    end if
end function

function SPLITLOG$_{\text{IMc}}(L, (\oplus, \Sigma_1, \ldots, \Sigma_n))$
    return SPLITLOG$_{\text{IMF}}$
end function

function FALLTHROUGH$_{\text{IMc}}(L)$
    $ft \leftarrow$ EMPTYTRACES(L)
    if $ft \neq \square$ **then return** ft

else return FLOWERMODEL(L)
    end if
end function

IM has a run time of $O(|L| \cdot |\Sigma(L)|^4)$, to which IMc adds an SMT step that is exponential in the number of activities, which is executed at most $|\Sigma(L)|$ times (once in each recursive step). $O(|L| \cdot |\Sigma(L)|^4 + |\Sigma| \cdot |2^{\Sigma}(L)|)$.

6.3.3 Example

As an example, consider the log

$$L_{100} = [\langle c, d, e, f, d, e, f, d, e \rangle, \langle b, a, d, e \rangle, \langle a, b, d, e, f, d, e \rangle, \langle c, g \rangle]$$

If IMc is applied to L_{100}, it first searches for the most likely cut, which is $(\rightarrow, \{a, b, c\}, \{d, e, f, g\})$, with a $p_\rightarrow$ of about 0.64. The choice for $\rightarrow$ is recorded, and L_{100} is split into

$$L_{101} = [\langle c \rangle^2, \langle b, a \rangle, \langle a, b \rangle]$$
$$L_{102} = [\langle d, e, f, d, e, f, d, e \rangle, \langle d, e \rangle, \langle d, e, f, d, e \rangle, \langle g \rangle]$$

Then, IMc recurses on both these sublogs. Figure 6.10 shows the recursive steps that are taken by IMc, which continues as IM. The final result is

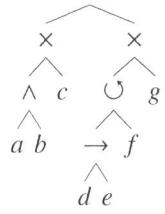

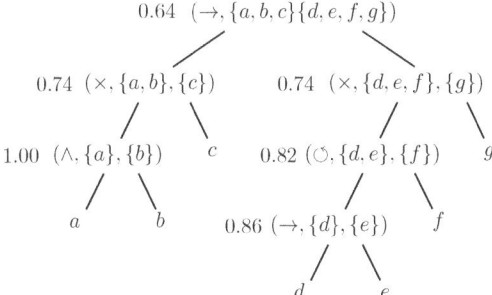

Fig. 6.10: Example of IMc applied to a log. As a first step, the cut with highest $p_\oplus$ is $(\rightarrow, \{a, b, c\}, \{d, e, f, g\})$, with $p_\oplus = 0.64$. Then, IMc recurses as shown.

6.3.4 Guarantees

We first discuss why IMc does not guarantee fitness, after which we show that IMc provides rediscoverability.

6.3.4.1 Fitness

In contrast with IM, IMc does not guarantee fitness. IMc cannot guarantee fitness, as a cut with the highest probability does not necessarily honour all observed activity relations.

For instance, consider the process tree $M_{103} =$ 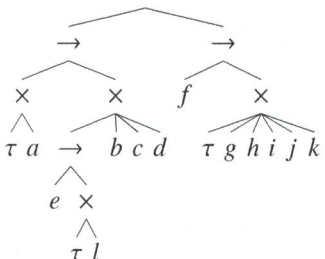 , and a

log L_{104} consisting of 1000 traces of M_{103} and an extra trace $\langle e, l \rangle$. Figure 6.11 selects the directly follows graph of L_{104}. Using this directly follows graph, IMc discovers the cut $(\times, \{a, b, c, d, e, l\}, \{f, g, h, i, j, k\})$. Then, splitting e.g. the trace $\langle f, l \rangle$ results in either the f or l event to be excluded. Consequently, the resulting model does not fit L_{104}.

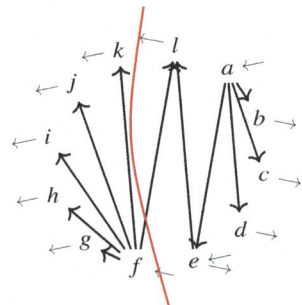

Fig. 6.11: $\twoheadrightarrow(L_{104})$. The red line denotes the cut that is selected by IMc.

6.3.4.2 Rediscoverability.

Similar to IMF, the fact that IMc as a first step applies IM cut detection, IMF log splitting and IM base cases, already provides rediscoverability:

Theorem 6.8 (IMc rediscoverability) *Take a system $S \in C_B$ and a log L such that* $set(L) \subseteq \mathcal{L}(S)$ *and* $\twoheadrightarrow(L) = \twoheadrightarrow(S)$. *Then,* $\mathcal{L}(IMc(L)) = \mathcal{L}(S)$.

However, in order to show that the new cut detection is consistent with IM, we prove rediscoverability also for the IMc SMT-cut detection in isolation, i.e. let FINDCUT$_{IMc}$ be FINDCUT$_{IMc}$ without the FINDCUT$_{IM}$ step.

Then, in order to prove rediscoverability, we reuse the results of Lemma 6.11, which shows that IM is abstraction preserving, and only need to prove Requirement AP.4, which states that each cut returned by FINDCUT$_{IMc}$ should conform to S. The following lemma proves that Requirement AP.4 holds for IMc. This lemma makes an additional assumption, i.e. that the activities of S appear in L at least a certain number of times. Let $least(L)$ denote the number of times the least occurring activity occurs in a log L (we will give a characterisation of a lower bound for $least(L)$ later).

Lemma 6.13 (IMc cut conformance) *Let $S \in C_B$ be a reduced system. Then, there exists a $l \in \mathbb{N}$ such that for all logs L with $set(L) \subseteq \mathcal{L}(S)$, $\twoheadrightarrow(L) = \twoheadrightarrow(S)$ and $least(L) \geq l$, and for which BASECASE$_{IMc}(L)$ does not apply, it holds that* FINDCUT$_{IMc}(L)$ *conforms to S (Definition 5.4).*

We prove this lemma as follows: we first show that IMc selects the correct root operator (Lemma 6.14), and second that IMc selects a partition corresponding to S (Lemma 6.15).

In the first lemma, we prove that for each log for which $least$ is sufficiently large, IMc selects the correct root operator.

Lemma 6.14 (IMc selects the correct root operator) *Assume a model $S = \oplus(S_1, \ldots, S_n)$ reduced according to Definition 5.1. Then there exists a $l \in \mathbb{N}$ such that for all logs L with $set(L) \subseteq \mathcal{L}(S)$, $\twoheadrightarrow(L) = \twoheadrightarrow(S)$ and $least(L) \geq l$, it holds that IMc(L) selects $\oplus$.*

Proof IMc selects binary cuts, while S can have an arbitrary number of children. Without loss of generality, assume that $c = (\oplus, \Sigma_1, \Sigma_2)$ is a binary cut conforming to S. Let $c' = (\otimes, \Sigma'_1, \Sigma'_2)$ be an arbitrary cut of S, with $\otimes \neq \oplus$. We need to prove that $p_\oplus(\Sigma_1, \Sigma_2) > p_\otimes(\Sigma'_1, \Sigma'_2)$, which we do by computing a lower bound for $p_\oplus(\Sigma_1, \Sigma_2)$ and an upper bound for $p_\otimes(\Sigma'_1, \Sigma'_2)$ and then comparing these two bounds. Apply case distinction on whether $\oplus = \circlearrowright$:

$\oplus \neq \circlearrowright$ We start with the lower bound for $p_\oplus(\Sigma_1, \Sigma_2)$. By Definition 6.1,

$$p_\oplus(\Sigma_1, \Sigma_2) = \frac{\sum_{a \in \Sigma_1, b \in \Sigma_2} p_\oplus(a, b)}{|\Sigma_1| \cdot |\Sigma_2|}$$

By semantics of process trees, the relations defined in Figure 6.9, the probabilities defined in Table 6.2, $set(L) \subseteq \mathcal{L}(S)$, and $\twoheadrightarrow(L) = \twoheadrightarrow(S)$, for each activity pair (a, b) that crosses c, $\oplus(a, b)$ holds. For each such pair, we defined $p_\oplus(a, b) \geq 1 - \frac{1}{z(a,b)+1}$ (notice that this is an equality for all operators except $p_\wedge$). Thus,

$$p_\oplus(\Sigma_1, \Sigma_2) \geq \frac{\sum_{a \in \Sigma_1, b \in \Sigma_2} 1 - \frac{1}{z(a,b)+1}}{|\Sigma_1| \cdot |\Sigma_2|}$$

For all a and b, $z(a, b) = \frac{|a|+|b|}{2} \geq \min(|a|, |b|) \geq least(L)$. Thus,

$$p_\oplus(\Sigma_1, \Sigma_2) \geq \frac{\sum_{a \in \Sigma_1, b \in \Sigma_2} 1 - \frac{1}{least(L)+1}}{|\Sigma_1| \cdot |\Sigma_2|}$$

$$p_\oplus(\Sigma_1, \Sigma_2) \geq 1 - \frac{1}{least(L) + 1} \tag{6.1}$$

Next, we prove an upper bound for $p_\otimes(\Sigma_1', \Sigma_2')$. By Definition 6.1,

$$\frac{\sum_{a \in \Sigma_1', b \in \Sigma_2'} p_{\underline{\otimes}}(a, b)}{|\Sigma_1'| \cdot |\Sigma_2'|} = p_\otimes(\Sigma_1', \Sigma_2')$$

Let (u, v) be a pair partitioned by both Σ_1, Σ_2 and Σ_1', Σ_2'. By Lemma 5.8, such a pair exists. For all other $(a, b) \neq (u, v)$, it holds that $p_{\underline{\otimes}}(a, b) \leq 1$ (abusing notation a bit by combining $\cup_i$ and $\cup_s$), and there are $|\Sigma_1| \cdot |\Sigma_2| - 1$ of those pairs.

$$\frac{(|\Sigma_1'| \cdot |\Sigma_2'| - 1) \cdot 1 + 1 \cdot p_{\underline{\otimes}}(u, v)}{|\Sigma_1'| \cdot |\Sigma_2'|} \geq p_\otimes(\Sigma_1', \Sigma_2')$$

As (u, v) crosses c, $\oplus(u, v)$ holds. Then by inspection of Table 6.2, $p_{\underline{\otimes}}(u, v) \leq \frac{1}{z(u,v)+1}$. Define y to be $|\Sigma_1'| \cdot |\Sigma_2'|$.

$$\frac{(y - 1) + \frac{1}{z(u,v)+1}}{y} \geq p_\otimes(\Sigma_1', \Sigma_2')$$

From $z(a, b) = \frac{|a|+|b|}{2} \geq 1$ follows that $\frac{1}{z(u,v)+1} \leq \frac{1}{2}$. Thus,

$$\frac{(y - 1) + \frac{1}{2}}{y} \geq p_\otimes(\Sigma_1', \Sigma_2') \tag{6.2}$$

Using the two bounds (6.1) and (6.2), we need to prove that

$$1 - \frac{1}{least(L) + 1} > \frac{(y - 1) + \frac{1}{2}}{y} \tag{6.3}$$

Note that y is at most $\lfloor \Sigma(S)/2 \rfloor \cdot \lceil \Sigma(S)/2 \rceil$, which allows us to choose l such that $l > 2y - 1$. By initial assumption $least(L) \geq l$, and therefore (6.3) holds. Hence, $p_\oplus(\Sigma_1, \Sigma_2) > p_\otimes(\Sigma_1', \Sigma_2')$.

$\oplus = \circlearrowleft$ Using reasoning similar to the $\oplus \neq \circlearrowleft$ case, while taking S_2, E_2 and the difference between $\circlearrowleft_s$ and $\circlearrowleft_i$ into account, we derive (6.1). We directly reuse (6.2) to arrive at (6.3) and conclude that $p_\oplus(\Sigma_1, \Sigma_2) > p_\otimes(\Sigma_1', \Sigma_2')$.

Thus, $p_\oplus(\Sigma_1, \Sigma_2) > p_\otimes(\Sigma_1', \Sigma_2')$ holds for all $\oplus$. As IMc selects the cut with highest $p_\oplus$, IMc selects $\oplus$. $\qquad\square$

Next, we prove that for a log L, if $least(L)$ is sufficiently large, then IMc will select a partition conforming to S.

Lemma 6.15 (IMc selects a correct partition) *Assume a model $S = \oplus(S_1, \ldots, S_n)$ in normal form. Let $c = (\oplus, \Sigma_1, \Sigma_2)$ be a cut conforming to S, and let $c' = (\oplus, \Sigma_1', \Sigma_2')$ be a cut not conforming to S. Then there exists a $l \in \mathbb{N}$ such that for all logs L with $set(L) \subseteq \mathcal{L}(S)$, $\twoheadrightarrow(L) = \twoheadrightarrow(S)$ and $least(L) \geq k$, holds that $p_\oplus(\Sigma_1, \Sigma_2) > p_\oplus(\Sigma_1', \Sigma_2')$.*

Proof We follow a similar reasoning as in the proof of Lemma 6.14 to prove that $p_\oplus(\Sigma_1, \Sigma_2) > p_\oplus(\Sigma_1', \Sigma_2')$: we prove a lower bound for $p_\oplus(\Sigma_1, \Sigma_2)$, an upper bound for $p_\oplus(\Sigma_1', \Sigma_2')$ and compare these two. Apply case distinction on whether $\oplus = \circlearrowleft$.

$\oplus \neq \circlearrowleft$ Obviously, (6.1) holds in this case as well. For the upper bound for $p_\oplus(\Sigma_1', \Sigma_2')$, we start with

$$\frac{\sum_{a \in \Sigma_1', b \in \Sigma_2'} p_\oplus(a, b)}{|\Sigma_1'| \cdot |\Sigma_2'|} = p_\oplus(\Sigma_1', \Sigma_2') \tag{6.4}$$

As the cut $c' = (\oplus, \Sigma_1', \Sigma_2')$ does not conform to S, there is a $\Sigma(S_i)$ partitioned by c': $\Sigma_1' \cap \Sigma(S_i) \neq \emptyset$ and $\Sigma_2' \cap \Sigma(S_i) \neq \emptyset$. Consider this $S_i = \otimes(\ldots)$, then $c_{S_i}' = (\otimes, (\Sigma(S_i) \cap \Sigma_1'), (\Sigma(S_i) \cap \Sigma_2'))$ is a cut of S_i. Take an arbitrary cut c_{S_i} that conforms to S_i. By Lemma 5.8, at least one activity pair (u, v) is partitioned by both c_{S_i} and c_{S_i}'. For all other $(a, b) \neq (u, v)$, by Table 6.2, it holds that $p_\oplus(a, b) \leq 1$, and there are $|\Sigma_1| \cdot |\Sigma_2| - 1$ of those pairs. Applying this to (6.4), we derive:

$$\frac{(|\Sigma_1'| \cdot |\Sigma_2'| - 1) \cdot 1 + 1 \cdot p_\oplus(u, v)}{|\Sigma_1'| \cdot |\Sigma_2'|} \geq p_\oplus(\Sigma_1', \Sigma_2')$$

As c_{S_i} conforms to S_i and $\oplus \neq \circlearrowleft$, we conclude that $\otimes(u, v)$ holds. As S is in normal form $\otimes \neq \oplus$, and therefore $\oplus(u, v)$ does not hold. Then, by Table 6.2, $p_\oplus(u, v) \leq \frac{1}{z(u,v)+1}$. From $z(u, v) = \frac{|u|+|v|}{2} \geq 1$ follows that $p_\oplus(u, v) \leq \frac{1}{z(u,v)+1} \leq \frac{1}{2}$. Define y to be $|\Sigma_1'| \cdot |\Sigma_2'|$.

$$\frac{(y - 1) + \frac{1}{2}}{y} = \frac{(|\Sigma_1'| \cdot |\Sigma_2'| - 1) \cdot 1 + \frac{1}{2}}{|\Sigma_1'| \cdot |\Sigma_2'|} \geq p_\oplus(\Sigma_1', \Sigma_2') \tag{6.5}$$

Similar to the proof of Lemma 6.14, from (6.1), (6.5) and choosing $l > 2y - 1$, follows that $p_\oplus(\Sigma_1, \Sigma_2) > p_\oplus(\Sigma_1', \Sigma_2')$.

$\oplus = \circlearrowright$ We follow a reasoning similar to the proof of Lemma 6.14, and derive the lower bound (6.1) again. For the upper bound for $p_\oplus(\Sigma_1', \Sigma_2')$, similar to the proof of Lemma 6.14, we derive

$$\frac{\sum_{a \in \Sigma_1', b \in \Sigma_2'} p_\oplus(a, b)}{|\Sigma_1'| \cdot |\Sigma_2'|} \geq p_\oplus(\Sigma_1', \Sigma_2')$$

As $\twoheadrightarrow(L) = \twoheadrightarrow(S)$ and by semantics of $\circlearrowright$, $\twoheadrightarrow^+$ holds for all activity pairs. Thus, $\circlearrowright_s \cup \circlearrowright_i \cup \wedge$ contains all activity pairs. By Requirement $\circlearrowright.1$, $Start(S) = Start(\overline{S_1}) \subseteq \Sigma_1$ and $Start(S_1) \subseteq \Sigma_1'$.

c' separates at least a $\Sigma(S_i)$. Let (u, v) be a pair of activities of $\Sigma(S_i)$ separated by c'. Prove by case distinction on whether $\Sigma(S_i) = \Sigma_1$ that at least one pair (u, v) is counted wrongly.

$\Sigma(S_i) = \Sigma_1$ Towards contradiction, assume no misclassified pair exists in $\Sigma(S_1)$. Take an arbitrary $a_k \in \Sigma(S_1)$. Apply case distinction on whether a_k is a start or an end activity.

- If $a_k \in Start(S)$ or $a_k \in End(S)$, by Requirement $\circlearrowright.1$, $a_k \in \Sigma_1'$.
- Consider two $\twoheadrightarrow$-paths: one path from a start activity to a_k: $a_1 \ldots a_k$ such that $a_1 \in Start(S)$ and $\forall_{a_{j>1}}\, a_j \notin Start(S) \cup End(S)$, and one from a_k to an end activity: $a_k \ldots a_l$ such that $a_l \in End(S)$ and $\forall_{a_{j<l}}\, a_j \notin Start(S) \cup End(S)$ (see Figure 6.12). Apply case distinction on whether such paths exist.

  $\exists a_1 \ldots a_k$ Then some pair (a_p, a_q), on this path crosses c'. As (a_p, a_q) is on a $\twoheadrightarrow$-path, $a_p \twoheadrightarrow a_q$ holds, so either $\circlearrowright_s(a_p, a_q)$ or $\wedge(a_p, a_q)$. Activity a_q is not a start activity and a_p is not an end activity, so (a_p, a_q) contributes as $\circlearrowright_i$ towards $p_\circlearrowright(c')$.

  $\exists a_k \ldots a_l$ Similar.

  $\nexists a_1 \ldots a_k \wedge \nexists a_k \ldots a_l$ Then a_k must be on a $\twoheadrightarrow$-path; let this path be $a_l \ldots a_k \ldots a_1$ with $a_1 \in Start(S)$ and $a_k \in End(S)$. As $S_1 \neq \circlearrowright$, this can only happen if $S_1 = \times$, which means that there is a $a_1' \in Start(S)$ such that no $\twoheadrightarrow$-path $a_k \ldots a_1'$ exists. Then, $\circlearrowright_i(a_k, a_1')$, but (a_k, a_1') contributes as $\circlearrowright_s$.

$\Sigma(S_i) \neq \Sigma_1$ As S is reduced, $S_i \neq \oplus(\ldots)$ and thus, the $\twoheadrightarrow$-graph of S_i is connected. By semantics of process trees, there is at least a start or end activity that can be executed before/after both u and v: either $\langle s \cdots u \rangle$ and $\langle s \cdots v \rangle$ or $\langle u \cdots e \rangle$ and $\langle v \cdots e \rangle$, with $s \in Start(S_i)$ and $e \in End(S_i)$. Without loss of generality, assume that two $\twoheadrightarrow$-paths $\langle s \cdots u \rangle$ and $\langle s \cdots v \rangle$ exist in the $\twoheadrightarrow$-graph, such that $s \in Start(S_i)$. The pair (u, v) crosses c', so one of these paths must cross c' as well. Let (x, y) be such a crossing pair in the $\twoheadrightarrow$-graph. As $\twoheadrightarrow(x, y)$, either $\wedge(x, y)$ or $\circlearrowright_s(x, y)$. Neither x nor y are start or end activities of S, so the pair (x, y) contributes as $\circlearrowright_i$ to the average $p_\circlearrowright(c')$.

Hence, at least one pair (u, v) is counted wrongly. Left to prove: this pair has a large enough influence on the final accumulated probability, thus the lower bound for $p_\oplus(\Sigma_1, \Sigma_2)$ and the upper bound for $p_\oplus(\Sigma'_1, \Sigma'_2)$ are separated. The remaining part of this case is similar to the case $\oplus \neq \circlearrowleft$.

Hence, $p_\oplus(\Sigma_1, \Sigma_2) > p_\oplus(\Sigma'_1, \Sigma'_2)$, so IMc will select a partition conforming to S.   □

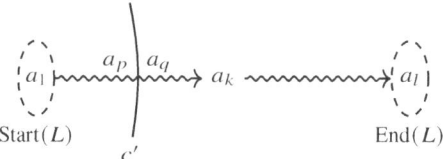

Fig. 6.12: Illustration of paths used in the proof of Lemma 6.15.

From these lemmas and the fact that for each tree there is a language-equivalent binary tree (applying reduction rules of Definition 5.1 in reverse if necessary), rediscoverability of IMc follows directly, which shows that the cut detection methods introduced in this section are sufficient to provide rediscoverability, under assumption that the log is directly follows complete and has no deviations with respect to the system, and assuming that the event log contains each activity at least l times.

In the proofs of lemmas 6.14 and 6.15, we chose $l > 2 \cdot \lfloor \Sigma(S)/2 \rfloor \cdot \lceil \Sigma(S)/2 \rceil - 1$. This gives an upper bound for the minimum $least(L)$ required, and a characterisation of sufficiency:

Corollary 6.2 (bound for $least$) *A bound for k and $least(L)$ as used in lemmas 6.14 and 6.15 is determined by the size of the alphabet: $least(L) \geq l \geq 2 \cdot \lfloor |\Sigma(S)|/2 \rfloor \cdot \lceil |\Sigma(S)|/2 \rceil$.*

In Chapter 8, we investigate the influence of incompleteness on rediscoverability.

Last, the unsolved question remaining is whether directly follows completeness of a log implies that the log is sufficiently large, and that a generalised version of Lemma 6.13 holds:

Conjecture 6.1 Assume a model $S \in C_B$ and a log L such that $set(L) \subseteq \mathcal{L}(S)$ and $\twoheadrightarrow(L) = \twoheadrightarrow(S)$. Then $\mathcal{L}(\text{IMc}(L)) = \mathcal{L}(S)$.

We finish this section with a more detailed description how IMc finds the cut with the highest accumulated probability. In the next section, we extend IM and IMF to handle more process tree constructs: τ, $\vee$ and $\leftrightarrow$.

6.3.5 Finding Cuts: Translation to SMT

In the previous parts of this section, we introduced IMc, an algorithm that handles incompleteness of behaviour. Furthermore, we showed that IMc guarantees redis-coverability. In this section, we describe the cut detection step of IMc in more detail, i.e. we describe how the problem of finding the most probable cut is translated to several SMT problems.

6.3.5.1 Translating ×, →, ∧

Cut searches for ×, → and ∧ are translated straightforwardly to optimisation problems by maximising the average probability of edges crossing the cut.

For $\oplus \in \{\times, \rightarrow, \wedge\}$, $p_\oplus = \frac{\sum_{a_1 \in \Sigma_1, a_2 \in \Sigma_2} p_\oplus(a_1, a_2)}{|\Sigma_1| \cdot |\Sigma_2|} = \frac{k}{l}$. Let n be $|\Sigma(L)|$. For the commutative × and ∧, we vary l from 1 to $n/2$, for the non-commutative → we vary l from 1 to $n - 1$. The basic decision to be made by the SMT solver is how to divide the activities in two sets: the ones on one side of the cut ($cut(a)$) and the ones on the other side of the cut ($\neg cut(a)$), such that $p_\oplus$ is maximised. As divisions cannot be translated to SMT directly, we solve multiple SMT problems with varying l. Each of these SMT problems will return a most probable cut for the l considered, and the most likely cut over all l is returned. We give the translation to SMT for the non-commutative →; the commutative × and ∧ are similar.

For a chosen l, the number of nodes on the left-hand side of the cut shall be l, so we add the constraint

$$|\{a | cut(a)\}| = l$$

$p_\oplus$ is defined on pairs of activities, so for each pair (a_1, a_2) we introduce a helper variable $crosses(a_1, a_2)$, denoting whether (a_1, a_2) crosses the cut:

$$crosses(a_1, a_2) \Leftrightarrow (cut(a_1) \wedge \neg cut(a_2))$$

The objective function to be maximised is the weighted sum of the crossing edges:

$$obj = \sum_{a_1 \in \Sigma_1, a_2 \in \Sigma_2} crosses(a_1, a_2) \cdot p_\oplus(a_1, a_2)$$

For ∧, a constraint is added that both Σ_1 and Σ_2 contain both start and end activities.

The SMT problem consists of the conjunction of these formulae. Once an optimal solution for the SMT problem is found, $\frac{obj}{l}$ gives the probability of $p_\oplus$ for the given l. This procedure is repeated for l varying between 1 and $|\Sigma(L)| - 1$, i.e. $l - 1$ SMT problems are solved, and the cut with the highest $p_\oplus$ is returned.

6.3.5.2 Translating ↺

For $p_↺$, each pair (a, b) that crosses the cut is categorised as being either *indirect, single* or *reverse single*. These correspond to $↺_i$ (indirect), $↺_s$ (single and reverse single). They are defined as follows:

$$(a, b) \text{ single} \Leftrightarrow (a \in End_1 \wedge b \in Start_2) \vee$$
$$(a \in End_2 \wedge b \in Start_1)$$
$$(a, b) \text{ reverse single} \Leftrightarrow (b, a) \text{ single}$$
$$(a, b) \text{ indirect} \Leftrightarrow \neg(a, b) \text{ single} \wedge \neg(b, a) \text{ single}$$

We give an example using Figure 6.13, which shows a directly follows graph. In this example, $\Sigma_1 = \{u, v\}$, $\Sigma_2 = \{w, x\}$, $Start_1 = \{u\}$, $End_1 = \{v\}$, $Start_2 = \{x\}$, and $End_2 = \{w\}$. Pairs (u, x), (v, w), (x, u) and (w, v) are indirect, (w, u) and (v, x) are single and (u, w) and (x, v) are reverse single.

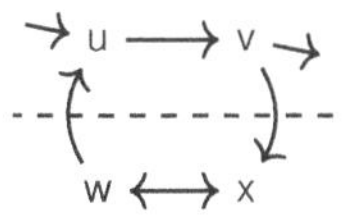

Fig. 6.13: Example ↠-graph. The dashed line denotes a cut.

The optimisation searches for assignments to Σ_1, Σ_2, $Start_2$ and End_2, and a classification of all edges that maximises the average probability of single and indirect edges. $Start_1$ and End_1 are taken as-is from the log.

Conclusion.

In this section, we have shown how incomplete behaviour, i.e. the directly follows graph not containing all information, might prevent rediscovery of a system by IM. We introduced a new algorithm, the *Inductive Miner - incompleteness* (IMc), which searches for the most likely cut instead of searching for a perfect cut, by estimating pairwise probabilities for activities and activity relations. Using a translation to SMT, IMc searches for the most likely cut and returns it. We showed that IMc does not provide local log-precision or fitness preservation, i.e. IMc guarantees neither fitness nor precision. However, we proved that rediscoverability holds, i.e. if the log is part of the language of the system, the system is of class C_B and the directly follows graph of the log is equal to the directly follows graph of the system, then IMc will rediscover the language of the system. Obviously, in such a case the discovered model and the log are fitting. This illustrates the flexibility of the IM framework: without changing its structure, we introduced an algorithm to handle infrequent and deviating behaviour (IMF) and an algorithm to handle incompleteness (IMC). In Chapter 8, we will evaluate the true gains in incompleteness handling provided by IMC.

All three algorithms introduced in the previous parts of chapter were restricted to the four operators $\times$, $\rightarrow$, $\wedge$ and $\circlearrowleft$, and no silent steps (τ) were allowed. In the next section, we introduce an algorithm that can handle all process tree constructs that were introduced in Section 2.2.5, i.e. we add τ leafs, $\vee$ operators and $\leftrightarrow$ operators.

6.4 Handling More Constructs: τ, $\leftrightarrow$ and $\vee$

In the previous section, we have introduced a process discovery algorithm for the operators $\times$, $\rightarrow$, $\wedge$ and $\circlearrowleft$. In this section, we study extensions for three more tree constructs: the silent activity τ, the interleaved operator $\leftrightarrow$ and the inclusive choice operator $\vee$. We also introduce a new discovery algorithm: *Inductive Miner - all operators* (IMA). We study the influence of these additions on rediscoverability, local log-precision preservation and local fitness preservation.

The interleaved operator has a distinctive footprint in the directly follows graph, which was introduced in Lemma 5.10. The algorithms introduced in this section use this footprint to detect interleaved behaviour. We will show that rediscoverability and fitness are guaranteed.

As shown in Section 5.6, τ, $\vee$ and $\wedge$ constructs do not have unique directly follows graphs and hence cannot be detected reliably by pure directly follows based discovery algorithms, as algorithms based on directly follows graphs would guarantee neither fitness nor rediscoverability. However, in Section 5.6 we identified three coo relations that allow to distinguish τ, $\vee$ and $\wedge$: concurrency $\overline{\wedge}$, concurrent optionality $\overline{\wedge?}$ and interchangeability $\overline{\vee}$.

In this section, we will introduce two algorithms that use these three coo relations to distinguish and discover $\vee$ and $\wedge$ operators. One of these algorithms, IMA, is a basic variant that guarantees fitness and corresponds to IM. Another algorithm filters infrequent and deviating behaviour, *Inductive Miner - infrequent - all operators* (IMFA) which makes it suitable for e.g. discovering 80% models, i.e. analogue to IMF. We start with an example in Section 6.4.1, after which we introduce the two algorithms in sections 6.4.2 and 6.4.3. We finish the section with a discussion of the guarantees provided by both new algorithms in Section 6.4.4.

6.4.1 Example

Consider the event log

$L_{105} = [\langle a,d\rangle, \quad \langle a,b,d\rangle, \quad \langle a,b,c,d\rangle, \quad \langle a,e\rangle, \quad \langle a,b,e\rangle,$

$\quad \langle a,b,c,e\rangle, \quad \langle a,d,e\rangle, \quad \langle a,b,d,e\rangle, \quad \langle a,b,c,d,e\rangle, \quad \langle d,a\rangle,$

$\quad \langle b,a,d\rangle, \quad \langle c,a,d,b\rangle, \quad \langle e,a\rangle, \quad \langle e,b,a\rangle, \quad \langle e,a,c,b\rangle,$

$\quad \langle d,c,b,a\rangle, \quad \langle a,e,d\rangle, \quad \langle e,c,b,a\rangle]$

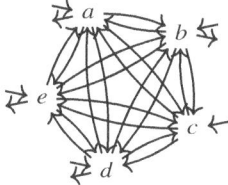

Fig. 6.14: $\twoheadrightarrow$ of L_{105}.

IMA implements the IM framework, so it considers base cases, cuts and fall throughs. On this event log, no base case applies as the log contains multiple activities. Second, IMA searches for a cut by considering the directly follows graph, which is shown in Figure 6.14, in which the footprint of both $\vee$ and $\wedge$ is present, with the activity partition $(\{a\}, \{b\}, \{c\}, \{d\}, \{e\})$. Hence, the question is which of these operators should be chosen. To decide this, IMA applies a bottom-up procedure that starts from the sets of activities identified in the cut. Then, it finds out whether a set of activities is optional, and whether two sets of activities are in an $\wedge$, an $\overline{\wedge?}$ or in an $\overline{\vee}$ relation. Any two sets of activities that are in such a relation are merged and the procedure is repeated until two sets are left. Finally, the relation between these sets determines the tree operator to be chosen.

In our example, IMA starts with the following activity sets, i.e. the activity partition of the identified cut:

$$P_1 \leftarrow \{\{a\}, \{b\}, \{c\}, \{d\}, \{e\}\}$$

Using these sets and L_{105}, IMA constructs the *activity set log*, expressing the language over the activity partition:

$T_1 \leftarrow \{\langle\{a\}, \{d\}\rangle, \quad \langle\{a\}, \{b\}, \{d\}\rangle, \quad \langle\{a\}, \{b\}, \{c\}, \{d\}\rangle,$

$\quad \langle\{a\}, \{e\}\rangle, \quad \langle\{a\}, \{b\}, \{e\}\rangle, \quad \langle\{a\}, \{b\}, \{c\}, \{e\}\rangle,$

$\quad \langle\{a\}, \{d\}, \{e\}\rangle, \quad \langle\{a\}, \{b\}, \{d\}, \{e\}\rangle, \quad \langle\{a\}, \{b\}, \{c\}, \{d\}, \{e\}\rangle\}$

Next, IMA considers which coo-relations hold for the activity sets in T_1. In our example, $\{c\}$ is optional, as each trace that occurs with a $\{c\}$ also occurs without it. Furthermore, in every trace in which $\{c\}$ occurs, $\{b\}$ occurs as well, thus $\{c\} \overline{\wedge?} \{b\}$. For every trace in which either $\{d\}$ or $\{e\}$ occurs, all variants of that trace with only $\{d\}$, only $\{e\}$ and both $\{d\}$ and $\{e\}$ occur as well, thus $\{d\} \overline{\vee} \{e\}$.

$$\{c\} \overline{\wedge?} \{b\}$$

$$\{d\} \overline{\vee} \{e\}$$

To move up in the hierarchy, one of the detected relations is chosen and all related activities are merged into one set. Subsequently, coo-relations between this new set and the other sets of activities can be detected. In our example, IMA arbitrarily chooses the first one, merges c and b and constructs a new activity set log:

$$P_2 \leftarrow \{\{a\}, \{b, c\}, \{d\}, \{e\}\}$$
$$T_2 \leftarrow \{\langle\{a\}, \{d\}\rangle, \quad\quad \langle\{a\}, \{b, c\}, \{d\}\rangle, \quad \langle\{a\}, \{e\}\rangle,$$
$$\langle\{a\}, \{b, c\}, \{e\}\rangle, \quad \langle\{a\}, \{d\}, \{e\}\rangle, \quad \langle\{a\}, \{b, c\}, \{d\}, \{e\}\rangle\}$$

In this log, the following coo relations hold:

$$\{b, c\} \overline{\wedge?} \{a\}$$

$$\{d\} \overline{\vee} \{e\}$$

IMA arbitrarily chooses the first one, merges a, b and c and constructs a new activity set log:

$$P_3 \leftarrow \{\{a, b, c\}, \{d\}, \{e\}\}$$
$$T_3 \leftarrow \{\langle\{a, b, c\}, \{d\}\rangle, \langle\{a, b, c\}, \{e\}\rangle, \langle\{a, b, c\}, \{d\}, \{e\}\rangle\}$$

In $T_{1,3}$, the following coo relation holds:

$$\{d\} \overline{\vee} \{e\}$$

IMA merges d and e and obtains

$$P_4 \leftarrow \{\{a, b, c\}, \{d, e\}\}$$
$$T_4 \leftarrow \{\langle\{a, b, c\}, \{d, e\}\rangle\}$$

In T_4, the coo relation $\{a, b, c\} \overline{\wedge} \{d, e\}$ holds. As there are only two components remaining, the coo process ends and IMA selects the cut $(\wedge, \{a, b, c\}, \{d, e\})$.

Using this cut, the event log L_{105} is split into two sublogs, and recursion continues on these sublogs as usual:

$$L_{106} = [\langle a\rangle^6, \langle a, b\rangle^6, \langle a, b, c\rangle^6]$$
$$L_{107} = [\langle d\rangle^7, \langle e\rangle^7, \langle d, e\rangle^4]$$

That is, on L_{106}, IMA identifies a nested sequence → 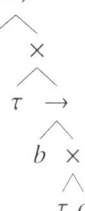 . On L_{107}, IMA

discovers an inclusive choice between d and e.

To illustrate the difference between ∨ and ∧, we show a second log with the same directly follows graph, however with the ∨-operator as root: Consider the event log

$$L_{108} = [\langle a, d\rangle, \qquad \langle a, b, d\rangle, \qquad \langle a, b, c, d\rangle, \quad \langle a, e\rangle, \qquad \qquad \langle a, b, e\rangle,$$
$$\langle a, b, c, e\rangle, \quad \langle a, d, e\rangle, \qquad \langle a, b, d, e\rangle, \quad \langle a, b, c, d, e\rangle, \quad \langle d, a\rangle,$$
$$\langle b, a, d\rangle, \qquad \langle c, a, d, b\rangle, \quad \langle e, a\rangle, \qquad \qquad \langle e, b, a\rangle, \qquad \langle e, a, c, b\rangle,$$
$$\langle d, c, b, a\rangle, \quad \langle a, e, d\rangle, \qquad \langle e, c, b, a\rangle, \quad \langle d\rangle, \qquad \qquad \langle e\rangle,$$
$$\langle d, e\rangle, \qquad \quad \langle a\rangle, \qquad \qquad \langle a, b\rangle, \qquad \quad \langle a, b, c\rangle, \qquad \langle b, a\rangle,$$
$$\langle c, a, b\rangle, \qquad \langle a, c, b\rangle, \qquad \langle c, b, a\rangle]$$

This log has the same directly follows graph as log L_{105}, which is shown in Figure 6.14. The steps taken by IMA are:

$P_{10} \leftarrow \{\{a\},\{b\},\{c\},\{d\},\{e\}\}$

$T_{10} \leftarrow \{\langle\{a\},\{d\}\rangle,$	$\langle\{a\},\{b\},\{d\}\rangle,$	$\langle\{a\},\{b\},\{c\},\{d\}\rangle,$
$\langle\{a\},\{e\}\rangle,$	$\langle\{a\},\{b\},\{e\}\rangle,$	$\langle\{a\},\{b\},\{c\},\{e\}\rangle,$
$\langle\{a\},\{d\},\{e\}\rangle,$	$\langle\{a\},\{b\},\{d\},\{e\}\rangle,$	$\langle\{a\},\{b\},\{c\},\{d\},\{e\}\rangle$
$\langle\{a\}\rangle,$	$\langle\{a\},\{b\}\rangle,$	$\langle\{a\},\{b\},\{c\}\rangle,$
$\langle\{d\}\rangle,$	$\langle\{e\}\rangle,$	$\langle\{d\},\{e\}\rangle\}$

coo relations: $\{c\}\,\overline{\wedge?}\,\{b\}$ $\{d\}\,\overline{\vee}\,\{e\}$

$P_{11} \leftarrow \{\{a\},\{b,c\},\{d\},\{e\}\}$

$T_{11} \leftarrow \{\langle\{a\},\{d\}\rangle,$	$\langle\{a\},\{b,c\},\{d\}\rangle,$	$\langle\{a\},\{e\}\rangle,$
$\langle\{a\},\{b,c\},\{e\}\rangle,$	$\langle\{a\},\{d\},\{e\}\rangle,$	$\langle\{a\},\{b,c\},\{d\},\{e\}\rangle,$
$\langle\{a\}\rangle,$	$\langle\{a\},\{b,c\}\rangle,$	$\langle\{d\}\rangle,$
$\langle\{e\}\rangle,$	$\langle\{d\},\{e\}\rangle\}$	

coo relations: $\{b,c\}\,\overline{\wedge?}\,\{a\}$ $\{d\}\,\overline{\vee}\,\{e\}$

$P_{12} \leftarrow \{\{a,b,c\},\{d\},\{e\}\}$

$T_{12} \leftarrow \{\langle\{a,b,c\},\{d\}\rangle,$	$\langle\{a,b,c\},\{e\}\rangle,$	$\langle\{a,b,c\},\{d\},\{e\}\rangle,$
$\langle\{a,b,c\}\rangle,$	$\langle\{d\}\rangle,$	$\langle\{e\}\rangle,$
$\langle\{d\},\{e\}\rangle\}$		

coo relations: $\{a,b,c\}\,\overline{\vee}\,\{d\}$ $\{a,b,c\}\,\overline{\vee}\,\{e\}$ $\{e\}\,\overline{\vee}\,\{f\}$

$P_{13} \leftarrow \{\{a,b,c\},\{d,e\}\}$

$T_{13} \leftarrow \{\langle\{a,b,c\},\{d,e\}\rangle,$	$\langle\{a,b,c\}\rangle,$	$\langle\{d,e\}\rangle\}$

coo relations: $\{a,b,c\}\,\overline{\vee}\,\{d,e\}$

Finally, the cut $(\vee,\{a,b,c\},\{d,e\})$ is selected.

In the remaining part of this section, we introduce IMA in more detail, as well as a variant that filters deviating and infrequent behaviour. Furthermore, we discuss the guarantees provided by these algorithms.

6.4.2 Inductive Miner - all operators (IMA)

To deal with τ, $\leftrightarrow$ and $\vee$, we extend the basic IM algorithm and introduce the *Inductive Miner - all operators* (IMA). We discuss in detail how IMA searches for cuts, splits the log, and which base cases and fall throughs it supports. As in Section 5.6, we focus on the discovery of $\leftrightarrow$ and $\vee$, as τ will be handled by a fall through.

6.4.2.1 Cut Detection

In this section, we introduce the cut detection mechanisms of IMA. As shown in Section 5.6, compared to the previously introduced algorithms, four cut detection functions need to be addressed: sequence, interleaved, concurrency and inclusive choice.

Sequence.

As discussed in Section 5.6.4, some changes to cut detection are necessary to detect sequence cuts if optionality might be present. Before we introduce the main cut detection function, we show a helper function that returns whether it is possible to skip an activity in an event log, in which p is the index of the set of activities in $C_1 \ldots C_n$ for which skippability is to be computed, and D is a directly follows graph.

> **function** SKIPPABLE($p, C_1 \ldots C_n, D$)   ▷ whether C_p can be skipped
>   **return** $\exists_{1 \le i < p < j \le n} \exists_{a_i \in C_i, a_j \in C_j} a_i \twoheadrightarrow_D a_j \vee$ ▷ by a $\twoheadrightarrow$-edge
> $\exists_{p < i \le n} \exists_{a_i \in C_i} a_i \in Start(D) \vee$ ▷ by a start activity after C_p
> $\exists_{1 \le i < p} \exists_{a_i \in C_i} a_i \in End(D)$ ▷ by an end activity before C_p
> **end function**

The main cut detection function consists of three parts: first, the normal sequence cut is computed. Second, for the activity sets of the partition of the cut, two relations are computed: one that describes for each activity set X the first activity set in the sequent cut that has a directly follows edge to X, and one that denotes the last activity set to which there is a directly follows edge from X. Third, the algorithm searches for a suitable pivot and maximises its scope by probing for the edges of the scope.

> **function** SEQUENCECUTSTRICT(L)
> $C \leftarrow C_1, \ldots C_n \leftarrow$ partition of SEQUENCECUT(L) (if none: fail)
>   ▷ $minFrom[x]$ the earliest activity set with an outgoing $\twoheadrightarrow$-edge to x
>   ▷ $maxTo[x]$ the latest activity set from with an incoming $\twoheadrightarrow$-edge from x
>   **for** $1 \le x \le n$ **do** $minFrom[x] \leftarrow \infty$; $maxTo[x] \leftarrow -\infty$ **end for**
>   **for** $\top \twoheadrightarrow a, a \in C_i$ **do** $minFrom[i] \leftarrow -\infty$ **end for**
>   **for** $b \twoheadrightarrow \bot, b \in C_j$ **do** $maxTo[j] \leftarrow \infty$ **end for**
>   **for** $a \twoheadrightarrow b, a \in C_i, b \in C_j$ **do**
> $minFrom[j] \leftarrow \min(minFrom[j], i)$
> $maxTo[i] \leftarrow \max(maxTo[i], j)$
>   **end for**
>   ▷ find a pivot and maximise its scope
>   **for** $p \leftarrow 1 \ldots n$ **do**
>   **if** SKIPPABLE(p, C, L) **then**
> $q \leftarrow p - 1$
>   **while** $maxTo[q] \le p$ **do**   ▷ C_q is in a pivot scope with p
>   remove C_q and add it to C_p in C
> $q \leftarrow q - 1$

           end while

           $q \leftarrow p + 1$

           while $minFrom[q] \geq p$ **do**   ▷ C_q is in a pivot scope with p

               remove C_q and add it to C_p in C

               $q \leftarrow q + 1$

           end while

        end if

      end for

      return $(\rightarrow, C)$

    end function

Interleaved.

To detect an interleaved cut, the following steps are performed:

    function INTERLEAVEDCUT(L)

      $P \leftarrow \{\{a\} | a \in \Sigma(\rightarrow)\}$

      for all $a \notin$ Start(L), $b \in \Sigma(L)$ such that $b \twoheadrightarrow a$ in L **do**

        let $a \in P_x$ and $b \in P_y$, then $P \leftarrow P \setminus \{P_x, P_y\} \cup \{P_x \cup P_y\}$

      end for

      for all $a \notin$ End(L), $b \in \Sigma(L)$ such that $a \twoheadrightarrow b$ in L **do**

        let $a \in P_x$ and $b \in P_y$, then $P \leftarrow P \setminus \{P_x, P_y\} \cup \{P_x \cup P_y\}$

      end for

      for all $a \in$ Start(L), $b \in$ End(L) such that $a \not\twoheadrightarrow b$ in L **do**

        let $a \in P_x$ and $b \in P_y$, then $P \leftarrow P \setminus \{P_x, P_y\} \cup \{P_x \cup P_y\}$

      end for

      for all $a, b \in \Sigma(L), a \neq b$ such that $a \oslash b \vee b \oslash a$ in L **do**

        let $a \in P_x$ and $b \in P_y$, then $P \leftarrow P \setminus \{P_x, P_y\} \cup \{P_x \cup P_y\}$

      end for

      for all $t \in L$ **do**   ▷ guarantee fitness

        if $\exists_{P_1, P_2 \in P} \exists_{a, c \in P_1, b \in P_2} t = \langle \ldots a \ldots b \ldots c \ldots \rangle$ **then**

          $P \leftarrow P \setminus \{P_1, P_2\} \cup \{P_1 \cup P_2\}$ merge P_1 and P_2 in P

        end if

      end for

      return $(\leftrightarrow, P_1 \ldots P_n)$

    end function

Notice that in order to locally preserve fitness, the INTERLEAVEDCUT requires an extra pass through the event log. Without the part denoted with "guarantee fitness", for some event logs, a non-fitting interleaved cut could be detected. For instance, consider the event log in Figure 6.15. For this log L_{109}, INTERLEAVEDCUT(L) will return the cut $c = (\leftrightarrow, \{a\}, \{b, c\})$ (and eventually the tree $\leftrightarrow(a, \rightarrow(b, c))$). However, this cut cannot lead to a model to which the first trace fits, i.e. $\langle a, b, c, a \rangle \notin \leftrightarrow_L(S(L, c))$, irrespective of the log splitting function S. Intuitively, the semantics of the $\leftrightarrow$-operator do not allow for executing the tree that will be discovered for $\{a\}$, leaving that tree and later executing it again. Hence, without the extra pass through the event log, the $\leftrightarrow$ cut detection method would not preserve fitness locally.

$$L_{109} = [\langle a, b, c, a \rangle, \langle b, c, a \rangle, \langle a, b, c \rangle]$$

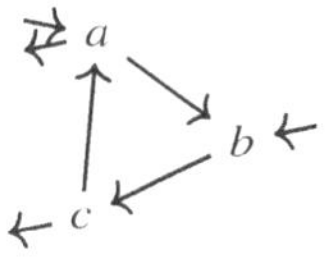

Fig. 6.15: A log and its directly follows graph.

Another method to guarantee local fitness preservation would be to treat the $\leftrightarrow$ cut as a "maybe"-interleaving cut, and divide the traces based on their first activity. Recursion continues, and if all discovered subtrees have a particular shape [6, S5], the "maybe" interleaved operator is replaced with a proper $\leftrightarrow$-operator. Otherwise, the "maybe" interleaved operator is considered an $\times$, which makes the log split fitting [6, S6]. This solution does not adhere to the IM framework, so we do not describe it in more detail. Nevertheless, it shows the flexibility and modularity of the IM framework: insertion of such a step does not influence the discovery of other operators.

Using this extension, we prove that all cuts returned by INTERLEAVEDCUT adhere to lemmas 5.3 and 5.14.

Lemma 6.16 (INTERLEAVEDCUT returns $\leftrightarrow$-cuts) *For any log L such that $\epsilon \notin L$, if* INTERLEAVEDCUT(*L*) *returns a cut, this is an $\leftrightarrow$-cut according to lemmas 5.3 and 5.14.*

Proof Requirement $\leftrightarrow$.1 can be split in three parts, which coincide with the three for-loops in INTERLEAVEDCUT: the first two express that all connections must go through a start or an end activity, while the third for-loop expresses that all such connections must be present. The fourth for-loop coincides with Requirement $\wedge \leftrightarrow$.1. The locally fitness-preserving part only limits cuts further, thus all cuts returned by INTERLEAVEDCUT adhere to Lemma 5.3. □

For local log-precision preservation, an argument similar to SEQUENCECUT holds: an extension is necessary to preserve log precision locally.

Inclusive Choice & Concurrency.

To detect an inclusive choice or concurrent cut, some changes are necessary. As described in Section 5.6, the directly follows detection algorithm CONCURRENTCUT returns the activity sets P of the non-coo subtrees of the event log L. In the example given in Section 6.4.1, this was $\{\{a\}, \{b\}, \{c\}, \{d\}, \{e\}\}$. Using the three coo relations $\overline{\wedge}$, $\overline{\wedge?}$ and $\overline{\vee}$, elements of the set P are merged until only two sets of activities remain. Finally, these two remaining sets of activities are returned as a cut, the operator of the cut depends on the remaining holding coo relation. If at some point in this procedure no coo relation would hold, we choose to return a cut using $\wedge$ and the remaining activity sets C, as $\wedge$ preserves the behaviour in the event log, and has less behaviour than $\vee$ (and hence a higher precision).

To compute the activity set log T for activity sets P and a log L, iterate over the traces, and for each trace denote which activity sets have a corresponding event in the trace (see Definition 5.16). For instance, the activity set log of log $[\langle a, b, c \rangle]$ and activity set $\{\{a, b\}, \{c, d\}, \{e, f\}\}$ is $\{\langle\{a, b\}, \{c, d\}\rangle\}$.

function COOCUT(L)
    $P = \{P_1 \ldots P_n\} \leftarrow$ partition of CONCURRENTCUT(L) (if none: fail)
    $T \leftarrow$ activity set log of L given C
    while true **do**
                                           ▷ compute coo relations on C as in Definition 5.18
        if $|P| > 2 \wedge \exists_{P_x, P_y \in P} \overline{\vee}(P_x, P_y) \vee \overline{\wedge}(P_x, P_y) \vee \overline{\wedge?}(P_x, P_y)$ **then**
            $P \leftarrow P \setminus \{P_x, P_y\} \cup \{P_x \cup P_y\}$
            $T \leftarrow$ activity set log of P
        else if $\exists_{P_x, P_y \in P} \overline{\vee}(P_x, P_y)$ **then**
            return $(\vee, P_1 \ldots P_n)$
        else
            return $(\wedge, P_1 \ldots P_n)$
        end if
    end while
end function

6.4.2.2 Log Splitting

In the log splitting functions, IMA does not require changes for the operators of IM. However, for the added operators $\leftrightarrow$ and $\vee$, new log splitting functions are necessary:

Interleaved.

To split a log L according to an interleaved cut, IM divides events over their corresponding subtraces:

function INTERLEAVEDSPLIT($L, (\leftrightarrow, \Sigma_1, \ldots, \Sigma_n)$)
    $\forall i : L_i \leftarrow [t|_{\Sigma_j} | t \in L]$
    return $L_1, \ldots, L_n$
end function

For instance, the log $L = [\langle a, b, c \rangle, \langle c, a, b \rangle]$ would be split using the cut $(\leftrightarrow, \{a, b\}, \{c\})$ into $[\langle a, b \rangle^2], [\langle c \rangle^2]$.

Lemma 6.17 (INTERLEAVEDSPLIT is locally fitness preserving) *Let L be a log. Then $set(L) \subseteq \leftrightarrow_{\mathcal{L}}(INTERLEAVEDSPLIT(L, c))$ for any cut c.*

Proof See the proof of Lemma 6.8.   □

Inclusive Choice.

To split a log L according to an inclusive choice cut, IM divides events over their corresponding subtraces:

function ORSPLIT$(L, (\vee, \Sigma_1, \ldots, \Sigma_n))$
  $\forall i : L_i \leftarrow [t|_{\Sigma_i} | t \in L \wedge t|_{\Sigma_i} \neq \epsilon]$
  return $L_1, \ldots, L_n$
end function

For instance, the log $L = [\langle a, b, c \rangle, \langle c \rangle, \langle a, b \rangle]$ would be split using the cut $(\vee, \{a, b\}, \{c\})$ into $[\langle a, b \rangle^2], [\langle c \rangle^2]$.

We prove local fitness preservation using the assumption that $\epsilon \notin L$. This assumption is satisfied in IMA by the base case *emptyTraces*.

Lemma 6.18 (cooCut & orSplit is locally fitness preserving) *Let L be a log such that $\epsilon \notin L$. Then $set(L) \subseteq \vee_{\mathcal{L}}(\text{orSplit}(L, c))$ for any cut c.*

Proof As $\epsilon \notin L$, see the proof of Lemma 6.8. □

Corollary 6.3 (cooCut & concurrentSplit is locally fitness preserving) *Let L be a log. Then $set(L) \subseteq \vee_{\mathcal{L}}(\text{concurrentSplit}(L, c))$ for any cut c.*

Local Guarantees

Fitness and log-precision preservation are defined on combinations of cut finders and log splitters (Definition 4.1). The following table summarises the local guarantee lemmas and the descriptions:

	locally fitness preserving	locally log precision preserving
xorCut & Split	yes (Lemma 6.5)	yes (Lemma 6.6)
sequenceCutStrict & Split	yes (Lemma 6.7)	when extended
cooCut & orSplit	yes (Lemma 6.18)	when extended
cooCut & concurrentSplit	yes (Corollary 6.3)	when extended
interleavedCut & Split	yes (Lemma 6.17)	when extended

6.4.2.3 Base Cases

As none of the constructs requires a base-case extension over IM, we reuse its base cases.

6.4.2.4 Fall Throughs

As none of the constructs requires any other fall through than IM, we reuse its base cases. Notice that the EMPTYTRACES fall through takes care of discovering $\times(\tau, \ldots)$ constructs.

6.4.2.5 Summary

To summarise, the *Inductive Miner - all operators* (IMA) implements the functions of the IM framework as follows:

function BASECASE$_\text{IMA}(L)$
    return BASECASE$_\text{IM}$
end function
function FINDCUT$_\text{IMA}(L)$
    if $\epsilon \notin L$ **then**
        $(\oplus, \Sigma_1 \ldots \Sigma_k) \leftarrow$ XORCUT$(\twoheadrightarrow(L))$
        if $k \leq 1$ **then** $(\oplus, \Sigma_1 \ldots \Sigma_k) \leftarrow$ SEQUENCECUTSTRICT$(\twoheadrightarrow(L))$ **end if**
        if $k \leq 1$ **then** $(\oplus, \Sigma_1 \ldots \Sigma_k) \leftarrow$ INTERLEAVEDCUT(L) **end if**
        if $k \leq 1$ **then** $(\oplus, \Sigma_1 \ldots \Sigma_k) \leftarrow$ COOCUT(L) **end if**
        if $k \geq 2$ **then return** $(\oplus, \Sigma_1 \ldots \Sigma_k)$
        else return LOOPCUT$(\twoheadrightarrow(L))$
        end if
    end if
    return $\square$
end function
function SPLITLOG$_\text{IMA}(L, (\oplus, \Sigma_1, \ldots, \Sigma_n))$
    if $\oplus = \times$ **then return** XORSPLIT$(L, (\oplus, \Sigma_1, \ldots, \Sigma_n))$
    else if $\oplus = \rightarrow$ **then return** SEQUENCESPLIT$(L, (\oplus, \Sigma_1, \ldots, \Sigma_n))$
    else if $\oplus = \vee$ **then return** ORSPLIT$(L, (\oplus, \Sigma_1, \ldots, \Sigma_n))$
    else if $\oplus = \wedge$ **then return** CONCURRENTSPLIT$(L, (\oplus, \Sigma_1, \ldots, \Sigma_n))$
    else if $\oplus = \leftrightarrow$ **then return** INTERLEAVEDSPLIT$(L, (\oplus, \Sigma_1, \ldots, \Sigma_n))$
    else if $\oplus = \circlearrowleft$ **then return** LOOPSPLIT$(L, (\oplus, \Sigma_1, \ldots, \Sigma_n))$
    end if
end function
function FALLTHROUGH$_\text{IMA}(L)$
    return FALLTHROUGH$_\text{IM}$
end function

In this section, we introduced IMA, which extends IM with handling of τ, $\vee$ and $\leftrightarrow$. Notice that IMA differs from IMF and IMC, as it refines IM with more advanced cut detection mechanisms, while IMF and IMC saw extra checks introduced in case IM failed to discover a cut.

6.4.3 Inductive Miner - infrequent - all operators (IM$_{FA}$)

In the previous section, we introduced IM$_A$, which extends IM with the process tree constructs τ, $\leftrightarrow$ and $\vee$. However, IM$_A$ is unable to handle deviating and infrequent behaviour. Therefore, in this section we adapt IM$_A$ in a new process discovery algorithm *Inductive Miner - infrequent - all operators* (IM$_{FA}$). IM$_{FA}$ reuses the base cases and fall throughs of IM$_F$, and combines cut detection functions of IM$_F$ with the $\wedge$ and $\vee$ detection of IM$_A$. That is, INTERLEAVEDCUTFILTERING and SEQUENCECUT-STRICTFILTERING first apply FILTER before applying respectively INTERLEAVEDCUT and SEQUENCECUTSTRICT. Similar to $\wedge$, for $\vee$ no behaviour can be identified as deviating and hence no new log splitting function is necessary. Consequentially, the only new function necessary is a log splitting function that splits the event log using non-conforming interleaved cuts: INTERLEAVEDSPLITFILTERING.

6.4.3.1 Interleaved Log Splitting

To split a log L according to an interleaved cut IM$_{FA}$ aims to minimise the number of events that are classified as deviating. Similar to SEQUENCESPLITFILTERING, the interleaved log splitting function searches for a split point, such that the number of filtered deviating events is minimised. An additional challenge is that, due to the semantics of $\leftrightarrow$, the sets of activities of the partition can appear in any order. These interleaved split points are detected using a divide-and-conquer strategy: first, on the entire trace, the *two* most efficient split points are detected, such that the trace in between them is the maximal subtrace that can be attributed to a single Σ_i. The two split points divide the trace in three subtraces: two outer ones and an inner one. The inner one is added to the sublog corresponding to Σ_i (events not of Σ_i are removed), the two outer ones are recursed upon (events of Σ_i are removed).

    For instance, consider the trace $t = \langle a, a, b, a, a, b, b, b, b, c \rangle$ and the cut $(\leftrightarrow, \{a\}, \{b\}, \{c\})$. In the first recursion, two optimal split points are discovered: $\langle a, a, b, a, a, | b, b, b, b, | c \rangle$, which belong to $\Sigma_2 = \{b\}$. Using the two split points, the trace is split into three parts: $t_1 = \langle a, a, a, a \rangle$ (notice that the b occurring before the first split point is of Σ_2 and is removed), $t_2 = \langle b, b, b, b \rangle$ and $t_3 = \langle c \rangle$. Subtrace t_2 is added to the sublog of Σ_2, and recursion continues on t_1 and t_3. In these recursions, $\langle a, a, a, a, \rangle$ is added to the sublog of Σ_1 and $\langle c \rangle$ is added to the sublog of Σ_3.

function INTERLEAVEDSPLITFILTERING($L, (\rightarrow, \Sigma_1, \ldots, \Sigma_n)$)
    $\forall_{1 \leq i \leq n} \, sublogs[\Sigma_i] \leftarrow [\;]$
    for $t \in L$ **do**
        $\forall_{1 \leq i \leq n} \, subtraces[\Sigma_i] \leftarrow \{\epsilon\}$
        $subtraces \leftarrow$ SPLITTRACE($L, t, \Sigma_1, \ldots \Sigma_n$)
        $\forall_{1 \leq i \leq n} \, sublogs[\Sigma_i] \leftarrow sublogs[\Sigma_i] \cup \{subtraces[\Sigma_i]\}$
    end for
end function
function SPLITTRACE($L, t, \Sigma_1, \ldots \Sigma_n, subtraces$)

$\sigma, start, end \leftarrow \text{FINDSPLITPOINTS}(t, \Sigma_1, \ldots \Sigma_n)$

$t_1 \leftarrow t[1 : start - 1]|_{\Sigma(L) \backslash \sigma}$

$t_2 \leftarrow t[start : end - 1]|_\sigma$

$t_3 \leftarrow t[end : |t|]|_{\Sigma(L) \backslash \sigma}$

$subtraces \leftarrow \text{SPLITTRACE}(L, t_1, \Sigma_1, \ldots \Sigma_n, subtraces)$

$subtraces[s] \leftarrow t_2$ such that $\sigma = \Sigma_s$

$subtraces \leftarrow \text{SPLITTRACE}(L, t_3, \Sigma_1, \ldots \Sigma_n, subtraces)$

return $subtraces$

end function

In the final function FINDSPLITPOINTS, the split points are computed: $start[i]$ holds the start of the best sequence for a Σ_i, while $values[i]$ holds the score, i.e. the number of events in Σ_i minus the number of events not in Σ_i since $start[i]$.

function FINDSPLITPOINTS$(t, \Sigma_1, \ldots \Sigma_n)$

    $maxStart \leftarrow -1$

    $maxEnd \leftarrow -1$

    $maxSigma \leftarrow \emptyset$

    $maxValue \leftarrow -1$

    $\forall_{1 \leq i \leq n} values[i] \leftarrow 0$

    $\forall_{1 \leq i \leq n} start[i] \leftarrow 0$

    $component \leftarrow -1$

    for $e \in t$ in order **do**

        $s \leftarrow i$ such that $e \in \Sigma_i$

        if $values[s] < 0$ **then**

            $values[s] \leftarrow 1$

            $start[s] \leftarrow$ position of e in t

        else

            $values[s] \leftarrow values[s] + 1$

        end if

        for $1 \leq i \leq n, i \neq s$ **do**

            $values[i] \leftarrow values[i] - 1$

        end for

        if $values[s] > maxValue$ **then**

            $maxSigma \leftarrow \Sigma_s$

            $maxStart \leftarrow start[s]$

            $maxEnd \leftarrow$ position of t in t

            $maxValue \leftarrow values[s]$

        end if

    end for

    return $maxSigma, maxStart, maxEnd$

end function

6.4.3.2 Summary

To summarise, the *Inductive Miner - infrequent - all operators* (IMFA) follows a similar strategy as IMF, i.e. applies FINDCUT$_{\text{IMA}}$ and if that fails, applies noise filtering. That is, IMFA implements the functions of the IM framework as follows:

function BASECASE$_{\text{IMFA}}(L)$
    return BASECASE$_{\text{IM}}$
end function
function FINDCUT$_{\text{IMFA}}(L)$
    if $\epsilon \notin L$ **then**
        $(\oplus, \Sigma_1 \ldots \Sigma_k) \leftarrow$ FINDCUT$_{\text{IMA}}(L)$
        if $k \leq 1$ **then** $(\oplus, \Sigma_1 \ldots \Sigma_k) \leftarrow$ XORCUTFILTERING$(\twoheadrightarrow(L))$ **end if**
        if $k \leq 1$ **then** $(\oplus, \Sigma_1 \ldots \Sigma_k) \leftarrow$ SEQUENCECUTSTRICTFILTERING$(\twoheadrightarrow(L))$
    end if
        if $k \leq 1$ **then** $(\oplus, \Sigma_1 \ldots \Sigma_k) \leftarrow$ INTERLEAVEDCUTFILTERING(L) **end if**
        if $k \leq 1$ **then return** LOOPCUTFILTERING$(\twoheadrightarrow(L))$
        else return $(\oplus, \Sigma_1 \ldots \Sigma_k)$
        end if
    end if
    return $\square$
end function
function SPLITLOG$_{\text{IMFA}}(L, (\oplus, \Sigma_1, \ldots, \Sigma_n))$
    if $\oplus = \times$ **then return** XORSPLITFILTERING$(L, (\oplus, \Sigma_1, \ldots, \Sigma_n))$
    else if $\oplus = \rightarrow$ **then return** SEQUENCESPLITFILTERING$(L, (\oplus, \Sigma_1, \ldots, \Sigma_n))$
    else if $\oplus = \vee$ **then return** ORSPLIT$(L, (\oplus, \Sigma_1, \ldots, \Sigma_n))$
    else if $\oplus = \wedge$ **then return** CONCURRENTSPLIT$(L, (\oplus, \Sigma_1, \ldots, \Sigma_n))$
    else if $\oplus = \leftrightarrow$ **then return** INTERLEAVEDSPLITFILTERING$(L, (\oplus, \Sigma_1, \ldots, \Sigma_n))$
    else if $\oplus = \circlearrowleft$ **then return** LOOPSPLITFILTERING$(L, (\oplus, \Sigma_1, \ldots, \Sigma_n))$
    end if
end function
function FALLTHROUGH$_{\text{IMFA}}(L)$
    return FALLTHROUGH$_{\text{IM}}$
end function

6.4.4 Guarantees

In this section, we discuss the guarantees provided by IMA and IMFA. Similar to IM and IMF, IMA guarantees fitness while IMFA does not:

Corollary 6.4 (IMA guarantees fitness) *As all steps of IMA are locally fitness preserving, by Corollary 4.1 for any log L it holds that $set(L) \subseteq \mathcal{L}(IMA(L))$.*

Next, we show rediscoverability for IMA and IMFA, i.e. we show that if a system model S is of class C_{coo}, and a log L is given to IM that is fitting to S and has the same directly follows graph and coo relations, then IMA and IMFA will return a model that is language equivalent to S. In order to prove this, we first prove that the log splitting of IMA preserves the log assumptions (Lemma 6.19), and second that IMA is abstraction preserving (Lemma 6.20).

In the following, let abstraction A_{IMA} denote the set of relations that combines the directly follows graph $\twoheadrightarrow$ and the coo relations $\overline{\triangle}$. The log assumptions function LA_{IMA} entails that the df and $\overline{\triangle}$ graphs are correct and complete, i.e. $L \in LA_{\text{IMA}}$ if and only if $\twoheadrightarrow(L) = \twoheadrightarrow(S)$ and $\overline{\triangle}(L) = \overline{\triangle}(S)$. However, we add a leniency: let S' be the reduced version of S according to Definition 5.1. Then, $\top \twoheadrightarrow_L \bot$ is only required if $S' = \times(\tau, \ldots)$. For instance, for the process tree , $\top \twoheadrightarrow \bot$ does not need

to be present in a log in order for that log to satisfy the log assumptions, however it may be present nevertheless.

Lemma 6.19 (IMA: log splitting preserves log assumptions) *Let $S = \oplus(S_1, \ldots S_n)$ with $S \in C_{coo}$, let $c = (\oplus, \Sigma_1, \ldots \Sigma_m)$ be a cut conforming to S, and let $L_1 \ldots L_m = \textsc{splitLog}_{\text{IMA}}(L, c)$. Then, there exist subtrees $M_1 \ldots M_m$ such that $\twoheadrightarrow(\oplus(M_1, \ldots M_m)) = \twoheadrightarrow(S)$, $\overline{\triangle}(\oplus(M_1, \ldots M_m)) = \overline{\triangle}(S)$ and $\forall_{1 \le i \le m} L_i \in LA_{\text{IMA}}(M_i)$.*

Proof We prove this lemma by constructing trees $M_1 \ldots M_m$ corresponding to $S_1 \ldots S_n$ as in Lemma 6.10 and showing that the log assumptions hold for these $M_1 \ldots M_m$, i.e. that the sublogs returned by $\textsc{splitLog}_{\text{IMA}}$ are fitting to their respective M_i and have the same directly follows graph and coo relations. As $\mathcal{L}(\oplus(M_1, \ldots M_m)) = \mathcal{L}(S)$, $\overline{\triangle}(\oplus(M_1, \ldots M_m)) = \overline{\triangle}(S)$.

Apply case distinction on $\oplus$:

$\oplus = \times$ As in Lemma 6.10, $\forall_{1 \le i \le m} set(L_i) \subseteq \mathcal{L}(M_i)$ and $\twoheadrightarrow(L_i) = \twoheadrightarrow(M_i)$.
Let $A \overline{\otimes}_S B$ be a coo relation that holds in S for sets of activities A and B. By Definition 5.18 and semantics of $\times$, $A, B \subseteq \Sigma(M_i)$ for some i. Thus, $\overline{\triangle}(S) = \cup_{1 \le k \le m} \overline{\triangle}(M_k)$. As $L \in LA_{\text{IMA}}(S)$, $A \overline{\otimes}_L B$ holds and consequently as $set(L_i) \subseteq set(L)$, $A \overline{\otimes}_{L_i} B$ holds for some i. Hence, $\overline{\triangle}(M_i) = \overline{\triangle}(L_i)$. As this holds for all i, $\forall_{1 \le i \le m} L_i \in LA_{\text{IMA}}(M_i)$.

$\oplus = \rightarrow$ As in Lemma 6.10, $\forall_{1 \le i \le m} set(L_i) \subseteq \mathcal{L}(M_i)$ and $\twoheadrightarrow(L_i) = \twoheadrightarrow(M_i)$.
Let $A \overline{\otimes}_S B$ be a coo relation that holds in S for sets of activities A and B. As $L \in LA_{\text{IMA}}(S)$, $\overline{\triangle}(L) = \overline{\triangle}(S)$ and $A \overline{\otimes}_L B$ holds. Perform case distinction on whether $A, B \subseteq \Sigma(M_i)$ for some i.

$A \cup B \subseteq \Sigma(M_i)$ By Definition 5.18 and semantics of $\rightarrow$, $A \overline{\otimes}_{M_i} B$ holds. By construction of $\textsc{sequenceSplit}$, $A \overline{\otimes}_{L_i} B$ holds.

$A \cup B \not\subseteq \Sigma(M_i)$ As A and B are not part of the same $\Sigma(M_i)$, then neither of the relations $A \overline{\otimes}_{M_i} B$ nor $A \overline{\otimes}_{L_i} B$ holds.

Hence, $\forall_{1 \le i \le m} L_i \in LA_{\text{IMA}}(M_i)$.

$\oplus = \leftrightarrow$ Let $1 \leq i \leq n$ and $t \in L_i$. By construction of INTERLEAVEDSPLIT, there must be a trace $t' = \langle \ldots t \ldots \rangle \in L$. As $L \in LA_{\mathrm{IMA}}(S)$, $t' \in \mathcal{L}(S)$. By Requirement $C_B.2$, the activities of t' in t can only be produced by M_i. Therefore, M_i must have produced t' and hence $set(L_i) \subseteq \mathcal{L}(M_i)$.

Left to prove: 1) $\twoheadrightarrow(L_i) = \twoheadrightarrow(M_i)$, which holds by an argument similar to the $\oplus = \rightarrow$ case in Lemma 6.10, and 2) $\forall_{1 \leq i \leq m} \overline{\wedge}(L_i) = \overline{\wedge}(S_i)$, which holds by an arguments similar to the $\oplus = \rightarrow$ case of this lemma.

$\oplus = \wedge$ As in Lemma 6.10, $\forall_{1 \leq i \leq m} set(L_i) \subseteq \mathcal{L}(M_i)$ and $\twoheadrightarrow(L_i) = \twoheadrightarrow(M_i)$.

Left to prove: $\forall_{1 \leq i \leq m} \overline{\wedge}(L_i) = \overline{\wedge}(M_i)$. Let $1 \leq i \leq m$ and let $A \overline{\oplus}_{L_i} B$ be a coo relation holding for sets of activities A and B. By CONCURRENTSPLIT, $A \overline{\oplus}_L B$. As $L \in LA_{\mathrm{IMA}}(S)$, $A \overline{\oplus}_S B$. By semantics of $\wedge$, $A \overline{\oplus}_{M_i} B$. The reverse direction is similar, hence $\forall_{1 \leq i \leq m} L_i \in LA_{\mathrm{IMA}}(M_i)$.

$\oplus = \vee$ Similar to the $\oplus = \wedge$ case.

$\oplus = \circlearrowleft$ As in Lemma 6.10, $\forall_{1 \leq i \leq m} set(L_i) \subseteq \mathcal{L}(M_i)$ and $\twoheadrightarrow(L_i) = \twoheadrightarrow(M_i)$.

Left to prove: $\forall_{1 \leq i \leq m} \overline{\wedge}(L_i) = \overline{\wedge}(M_i)$. Let $1 \leq i \leq m$ and let $A \overline{\oplus}_{L_i} B$ be a coo relation holding for sets of activities $A, B \subseteq \Sigma(L_i)$. Perform case distinction on $\overline{\oplus}$ to prove that $A \overline{\oplus}_L B$:

$\overline{\oplus} \neq \overline{\wedge?}$ By construction of LOOPSPLIT, all activity set traces of L_i are also present in L. The definitions of $\overline{?}$, $\overline{\Rightarrow}$, $\overline{\vee}$ and $\overline{\wedge}$ only involve A and B, and are upward closed, i.e. adding behaviour cannot negate a relation. Hence, $A \overline{\oplus}_L B$.

$\overline{\oplus} = \overline{\wedge?}$ Towards contradiction, assume that $A \overline{\cancel{\wedge?}}_L B$. By reasoning similar to the $\overline{\oplus} \neq \overline{\wedge?}$ case, $\overline{?}_L A$ and $A \overline{\Rightarrow}_L B$. Hence, there must exist a $C \subseteq \Sigma \setminus (A \cup B)$ such that $B \overline{\cancel{\Rightarrow}} C \wedge C \overline{\Rightarrow} B \wedge \forall_{a \in A \cup B, c \in C} a \twoheadrightarrow c \wedge c \twoheadrightarrow a)$. As $A \overline{\cancel{\wedge?}}_{L_i} B$, such a $C \cap \Sigma(L_i) = \emptyset$, i.e. C consists of non-L_i activities. Perform case distinction on whether $i = 1$:

$i = 1$ By semantics of $\circlearrowleft$ and Requirement $C_B.3$ of C_{coo}, there is no back-and-forth connection between all activities of $A \cup B$ and C. That is, $\neg \forall_{a \in A \cup B, c \in C} a \twoheadrightarrow c \wedge c \twoheadrightarrow a$, which contradicts that there exists such a C.

$i > 1$ By semantics of $\circlearrowleft$, $C \overline{\cancel{\Rightarrow}} B$, which contradicts that there exists such a C.

There exist no such C, hence $A \overline{\wedge?}_L B$.

Hence, $A \overline{\oplus}_L B$. As $L \in LA_{\mathrm{IMA}}(S)$, $A \overline{\oplus}_S B$. By semantics of $\circlearrowleft$, $A \overline{\oplus}_{M_i} B$.

Hence, subtrees $M_1 \ldots M_m$ exist such that $A_{\mathrm{IMA}}(\oplus(M_1, \ldots M_m)) = A_{\mathrm{IMA}}(S)$ and $\forall_{1 \leq i \leq m} L_i \in LA_{\mathrm{IMA}}(M_i)$. $\square$

Lemma 6.20 (IMA is abstraction preserving) *IMA is abstraction preserving, i.e. the combination of the class of process trees C_{coo}, the directly follows abstraction $\twoheadrightarrow$, the combined relations A_{IMA}, the log assumptions function LA_{IMA}, and the algorithm IMA implementing the IM framework with BASECASE$_{IMA}$, FINDCUT$_{IMA}$, SPLITLOG$_{IMA}$ and FALLTHROUGH$_{IMA}$, is abstraction preserving.*

Proof We discuss the requirements of Definition 4.5:

AP.1 An activity base case preserves the abstraction.

See the proof of Lemma 6.11.

AP.2 A τ base case preserves the abstraction.

As $L \in LA_{\text{IMA}}(S)$ holds, $set(L) = \{\epsilon\}$. By code inspection, the fall through case EMPTYTRACES applies, which returns $\times(\tau, \text{IMA}(L'))$, with L' being the empty log, for which in a next recursion τ is discovered by the base case EMPTYLOG. Hence, $A(\text{BASECASE}_{\text{IMA}}(L)) = A(\tau) = A(\times(\tau, \tau) = A(\tau) = A(S)$.

AP.3 The base case parameter function preserves the abstraction.

If $S = \oplus(S_1, \ldots S_n)$, with $S \in C_{\text{coo}}$, then $\Sigma(S) \geq 2$. As $L \in LA_{\text{IMA}}(S)$ holds, by code inspection, no base case in BASECASE$_{\text{IMA}}$ applies. Therefore, the requirement holds.

AP.4 Every cut that is detected conforms to S.

As $L \in LA_{\text{IMA}}(S)$, $\twoheadrightarrow(S) = \twoheadrightarrow(L)$ and $\overline{\overline{\wedge}}(S) = \overline{\wedge}(L)$. By lemmas 5.16, 5.17, 5.18 and 5.19, $\twoheadrightarrow(L)$ and $\overline{\wedge}(L)$ contain a cut $c = \oplus(\Sigma(S_1), \ldots \Sigma(S_n))$. By Corollary 5.8, no other footprint is present in $\twoheadrightarrow(L)$ and $\overline{\wedge}(L)$. By code inspection of FINDCUT$_{\text{IMA}}$, this cut c is returned, hence FINDCUT$_{\text{IMA}}(L)$ conforms to S (Definition 5.4).

AP.5 Log splitting preserves the log assumptions.

This requirement follows from Lemma 6.19.

AP.6 A fall through preserves the abstraction.

By the previous requirements, the only systems S of C_{coo} for which neither BASECASE$_{\text{IMA}}$ nor FINDCUT$_{\text{IMA}}$ apply have $\tau \in \mathcal{L}(S)$ and $\top \twoheadrightarrow_L \bot$. For these systems, the fall through EMPTYTRACES applies. Let S' be the reduced version of S according to Definition 5.1. We consider two cases:

- S' is of the form $\times(\tau, S'')$. By construction of EMPTYTRACE, IMA$(L) = \times(\tau, \text{IMA}(L'))$ with $L' = L \setminus \{\epsilon\}$. By semantics of $\twoheadrightarrow$ and $\overline{\wedge}$, $L' \in LA_{\text{IMA}}(S'')$.
  By assumption of Requirement AP.6, $A_{\text{IMA}}(\text{IMA}(L')) = A_{\text{IMA}}(S'')$. Then, $A_{\text{IMA}}(\text{IMA}(L)) = A_{\text{IMA}}(\text{IMA}(L')) \cup \top \twoheadrightarrow \bot = A_{\text{IMA}}(S'') \cup \top \twoheadrightarrow \bot = A(S)$.
- S' is not of the form $\times(\tau, \ldots)$. By construction of EMPTYTRACE, IMA$(L) = \times(\tau, \text{IMA}(L'))$ with $L' = L \setminus \{\epsilon\}$. By construction of LA_{IMA}, $L' \in LA_{\text{IMA}}(S'')$.
  By assumption of Requirement AP.6, $A_{\text{IMA}}(\text{IMA}(L')) = A_{\text{IMA}}(S'')$. Then, $A_{\text{IMA}}(\text{IMA}(L)) = A_{\text{IMA}}(\text{IMA}(L')) \cup \top \twoheadrightarrow \bot = A_{\text{IMA}}(S'') \cup \top \twoheadrightarrow \bot = A(S)$.

Hence, $A_{\text{IMA}}(L) = A_{\text{IMA}}(S)$. □

We show that IMA is language-class preserving (Definition 4.6), i.e. that the discovered model is of C_{coo}:

Lemma 6.21 (IMA is language-class preserving) *For all systems $S \in C_{coo}$ and logs L such that $L \in LA_{IMA}(S)$, it holds that $IMA(L) \in C_{coo}$.*

Proof We consider the requirements of C_{coo} separately:

$C_{\text{B}}.2$ No duplicate activities.

This requirement is guaranteed by the cuts discovered by FINDCUT$_{\text{IM}}$, which guarantee that all Σ_i are disjoint, and SPLITLOG$_{\text{IMA}}$ being fitting (Requirement AP.5.

$C_B.3$ The body of a loop has disjoint start and end activities.

As $S \in C_{coo}$, the only FALLTHROUGH$_{IMA}$ function that is reached is EMPTYTRACES, i.e. if a loop operator is discovered, then this is discovered by the cut detection, and the log is split into sublogs $L_1 \ldots L_n$. By the previous requirements, FINDCUT$_{IMA}$ only selects a $\circlearrowright$ if $S = \oplus(S_1, \ldots S_n)$. As $S \in C_{coo}$, Start$(S_1) \cap$ End$(S_1) = \emptyset$. By Requirement AP.5 and the log assumptions LA_{IMA}, Start$(L_1) \cap$ End$(L_1) = \emptyset$. By lemmas 6.11 and 4.1, $\twoheadrightarrow(IMA(L_1)) = \twoheadrightarrow(IMA(S_1))$, hence Start$(IMA(L)) \cap$ End$(IMA(L)) = \emptyset$.

$C_{coo}.2$ No redo child of a loop can produce the empty trace.

As $S \in C_{coo}$, if a loop operator is discovered, then this is discovered by cut detection, and the log is split into sublogs $L_1 \ldots L_m$. By Lemma 6.19, there exist $M_1 \ldots M_m$ such that $\twoheadrightarrow(\circlearrowright(M_1, \ldots M_m)) = \twoheadrightarrow(S)$ and $\forall_{1 \leq i \leq m} L_i \in LA_{IMA}(M_i)$. Then, by Requirement $C_{coo}.2$, $\forall_{2 \leq i \leq m} \epsilon \notin \mathcal{L}(M_i)$, thus $\forall_{2 \leq i \leq m} \epsilon \notin L_i$. By Lemma 6.20, $\epsilon \notin IMA(L_i)$.

$C_{coo}.3$ Interleaving cannot be nested using optionality.

By code inspection, if an $\leftrightarrow$ is discovered, this happens by cut detection, and the log is split into sublogs $L_1, \ldots L_m$. By Lemma 6.19, there exist $M_1 \ldots M_m$ such that $\twoheadrightarrow(\circlearrowright(M_1, \ldots M_m)) = \twoheadrightarrow(S)$ and $\forall_{1 \leq i \leq m} L_i \in LA_{IMA}(M_i)$. Towards contradiction, without loss of generality, assume that M_1 is optional, i.e. $M_1 = \times(\tau, M')$, and that $M' = \leftrightarrow(M'_1, \ldots M'_n)$. Let $set(L') \subseteq \mathcal{L}(M')$ be the sublog created by IMA for the recursive step on L_1. Then, by reasoning similar to the previous requirement, $S \notin C_{coo}$, which contradicts the initial assumption. Hence, Requirement $C_{coo}.3$ holds for S.

$C_{coo}.4$ An inclusive choice child of an interleaving has at least one child with disjoint start and end activities,

$C_I.2$ An interleaving has at least one child with disjoint start and end activities,

$C_I.3$ An interleaving has no interleaved child, and

$C_I.4$ A concurrent child of an interleaving has at least one child with disjoint start and end activities. The proofs for these last four requirements are similar to the proof of the two requirements before them. $\square$

Then, by Theorem 4.2, IMA guarantees rediscoverability for C_{coo}. Rediscoverability is guaranteed by IMFA as well, as IMFA first applies the cut detection of IMA before attempting filtering (and this is not necessary in case the preconditions for rediscoverability hold).

Theorem 6.9 (IMA & IMFA rediscoverability) *Let L be a log and $S \in C_{coo}$ be a system such that $set(L) \subseteq \mathcal{L}(S) \wedge \twoheadrightarrow(S) = \twoheadrightarrow(L) \wedge \overline{\circledwedge}(L) = \overline{\circledwedge}(S)$. Then, $\mathcal{L}(IMA(L)) = \mathcal{L}(S)$ and $\mathcal{L}(IMFA(L)) = \mathcal{L}(S)$.*

In this section, we extended the IM algorithm to discover the remaining process-tree constructs τ, $\leftrightarrow$ and $\vee$. We introduced a basic version (IMA) and a second version (IMFA) to handle infrequent and deviating behaviour. We showed that IMA guarantees fitness, but to guarantee that needs a dedicated pass over the event log for

↔. Notice that in IMFA, we did not apply deviation filtering to coo-cut detection, i.e. detecting ∧ and ∨ behaviour. This might be an interesting area of future research:

Future work 6.10: Extend IMFA to apply deviation filtering to the detection of ∧ and ∨.

In the remainder of this chapter, we study two more properties of event logs for which new discovery techniques are necessary: non-atomic event logs and exceptionally large event logs. Furthermore, we give an overview of and a way to choose between the discovery algorithms presented in this chapter.

6.5 Handling Non-Atomic Event Logs

In some real-life event logs, activity executions take time and hence are non-atomic. As described in Section 2.3.2, we assume that each non-atomic execution of an activity consists of two events: one event denoting the start of the execution, and one denoting the completion of the execution. For instance, the trace $\langle a_s, b_s, b_c, a_c \rangle$ denotes that during execution of activity a, an execution of activity b occurred. We assume that each non-atomic trace in a non-atomic event log is consistent (Definition 2.13).

In this section, we first illustrate how the techniques presented before treat non-atomic traces and why they could fail (Section 6.5.1). Second, we introduce several algorithms that deal with non-atomicity by considering the life cycle information of events. We introduce three algorithms: a basic fitness guaranteeing one, *Inductive Miner - life cycle* (IMLC) in Section 6.5.2, one to handle infrequent and deviating behaviour, *Inductive Miner - infrequent - life cycle* (IMFLC) in Section 6.5.3, and one to handle incomplete behaviour, *Inductive Miner - incompleteness - life cycle* (IMCLC) in Section 6.5.3.

All these three algorithms implement the IM framework, and reuse parts of respectively the IM, IMF and IMC algorithms. The key step in the three new algorithms is the construction of the directly follows relation, which is computed while being aware of the life cycle information contained in events. These non-atomic directly follows graphs were introduced in detail in Section 5.7.3. We finish this section with some brief comments on the implementation of IMLC, IMFLC and IMCLC in Section 6.5.4, and a discussion of the guarantees provided by these algorithms in Section 6.5.5.

6.5.1 Non-Atomic Event Logs

In case the event log contains non-atomic behaviour, the previously introduced algorithms will misinterpret this behaviour. For instance, consider the event log consisting of a single trace

$$\tilde{L_{110}} = [\langle a_s, b_s, b_c, a_c \rangle]$$

The algorithms presented before, e.g. IM, are not aware of life-cycle information and will either ignore it ($L_{110}' = [\langle a, b, b, a \rangle]$) or treat the life-cycle information as part of the activity name ($[\langle a + start, b + start, b + complete, a + complete \rangle]$), depending on the chosen classifier (see Section 6.5.4). The directly follows graph of this log is shown in Figure 6.16a. When applied to L_{110}', IM performs the following steps:

$$IM(L_{110}') = \circlearrowleft(IM(L_{111}'), IM(L_{112}'))$$
$$L_{111}' = [\langle a \rangle^2]$$
$$L_{112}' = [\langle b, b \rangle]$$
$$IM(L_{111}') = a$$
$$IM(L_{112}') = \circlearrowleft(b, \tau)$$

and discovers the model $\circlearrowleft(a, \circlearrowleft(b, \tau))$.

(a) $\twoheadrightarrow$ of L_{110}'.   (b) $\tilde{\twoheadrightarrow}$ of $\tilde{L_{110}}$.   (c) $\parallel$ of $\tilde{L_{110}}$.

This model does not cover the original event log L_{110} well, which contains just two executions of activities: one of a and one of b. Alternatively, one could filter L_{110} by removing all start events, on which IM would discover the model $\rightarrow(b, a)$, which does not describe the event log well either.

The algorithms that are introduced in the next sections compute the non-atomic directly follows graph that is shown in Figure 6.16b, according to Definition 5.20, which states that $a \tilde{\twoheadrightarrow} b$ if there is no full activity instance between the completion of a and the start of b. In contrast, in the atomic directly follows graph (shown in Figure 6.16a), b is neither a start nor an end activity, as the trace in L_{110}' does not start or end with a b. From this graph, the default directly follows footprints of $\times$, $\rightarrow$ and $\circlearrowleft$ apply. For concurrency, the new algorithms use the extra information of the concurrency graph. In our example, the concurrency graph is shown in Figure 6.16c, which shows that a and b should be concurrent, as there is a connection $a \parallel b$ (which denotes that activity instances of a and b overlap in time).

Thus, in our example event log $\tilde{L_{110}}$, the concurrent footprint is present and IMLC would discover $\wedge(\tilde{a}, \tilde{b})$ and thus discover a model that resembles $\tilde{L_{110}}$ much better than the model discovered by IM.

In the remainder of this section, we describe three algorithms that are aware of the non-atomicity of activities: IMLC, IMFLC and IMCLC, after which we discuss the guarantees provided by them.

6.5.2 Inductive Miner - life cycle (IMLC)

In this section, we discuss how the IMLC algorithm implement the IM framework, i.e. we discuss their cut detection, log splitting, base cases and fall throughs. We discuss local fitness and log-precision preservation, as well as a new property that every log on which the algorithms recurse should be consistent (assuming that the input event log is consistent). We will discuss this property for fall throughs and combinations of cut detection functions and log splitters.

6.5.2.1 Cut Detection

For $\times$ and $\to$, cut detection of IMLc resembles cut detection of IM. That is, after construction of a non-atomic directly follows graph (see Section 5.7.3, the cut detection of IM is applied. For these functions, see Section 6.1.2.1.

The cut detection functions for $\wedge$, $\leftrightarrow$ and $\circlearrowleft$ however need some adjustments from IM.

Concurrency.

For detecting concurrent cuts for non-atomic event logs, the function takes both the directly follows graph (Definition 2.14) and the concurrency graph (Definition 5.21) into account. That is, in a concurrent cut, all activities should be connected by either a double directly follows edge or an edge in the concurrency graph:

function NONATOMICCONCURRENTCUT($\twoheadrightarrow$, $\parallel$)

$\quad P = \{P_1 \ldots P_n\} \leftarrow \{\{a\} | a \in \Sigma(\twoheadrightarrow)\}$

$\qquad\qquad\qquad\qquad\qquad\qquad$ ▷ merge not-fully connected or not-concurrent sets

$\quad$**for all** $a, b \in \Sigma(\twoheadrightarrow), a \neq b$ **do**

$\qquad$**if** $a \nparallel b \wedge (a \stackrel{\sim}{\nrightarrow} b \vee b \stackrel{\sim}{\nrightarrow} a)$ **then**

$\qquad\qquad$let $a \in P_x$ and $b \in P_y$, then $P \leftarrow P \setminus \{P_x, P_y\} \cup \{P_x \cup P_y\}$

$\qquad$**end if**

$\quad$**end for**

$\qquad\qquad\qquad\qquad\qquad\qquad\qquad$ ▷ merge sets without start or end activities

$\quad$**for all** $P_c \in P$ **do**

$\qquad$**if** $P_c \cap \text{Start}(\twoheadrightarrow) = \emptyset \vee P_c \cap \text{End}(\twoheadrightarrow) = \emptyset$ **then**

$\qquad\qquad$let $P_{x \neq c}$ be an arbitrary set in P, then $P \leftarrow P \setminus \{P_c, P_x\} \cup \{P_c \cup P_x\}$

$\qquad$**end if**

$\quad$**end for**

$\quad$**return** $(\wedge, P_1 \ldots P_n)$

end function

Interleaved.

For the interleaved cut detection function NONATOMICINTERLEAVEDCUT, we reuse the algorithmic idea of INTERLEAVEDCUT, however as the $\oslash$-relation is not defined for non-atomic event logs, its corresponding merge-loop is omitted.

Notice that NONATOMICINTERLEAVEDCUT implicitly takes the concurrency graph into account, as the fitness guarantee prevents concurrent activities from being divided. For instance, consider the trace $\langle a_s, b_s, b_c, a_c \rangle$, in which a and b are concurrent, i.e. $a \parallel b$. Then, a and b are merged into the same P_i by the last for-loop of the function. In this last for loop, we denote an event of a (either start or completion) with $a_?$.

> **function** NONATOMICINTERLEAVEDCUT($\tilde{L}$)
> $P \leftarrow \{\{a\} | a \in \Sigma(\tilde{\twoheadrightarrow})\}$
>   **for all** $a \notin \text{Start}(\tilde{L}), b \in \Sigma(\tilde{L})$ such that $b \tilde{\twoheadrightarrow} a$ in $\tilde{L}$ **do**
>   let $a \in P_x$ and $b \in P_y$, then $P \leftarrow P \setminus \{P_x, P_y\} \cup \{P_x \cup P_y\}$
>   **end for**
>   **for all** $a \notin \text{End}(\tilde{L}), b \in \Sigma(\tilde{L})$ such that $a \tilde{\twoheadrightarrow} b$ in $\tilde{L}$ **do**
>   let $a \in P_x$ and $b \in P_y$, then $P \leftarrow P \setminus \{P_x, P_y\} \cup \{P_x \cup P_y\}$
>   **end for**
>   **for all** $a \in \text{Start}(\tilde{L}), b \in \text{End}(\tilde{L})$ such that $a \tilde{\not\twoheadrightarrow} b$ in $\tilde{L}$ **do**
>   let $a \in P_x$ and $b \in P_y$, then $P \leftarrow P \setminus \{P_x, P_y\} \cup \{P_x \cup P_y\}$
>   **end for**
>   **for all** $\tilde{t} \in \tilde{L}$ **do**   ▷ guarantee fitness
>   **if** $\exists_{P_1,P_2 \in P} \exists_{a,c \in P_1, b \in P_2} \tilde{t} = \langle \ldots a_? \ldots b_? \ldots c_? \ldots \rangle$ **then**
> $P \leftarrow P \setminus \{P_1, P_2\} \cup \{P_1 \cup P_2\}$ merge P_1 and P_2 in P
>   **end if**
>   **end for**
>   **return** $(\leftrightarrow, P_1 \ldots P_n)$
> **end function**

Loop.

Furthermore, also the loop cut detection takes the concurrency graph footprint into account, i.e. activities in a concurrent relation cannot be split by a loop cut. In the following algorithm, the start and end activities are taken separate in the body P_1 (Requirement $\circlearrowleft$.1), and the remaining activities are divided such that only unconnected parts remain (Requirement $\circlearrowleft$.3). Second, the divided sets that cannot be redo parts are merged with the body. Third, the divided sets that do not have all required connections (i.e. from end and to start) are merged with the body (Requirement $\circlearrowleft$.4).

> **function** NONATOMICLOOPCUT($\tilde{\twoheadrightarrow}, \parallel$)
> $P_1 \leftarrow \text{Start}(\tilde{\twoheadrightarrow}) \cup \text{End}(\tilde{\twoheadrightarrow})$
> $P_2 \ldots P_n \leftarrow$ partition of $\Sigma(\tilde{\twoheadrightarrow}) \setminus P_1$ such that $\forall_{2 \leq i < j \leq n, a \in P_i, b \in P_j} a \tilde{\not\twoheadrightarrow} b \wedge b \tilde{\not\twoheadrightarrow} a \wedge a \not\parallel b$ and $\forall_{2 \leq i \leq n} \forall_{a,b \in P_i} a \leftrightsquigarrow b$

▷ exclude sets that are connected from a start activity
for all $a \in \text{Start}(\twoheadrightarrow) \setminus \text{End}(\twoheadrightarrow)$ **do**
   for all b such that $a \twoheadrightarrow b \vee a \parallel b$ **do**
      $P_1 \leftarrow P_1 \cup$ set of b
   end for
end for

▷ exclude sets that are connected to an end activity
for all $b \in \text{End}(\twoheadrightarrow) \setminus \text{Start}(\twoheadrightarrow)$ **do**
   for all a such that $a \twoheadrightarrow b \vee a \parallel b$ **do**
      $P_1 \leftarrow P_1 \cup$ set of a
   end for
end for

▷ sets should have all connections
for all $2 \leq i \leq n, a \in P_i$ **do**
   if $\exists_{b \in \text{Start}(\twoheadrightarrow)} \, a \twoheadrightarrow b \wedge \neg \forall_{b \in \text{Start}(\twoheadrightarrow)} \, a \twoheadrightarrow b$ **then**
      $P_1 \leftarrow P_1 \cup$ set of a
   end if
   if $\exists_{b \in \text{End}(\twoheadrightarrow)} \, b \twoheadrightarrow a \wedge \neg \forall_{b \in \text{End}(\twoheadrightarrow)} \, b \twoheadrightarrow a$ **then**
      $P_1 \leftarrow P_1 \cup$ set of a
   end if
end for
return $(\circlearrowright, P_1, \ldots P_n)$
end function

6.5.2.2 Log Splitting

Given a cut and an event log, the life cycle algorithm IMLC splits the event logs
into sublogs. For these log splitting functions, we reuse the log splitting functions of
IMFA without modifications.

Local Guarantees for IMLC.

Fitness and log-precision preservation are defined on combinations of cut finders
and log splitters (Definition 4.1). Given a cut $(\oplus, \Sigma_1, \ldots \Sigma_n)$, for each event these log
splitting functions consider the Σ_i belonging to the event. Thus, for the log splitting
functions it is only relevant in which Σ_i the event belongs and hence, the local fitness
and log precision preservation result also hold for non-atomic event logs and process
trees, using that NONATOMICCONCURRENTCUT, NONATOMICINTERLEAVEDCUT and
NONATOMICLOOPCUT return cuts according to the footprints of Lemma 5.3.
   The following table summarises these local guarantees:

	locally fitness preserving	locally log precision preserving
XORCUT & SPLIT	yes (Lemma 6.5)	yes (Lemma 6.6)
SEQUENCECUT & SPLIT	yes (Lemma 6.7)	when extended
NONATOMICCONCURRENTCUT & CONCURRENTSPLIT	yes (Lemma 6.8)	when extended
NONATOMICINTERLEAVEDCUT & INTERLEAVEDSPLIT	yes (Lemma 6.17)	when extended
NONATOMICLOOPCUT & LOOPSPLIT	yes (Lemma 6.9)	no (see Section 4.1.4.2)

Another desirable local guarantee is that the sublogs that result from log splitting should be consistent. For $\times$, $\rightarrow$, $\wedge$ and $\leftrightarrow$, this property follows from local fitness preservation, as log splitting using cuts according to Lemma 5.3 does not remove or reorder events, and consistency is defined on events of the same activity (see Section 2.3.2). Hence, if the input event log is consistent, all sublogs are consistent as well.

For $\circlearrowleft$, an inconsistent sublog can only appear if a start event and its corresponding completion event get separated in different subtraces, e.g. the trace $\langle a_s, b_s, b_c, a_c \rangle$ would be split using the cut $(\circlearrowleft, \{a\}, \{b\})$ into $L_1 = [\langle a_s \rangle, \langle a_c \rangle]$ and $L_2 = [\langle b_s, b_c \rangle]$. Both traces in L_1 are inconsistent. However, notice that in such cases, the activities a and b are in a concurrent relation $(a \parallel b)$, and therefore NONATOMICLOOPCUT will put both a and b in Σ_1 and not discover a loop cut. Therefore, also for $\circlearrowleft$ cuts, if the input log is consistent, all sublogs are consistent as well.

6.5.2.3 Base Cases

The base cases of IMlc resemble the base cases of IM: SINGLENONATOMICACTIVITY applies when the event log contains only traces having a single activity instance, i.e. the combination of a start event and a corresponding completion event. Similar to the atomic base case SINGLEACTIVITY in IM, the non-atomic base case SINGLEN-ONATOMICACTIVITY in IMlc is locally fitness preserving and locally log-precision preserving.

	locally fitness preserving	locally log precision preserving
SINGLENONATOMICACTIVITY	yes	yes

Notice that the base cases do not recurse and thus consistency preservation of the event logs is irrelevant.

6.5.2.4 Fall Throughs

We discuss the fall throughs of IM, i.e. we argue whether they apply and what changes are necessary when these fall throughs are applied to non-atomic event logs:

- EMPTYTRACES applies without modifications to IMLC: non-atomic and atomic empty traces are equivalent.
- NONATOMICACTIVITYONCEPERTRACE applies when an activity a has a single activity instance, i.e. a pair of start and completion events (instead of single events) in every trace of the event log.
- ACTIVITYCONCURRENT applies without modifications.
- STRICTTAULOOP and TAULOOP apply when looping behaviour is present, i.e. the occurrence of an end activity followed by a start activity. These fall throughs need modification to be applicable to non-atomic event logs, as the IMLC algorithms assume that the event log is consistent, and all sublogs on which they are applied recursively need to be consistent as well. For instance, consider the trace $t_1 = \langle a_s, b_s, b_c, c_s, c_c, b_s, b_c, c_s, c_c, a_c \rangle$, which is consistent. Splitting this trace after each end activity followed by a start activity would yield $t_2 = \langle a_s, b_s, b_c, c_s, c_c \rangle$ and $t_3 = \langle b_s, b_c, c_s, c_c, a_c \rangle$, both of which are not consistent, as a_s and a_c have been separated over different sub traces. Therefore, STRICTNONATOMICTAULOOP and NONATOMICTAULOOP only split traces if this would not introduce inconsistencies, i.e. each start event before the split point has a corresponding completion event before the split point.
- FLOWERMODEL is the last resort of the fall throughs and applies to all event logs. However, as described in Section 5.7.2, a non-atomic flower model might not describe all behaviour of an event log. For instance, the trace $\langle a_s, b_s, a_s, b_c, a_c, a_c \rangle$ does not fit the flower model (a is executed concurrently with itself), even

though all activities of the trace are contained in the flower model.

Therefore, we introduce the CONCURRENTFLOWERMODEL, which counts the maximum number of concurrent activities in an event log, and returns a concurrent model accordingly. In our example, this fall through would return  ,

i.e. two repeatable a's and one repeatable b concurrently, and this model fits the trace $\langle a_s, b_s, a_s, b_c, a_c, a_c \rangle$.

Notice that all activities in these fall throughs are non-atomic.

Local Guarantees.

The following table restates the preservation guarantees of these fall throughs:

	locally fitness preserving	locally log precision preserving
EMPTYTRACES	yes	yes
NONATOMICACTIVITYONCEPERTRACE	yes	when extended
ACTIVITYCONCURRENT	yes	when extended
STRICTNONATOMICTAULOOP	yes	no
NONATOMICTAULOOP	yes	no
CONCURRENTFLOWERMODEL	yes	no

All fall throughs preserve consistency: EMPTYTRACES does not alter any nonempty trace, NONATOMICACTIVITYONCEPERTRACE and ACTIVITYCONCURRENT remove an activity from each trace and thus do not influence consistency of the remaining activities, and NONATOMICTAULOOP and CONCURRENTFLOWERMODEL have been discusses before.

6.5.2.5 Summary

We summarise IMLC, which implements the functions of the IM framework as follows:

function BASECASE$_{IMLC}$($\tilde{L}$)
    if $\epsilon \notin L$ **then**
        $bc \leftarrow$ EMPTYLOG(L)
        if $bc = \square$ **then** $bc \leftarrow$ SINGLENONATOMICACTIVITY(L) **end if**
        if $bc \neq \square$ **then return** bc **end if**
    end if
    return $\square$
end function
function FINDCUT$_{IMLC}$($\tilde{L}$)
    if $\epsilon \notin L$ **then**
        $(\oplus, \Sigma_1 \dots \Sigma_k) \leftarrow$ XORCUT($\tilde{\twoheadrightarrow}(\tilde{L})$)
        if $k \leq 1$ **then** $(\oplus, \Sigma_1 \dots \Sigma_k) \leftarrow$ SEQUENCECUT($\tilde{\twoheadrightarrow}(\tilde{L})$) **end if**
        if $k \leq 1$ **then** $(\oplus, \Sigma_1 \dots \Sigma_k) \leftarrow$ NONATOMICCONCURRENTCUT($\tilde{\twoheadrightarrow}(\tilde{L}), \|(L)$)
end if
        if $k \leq 1$ **then** $(\oplus, \Sigma_1 \dots \Sigma_k) \leftarrow$ NONATOMICINTERLEAVEDCUT($\tilde{L}$) **end if**
        if $k \leq 1$ **then** $(\oplus, \Sigma_1 \dots \Sigma_k) \leftarrow$ NONATOMICLOOPCUT($\tilde{\twoheadrightarrow}(\tilde{L}), \|(L)$) **end**
if
        if $k \geq 2$ **then return** $(\oplus, \Sigma_1 \dots \Sigma_k)$ **end if**
    end if
    return $\square$
end function
function SPLITLOG$_{IMLC}$($\tilde{L}, (\oplus, \Sigma_1, \dots, \Sigma_n)$)
    if $\oplus = \times$ **then return** XORSPLIT($\tilde{L}, (\oplus, \Sigma_1, \dots, \Sigma_n)$)
    else if $\oplus = \rightarrow$ **then return** SEQUENCESPLIT($\tilde{L}, (\oplus, \Sigma_1, \dots, \Sigma_n)$)
    else if $\oplus = \wedge$ **then return** CONCURRENTSPLIT($\tilde{L}, (\oplus, \Sigma_1, \dots, \Sigma_n)$)

else if $\oplus = \leftrightarrow$ **then return** INTERLEAVEDSPLIT$(\tilde{L}, (\oplus, \Sigma_1, \ldots, \Sigma_n))$
else if $\oplus = \circlearrowleft$ **then return** LOOPSPLIT$(\tilde{L}, (\oplus, \Sigma_1, \ldots, \Sigma_n))$
end if
end function

function FALLTHROUGH$_{\text{IMLC}}(\tilde{L})$
  $ft \leftarrow$ EMPTYTRACES$(\tilde{L})$
  if $ft = \square$ **then** $ft \leftarrow$ NONATOMICACTIVITYONCEPERTRACE$(\tilde{L})$ **end if**
  if $ft = \square$ **then** $ft \leftarrow$ ACTIVITYCONCURRENT$(\tilde{L})$ **end if**
  if $ft = \square$ **then** $ft \leftarrow$ STRICTNONATOMICTAULOOP$(\tilde{L})$ **end if**
  if $ft = \square$ **then** $ft \leftarrow$ NONATOMICTAULOOP$(\tilde{L})$ **end if**
  if $ft \neq \square$ **then return** ft
  else return CONCURRENTFLOWERMODEL$(\tilde{L})$
  end if
end function

6.5.3 Inductive Miner - infrequent - life cycle (IMFLC) & Inductive Miner - incompleteness - life cycle (IMCLC)

In the previous section, we introduced IMLC, that takes non-atomicity of activity executions into account and guarantees fitness. In this section, we adapt IMLC to handle two types of challenges: infrequent and deviating behaviour, and incomplete behaviour. For each of these challenges, we introduce a new discovery algorithm: *Inductive Miner - infrequent - life cycle* (IMFLC) to handle deviating and infrequent behaviour, and *Inductive Miner - incompleteness - life cycle* (IMCLC) to handle incomplete behaviour. We describe the changes in the four function parameters of the IM framework: cut detection, log splitting, base cases and fall throughs.

6.5.3.1 Cut Detection

Cut detection for IMFLC and IMCLC resembles the cut detection functions of IMFA and IMC. That is, both first try to detect a cut using the cut detection of IMLC. If that fails, then IMFLC filters the non-atomic directly follows graph using the function FILTER, after which again the cut detection of IMLC is applied. If cut detection fails for IMCLC, then the cut with the highest probability is selected. We adapt the activity relations defined in Section 5.3 to take the concurrency graph $\parallel$ into account, i.e. if for two activities a and b it holds that $a \parallel b$, they are considered to be in a concurrent activity relation: $a \wedge b$.

6.5.3.2 Log Splitting

Given a cut and an event log, the life cycle algorithms IM$_{FLC}$ and IM$_{CLC}$ split the event logs into sublogs. For these log splitting functions, we reuse the log splitting functions of IM$_{FA}$ without modifications.

Local Guarantees.

Similar to IM$_F$ and IM$_C$, neither IM$_{FLC}$ nor IM$_{CLC}$ guarantee local fitness or log-precision preservation for cut detection and log splitting. As cut detection and log splitting might classify events as deviating and remove them, neither algorithm guarantees that all sublogs are consistent. Therefore, we apply a post-processing step to these sublogs that makes the sublogs consistent. The post-processing step was described in Section 2.3.2, and introduces a completion event right after each start event without a corresponding completion event. Notice that this post-processing step is only necessary for →, ↔ and ↺, as × will only remove deviating start and completion events together, and ∧ does not remove any deviating event.

6.5.3.3 Base Cases

IM$_{FLC}$ combines techniques from IM$_F$ and IM$_{LC}$, i.e. the base case SINGLENONATOMICACTIVITYFILTERING applies the base case if "enough" traces consisting of a single activity execution (i.e. a start and a completion event of the same activity) are present (see SINGLEACTIVITYFILTERING in Section 6.2). This base case is neither locally fitness nor log-precision preserving:

	locally fitness preserving	locally log precision preserving
SINGLENONATOMICACTIVITYFILTERING	no	no

Notice that this base case does not recurse and thus consistency preservation of the event logs is irrelevant.

6.5.3.4 Fall Throughs

IM$_{CLC}$ and IM$_{FLC}$ reuse the fall throughs of IM$_{LC}$ and IM$_F$. Notice that EMPTYTRACESFILTERING applies without modification: non-atomic and atomic empty traces are equivalent. This fall through is neither locally fitness nor log-precision preserving, however it preserves consistency: no nonempty trace is altered.

	locally fitness preserving	locally log precision preserving
EMPTYTRACESFILTERING	no	yes

6.5.3.5 Summary

We summarise how IMFLC and IMCLC implement the IM framework. Notice that both combine IMLC with IMF and IMC.

IMFLC.

We start with IMFLC, which filters infrequent and deviating behaviour:

function BASECASE$_{\text{IMFLC}}(\tilde{L})$
    if $\epsilon \notin \tilde{L}$ **then**
        $bc \leftarrow$ EMPTYLOG$(\tilde{L})$
        if $bc = \square$ **then**
            $bc \leftarrow$ SINGLENONATOMICACTIVITYFILTERING$(\tilde{L})$
        end if
        return bc
    end if
    return $\square$
end function
function FINDCUT$_{\text{IMFLC}}(\tilde{L})$
    if $\epsilon \notin \tilde{L}$ **then**
        $(\oplus, \Sigma_1 \ldots \Sigma_k) \leftarrow$ FINDCUT$_{\text{IM}}(\overset{\sim}{\twoheadrightarrow}(\tilde{L}))$
        if $k \leq 1$ **then** $(\oplus, \Sigma_1 \ldots \Sigma_k) \leftarrow$ XORCUTFILTERING$(\overset{\sim}{\twoheadrightarrow}(\tilde{L}))$ **end if**
        if $k \leq 1$ **then**
            $(\oplus, \Sigma_1 \ldots \Sigma_k) \leftarrow$ SEQUENCECUTFILTERING$(\overset{\sim}{\twoheadrightarrow}(\tilde{L}))$
        end if
        if $k \leq 1$ **then**
            $(\oplus, \Sigma_1 \ldots \Sigma_k) \leftarrow$ NONATOMICCONCURRENTCUTFILTERING$(\overset{\sim}{\twoheadrightarrow}(\tilde{L}))$
        end if
        if $k \leq 1$ **then**
            $(\oplus, \Sigma_1 \ldots \Sigma_k) \leftarrow$ NONATOMICINTERLEAVEDCUTFILTERING$(\overset{\sim}{\twoheadrightarrow}(\tilde{L}))$
        end if
        if $k \leq 1$ **then**
            return NONATOMICLOOPCUTFILTERING$(\overset{\sim}{\twoheadrightarrow}(\tilde{L}))$
        else
            return $(\oplus, \Sigma_1 \ldots \Sigma_k)$
        end if
    end if
    return $\square$
end function
function SPLITLOG$_{\text{IMFLC}}(\tilde{L}, (\oplus, \Sigma_1, \ldots, \Sigma_n))$
    if $\oplus = \times$ **then return** XORSPLITFILTERING$(\tilde{L}, (\oplus, \Sigma_1, \ldots, \Sigma_n))$
    else if $\oplus = \rightarrow$ **then return**
    SEQUENCESPLITFILTERING$(\tilde{L}, (\oplus, \Sigma_1, \ldots, \Sigma_n))$ made consistent

  else if $\oplus = \wedge$ **then return** CONCURRENTSPLIT$(\tilde{L}, (\oplus, \Sigma_1, \ldots, \Sigma_n))$
  else if $\oplus = \rightarrow$ **then return**
   INTERLEAVEDSPLITFILTERING$(\tilde{L}, (\oplus, \Sigma_1, \ldots, \Sigma_n))$ made consistent
  else if $\oplus = \circlearrowright$ **then return** LOOPSPLITFILTERING$(\tilde{L}, (\oplus, \Sigma_1, \ldots, \Sigma_n))$ made
consistent
  end if
 end function
 function FALLTHROUGH$_{\mathrm{IM_{FLC}}}(\tilde{L})$
  $bc \leftarrow$ EMPTYTRACESFILTERING$(\tilde{L})$
  if $bc \neq \square$ **then return** bc
  else return FALLTHROUGH$_{\mathrm{IM_{LC}}}(\tilde{L})$
  end if
 end function

IM$_{\mathrm{CLC}}$.

Then, we summarise IM$_{\mathrm{CLC}}$, which handles incomplete behaviour:
 function BASECASE$_{\mathrm{IM_{CLC}}}(\tilde{L})$
  return BASECASE$_{\mathrm{IM_{LC}}}$
 end function
 function FINDCUT$_{\mathrm{IM_{CLC}}}(\tilde{L})$
  if $\epsilon \notin \tilde{L}$ **then**
   return cut $(\oplus, \Sigma_1, \Sigma_2)$ of $\Sigma(L)$ with highest $p_{\oplus}(\Sigma_1, \Sigma_2); \oplus \in \{\times, \rightarrow, \wedge, \circlearrowright\}$
  end if
 end function
 function SPLITLOG$_{\mathrm{IM_{CLC}}}(\tilde{L}, (\oplus, \Sigma_1, \ldots, \Sigma_n))$
  return SPLITLOG$_{\mathrm{IM_{FLC}}}$
 end function
 function FALLTHROUGH$_{\mathrm{IM_{CLC}}}(\tilde{L})$
  $ft \leftarrow$ EMPTYTRACES$(\tilde{L})$
  if $ft \neq \square$ **then return** ft
  else return CONCURRENTFLOWERMODEL$(\tilde{L})$
  end if
 end function

6.5.4 Implementation

We finish the description of IM$_{\mathrm{LC}}$, IM$_{\mathrm{FLC}}$ and IM$_{\mathrm{CLC}}$ with a brief description of their implementation in the ProM framework. At the moment of writing, IM$_{\mathrm{LC}}$ and IM$_{\mathrm{FLC}}$ have been implemented.

In the descriptions of IM$_{LC}$, IM$_{FLC}$ and IM$_{CLC}$ in this section, it was assumed that each trace in the event log is consistent, i.e. each start event in the trace corresponds to a single completion event and vice versa (see Section 5.7.1). However, the implementation is more lenient: in the implementation, we pose this requirement only on the start events, i.e. we require that for each start event there is a corresponding completion event, but we allow completion events without corresponding start events. This leniency allows the implemented algorithms to handle both non-atomic and atomic event logs without preprocessing steps, as the atomic events in atomic event logs are typically considered to be (or explicitly defined to be) completion events.

The output of the implementations of IM$_{LC}$ and IM$_{FLC}$ is a process tree. Thus, there is no observable difference between an atomic activity a in a process tree returned by an atomic-log algorithm and a non-atomic activity $\tilde{a}$ in a process tree returned by a non-atomic-log algorithm, due to a lack of compatible suitable output formats in the ProM framework (and hence there are no other tools to interpret and use such models correctly). Therefore, unfortunately, it is up to the user of the model to interpret the activities in the correct way.

We provide a plug-in, "Expand collapsed process tree", that transforms a process tree into a process tree with explicit non-atomicity, e.g. this plug-in transforms the process tree ╳ into ╳ , such that it can
∧
a b

$\rightarrow$ $\rightarrow$

a+start a+complete b+start b+complete

be processed further in techniques that are not non-atomic aware. In ProM, for further processing users should ensure to use an XEventClassifier that adds the +start/+complete parts. Ideally, conformance checking, log enhancement and other process mining techniques would support non-atomic event logs and process trees without the need for manually expanding models and selecting classifiers.

In the Inductive visual Miner, users need not to be aware of this, as it supports non-atomic event logs and process trees in a transparent manner (see Section 9.1).

6.5.5 Guarantees

As all parameter functions of IM$_{LC}$ are locally fitness preserving, IM$_{LC}$ itself guarantees fitness by Corollary 4.1:

Corollary 6.5 (IM$_{LC}$ guarantees fitness) *For any consistent log L it holds that* $set(L) \subseteq \mathcal{L}(IM_{LC}(L))$.

Not all parameter functions of IM$_{FLC}$ and IM$_{CLC}$ are locally fitness preserving, so these algorithms do not guarantee fitness. Similar to the atomic algorithms, neither IM$_{LC}$ nor IM$_{FLC}$ nor IM$_{CLC}$ guarantees log-precision.

As all parameter functions guarantee consistency preservation, we have shown that if the input event log is consistent, then in every recursion of these three algorithms, the event log under consideration is consistent.

Next, we prove rediscoverability, using the framework introduced in Section 4.2.2. That is, we first show that log splitting preserves the directly follows graph and concurrency graph abstractions (Lemma 6.22). Second, we show that IMLC preserves these abstractions (Lemma 6.23). Third, we prove rediscoverability (Theorem 6.11).

Lemma 6.22 (IMLC: log splitting preserves log assumptions) *Let* $S = \oplus(S_1, \ldots S_n)$ *with* $S \in C_{lc}$, *let* $c = (\oplus, \Sigma_1, \ldots \Sigma_m)$ *be a cut conforming to* S, *and let* $L_1 \ldots L_m = $ SPLITLOG(L, c). *Then, there exist subtrees* $M_1 \ldots M_m$ *such that* $\twoheadrightarrow(\oplus(M_1, \ldots M_m)) = \twoheadrightarrow(S)$, $\|(\oplus(M_1, \ldots M_m)) = \|(S)$ *and* $\forall_{1 \leq i \leq m} L_i \in LA_{IMLC}(M_i)$.

Proof We prove this lemma by constructing trees $M_1 \ldots M_m$ corresponding to $S_1 \ldots S_n$ as in Lemma 6.10 and showing that the log assumptions hold for these $M_1 \ldots M_m$, i.e. that the sublogs returned by SPLITLOG$_{IMLC}$ are fitting to their respective M_i and have the same directly follows graph and concurrency graph (similar to Figure 6.4). By construction, $\mathcal{L}(\oplus(M_1, \ldots M_m)) = \mathcal{L}(S)$, therefore $\twoheadrightarrow(\oplus(M_1, \ldots M_m)) = \twoheadrightarrow(S)$ and $\|(\oplus(M_1, \ldots M_m)) = \|(S)$.

As in Lemma 6.10, $\forall_{1 \leq i \leq m}$ $set(L_i) \subseteq \mathcal{L}(M_i)$ and $\twoheadrightarrow(L_i) = \twoheadrightarrow(M_i)$. Left to prove: $\forall_{1 \leq i \leq m}$ $\|(L_i) = \|(S_i)$. Take such an L_i and M_i, and a pair of activities $a, b \in \Sigma(L_i)$ such that $a \parallel_{L_i} b$. Then, there is a trace t in L_i such that an execution of a and an execution of b overlap in t. By construction of the split functions and semantics of the process tree operators, this overlap is present in a trace t' in L, hence $a \parallel_L b$. As $L \in LA_{IMLC}(S)$, $t' \in \mathcal{L}(S)$ and therefore $a \parallel_S b$. Similarly, $t \in M_i$, hence $a \parallel_{M_i} b$.

To prove the other direction, consider an M_i and activities $a, b \in \Sigma(M_i)$ such that $a \parallel_{M_i} b$. By construction of M_i and semantics of the process tree operators, $a \parallel_S b$. As $L \in LA_{IMLC}(S)$, $a \parallel_L b$. Then, there must be a trace $t \in \mathcal{L}(L)$ such that some executions of a and b overlap in t. Without loss of generality, assume that $t = \langle \ldots a_s, b_s, \ldots_1 a_c, b_c, \ldots \rangle$. For all activities d that have an execution (i.e. an event) in $\ldots_1$, it holds that $a \parallel_L d$ and $b \parallel_L d$. As cut c conforms to S, by Lemma 5.24, $d \in \Sigma(L_i)$. Hence, for the process tree operators $\times$, $\rightarrow$, $\wedge$ and $\circlearrowleft$, $t \in L_i$ and hence $a \parallel_{L_i} b$.

Hence, subtrees $M_1 \ldots M_m$ exist such that $\twoheadrightarrow(\oplus(M_1, \ldots M_m)) = \twoheadrightarrow(S)$, $\|(\oplus(M_1, \ldots M_m)) = \|(S)$ and $\forall_{1 \leq i \leq m} L_i \in LA_{IMLC}(M_i)$. □

Lemma 6.23 (IMLC, IMFLC and IMCLC are abstraction preserving) *IMLC, IMFLC and IMCLC are abstraction preserving, i.e. the combination of the class of process trees* C_{lc}, *the directly follows abstraction* $\twoheadrightarrow$, *the log assumptions function* $L \in LA_{IMLC}(S) \equiv set(L) \subseteq \mathcal{L}(S) \wedge \twoheadrightarrow(S) = \twoheadrightarrow(L) \wedge \|(S) = \|(L)$, *and the algorithms IMLC implementing the IM framework with* BASECASE$_{IMLC}$, FINDCUT$_{IMLC}$, SPLITLOG$_{IMLC}$ *and* FALLTHROUGH$_{IMLC}$ *is abstraction preserving.*

Proof We show that each requirement of Definition 4.5 holds:

AP.1 i.e. an activity base case preserves the abstraction. As $L \in LA_{IM}(S)$ holds, $set(L) = \{\langle a_s, a_c \rangle\}$. By code inspection, in IMLC the base case SINGLENONA-TOMICACTIVITY applies, which returns $\tilde{a}$.
Hence, $\twoheadrightarrow($BASECASE$_{IMLC}(L)) = \twoheadrightarrow(\tilde{a})$.

AP.2 i.e. a τ base case preserves the abstraction. As $\tau \notin C_{\text{LC}}$, this case cannot occur and the requirement holds.

AP.3 i.e. the base case parameter function preserves the abstraction. If $S = \oplus(S_1, \ldots S_n)$, with $S \in C_{\text{LC}}$, then $\Sigma(S) \geq 2$. As $L \in LA_{\text{IMLC}}(S)$ holds, by code inspection, no base case in BASECASE$_{\text{IMLC}}$ applies and the requirement trivially holds.

AP.4 i.e. every cut that is detected conforms to S. As $L \in LA_{\text{IMLC}}(S)$, $\rightarrow\!\!\!\rightarrow(S) = \rightarrow\!\!\!\rightarrow(L)$ and $\|(S) = \|(L)$. By lemmas 5.10 and 5.24, $\rightarrow\!\!\!\rightarrow(L)$ and $\|(L)$ contain a cut $c = \oplus(\Sigma(S_1), \ldots \Sigma(S_n))$. By Corollary 5.9, no other footprint is present in the combination of $\rightarrow\!\!\!\rightarrow(L)$ and $\|(S)$. By code inspection of FINDCUT$_{\text{IMLC}}$, this cut c is returned, hence these three cut detection functions conform to S (Definition 5.4).

AP.5 i.e. log splitting preserves the log assumptions. This requirement follows from Lemma 6.22.

AP.6 i.e. fall throughs preserve the abstraction. By the previous requirements and Lemma 5.3, for all systems $S \in C_{\text{LC}}$, either BASECASE$_{\text{IMLC}}$ or FINDCUT$_{\text{IMLC}}$ applies, i.e. FALLTHROUGH$_{\text{IMLC}}$ is never reached for $S \in C_{\text{LC}}$. Therefore, this case cannot occur and the requirement holds. $\square$

We show that IMLC is language-class preserving (Definition 4.6), i.e. that the discovered model is of C_{LC}:

Lemma 6.24 (IMLC is language-class preserving) *For all non-atomic systems $S \in C_{\text{LC}}$ and logs $\tilde{L}$ such that $\tilde{L} \in LA_{\text{IMLC}}(S)$, it holds that $IM_A(\tilde{L}) \in C_{\text{LC}}$.*

The proof of this lemma is similar to the proofs of lemmas 6.12 and 6.21, i.e. for each requirement of C_{LC} it is shown that there exists subtrees $M_1 \ldots M_m$ such that each of the sublogs L_i adheres to the log assumptions of M_i (Lemma 6.22). By Lemma 6.23, the non-atomic directly follows graph is preserved by IMLC, and as the requirement of C_{LC} holds for M_i, this requirement holds for S as well.

Finally, by Theorem 4.2, IMLC guarantees rediscoverability for C_{LC}: Rediscoverability is guaranteed by IMFLC and IMCLC as well, as these algorithms first apply the cut detection of IMLC before attempting filtering (and this is not necessary in case the preconditions for rediscoverability hold).

Theorem 6.11 (IMLC, IMFLC & IMCLC rediscoverability) *Let L be a log and $S \in C_{\text{LC}}$ be a system such that $set(L) \subseteq \mathcal{L}(S) \wedge \rightarrow\!\!\!\rightarrow(L) = \rightarrow\!\!\!\rightarrow(S)$. Then, $\mathcal{L}(IM_{LC}(L)) = \mathcal{L}(IM_{FLC}(L)) = \mathcal{L}(IM_{CLC}(L)) = \mathcal{L}(S)$.*

In this section, we introduced three algorithms that use life cycle information in event logs, i.e. in non-atomic event logs, to distinguish concurrency and looping behaviour. One of these algorithms is the basic fitness-guaranteeing IMLC, one the infrequent and deviating behaviour handling IMFLC, and the third, IMCLC, handles incomplete behaviour. We have shown that all three algorithms provide rediscoverability on process trees of C_{LC}. Furthermore, we have shown that IMLC guarantees fitness. Both IMLC and IMFLC have been implemented in the ProM framework [4] (see Section 6.7). Implementing and evaluating IMCLC remains part of future work:

Future work 6.12: Implement and evaluate IMCLC.

An interesting area of further research is to combine the life cycle handling capabilities of IMʟᴄ with the concurrency detection using the ⌾-relation of IM, and to extend support for the interleaved operator ↔.

Future work 6.13: Combine life cycle handling capabilities with the minimum self-distance relation ⌾.

Future work 6.14: Include support for ↔ in IMʟᴄ and IMꜰʟᴄ.

6.6 Handling Large Event Logs

In the previous sections, we introduced several algorithms that use the IM framework to discover process models. In Chapter 8, we will show that these algorithms (except IMᴄ) can be applied to event logs with millions of events and thousands of traces. Handling even larger event logs, the algorithms of the IM framework face the problem that they need to pass through the event log in every recursion and copy the event log in each recursion, which increases run time and memory consumption by the algorithms that use the IM framework. In this section, we will introduce a new family of algorithms that is able to handle logs with tens of millions of events and thousands of activities. These new algorithms pass through the event log once to construct a directly follows graph, and then recurse on this directly follows graph only instead of on the event log, thereby avoiding to copy the event log. Therefore, we adapt the IM framework to pass over the event log once: four steps are applied to the directly follows graph, similar to the IM framework: it detects a cut of the graph, splits the directly follows graph in subgraphs, and recurses on these subgraphs until it encounters a base case, and if no cut can be detected it chooses a fall through. We refer to this adapted framework as the *Inductive Miner - directly follows based framework* (IMᴅ framework).

We introduce three algorithms that implement the IMᴅ framework: a basic one called *Inductive Miner - directly follows* (IMᴅ), one to handle infrequent and deviating behaviour, called *Inductive Miner - infrequent - directly follows* (IMꜰᴅ), and one to handle incomplete behaviour *Inductive Miner - incompleteness - directly follows* (IMᴄᴅ).

We start with an example, after which we introduce the IMᴅ framework, and the three algorithms implementing it (IMᴅ, IMꜰᴅ and IMᴄᴅ). We conclude the section with a discussion of the guarantees provided by the IMᴅ framework and the introduced algorithms.

6.6.1 Example

We illustrate the IMᴅ framework and the basic algorithm IMᴅ with an example, using the event log

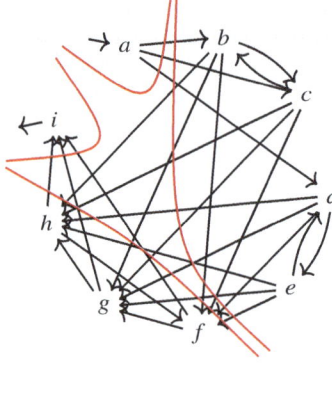

(a) ⟶ graph.

$\overset{?}{\twoheadrightarrow}$	a	b	c	d	e	f	g	h	i	⊥
⊤	9									
a			3	3	3					
b				3			1	1	1	
c			3				1	1	1	
d					3	1	1	1		
e					3					
f								6		3
g									6	3
h							6			3
i										9

(b) Cardinalities of the ⟶ graph.

Fig. 6.17: Directly follows graph of L_{113}, with cardinalities denoted in a table. The red lines denote the cut $(\rightarrow, \{a\}, \{b, c, d, e\}, \{f, g, h\}, \{i\})$.

$L_{113} = [\langle a, b, c, f, g, h, i \rangle, \quad \langle a, b, c, g, h, f, i \rangle \quad \langle a, b, c, h, f, g, i \rangle,$
$\qquad\quad \langle a, c, b, f, g, h, i \rangle, \quad \langle a, c, b, g, h, f, i \rangle, \quad \langle a, c, b, h, f, g, i \rangle,$
$\qquad\quad \langle a, d, f, g, h, i \rangle, \quad \langle a, d, e, d, g, h, f, i \rangle, \quad \langle a, d, e, d, e, d, h, f, g, i \rangle]$

The directly-follows graph of L_{113} is shown in Figure 6.17.

To this graph, IMD first tries to apply a base case, which does not apply. Second, the IMD applies cut detection, and detects the cut $(\rightarrow, \{a\}, \{b, c, d, e\}, \{f, g, h\}, \{i\})$. Using this cut, the directly follows graph is split into four subgraphs, shown in Figure 6.18, by, for each part of the cut, taking the nodes and edges within that part, and converting the inter-edges into start and end activities (depending on the process tree operator). Notice the difference with IM, which would split the event log, and in the recursion recompute the subgraphs; IMD avoids this step. The intermediate result is recorded:

$$\text{IMD}(\rightarrow(L_{113})) = \rightarrow(\text{IMD}(D_{114}), \text{IMD}(D_{115}), \text{IMD}(D_{116}), \text{IMD}(D_{117}))$$

On these four subgraphs, IMD recurses. Two of these subgraphs, i.e. D_{114} and D_{117}, are base cases:

$$\text{IMD}(D_{114}) = a$$
$$\text{IMD}(D_{117}) = i$$

Once IMD recurses on D_{115}, it detects the cut $(\times, \{b, c\}, \{d, e\})$, splits the graph into D_{118} and D_{119} (see Figure 6.19), and records the intermediate result

$$\text{IMD}(D_{115}) = \times(\text{IMD}(D_{118}), \text{IMD}(D_{119}))$$

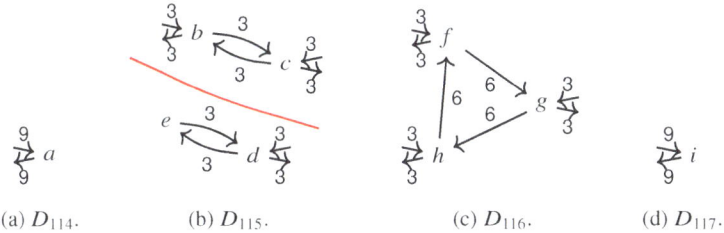

(a) D_{114}.   (b) D_{115}.   (c) D_{116}.   (d) D_{117}.

Fig. 6.18: Sub $\twoheadrightarrow$ graphs of L_{113}. The red line denotes the cut $(\times, \{b,c\}, \{d,e\})$.

For D_{118}, IMD discovers the tree $\wedge(b,c)$, while for D_{119}, IMD discovers the tree $\circlearrowright(d,e)$.

Recursing on D_{116}, having nodes f, g and h (see Figure 6.18), IMD detects neither a base case nor a cut. Therefore, a fall through must be used: IMD performs a fall through similar to STRICTTAULOOP of IM. Where STRICTTAULOOP splits traces on every transition from an end to a start activity, IMD removes all edges from an end to a start activity. On our example, IMD thus obtains the directly follows graph D_{120}, and records the intermediate result

$$\text{IMD}(D_{116}) = \circlearrowright(\text{IMD}(D_{120}, \tau))$$

On D_{120}, IMD discovers the model $\times(f,g,h)$. The end result of IMD applied to $\twoheadrightarrow(L_{113}))$ is

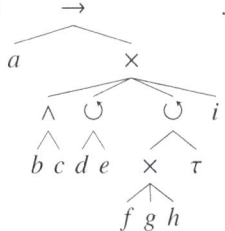

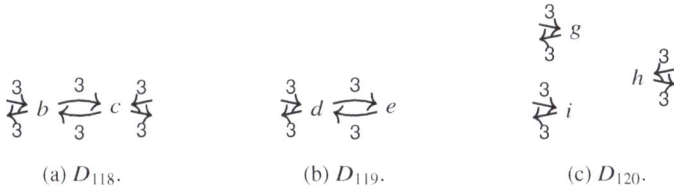

(a) D_{118}.   (b) D_{119}.   (c) D_{120}.

Fig. 6.19: Further sub $\twoheadrightarrow$ graphs.

6.6.2 Inductive Miner - directly follows based framework (IMᴅ framework)

In this section, we introduce the IMᴅ framework, which consists of four steps: first, a cut is detected, second, the directly follows graph is split into smaller subgraphs and third, the IMᴅ framework recurses on these subgraphs until a base case is encountered. If no cut can be found, a *fall through* is returned, i.e. a process tree is discovered such that recursion can continue. These four steps are parameters of the IMᴅ framework and have to be provided as plug-ins by a process discovery algorithm: each algorithm that implements the IMᴅ framework should provide each of these four functions. That is, for a directly follows graph $\twoheadrightarrow$, the parameter function BASECASE detects base cases of the recursion. The parameter function FINDCUT searches for a cut, i.e., FINDCUT($\twoheadrightarrow$) searches for a cut in directly follows graph $\twoheadrightarrow$ and returns that cut if it exists. The parameter function SPLITDFG splits the directly follows graph into smaller directly follows graphs: SPLITDFG($\twoheadrightarrow$, c) splits $\twoheadrightarrow$ according to cut c and returns the remaining subgraphs. The parameter function FALLTHROUGH($\twoheadrightarrow$) returns a fall through for $\twoheadrightarrow$. This function must not fail and always return a process tree.

Formally, let D be a directly follows relation:

function IMᴅ FRAMEWORK$_{\text{BASECASE,FINDCUT,SPLITDFG,FALLTHROUGH}}(D)$
    $bc \leftarrow$ BASECASE(D)
    if $bc \neq \square$ **then**
        return bc
    end if
    $(\oplus, \Sigma_1, \ldots, \Sigma_n) \leftarrow$ FINDCUT(D)
    if $(\oplus, \Sigma_1, \ldots, \Sigma_n) \neq \square$ **then**
        $D_1 \ldots D_n \leftarrow$ SPLITDFG$(D, (\oplus, \Sigma_1, \ldots, \Sigma_n))$
        return $\oplus(\text{IMᴅ} framework(D_1), \ldots, \text{IMᴅ} framework(D_n))$
    else
        return FALLTHROUGH(D)
    end if
end function

In the remainder of this section, we introduce three algorithms that implement this framework.

6.6.3 Inductive Miner - directly follows (IMᴅ)

The first algorithm we introduce is a basic one: the *Inductive Miner - directly follows* (IMᴅ) algorithm. This algorithm resembles the IM algorithm, i.e. uses the same concepts. We discuss its four parameter functions: cut detection, directly follows graph splitting, base cases and fall throughs.

6.6.3.1 Cut Detection

Cut detection of the IMd algorithm is identical to cut detection of IM. That is, XORCUT, SEQUENCECUT, CONCURRENTCUT and LOOPCUT of IM are applied until one returns a cut.

6.6.3.2 Directly Follows Graph Splitting

After finding a cut, the IMd framework splits the directly follows graph into several subgraphs, on which recursion continues. The IMd algorithm uses several log splitting functions. The idea is to keep the internal structure of each of the clusters of the cut by projecting a graph on the cluster.

Exclusive Choice and Concurrency.

For exclusive choice and concurrency, a simple projection suffices. That is, all inter-edges of a cluster are kept, and all intra-cluster edges are removed. For instance, consider the directly follows relation shown in Figure 6.20: starting with D_{121}, this graph is split using the cut $(\times, \{a\}, \{b, c\})$ into D_{122} and D_{123}. Similarly, Figure 6.21 shows a directly follows graph, the cut $(\wedge, \{a\}, \{b, c\})$ and the directly follows graphs that result from splitting.

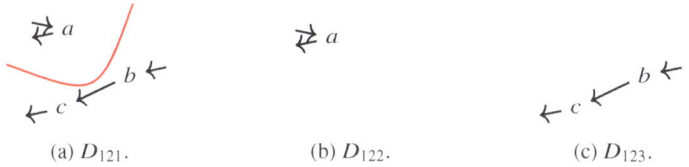

(a) D_{121}.   (b) D_{122}.   (c) D_{123}.

Fig. 6.20: A directly follows graph, split by the cut denoted by a red line.

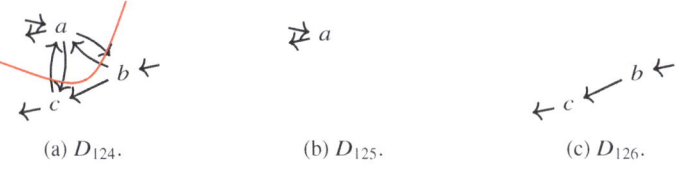

(a) D_{124}.   (b) D_{125}.   (c) D_{126}.

Fig. 6.21: A directly follows graph, split by the cut denoted by a red line.

Formally:
function SIMPLEDFGSPLIT$(D, \oplus, \Sigma_1 \ldots \Sigma_n)$

$\forall i : D_i \leftarrow [a \twoheadrightarrow b | a, b \in \Sigma_i \cup \{\top, \bot\} \wedge a \twoheadrightarrow_D b]$
  return $D_1, \ldots, D_n$
end function

Sequence and Loop.

For $\circlearrowright$ and $\rightarrow$, the start and end activities of a child might be different from the start and end activities of its parent. Therefore, every edge that enters a cluster is counted as a start activity, and an edge leaving a cluster is counted as an end activity. For $\rightarrow$, an edge bypassing a cluster is counted as an empty trace ($\top \twoheadrightarrow \bot$, denoted as ϵ in the directly follows graphs). For instance, the directly follows graph D_{127} in Figure 6.22 is split using the cut $(\rightarrow, \{a\}, \{b\}, \{c\})$ in the graphs D_{128}, D_{129} and D_{130}.

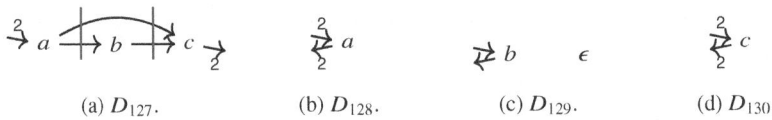

| (a) D_{127}. | (b) D_{128}. | (c) D_{129}. | (d) D_{130}. |

Fig. 6.22: A directly follows graph, split by the cut denoted by the red lines.

Formally:
function SEQUENCEDFGSPLIT$(D, \oplus, \Sigma_1 \ldots \Sigma_n)$
 $\forall i : D_i \leftarrow [a \twoheadrightarrow b | a, b \in \Sigma_i \wedge a \twoheadrightarrow_D b]$
    $\uplus [\top \twoheadrightarrow b | b \in \Sigma_i \wedge a \in \Sigma_{j<i} \cup \{\top\} \wedge a \twoheadrightarrow_D b]$
    $\uplus [a \twoheadrightarrow \bot | a \in \Sigma_i \wedge b \in \Sigma_{j>i} \cup \{\bot\} \wedge a \twoheadrightarrow_D b]$
    $\uplus [\top \twoheadrightarrow \bot | a \in \Sigma_{j<i} \cup \{\top\} \wedge b \in \Sigma_{k>i} \cup \{\bot\} \wedge a \twoheadrightarrow_D b]$
  return $D_1, \ldots, D_n$
end function

The function for loop cuts is similar: it takes the clusters of the cuts and keeps the internal $\twoheadrightarrow$-edges of these clusters. For the first (body) cluster, empty traces are added in three cases, as illustrated in Figure 6.23: (a) a redo cluster contains a start activity, which means that execution of the body left no trace in the directly follows graph, thus an empty trace should be added to the body (b) similarly, for each end activity an empty trace is added, and (c) a directly follows edge between two redo clusters indicates that the execution of the body resulted in an empty trace. For the non-first (redo) clusters, start and end activities are added according to the directly follows edges between the body cluster and the redo cluster.

 Notice that for $\circlearrowright$, the cut detection routines of IMD, IMFD and IMCD guarantee that only case (c) might occur. Figure 6.24 shows an example: D_{131} is split into D_{132}, D_{133} and D_{134}. The edge $c \twoheadrightarrow d$ violates the red loop cut and appears an as empty trace in D_{132}.

 function LOOPDFGSPLIT$(D, \oplus, \Sigma_1 \ldots \Sigma_n)$

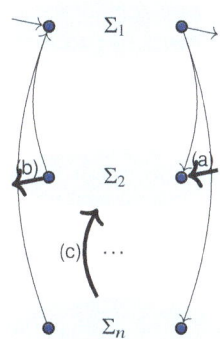

Fig. 6.23: Three possible bypasses of the body part in a loop.

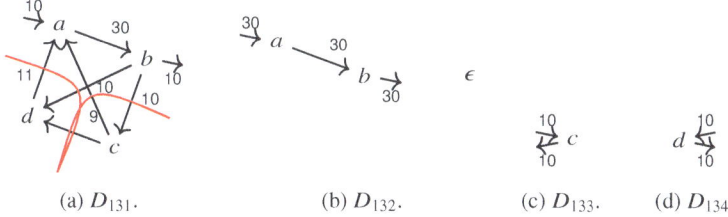

(a) D_{131}.   (b) D_{132}.   (c) D_{133}.   (d) D_{134}.

Fig. 6.24: Example of directly follows graph splitting. The red lines denote the loop cut ($\circlearrowleft, \{a, b\}, \{c\}, \{d\}$).

$$D_1 \leftarrow [a \twoheadrightarrow b | a, b \in \Sigma_1 \cup \{\top, \bot\} \wedge a \twoheadrightarrow_D b \wedge \neg(a = \top \wedge b = \bot)]$$

$$\uplus \; [\top \twoheadrightarrow \bot | a \in \bigcup_{1 < i \leq n} \Sigma_i \cup \{\top\} \wedge b \in \bigcup_{1 < j \leq n, i \neq j} \Sigma_i \cup \{\bot\} \wedge a \twoheadrightarrow_D b]$$

for $2 \leq i \leq n$ **do**
    $D_i \leftarrow [a \twoheadrightarrow b | a, b \in \Sigma_i \wedge a \twoheadrightarrow_D b]$
        $\uplus \; [\top \twoheadrightarrow b | a \in \Sigma_1 \cup \{\top\} \wedge b \in \Sigma_i \wedge a \twoheadrightarrow_D b]$
        $\uplus \; [a \twoheadrightarrow \bot | a \in \Sigma_i \wedge b \in \Sigma_1 \cup \{\bot\} \wedge a \twoheadrightarrow_D b]$
end for
return $D_1, \ldots, D_n$
end function

6.6.3.3 Base Cases

We identified two base cases for IMD: no activities and single activities.

The EMPTYDFG base case applies when the directly follows graph contains no activities. In that case, τ is returned.

The SINGLEACTIVITYDFG base case applies when the directly follows graph contains a single activity, and that activity has no self-edges. In case the activity has self-edges, the base case does not apply, as a self-edge implies that the log underlying the directly follows graph contains a trace with more than one execution of the activity, e.g. $\langle a, a \rangle$ (this case is handled by the fall through STRICTDFGTAULOOP that will be introduced below).

6.6.3.4 Fall Throughs

The IMD algorithm uses fewer fall throughs than the IM algorithm. For each of the fall throughs of IM, we discuss whether they are used by IMD as well, and how they were adapted to directly follows relations:

- The IM fall through EMPTYTRACES applies to directly follows relations, though needs to be adapted: instead of probing the event log for empty traces, it probes a directly follows relation D for relations $\top \rightarrow_D \bot$. We refer to this adapted fall through as EMPTYTRACESDFG.
- The IM fall through ACTIVITYONCEPERTRACE does not apply to directly follows relations, as from such a relation, it cannot always be derived whether an activity was executed once in each trace.
- The IM fall through ACTIVITYCONCURRENT could be adapted to directly follows relations by removing an activity from the relation and trying whether in the remaining graph a cut is present. However, given the time-consuming nature of this fall through and the focus of IMD on scalability, we chose not to include this fall through in IMD.
- The IM fall through STRICTTAULOOP can be adapted to directly follows relations: every edge from an end activity to a start activity is removed. We refer to the adapted fall through as STRICTDFGTAULOOP. The fall through applies if this step actually removes at least an edge. For instance, Figure 6.25 shows a directly follows relation and the filtered relation on which STRICTDFGTAULOOP recurses. That is, STRICTDFGTAULOOP returns $\circlearrowleft(\text{IMD}(D_{136}), \tau)$.

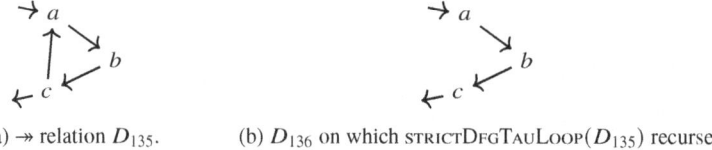

(a) $\twoheadrightarrow$ relation D_{135}.   (b) D_{136} on which STRICTDFGTAULOOP(D_{135}) recurses.

Fig. 6.25: A $\twoheadrightarrow$ relation, and the filtered $\twoheadrightarrow$ relation on which STRICTDFGTAULOOP recurses.

- The IM fall through TAULOOP can be adapted to directly follows relations, in a way similar to STRICTDFGTAULOOP, to obtain DFGTAULOOP. That is, of a directly follows graph D, all incoming edges of start activities are removed to obtain a new directly follows graph D', i.e. $\text{IM}_D(D) = \circlearrowright(\text{IM}_D(D'), \tau)$.

- The IM fall through FLOWERMODEL applies to directly follows relations without changes.

6.6.3.5 Summary

To summarise, the IM_D algorithm implements the parameter functions of the IM_D framework as follows:

function BASECASE$_{\text{IM}_D}(D)$
    if $\top \twoheadrightarrow \bot \notin D$ **then**
        $bc \leftarrow$ EMPTYDFG(D)
        if $bc = \square$ **then** $bc \leftarrow$ SINGLEACTIVITYDFG(D) **end if**
        if $bc \neq \square$ **then return** bc **end if**
    end if
    return $\square$
end function
function FINDCUT$_{\text{IM}_D}(D)$
    if $\top \twoheadrightarrow \bot \notin D$ **then**
        $(\oplus, \Sigma_1 \ldots \Sigma_k) \leftarrow$ XORCUT(D)
        if $k \leq 1$ **then** $(\oplus, \Sigma_1 \ldots \Sigma_k) \leftarrow$ SEQUENCECUT(D) **end if**
        if $k \leq 1$ **then** $(\oplus, \Sigma_1 \ldots \Sigma_k) \leftarrow$ CONCURRENTCUT$(D, \emptyset)$ **end if**
        if $k \leq 1$ **then** $(\oplus, \Sigma_1 \ldots \Sigma_k) \leftarrow$ LOOPCUT(D) **end if**
        if $k \geq 2$ **then return** $(\oplus, \Sigma_1 \ldots \Sigma_k)$ **end if**
    end if
    return $\square$
end function
function SPLITDFG$_{\text{IM}_D}(D, (\oplus, \Sigma_1, \ldots, \Sigma_n))$
    if $\oplus = \times$ **then return** SIMPLEDFGSPLIT$(D, (\oplus, \Sigma_1, \ldots, \Sigma_n))$
    else if $\oplus = \rightarrow$ **then return** SEQUENCEDFGSPLIT$(D, (\oplus, \Sigma_1, \ldots, \Sigma_n))$
    else if $\oplus = \wedge$ **then return** SIMPLEDFGSPLIT$(D, (\oplus, \Sigma_1, \ldots, \Sigma_n))$
    else if $\oplus = \circlearrowright$ **then return** LOOPDFGSPLIT$(D, (\oplus, \Sigma_1, \ldots, \Sigma_n))$
    end if
end function
function FALLTHROUGH$_{\text{IM}_D}(D)$
    $ft \leftarrow$ EMPTYTRACESDFG(D)
    if $ft = \square$ **then** $ft \leftarrow$ STRICTDFGTAULOOP(D) **end if**
    if $ft = \square$ **then** $ft \leftarrow$ DFGTAULOOP(D) **end if**
    if $ft \neq \square$ **then return** ft
    else return FLOWERMODEL(D)
    end if
end function

In Section 6.6.6, we will discuss the guarantees offered by IMD: rediscoverability, but not fitness.

6.6.4 Inductive Miner - infrequent - directly follows (IMFD)

In the previous sections, we introduced the IMD algorithm, which splits directly follows graphs recursively. In this section, we combine the concepts of the IMD and IMF algorithms to handle infrequent and deviating behaviour. We introduce a new algorithm, *Inductive Miner - infrequent - directly follows* (IMFD), and describe how it implements the four parameter functions of the IMD framework, i.e. cut detection, directly follows graph splitting, base cases and fall throughs.

6.6.4.1 Cut Detection

Cut detection of the IMFD resembles cut detection of IMF and IMD, i.e. first the cut detection of the basic IMD is attempted. If that fails, the graph is filtered using the FILTER function (see Section 6.2.2), after which the IMD cut detection is applied again.

6.6.4.2 Directly Follows Graph Splitting

As fitness is not guaranteed by the basic IMD algorithm, its directly follows graph splitting functions were designed to handle non-perfect cuts. Therefore, these log splitting functions of the IMD algorithm suffice for IMFD.

6.6.4.3 Base Cases

We consider how to adapt the two base cases of IMF to IMFD: EMPTYDFG applies to empty logs, and there is not much to filter, so this fall through is included unchanged. The SINGLEACTIVITYDFG fall through is sensitive to only one type of infrequent behaviour: self edges. That is, if a directly follows graph consists of a single activity a, then the only possible extra information is a self edge of a, indicating that a was executed multiple times in a trace. In such cases, the IMF algorithm needs to decide whether there are enough traces to justify the model $\circlearrowleft(a, \tau)$ or the more precise model a.

In IMF, the fall through SINGLEACTIVITYFILTERING assumes a geometric distribution with parameter p, which is estimated as $\widehat{p} = |L|/(||L|| + |L|)$, in which $|L|$ is the number of traces in log L and $||L||$ is the number of events in L. If the log contains only traces with a single a, then $\widehat{p} = 0.5$. If this $\widehat{p}$ is 'close enough' to 0.5, i.e. $|\widehat{p} - 0.5| \leq f$, the activity a is returned as a leaf. The IMFD algorithm applies a similar strategy with the SINGLEACTIVITYDFGFILTERING, and derives $|L|$ by the number of times a was a start activity, and $||L||$ by the total weight of incoming edges of a, i.e. the number of times a is a start activity and the weight of the self edge of a.

6.6.4.4 Fall Throughs

Most fall throughs of the IMD algorithm apply to IMFD without change. However, similar to the IMF algorithm, the fall through EMPTYTRACES is adapted to not apply when just a few empty traces are present. Thus, the fall through EMPTYTRACESD-FGFILTERING applies if "enough" empty traces, i.e. $|\top \twoheadrightarrow_D \bot| \geq |\text{Start}(D)| \times f$, the model $\times(\tau, \text{IMF}(D \text{ without } \top \twoheadrightarrow \bot))$ is returned and recursion continues on a directly follows graph without the empty traces. Otherwise, the empty traces are filtered out and recursion continues, i.e. IMFD(D without $\top \twoheadrightarrow \bot$).

6.6.4.5 Summary

To summarise, the IMFD algorithm implements the parameter functions of the IMD framework as follows:

function BASECASE$_{\text{IMFD}}(D)$
    if $\top \twoheadrightarrow \bot \notin D$ **then**
        $bc \leftarrow$ EMPTYDFG(D)
        if $bc = \square$ **then** $bc \leftarrow$ SINGLEACTIVITYDFGFILTERING(D) **end if**
        if $bc \neq \square$ **then return** bc **end if**
    end if
    return $\square$
end function
function FINDCUT$_{\text{IMFD}}(D)$
    if $\top \twoheadrightarrow \bot \notin D$ **then**
        $(\oplus, \Sigma_1 \ldots \Sigma_k) \leftarrow$ FINDCUT$_{\text{IMD}}(D)$
        if $k \leq 1$ **then** $(\oplus, \Sigma_1 \ldots \Sigma_k) \leftarrow$ XORCUTFILTERING(D) **end if**
        if $k \leq 1$ **then** $(\oplus, \Sigma_1 \ldots \Sigma_k) \leftarrow$ SEQUENCECUTFILTERING(D) **end if**
        if $k \leq 1$ **then** $(\oplus, \Sigma_1 \ldots \Sigma_k) \leftarrow$ CONCURRENTCUTFILTERING$(D, \emptyset)$ **end if**
        if $k \leq 1$ **then** $(\oplus, \Sigma_1 \ldots \Sigma_k) \leftarrow$ LOOPCUTFILTERING(D) **end if**
        if $k \geq 2$ **then return** $(\oplus, \Sigma_1 \ldots \Sigma_k)$ **end if**
    end if
    return $\square$
end function
function SPLITDFG$_{\text{IMFD}}(D, (\oplus, \Sigma_1, \ldots, \Sigma_n))$
    if $\oplus = \times$ **then return** SIMPLEDFGSPLIT$(D, (\oplus, \Sigma_1, \ldots, \Sigma_n))$
    else if $\oplus = \rightarrow$ **then return** SEQUENCEDFGSPLIT$(D, (\oplus, \Sigma_1, \ldots, \Sigma_n))$
    else if $\oplus = \wedge$ **then return** SIMPLEDFGSPLIT$(D, (\oplus, \Sigma_1, \ldots, \Sigma_n))$

       else if $\oplus = \circlearrowleft$ **then return** LOOPDFGSPLIT$(D, (\oplus, \Sigma_1, \ldots, \Sigma_n))$
       end if
   end function
   function FALLTHROUGH$_{\text{IM}_{\text{FD}}}(D)$
       $ft \leftarrow$ EMPTYTRACESDFGFILTERING(D)
       if $ft = \square$ **then** $ft \leftarrow$ STRICTTAULOOP(D) **end if**
       if $ft = \square$ **then** $ft \leftarrow$ TAULOOP(D) **end if**
       if $ft \neq \square$ **then return** ft
       else return FLOWERMODEL(D)
       end if
   end function

6.6.5 Inductive Miner - incompleteness - directly follows (IMCD)

In the previous sections, we introduced the IMD framework to increase scalability, and the IMD and IMF algorithms. In this section, we describe how the concepts of the incompleteness-handling algorithm IMc can be applied in the IMD framework, i.e. we introduce a new algorithm *Inductive Miner - incompleteness - directly follows* (IMCD) that handles incomplete behaviour. We discuss cut detection, log splitting, base cases and fall throughs, after which we summarise the algorithm. Notice that even though the IMD framework focuses on speed, the IMCD algorithm that is presented in this section is exponential in the number of activities. We nevertheless included it as it illustrates the flexibility of the IM framework and IMD framework: without much effort, we can reuse existing techniques, implementations and proofs.

6.6.5.1 Cut Detection

The activity relations described in Section 5.3 are defined using a directly follows graph. Hence, IMCD can reuse the cut detection of IMc with a small adjustment, i.e. it first constructs the activity relations and computes probabilities for them. In these probabilities, the number of occurrences of an activity is used, which is derived from the sum of incoming edges in the directly follows graph. For instance, if activity a has several incoming directly follows edges whose weights sum up to 10, activity a was executed 10 times. Second, several SMT problems are constructed whose solutions correspond to the cuts with maximum accumulated probabilities, and the cut with the highest probability is returned. For more details, please refer to Section 6.3.5.

6.6.5.2 Directly Follows Graph Splitting

As fitness is not guaranteed by the basic IM_D algorithm, its directly follows graph splitting functions were designed to handle non-perfect cuts. Therefore, these log splitting functions of the IM_D algorithm suffice for IM_{CD}.

6.6.5.3 Base Cases

Similar to IM_C, which reuses the base cases of IM, IM_{CD} reuses the non-filtering base cases of IM_D.

6.6.5.4 Fall Throughs

Similar to IM_C, IM_{CD} always discovers a cut, thus a fall through is only necessary if the event log consists of a single activity, or if the event log contains empty traces ϵ. That is, EMPTYTRACESDFG and FLOWERMODEL.

6.6.5.5 Summary

To summarise, the IM_{CD} algorithm implements the parameter functions of the IM_D framework as follows:

> **function** BASECASE$_{IM_{CD}}(D)$
> > **return** BASECASE$_{IM_D}(D)$
>
> **end function**
> **function** FINDCUT$_{IM_{CD}}(D)$
> > **if** $\top \twoheadrightarrow \bot \notin D$ **then**
> > > **return** cut $(\oplus, \Sigma_1, \Sigma_2)$ of $\Sigma(D)$ with highest $p_\oplus(\Sigma_1, \Sigma_2)$; $\oplus \in \{\times, \rightarrow, \wedge, \circlearrowright\}$
> >
> > **end if**
>
> **end function**
> **function** SPLITDFG$_{IM_{CD}}(D, (\oplus, \Sigma_1, \dots, \Sigma_n))$
> > **return** SPLITDFG$_{IM_D}(D)$
>
> **end function**
> **function** FALLTHROUGH$_{IM_{CD}}(D)$
> > $ft \leftarrow$ EMPTYTRACESDFG(D)
> > **if** $ft \neq \square$ **then return** ft
> > **else return** FLOWERMODEL(L)
> > **end if**
>
> **end function**

6.6.6 Guarantees

In the previous sections, we introduced the IMD framework that discovers process trees by recursing on directly follows graphs, and three algorithms implementing it, i.e. the basic IMD, the the infrequent and deviating behaviour handling IMFD, and the incompleteness handling IMCD. In this section, we discuss the guarantees provided by the IMD framework and the three algorithms. We start with an explanation why we chose to *not* guarantee fitness. Second, we lift the rediscoverability framework of the IM framework to the IMD framework. Third, we show that all three algorithms provide rediscoverability for selected subclasses of models.

6.6.6.1 Fitness

A desirable property of discovery algorithms is the ability to return a fitting model, i.e. guarantee fitness. For directly follows based algorithms without recursion, this is challenging. For instance, Figure 6.26 shows the directly follows graph of the process tree $M_{137} = \begin{array}{c} \wedge \\ \wedge \\ \rightarrow \quad c \\ \wedge \\ a \; b \end{array}$. However, it is also the directly follows graph of the event log $L_{138} = [\langle a, c, b, c, a, b \rangle, \langle c \rangle]$. Thus, any directly follows algorithm that aims to return a fitting model should return a model that includes the behaviour of M_{137} as well as the behaviour of L_{138}, as the directly follows graph that the algorithm considers cannot distinguish between the behaviour of M_{137} and L_{138}.

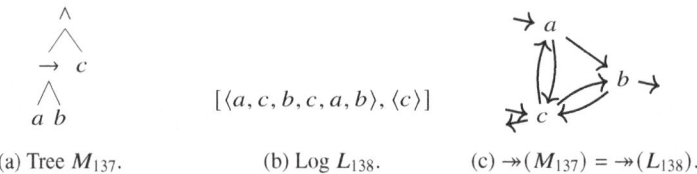

(a) Tree M_{137}.   (b) Log L_{138}.   (c) $\twoheadrightarrow(M_{137}) = \twoheadrightarrow(L_{138})$.

Fig. 6.26: Example of a tree and non-fitting log with the same $\twoheadrightarrow$ graph.

Hence, a directly follows based algorithm that guarantees fitness can never return tree M_{137}. Specifically, a fitness-guaranteeing algorithm would need to return a model that allows for any possible path through the directly follows graph. Therefore, such an algorithm has to seriously underfit/generalise, and we chose the basic algorithm IMD to not guarantee fitness, in contrast to the basic IM, which is aided by its log splitting in detecting the difference between M_{137} and L_{138}. Furthermore, this example shows that a fitness-guaranteeing directly follows-based discovery algorithm cannot guarantee rediscoverability on any class of process models that includes M_{137}, e.g. C_B.

This challenge holds for any pure directly-follows based process discovery algorithm, such as IMD and α. The algorithms of the IM framework do not suffer from this as due to their recursion, these algorithms use more information from the event log. The trees returned by IM and IMD for M_{137} and L_{138} are as follows:

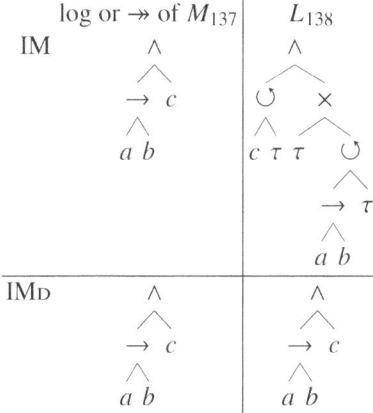

6.6.6.2 Rediscoverability in the IMD framework

The IMD framework provides several guarantees: the model returned is a process tree and hence is sound, and furthermore algorithms that implement the IMD framework can guarantee rediscoverability.

In Section 4.2.2, we introduced a formal framework for rediscoverability and expressed its requirements in terms of the IM framework functions. As the IMD framework differs from the IM framework, we adapt the rediscoverability framework as well. In the rediscoverability framework we used a discovery algorithm $\Diamond$, a log-assumption function LA, a language abstraction A and a class of models C. We defined the property *abstraction preservation*, which expresses that the algorithm chooses sensible cuts, base cases and fall throughs, and never performs a step that causes the discovered model to have a different abstraction than the system model (see Definition 4.5). Furthermore, we defined the property *language-class preservation*, which expresses that the output model should be of class C. For both properties, one may assume that the system is of class C, and that the log adheres to the log-assumption function LA. From these two properties, rediscoverability follows, as proven in Theorem 4.2.

For the IMD framework, we adapt this formal framework as follows. In order not to repeat large parts of Section 4.2.2, we restate properties and requirements of the IMD framework while referring to similar proofs of the IM framework. We introduce two properties: one corresponding to abstraction preservation and one similar to language-class preservation.

Definition 6.3 (directly follows preservation) Let C be a class of process trees, let $DA : C \to 2^{\text{directly follows relation}}$ be a directly follows relation assumption function, let D be a directly follows relation, and let $\Diamond$ be a discovery algorithm implementing the IM$_D$ framework with BASECASE$_\Diamond$, FINDCUT$_\Diamond$, SPLITDFG$_\Diamond$ and FALLTHROUGH$_\Diamond$. Then, $\Diamond$ is *abstraction preserving* if for every tree $S \in C$:

DAP.1 The directly follows relation of an activity is preserved: for all reduced systems $a \in C$ such that a is an activity, and for all directly follows relations $D \in DA(a)$, it holds that $\twoheadrightarrow(\text{BASECASE}_\Diamond(D)) = \twoheadrightarrow(a)$.

DAP.2 The directly follows relation of a τ step is preserved: for the reduced system $\tau \in C$ and for all directly follows relations $D \in DA(\tau)$, it holds that $\twoheadrightarrow(\text{BASECASE}_\Diamond(D)) = \twoheadrightarrow(\tau)$.

DAP.3 If the algorithm applies a base case, then the directly follows relation is preserved: let $S = \oplus(S_1, \ldots S_n)$, with $S \in C$ be a reduced system, and let $D \in DA(S)$ be a directly follows relation. Assume that for all S' such that $|S'| \le |S|$ and $D' \in DA(S')$, it holds that $\twoheadrightarrow(\text{BASECASE}_\Diamond(D')) = \twoheadrightarrow(S')$. Then, if BASECASE$_\Diamond(D)$ applies, then $\twoheadrightarrow(\text{BASECASE}_\Diamond(D)) = \twoheadrightarrow(S)$.

DAP.4 If the algorithm detects a cut, then this cut conforms to the system: for all reduced systems $S = \oplus(S_1, \ldots S_n)$ with $S \in C$ and for all directly follows relations $D \in DA(S)$ for which BASECASE$_\Diamond(D)$ does not apply, it holds that FINDCUT$_\Diamond(D)$ conforms to S (Definition 5.4).

DAP.5 If a conforming cut is found, then the directly follows assumptions hold for the subgraphs (for the next recursive step): for all reduced systems $S = \oplus(S_1, \ldots S_n)$ with $S \in C$, and all cuts $c = (\otimes, \Sigma_1, \ldots \Sigma_m)$ that conform to S (Definition 5.4, notice that n and m may be different due to the reduction rules of Definition 5.1), let $D_1 \ldots D_m = \text{SPLITDFG}(D, c)$, then there exist trees $M_1 \ldots M_m$ such that $\twoheadrightarrow(\oplus(M_1, \ldots M_m)) = \twoheadrightarrow(S)$, and $\forall_{1 \le i \le m} \ D_i \in DA(M_i)$.

DAP.6 If the algorithm uses a fall through, the directly follows relation is preserved: let $S = \oplus(S_1, \ldots S_n)$ with $S \in C$ be a reduced system, and let $D \in DA(S)$ be a directly follows relation, but neither BASECASE$_\Diamond(D)$ nor FINDCUT$_\Diamond(D)$ applies. Assume that for all S' such that $|S'| \le |S|$ and $D' \in DA(S')$, it holds that $\twoheadrightarrow(\Diamond(D')) = \twoheadrightarrow(S')$.
Then, $\twoheadrightarrow(\text{FALLTHROUGH}_\Diamond(D)) = \twoheadrightarrow(S)$.

An algorithm having this property preserves the abstraction of a system, i.e.

Lemma 6.25 (Directly follows rediscoverability of the IM$_D$ framework) *Let C be a class of process trees, DA be a directly follows assumption function, and let $\Diamond$ be an algorithm that implements the IM$_D$ framework with* BASECASE$_\Diamond$, FINDCUT$_\Diamond$, SPLITDFG$_\Diamond$ *and* FALLTHROUGH$_\Diamond$, *such that $\Diamond$ is directly follows preserving (Definition 6.3). Then, for all reduced systems $S \in C$ and directly follows relations $D \in DA(S)$, it holds that $\twoheadrightarrow(\Diamond(D)) = \twoheadrightarrow(S)$.*

The proof of this lemma is similar to the proof of Lemma 4.1.

The second property expresses that the discovered model should be of class C, in terms of C and LA:

Definition 6.4 (language-class preservation) A combination of a class of process trees C, a directly follows assumption function DA and an algorithm $\Diamond$ is *language-class preserving* if and only if for all reduced systems $S \in C$ and directly follows relations $D \in DA(S)$, it holds that $\Diamond(D) \in C$.

Finally, we prove the main theorem, i.e. an algorithm that is abstraction preserving and language-class preserving has rediscoverability:

Theorem 6.15 *Let C be a class of process trees, DA be a directly follows assumption function, and $\Diamond = IM_D\ FRAMEWORK_{BASECASE_{\Diamond},FINDCUT_{\Diamond},SPLITDFG_{\Diamond},FALLTHROUGH_{\Diamond}}$, such that the combination of C, DA and $\Diamond$ is directly follows preserving (Definition 6.3), such that the combination of C, DA and $\Diamond$ is language-class preserving (Definition 6.4), and such that the combination of $\twoheadrightarrow$ and the set of languages represented by C is language unique (Definition 4.3).*

Let SM and S be process trees such that $\mathcal{L}(SM) = \mathcal{L}(S)$, $SM \in C$ and $S \in C$. Then, $\Diamond$ has rediscoverability (Definition 4.2): for each directly follows relation $D \in DA(S)$, it holds that $\mathcal{L}(SM) = \mathcal{L}(\Diamond(D))$.

For the proof of this theorem, we refer to the proof of Theorem 4.2.

6.6.6.3 Rediscoverability of IMD, IMFD and IMCD

Theorem 6.16 (IMD rediscoverability) *Take a system $S \in C_B$ (Definition 5.2) and a log L such that $set(L) \subseteq \mathcal{L}(S)$ and $\twoheadrightarrow(L) = \twoheadrightarrow(S)$. Then, $\mathcal{L}(IM_D(\twoheadrightarrow(L))) = \mathcal{L}(S)$.*

Proof In order to prove the theorem, we prove that the directly follows relation is preserved (Definition 6.3), and that the model discovered by IMD is of class C_B (Definition 6.4). Then, by Theorem 6.15, the theorem holds.

We prove that the parameter functions of IMD are directly follows preserving: let $\twoheadrightarrow(L) = D \in DA_{IM_D}$ if and only if $\twoheadrightarrow(S) = D$.

DAP.1 Let $S = a$, with a being an activity. By DA_{IM_D}, $\twoheadrightarrow(L) = \twoheadrightarrow(S)$. By construction of $BASECASE_{IM_D}$, $\twoheadrightarrow(BASECASE_{IM_D}(\twoheadrightarrow(L))) = \twoheadrightarrow(a)$.

DAP.2 Let $S = \tau$. By DA_{IM_D}, $\twoheadrightarrow(L) = \twoheadrightarrow(S)$. By construction, neither $BASECASE_{IM_D}(\twoheadrightarrow(L))$ nor $FINDCUT_{IM_D}(\twoheadrightarrow(L))$ apply. Then, by construction of $BASECASE_{IM_D}$, $\twoheadrightarrow(BASECASE_{IM_D}(\twoheadrightarrow(L))) = \twoheadrightarrow(a)$.

DAP.3 This case never applies, i.e. if $S = \oplus(S_1, \ldots S_n)$, then $BASECASE_{IM_D}(\twoheadrightarrow(L))$ does not apply.

DAP.4 As IMD reuses the cut detection of IM, the proof of Requirement AP.4 of Lemma 6.11 applies.

DAP.5 Let $S = \oplus(S_1, \ldots S_n)$ with $S \in C_B$, let $c = (\oplus, \Sigma_1, \ldots \Sigma_m)$ be a cut conforming to S, and let $D_1 \ldots D_m = SPLITLOG(D, c)$. To prove: there exist subtrees $M_1 \ldots M_m$ such that $\twoheadrightarrow(\oplus(M_1, \ldots M_m)) = \twoheadrightarrow(S)$ and $\forall_{1 \leq i \leq m}\ D_i \in DA(S_i)$. We prove this by constructing trees $M_1 \ldots M_m$ corresponding to $S_1 \ldots S_n$ and showing that the log assumptions hold for these $M_1 \ldots M_m$, i.e. that the subgraphs returned by $SPLITDFG_{IM}$ are fitting to their respective M_i and have the same directly follows graph.

As c is conforming, each $\Sigma_1 \ldots \Sigma_m$ is the union of one or more $\Sigma(S_i)$. Let each $M_1 \ldots M_m$ be the trees corresponding to the subtrees S_i, combined with $\oplus$ if necessary. (for instance, if $S = \rightarrow(a, b, c)$ and $c = (\rightarrow, \{a, b\}, \{c\})$, then $M_1 = \rightarrow(a, b)$ and $M_2 = c$).

We prove the directly follows assumptions DA_{IM} for these subgraphs, i.e. $\forall_{1 \leq i \leq m} \twoheadrightarrow (M_i) = D_i$ by case distinction on $\oplus$. As of Requirement $C_B.1$, we do not need to consider $\top \twoheadrightarrow \bot$.

- $\oplus = \times$ and $\oplus = \wedge$ By construction of SIMPLEDFGSPLIT and Requirement $C_B.2$, for any activities $a, b \in \Sigma(D_i)$, $a \twoheadrightarrow_{D_i} b \wedge a \twoheadrightarrow_{M_i} b$. Furthermore, for a and b in different $\Sigma(D_i)$, $\Sigma(D_{j \neq i})$, by construction of SIMPLEDFGSPLIT, $a \not\twoheadrightarrow_{D_i} b$ and $a \not\twoheadrightarrow_{M_i} b$. Similarly, $\text{Start}(D_i) = \text{Start}(M_i)$ and $\text{End}(D_i) = \text{End}(M_i)$. Hence, $D_i = \twoheadrightarrow(M_i)$.

- $\oplus = \rightarrow$ and $\oplus = \circlearrowleft$ By reasoning similar to the $\times$ case, for any activities $a, b \in \Sigma(D_i)$, $a \twoheadrightarrow_{D_i} b = a \twoheadrightarrow_{M_i} b$ and for a and b in different $\Sigma(D_i)$, $\Sigma(D_{j \neq i})$, $a \not\twoheadrightarrow_{D_i} b$ and $a \not\twoheadrightarrow_{M_i} b$. Left to prove: $\text{Start}(D_i) = \text{Start}(M_i)$ and $\text{End}(D_i) = \text{End}(M_i)$. We consider the start activities (the end activities are symmetrical). For $i = 1$, $\text{Start}(M_i) = \text{Start}(S) = \text{Start}(D)$. For $i > 1$, take an activity $b \in \text{Start}(M_i)$, then as $\twoheadrightarrow(S) = D$, there exists an activity $a \in \Sigma_{i-1}$ such that $a \twoheadrightarrow_D b$. By construction of SEQUENCEDFGSPLIT and LOOPDFGS-PLIT, $b \in \text{Start}(D_i)$. Hence, $\text{Start}(D_i) = \text{Start}(M_i)$ and by a symmetrical argument, $\text{End}(D_i) = \text{End}(M_i)$.

Hence, subtrees $M_1 \ldots M_m$ exist such that $\twoheadrightarrow(\oplus(M_1, \ldots M_m)) = \twoheadrightarrow(S)$ and $\forall_{1 \leq i \leq m} D_i \in DA(S_i)$.

DAP.6 This case never applies, i.e. if $S = \oplus(S_1, \ldots S_n)$, then FINDCUT$_{\text{IMD}}(\twoheadrightarrow(L))$ applies and FALLTHROUGH$_{\text{IMD}}$ is never executed.

By reasoning similar to Lemma 6.12, IMD$(\twoheadrightarrow(L)) \in C_B$, i.e. IMD is abstraction preserving. Then by Theorem 6.15, IMD provides rediscoverability if $\twoheadrightarrow(L) = \twoheadrightarrow(S)$ and $S \in C_B$. $\square$

Both IMFD and IMCD as a first step apply IMD cut detection, IMD log splitting and IMD base cases before applying filtering and SMT-cut detection. Therefore, IMFD and IMCD provide rediscoverability as well:

Theorem 6.17 (IMFD & IMCD rediscoverability) *Take a system $S \in C_B$ and a log L such that $\text{set}(L) \subseteq \mathcal{L}(S)$ and $\twoheadrightarrow(L) = \twoheadrightarrow(S)$. Then, $\mathcal{L}(IMFD(\twoheadrightarrow(L))) = \mathcal{L}(IMCD(\twoheadrightarrow(L))) = \mathcal{L}(S)$.*

6.7 Tool Support

The algorithms described in this chapter have been implemented as plug-ins of the ProM framework The algorithms of the IM framework are accessible via the plug-ins "Mine Petri net with Inductive Miner" and "Mine process tree with Inductive Miner".

The algorithms of the IMᴅ framework are accessible via the plug-ins "Mine Petri net with Inductive Miner - directly follows" and "Mine process tree with Inductive Miner - directly follows". In this section, we provide a user manual and describe its architecture.

In this chapter, two types of algorithms were introduced: the algorithms of the IM framework, and the algorithms of the IMᴅ framework. The difference between these algorithms is their input: the IM framework algorithms take an event log as input, while the IMᴅ framework algorithms take a directly follows graph as input.

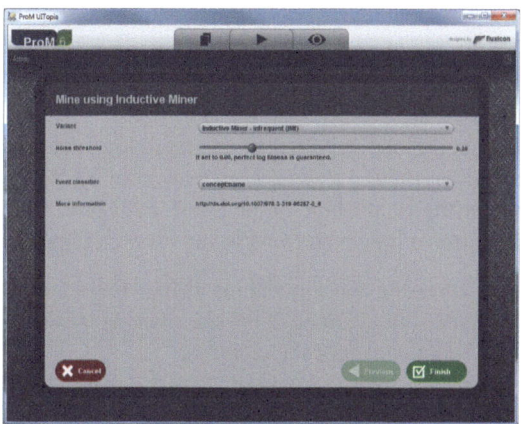

(a) Parameter settings for the IM framework algorithms in the ProM framework.

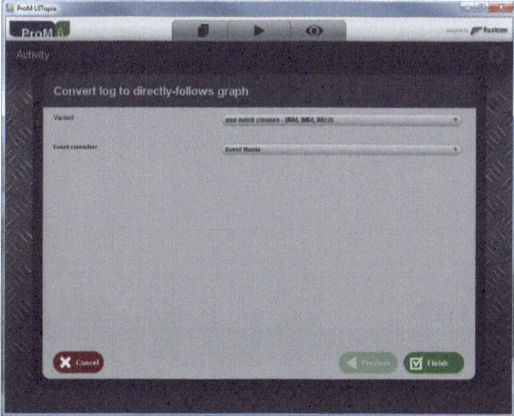

(b) Parameter settings for the conversion from log to directly follows graph.

Fig. 6.27: Parameter settings of Inductive Miner plug-ins.

For event logs, two plug-ins are available, that produce either a Petri net ("Mine Petri net with Inductive Miner") or a process tree ("Mine process tree with Inductive Miner"). A process tree is suitable for fast further processing using for instance the plug-in "Visualise deviations on process tree", while a Petri net is widely supported in other plug-ins. On activation, both plug-ins show a graphical user interface in which the settings for the miner can be set (see Figure 6.27a) for a screenshot. In this interface, the following parameters can be set:

1. The specific mining algorithm can be chosen ('Variant'), please refer to earlier in this chapter for a discussion on these algorithms.
2. If a filtering threshold is relevant for the mining algorithm, this threshold can be set ('Noise threshold').
3. The classifier selector controls what determines the event types (i.e. activities) of events: events in XES-logs can have several data attributes [5], and this selector determines which one of these data attributes determines the activities. One can choose either one of the classifiers defined in the event log, or use any combination of attributes of the event log.
   By default, the first defined classifier of the event log is chosen. If the log does not define a classifier, the *concept:name* extension of XES is used. If that is not present, then an arbitrary attribute of the event log is used.
4. For more information a (clickable) link to the relevant paper is provided.

Notice that when choosing the algorithms IM_{LC}, IM_{FLC} or IM_{CLC}, the "life-cycle:transition" attribute should not be in the classifier, as these algorithms take this attribute into account independent of the classifier.

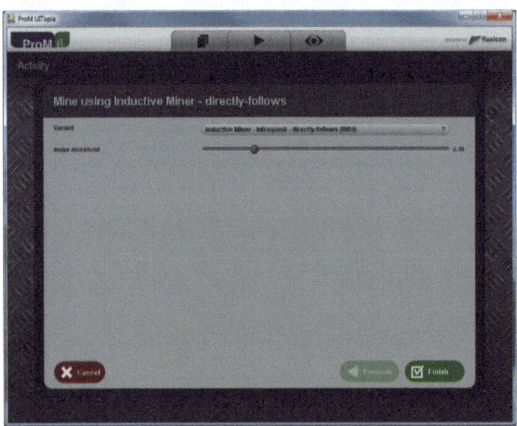

Fig. 6.28: Parameter settings for the IM_D framework algorithms in the ProM framework.

Directly follows graphs can be obtained by using the plug-in "Convert log to directly follows graph", a screenshot is shown in Figure 6.27b. The parameters of this plug-in entail a classifier (see before), and whether to take life cycle transitions into account. Given a directly follows graph, the plug-ins "Mine process tree with Inductive Miner - directly follows" and "Mine Petri net with Inductive Miner -

directly follows" provide access to the algorithms IM$_D$ and IM$_{FD}$. Figure 6.28 shows the graphical user interface in which the two parameters can be set, i.e. the variant (specific algorithm) and, if applicable, the noise threshold. Figure 6.29 shows the result of applying IM$_F$ to a real-life log of a road fine management process: as a process tree, as a process tree in the notation used by the Inductive visual Miner (see Section 9.1) and as a Petri net. In Chapter 8, we will evaluate this and other models.

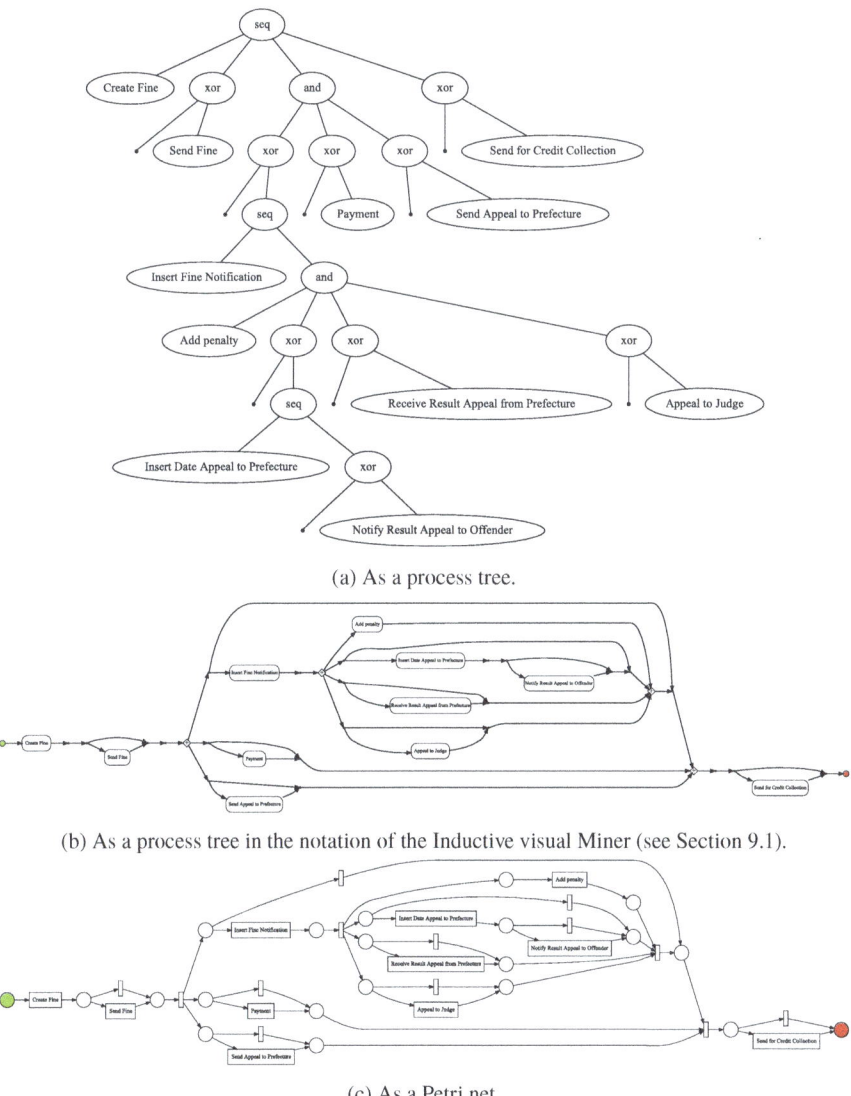

(a) As a process tree.

(b) As a process tree in the notation of the Inductive visual Miner (see Section 9.1).

(c) As a Petri net.

Fig. 6.29: The result of applying IM$_F$ to a real-life log [7].

The IM framework has been implemented as a plug-in of the ProM framework [11]. The implementation resembles the formal definition given in Chapter 4, i.e. a developer can provide new functions to implement new algorithms easily, i.e. the base cases, cut detection, log splitters and fall-throughs can be changed. Furthermore, there is a post-processing step that is executed on each node before that node is returned.

For developers, the Inductive Miner framework allows for easy extension and adaption: all steps taken by any algorithm can be adjusted using the MiningParameters class. It is possible to implement this interface, but we would recommend to simply create an instance of an existing algorithm (for instance, MiningParametersIMi), and change the parameters as necessary. Base cases, cut detection, log splitters, fall-throughs and post-processing steps all have their own self-explanatory interfaces.

The source code of the IM framework is available at `https://svn.win.tue.nl/repos/prom/Packages/InductiveMiner/Trunk`. The implementation of IMc uses the SAT4j SAT solver [1].

6.8 Summary: Choosing a Miner

In the previous sections, several algorithms have been introduced. In this section, we provide an overview and guidance which algorithm to choose in which situation. Figure 6.30 shows a flow chart containing each of the discovery algorithms that have been introduced in this chapter, and guides the choice. As described in Section 3.1, finding the best discovery algorithm might depend on the use case and the event log, and might be an iterative process. Therefore, the flowchart might guide this iterative process: if an algorithm does not work, one might take a different choice somewhere and try another algorithm[2].

Furthermore, Table 6.3 summarises the guarantees that the algorithms of this chapter provide, i.e. whether each algorithm guarantees fitness and for which class of process trees rediscoverability was proven in this chapter.

In this chapter, we have introduced several algorithms, focusing on and combining several dimensions. One such dimension is to guarantee fitness (IM, IMA, IMLc), to handle infrequent and deviating behaviour (IMF, IMFA, IMFLC, IMFD), or to handle incomplete behaviour (IMc, IMcLc, IMcD). Another dimension is the type of input data, i.e. atomic event logs (IM, IMF, IMc, IMA, IMFA), non-atomic event logs (IMLc, IMFLC, IMcLc) and directly follows graphs (IMD, IMFD, IMcD). These dimensions and the variety of algorithms introduced in this chapter illustrate the flexibility of the IM framework.

Even though the algorithms presented in this chapter apply different strategies, the IM framework allowed us to focus on strategies to find the *most important* behaviour in an event log (i.e. the root operator and root activity partition), instead of searching

---

[2] Recommender systems have been proposed that suggest process discovery techniques for a given event log, e.g. [9].

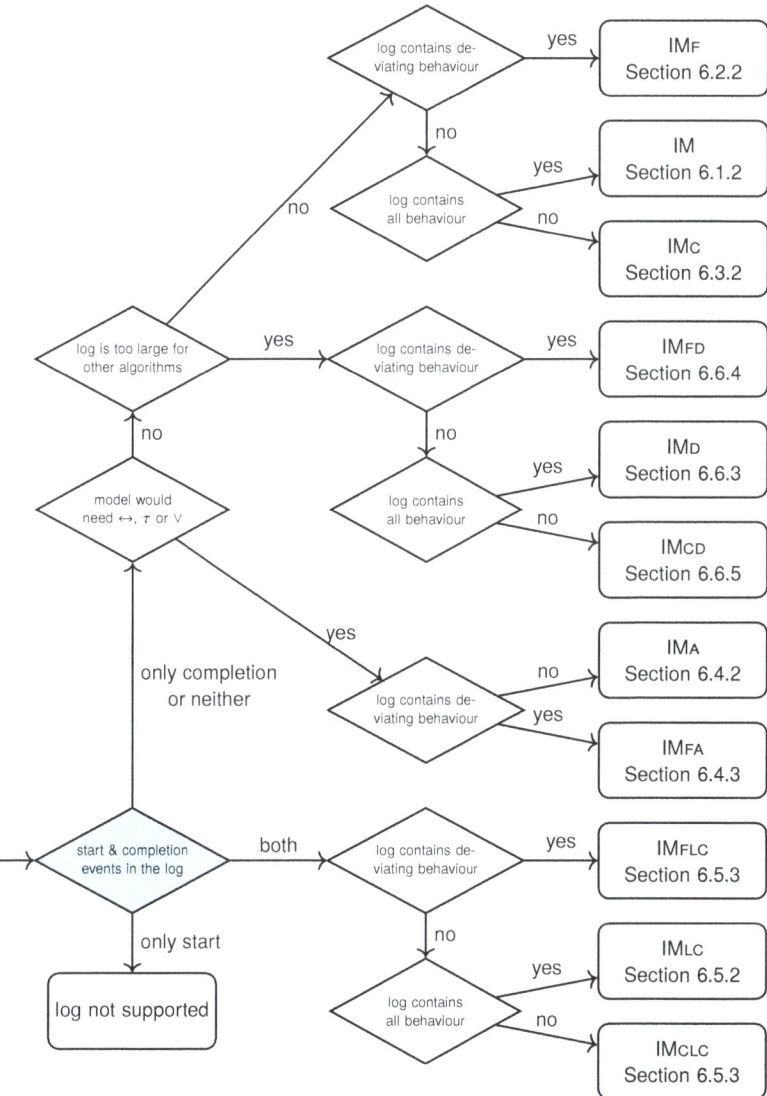

Fig. 6.30: Flowchart to choose an Inductive Miner.

Table 6.3: Guarantees provided by the algorithms introduced in this chapter.

	fitness guaranteed	rediscoverability proven for
IM	yes (Corollary 6.1)	C_B (Theorem 6.2)
IM$_F$	no	C_B (Theorem 6.5)
IM$_C$	no	C_B (Theorem 6.8)
IM$_A$	yes (Corollary 6.4)	C_{coo} (Theorem 6.9)
IM$_{FA}$	no	C_{coo} (Theorem 6.9)
IM$_{LC}$	yes (Corollary 6.5)	C_{LC} (Theorem 6.11)
IM$_{FLC}$	no	C_{LC} (Theorem 6.11)
IM$_{CLC}$	no	C_{LC} (Theorem 6.11)
IM$_D$	no	C_B (Theorem 6.16)
IM$_{FD}$	no	C_B (Theorem 6.17)
IM$_{CD}$	no	C_B (Theorem 6.17)

for the *entire* behaviour while worrying about soundness. In future work, many advanced techniques might be designed to handle specific events, specific use cases and unstructured behaviour. All of these techniques might benefit from the ideas of the IM framework.

For instance, a *hybrid process model* combine block-structured process trees and declarative models hierarchically in a single formalism: whenever a part of the system is structured enough, it is represented by process tree constructs, while if it is not structured enough, or the discovery technique cannot find this structure, it is represented as e.g. a Declare model. In this Declare model, certain structured parts can be represented by process trees again [8, 10]. Such an approach could use the fall through concepts of the IM framework.

Future work 6.18: Develop cut detection, log splitting, base case detection and fall through techniques further.

Furthermore, in this chapter we described the implementation of the described algorithms and provided a user manual. We could imagine a graphical user interface that would allow an end user to compose a discovery algorithm by manually choosing the four parameter functions, however it remains future work to make such an interface understandable and user friendly.

Future work 6.19: Engineer a do-it-yourself graphical user interface to compose an algorithm in the IM framework.

In the next chapter, we introduce our approach for conformance checking. In Chapter 8, we evaluate the newly introduced discovery and conformance checking techniques.

References

1. Berre, D.L., Parrain, A.: The sat4j library, release 2.2. JSAT **7**(2-3), 59–6 (2010). URL http://jsat.ewi.tudelft.nl/content/volume7/JSAT7_4_LeBerre.pdf

2. Bose, R.P.J.C., van der Aalst, W.M.P.: Trace clustering based on conserved patterns: To-wards achieving better process models. In: S. Rinderle-Ma, S.W. Sadiq, F. Leymann (eds.) Business Process Management Workshops, BPM 2009 International Workshops, Ulm, Germany, September 7, 2009. Revised Papers, *Lecture Notes in Business Information Processing*, vol. 43, pp. 170–181. Springer (2009). DOI 10.1007/978-3-642-12186-9_16. URL http://dx.doi.org/10.1007/978-3-642-12186-9_16

3. Buijs, J.C.A.M.: Flexible evolutionary algorithms for mining structured process models. Ph.D. thesis, Eindhoven University of Technology (2014)

4. van Dongen, B.F., de Medeiros, A.K.A., Verbeek, H.M.W., Weijters, A.J.M.M., van der Aalst, W.M.P.: The prom framework: A new era in process mining tool support. In: G. Ciardo, P. Darondeau (eds.) Applications and Theory of Petri Nets 2005, 26th International Conference, ICATPN 2005, Miami, USA, June 20-25, 2005, Proceedings, *Lecture Notes in Computer Science*, vol. 3536, pp. 444–454. Springer (2005). DOI 10.1007/11494744_25. URL http://dx.doi.org/10.1007/11494744_25

5. Günther, C., Verbeek, H.: XES v2.0 (2014). URL http://www.xes-standard.org/

6. Leemans, S.J.J., Fahland, D., van der Aalst, W.M.P.: Using life cycle information in process discovery. In: M. Reichert, H.A. Reijers (eds.) Business Process Management Workshops - BPM 2015, 13th International Workshops, Innsbruck, Austria, August 31 - September 3, 2015, Revised Papers, *Lecture Notes in Business Information Processing*, vol. 256, pp. 204–217. Springer (2015). DOI 10.1007/978-3-319-42887-1_17. URL http://dx.doi.org/10.1007/978-3-319-42887-1_17

7. de Leoni, M., Mannhardt, F.: Road traffic fine management process (2015). DOI dx.doi.org/10.1007/s00607-015-0441-1. URL http://dx.doi.org/10.1007/s00607-015-0441-1

8. Maggi, F.M., Slaats, T., Reijers, H.A.: The automated discovery of hybrid processes. In: S.W. Sadiq, P. Soffer, H. Völzer (eds.) Business Process Management - 12th International Conference, BPM 2014, Haifa, Israel, September 7-11, 2014. Proceedings, *Lecture Notes in Computer Science*, vol. 8659, pp. 392–399. Springer (2014). DOI 10.1007/978-3-319-10172-9_27. URL http://dx.doi.org/10.1007/978-3-319-10172-9_27

9. Ribeiro, J., Carmona, J.: RS4PD: A tool for recommending control-flow algorithms. In: L. Limonad, B. Weber (eds.) Proceedings of the BPM Demo Sessions 2014 Co-located with the 12th International Conference on Business Process Management (BPM 2014), Eindhoven, The Netherlands, September 10, 2014., *CEUR Workshop Proceedings*, vol. 1295, p. 66. CEUR-WS.org (2014). URL http://ceur-ws.org/Vol-1295/paper14.pdf

10. Slaats, T., Schunselaar, D.M.M., Maggi, F.M., Reijers, H.A.: The semantics of hybrid process models. In: C. Debruyne, H. Panetto, R. Meersman, T.S. Dillon, eva Kühn, D. O'Sullivan, C.A. Ardagna (eds.) On the Move to Meaningful Internet Systems: OTM 2016 Conferences - Confederated International Conferences: CoopIS, C&TC, and ODBASE 2016, Rhodes, Greece, October 24-28, 2016, Proceedings, *Lecture Notes in Computer Science*, vol. 10033, pp. 531–551 (2016). DOI 10.1007/978-3-319-48472-3_32. URL http://dx.doi.org/10.1007/978-3-319-48472-3_32

11. Verbeek, H.M.W., Buijs, J.C.A.M., van Dongen, B.F., van der Aalst, W.M.P.: XES, XESame, and ProM 6. In: P. Soffer, E. Proper (eds.) Information Systems Evolution - CAiSE Forum 2010, Hammamet, Tunisia, June 7-9, 2010, Selected Extended Papers, *Lecture Notes in Business Information Processing*, vol. 72, pp. 60–75. Springer (2010). DOI 10.1007/978-3-642-17722-4_5. URL http://dx.doi.org/10.1007/978-3-642-17722-4_5

12. Weijters, A.J.M.M., Ribeiro, J.T.S.: Flexible heuristics miner (FHM). In: Proceedings of the IEEE Symposium on Computational Intelligence and Data Mining, CIDM 2011, part of the IEEE Symposium Series on Computational Intelligence 2011, April 11-15, 2011, Paris, France, pp. 310–317. IEEE (2011). DOI 10.1109/CIDM.2011.5949453. URL http://dx.doi.org/10.1109/CIDM.2011.5949453

Chapter 7
Conformance Checking

Sander J. J. Leemans: Robust Process Mining with Guarantees, LNBIP 440, pp. 327–354, 2022
https://doi.org/10.1007/978-3-030-96655-3_7

Abstract In this chapter, we introduce a conformance checking framework focused on speed. The framework can be applied to any combination of logs and models. The framework projects the behaviour exhaustively on all k-sized subsets of activities, after which the results can be shown as a measure, being projected on a model, or inspected in tabular format. We also explore how unbounded and weakly unsound Petri nets may be handled.

In the previous chapters, a framework for process discovery algorithms was introduced, abstractions analysed and new algorithms introduced. In this chapter, we discuss conformance checking. As described in Chapter 3, conformance checking plays an important role in evaluating new process discovery techniques. Moreover, conformance checking is also used to check compliance of actual process executions with respect to some normative model. In such cases, an event log is compared to a given model. In other process mining projects, model-model comparison can be used for instance to detect concept drift, to detect relations between processes, and to retrieve process models from large collections (e.g. to ease implementation). Furthermore, a useful technique to evaluate process discovery algorithms is to take a process model, generate an event log from it, discover a model and compare the original and the discovered model. Using an appropriate measure for model-model equivalence, robustness of discovery algorithms against noise, infrequent behaviour and incompleteness can be tested, which will be done in Chapter 8. Figure 7.1 shows the context of log-model and model-model conformance checking; see Section 3.4 for more details.

In line with existing process and data mining literature, we consider two measures for each use case (as discussed in Section 3.2): in model-model conformance checking *recall* describes how much behaviour of the system is present in the discovered model and *system precision* describes how much behaviour of the discovered model is present in the system. In log-model conformance checking *fitness* describes how much behaviour of an event log is allowed by the model and *log precision* describes how much behaviour of the discovered model is present in the event log.

In sections 3.2.4 and 3.4.3, we discussed several formal and practical requirements for conformance checking techniques. For instance, a desirable property of log-conformance measures is that its measures *coincide* with language equivalence, i.e. fitness and log precision are both 1 if and only if model and log are language equivalent, and, in the model-model case, recall and system precision are both 1 if and only if both models are language equivalent (requirements CR3 and CR2). Notice that log precision can only be 1 if the model does not contain a loop. Furthermore, the measures should accept all weakly sound models (Requirement CR1), work fast on real-life models (Requirement CR4), provide insights on summarative, model and log level (Requirement CR5), be normalisable (Requirement CR6), and symmetric (Requirement CR7).

In this chapter, we introduce a technique, the Projected Conformance Checking framework (PCC framework), that supports both log-model and model-model comparisons and performs these comparisons on language. It applies to all process model formalisms with executable semantics, i.e. a language, and only requires that this language is regular (bounded). For clarity of presentation, we will show the PCC

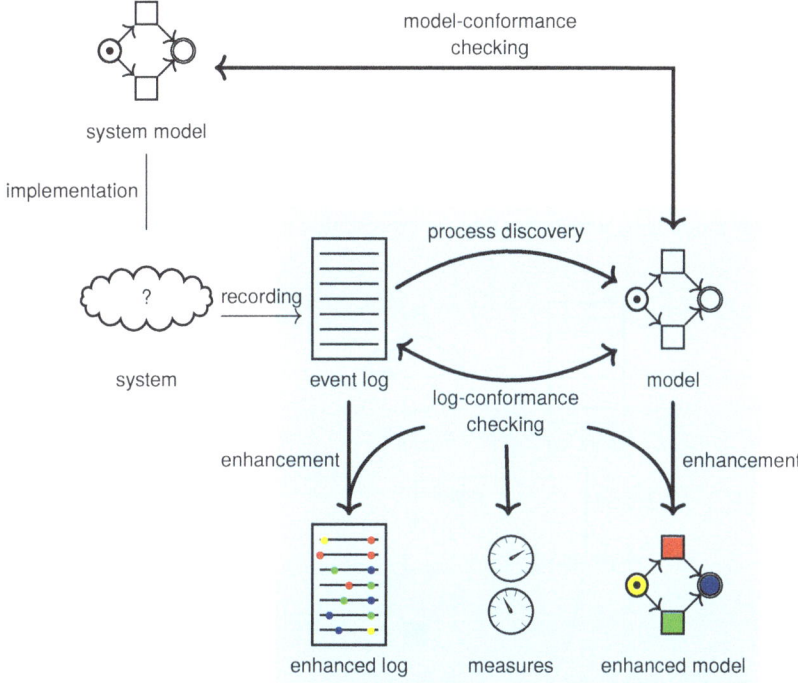

Fig. 7.1: Conformance checking in a process discovery context.

framework for process trees and Petri nets; the latter is included as most process discovery algorithms we discussed in Chapter 3 use it. Using the properties described in Chapter 5, we will prove for a particular class of models that the model-model measures of the PCC framework coincide with language equivalence.

In the remainder of this chapter, we first introduce the framework in Section 7.1, then give an extensive example and illustrate how intermediate steps of the framework might provide more insight into conformance in Section 7.2, and prove the language-equivalence property in Section 7.3. We finish the chapter with a description of the implementation (Section 7.4), conclude the chapter in Section 7.5 and describe ideas to extend the PCC framework to handle unbounded and weakly unsound Petri nets (Section 7.6).

7.1 Projected Conformance Checking Framework

Requirement CR4 states that a conformance checking technique should preferably be fast and able to handle large real-life event logs. In Chapter 8, we will illustrate that existing techniques have difficulties to handle real-life event logs in reasonable time and memory. Therefore, the PCC framework applies a divide-and-conquer strategy, shown in Figure 7.2: for each k-subset of activities (for a user-specified value of k), it projects the logs or models to the activities in the k-subset, and constructs a deterministic finite automaton for the behaviour of both. For each of these deterministic finite automata, precision and recall with respect to these k activities are computed; the final result of either measure is the average over all subsets. Due to the use of DFAs, we assume that the languages of all process models are regular languages, which can thus be represented by minimal DFAs [9].

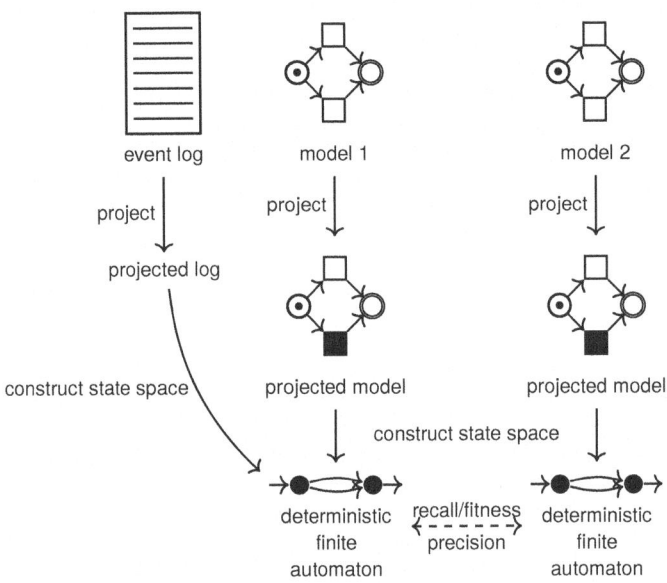

Fig. 7.2: The PCC framework: compare either a log or model 1 to a model 2.

Due to the divide-and-conquer strategy, the PCC framework avoids creating a state space of the *entire* model: it trades state explosion for a, in case of low k, limited number of smaller state spaces. For $k = 1$, the number of smaller state spaces is linear in the number of activities, for $k = 2$ quadratic and for variable k factorial ($O\binom{n}{k}$). A further reduction of computation time is achieved by all k-subsets being completely independent and therefore computing measures is highly parallelisable.

In the remainder of this section, we introduce each step of the PCC framework in detail, and provide formal definitions: we start with the projection of an event log, and transforming the projected log into a DFA (Section 7.1.1). Second, we describe how models can be projected and transformed to DFAs in Section 7.1.2. Third, in Section 7.1.3, we show how these DFAs are compared and measured. We finish the PCC framework with a description how the final recall/fitness and precision are computed (Section 7.1.4).

7.1.1 Log to Projected Log to DFA

A log can straightforwardly be projected on a set of activities A by removing all events that are not in A. For instance,

$$L_{139} = [\langle a, b \rangle, \langle b, a \rangle, \langle c, d, c \rangle, \langle c, d, c, d, c \rangle]$$
$$L_{139}|_{\{a,b\}} = [\langle a, b \rangle, \langle b, a \rangle, \epsilon^2]$$

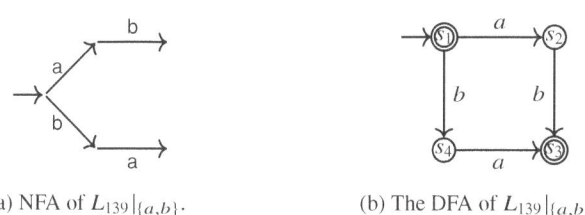

(a) NFA of $L_{139}|_{\{a,b\}}$.   (b) The DFA of $L_{139}|_{\{a,b\}}$.

Fig. 7.3

Second, the log is translated into an NFA by, for each trace in the log, extending the NFA with an explicit path accepting the trace. For our example log L_{139} and the subset $\{a, b\}$, this NFA is shown in Figure 7.3a. Third, the NFA is converted into a DFA and the resulting DFA is reduced; for our example this is shown in Figure 7.3b. Reducing the DFA ensures that the reduced DFA is language-unique [9].

7.1.2 Model to Projected Model to DFA

Any formalism with executable semantics can be transformed to DFA and be used by the PCC framework, as long as the resulting models have a regular language, i.e. their behaviour can be captured in a finite state space. In this section, we show this projection and translation for process trees and Petri nets.

7.1.2.1 Process Trees

A process tree can be projected on a set of activities $A = \{a_1 \ldots a_k\}$ by replacing every leaf that is not in A with τ (in which $\oplus$ is any process tree operator):

$$a|_A = \text{if } a \in A \text{ then } a \text{ else } \tau$$
$$\tau|_A = \tau$$
$$\oplus(M_1 \ldots M_n)|_A = \oplus(M_1|_A \ldots M_n|_A)$$

The projected process tree is likely full of τ leafs. Therefore, after projection a major problem reduction (and speedup) can be achieved by applying structural language-preserving reduction rules to the process tree, such as the rules described in Section 5.1. For instance, the process tree $M_{140} = \qquad\qquad \rightarrow \qquad$ projected on $\{a, b\}$

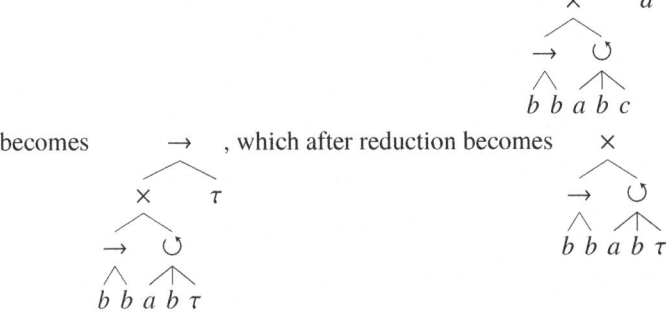

becomes $\qquad \rightarrow \qquad$, which after reduction becomes $\qquad \times \qquad$.

Translating to a DFA.

In Section 2.2.5, we defined the semantics of process trees using regular expressions, thus every process tree can straightforwardly be transformed into an NFA. Notice that for the translation of $\wedge$, we used the shuffle (interleaving) operator [8]. Second, a simple procedure transforms the NFA into a DFA [9], and minimises the DFA. For instance, applying this procedure to the tree $\quad \times \quad$ yields the DFA denoted in

Figure 7.4.

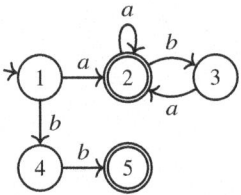

Fig. 7.4: The DFA of $M_{140} = \times(\rightarrow(b, b), \circlearrowleft(a, b, \tau))$.

7.1.2.2 Petri nets

A Petri net can be projected onto a subset of activities A, i.e. transitions, by removing every label not in A, i.e., replacing each transition that is not in A with a silent transition. In complex models and for smaller k values, this step introduces a lot of silent transitions in the projection. However, many of these τ-transitions might be removable without changing the language of the projected model. Therefore, language-preserving reduction rules could be applied to reduce the size of the Petri net, and hence the computation times required. For instance, a subset of the Murata rules [11] could be applied, or (adaptions of) the rules described in [6, 13].

A Petri net can be translated to an automaton using state space exploration. The more the net was reduced, the smaller the state space will be in this step. For instance, Figure 7.5 shows a Petri net, its projection on $\{a, b\}$ and the DFA of that projection.

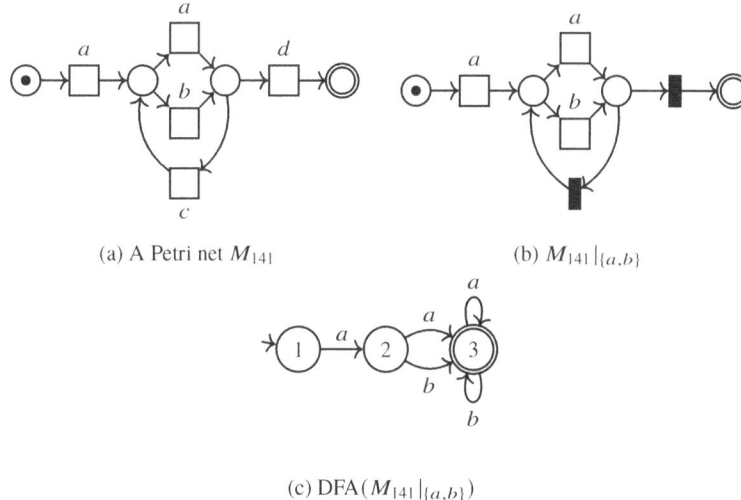

(a) A Petri net M_{141} (b) $M_{141}|_{\{a,b\}}$

(c) DFA($M_{141}|_{\{a,b\}}$)

Fig. 7.5: A Petri net, its projection and the DFA of that projection.

A necessary condition for the translation to a DFA is that the language of the Petri net can be described by a DFA, as described in Section 2.2.1. A definition of a language requires the notion of start and acceptance of traces, and a finite state space, thus the model needs to be bounded, and provide an initial marking and a set of final markings. As sound workflow nets are bounded [1] and have clear initial and final markings, all sound workflow nets can be handled by the PCC framework. However, for general Petri nets, a translation to a DFA is impossible as we could use such a translation to decide language inclusion, which is undecidable for general Petri nets [7]. Therefore, we decided to only support Petri nets having languages that can be described by DFAs, i.e. bounded Petri nets with initial and final markings. Nevertheless, in Section 7.6, we will describe some heuristics to handle Petri nets that do not satisfy these conditions.

7.1.3 Comparing DFAs & Measuring

Given two DFAs of a model and a log/model, i.e. recall, fitness, log precision and system precision, can be computed.

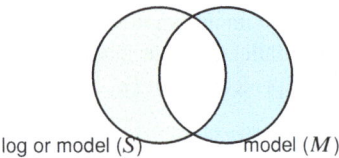

Fig. 7.6: A venn diagram illustrating precision: the green region denotes unfitting or unrecalled behaviour, the blue region denotes log- or model-imprecise behaviour.

As defined in Section 3.2.2, precise behaviour is the behaviour that is present in S and is also present in M. Thus, the measures *log- and model-precision* capture the amount of precise behaviour compared to the imprecise behaviour (the blue region in Figure 7.6), and precision is the part of the behaviour of M that is precise, i.e. "$|S \cap M|/|M|$". Similarly, we defined recall and fitness as the part of the behaviour of S that is in M, i.e. "$|S \cap M|/|S|$". In the remainder of this section, we show how the PCC framework uses these concepts to compute precision and recall measures by defining these informal formulae step by step. We first explain precision and recall, after which we explain fitness.

7.1.3.1 Log Precision, System Precision and Recall

We first define the conjunction of two DFAs ("$S \cap M$"), after which we map the states of the conjunction to corresponding states of S or M. Finally, we compute log-precision, model-precision and recall by counting corresponding states ("$|S \cap M|$" and "$|S|$"). In these explanations, we use our running example consisting of the process tree $M_{140} =$ 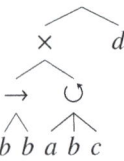 and the Petri net M_{141} given in Figure 7.5.

Conjunction.

The conjunction of two DFAs is a DFA that accepts any and all traces that both input DFAs accept.

Definition 7.1 (DFA conjunction) Let $D = (S, s_0, F, A)$ and $D' = (S', s'_0, F', A')$ be minimal deterministic finite automata. Then, $D \cap D'$ denotes the *conjunctive* DFA such that $\mathcal{L}(D \cap D') = \mathcal{L}(D) \cap \mathcal{L}(D')$.

A conjunctive DFA $D \cap D' = (S'', s''_0, F'', A'')$ could be constructed as follows:

$$S'' = S \times S'$$
$$s''_0 = (s_0, s'_0)$$
$$F'' = F \times F'$$
$$A'' : (S \times S') \times \Sigma \to S \times S'$$

such that $\forall_{s,p \in S, s', p' \in S', a \in \Sigma(D) \cap \Sigma(D')}$,

$$A''((s, s'), a) = (p, p') \Leftrightarrow A(s, a) = p \wedge A'(s', a) = p'$$

Without loss of generality, we assume that the conjunctive DFA is minimal. For instance, the conjunction could be minimised after construction using the algorithm described in [9].

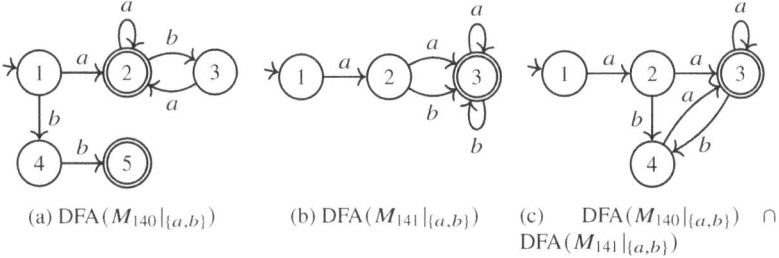

(a) DFA($M_{140}|_{\{a,b\}}$) (b) DFA($M_{141}|_{\{a,b\}}$) (c) DFA($M_{140}|_{\{a,b\}}$) $\cap$ DFA($M_{141}|_{\{a,b\}}$)

Fig. 7.7: DFAs for the models M_{140} and M_{141} of our example, projected to $\{a, b\}$ and reduced, and their conjunction.

For instance, Figure 7.7 revisits the DFAs of M_{140} and M_{141} used in our previous examples; Figure 7.7c shows their conjunctive DFA.

Mapping.

The states denoted in Figure 7.7c link the conjuctive DFA to states in M_{140} and M_{141}. While constructing the conjunction, such a state mapping is easily constructed using a little bookkeeping. However, as the conjunction is minimised, this mapping might get lost. Therefore, we introduce a function to recompute this mapping.

Given two minimal DFAs D and D' such that $\mathcal{L}(D) \subseteq \mathcal{L}(D')$, the function MAP returns the mapping between the states of the automata. In this function, m keeps track of the mapped states, starting with mapping the initial states. Using the head of the state queue q, all outgoing edges are followed in both DFAs and the resulting states (b and b') are mapped, until all combinations of states have been mapped.

While following the edges, we use the fact that D is a subset of D', i.e. D' can follow any step that D takes.

function MAP($D = (S, s_0, F, A), D' = (S', s'_0, F', A')$)
    $m \leftarrow \{(s_0, s'_0)\}$
    $q \leftarrow \{(s_0, s'_0)\}$
    while $q \neq \emptyset$ **do**
        $(s, s') \leftarrow$ remove and return an element of q
        for $b = A(s, a)$ **do**
            $b' \leftarrow A'(s', a)$
            if $(b, b') \notin m$ **then**
                $m \leftarrow m \cup \{(b, b')\}$
                $q \leftarrow q \cup \{(b, b')\}$
            end if
        end for
    end while
    return m
end function

For instance, when applying the function MAP to DFA($M_{140}|_{\{a,b\}}$)$\cap$DFA($M_{141}|_{\{a,b\}}$) and DFA($M_{141}|_{\{a,b\}}$), the following mapping is returned:

$$\{(1, 1), (2, 2), (3, 3), (4, 3)\}$$

Outgoing edges.

Let S be a log or a model, let M be a model and let A be a set of activities. In the PCC framework, precision is measured similarly to several existing precision metrics, such as in [4]. That is, we count the outgoing edges of all states in the projected automaton DFA($M|_A$), and compare that to the outgoing edges of corresponding states in the conjunctive automaton DFA($S|_A$) $\cap$ DFA($M|_A$).

Given our focus on end-to-end languages, the PCC framework takes the final states into account. That is, besides the steps that can be taken from a state in the automaton, we consider acceptance of a state to be an extra outgoing edge as well.

Definition 7.2 (post set) Let $D = (S, s_0, F, A)$ be a deterministic finite automaton, and let $s \in S$ be a state. Then, $s^\bullet$ denotes the *post set* of s, i.e. the steps that can be performed from s, appended with $\bot$ if s is an accepting state:

$$s^\bullet = \{(b, s')|A(s, b) = s'\} \cup \{\bot|s \in F\}$$

For instance, in our example DFA($M_{141}|_{\{a,b\}}$) of Figure 7.7b, the post set of state 3, i.e. $3^\bullet$, is $\{(a, 3), (b, 3), \bot\}$.

Precision.

Then, precision is computed by counting edges. Let S be an event log or (a model with) a regular language ($S \in \mathbb{B} \cup \mathbb{L}$), M be (a model with) a regular language ($M \in \mathbb{L}$), and X is a subset of activities ($X \subseteq \Sigma(S) \cup \Sigma(M)$).

Consider a state s in the automaton of the projected M. The post set of s is the behaviour that M allows from s. The mapping obtained using the MAP function maps s to zero or more states in the conjunction of projected S and M. Precision measures the behaviour of M that is also present in S, thus we compare the post set of s with the post sets of the mapped states. That is, we count the behaviour at s (the size of its post set) and the behaviour of the mapped states.

If s maps to multiple states in the conjunction, we count multiple occurrences of s in the conjunction accordingly ("count states in the mapping"). If s does not map to any state in the conjunction, we count s anyway, as precision covers the behaviour in M (and thus in s) that does not appear in S ("count states not in the mapping").

Table 7.1: System precision for $\{a, b\}$ on M_{140} and M_{141}.

state in DFA($M_{141}|_{\{a,b\}}$)	post-set size	state in conjunction	post-set size
1	1	1	1
2	2	2	2
3	3	3	3
3	3	4	1

function PRECISION(M_1, M_2, X)
    $D \leftarrow (S, s_0, F, A) \leftarrow \text{DFA}(M_1|_X) \cap \text{DFA}(M_2|_X)$
    $D' \leftarrow (S', s_0', F', A') \leftarrow \text{DFA}(M_2|_X)$
    $m \leftarrow \text{MAP}(D, D')$
    $c, c' \leftarrow 0, 0$
    for $s' \in S'$ **do**
        for $(s, s') \in m$ **do**   ▷ count states in the mapping
            $c' \leftarrow c' + |s'^{\bullet}|$
            $c \leftarrow c + |s^{\bullet}|$
        end for
        if $\neg \exists_{(s,s') \in m}$ **then**   ▷ count states not in the mapping
            $c' \leftarrow c' + |s'^{\bullet}|$
        end if
    end for
    if $c' = 0$ **then**
        return 1
    else
        return $\dfrac{c}{c'}$
    end if
end function

As the conjunctive DFA D is a subset of D', the precision measure is a number between 0 and 1 (both inclusive), which satisfies Requirement CR6, which favours normalised measures, such that measures on different models can be compared.

For instance, Table 7.1 shows the counting on our example projected process tree $M_{140}|_{\{a,b\}}$ and Petri net $M_{141}|_{\{a,b\}}$. In this example, state 3 of DFA(M_{141}) is counted twice, as it is mapped to both states 3 and 4 of the conjunction. Therefore, system precision is $\frac{1+2+3+1}{1+2+3+3} = 0.778$.

Recall.

For a model S, a model M and a set of activities X, recall is defined as the part of behaviour in $S|_X$ that is not in $M|_X$, i.e. the opposite of system precision:

function RECALL(S, M, X)

    return PRECISION(M, S, X)

end function

This satisfies the requirement that these measures should be symmetric (Requirement CR7). In our example projected process tree $M_{140}|_{\{a,b\}}$ and Petri net $M_{141}|_{\{a,b\}}$ (see Figure 7.7), the mapping computed by MAP is as follows:

$$\{(1, 1), (2, 2), (3, 2), (4, 3)\}$$

Recall for this example, as shown in Table 7.2, is $\frac{1+2+3+1}{2+3+3+1+1+1} = 0.636$. Notice that state 2 of DFA($M_{140}|_{\{a,b\}}$) has been included twice, as two states of the conjunction are mapped to it. Furthermore, states 4 and 5 of DFA($M_{140}|_{\{a,b\}}$) are not mapped to states in the conjunction, but are included nevertheless in the denominator, as they represent behaviour in $M_{140}|_{\{a,b\}}$.

Table 7.2: Recall for $\{a, b\}$ on M_{140} and M_{141}.

state in DFA($M_{140}|_{\{a,b\}}$)	post-set size	state in conjunction	post-set size
1	2	1	1
2	3	2	2
2	3	3	3
3	1	4	1
4	1	-	0
5	1	-	0

7.1.3.2 Fitness

The fitness measure could be measured like recall and precision, however this would have a downside: these measures do not take the frequency of traces into account. That is, the repeated occurrence of behaviour makes no difference in the DFAs, and a trace that occurs once would have the same influence as a trace that occurs 1,000,000 times. Therefore, even though we do not satisfy Requirement CR7, we apply a different strategy to measure fitness: instead of building automata and taking their conjunction, we construct the DFA of the projected model as described before. Next, we replay each projected trace of the event log on this DFA, and record whether the projected trace is accepted by the DFA. The reported fitness measure is the fraction of projected traces that is accepted. Notice that even though the fitness is computed as a 0/1 value on the projected traces: if the trace cannot be replayed due to even a single event, 0 is reported. However, the final result will be fine-grained nevertheless as the average over many projected traces is taken.

Definition 7.3 Let S be an event log, let M a model with a regular language and let X be a set of activities. Then,

$$\text{FITNESS}(S, M, X) = \frac{|[t \mid t \in S|_X \wedge t \in \mathcal{L}(\text{DFA}(M|_X))]|}{|[t \mid t \in S|_X]|}$$

Notice that typical fitness measures [12, 4] avoid taking full traces into account. For instance, consider a trace of 100 events, of which the first 99 events fit the model and the last event does not fit the model. Then, a fitness measure that considers full traces will classify the trace as non-fitting, while intuitively most of the trace corresponds to the model. Consequently, if the entire event log contains many such traces, a full-trace measure will report a fitness value that is intuitively too low. The PCC framework computes fitness over subsets of traces, thus already considers non-full-trace behaviour, so we deemed a more detailed approach not necessary. However, other techniques such as token-based replay [12] or alignments [4] could be used as well.

7.1.4 Measuring over All Activities

In the previous section, we defined how precision, recall and fitness between a model or log S and a model M are measures for a chosen subset of k activities. To measure precision, recall and fitness for entire S and M, the previous steps are repeated for each set of activities $\{a_1 \ldots a_k\} \subseteq \Sigma$ of size k and the results are averaged:

$$k^{\Sigma(S)} = \{A \mid A \subseteq \Sigma(S) \cup \Sigma(M) \wedge |A| = k\}$$

$$\text{PRECISION}(S, M, k) = \frac{\sum_{a_1 \dots a_k \in k^{\Sigma(S)}} \text{PRECISION}(S, M, a_1 \dots a_k)}{|k^{\Sigma(S)}|}$$

$$\text{with } S \in \mathbb{L} \cup \mathbb{E}$$

$$\text{RECALL}(S, M, k) = \text{PRECISION}(M, S, k)$$

$$\text{with } S \in \mathbb{L}$$

$$\text{FITNESS}(S, M, k) = \frac{\sum_{A \subseteq \Sigma(S) \cup \Sigma(M) \wedge |A| = k} \text{FITNESS}(S, M, a_1 \dots a_k)}{|\{A \mid A \subseteq \Sigma(S) \cup \Sigma(M) \wedge |A| = k\}|}$$

$$\text{with } S \in \mathbb{E}$$

As the intermediate measures all result in a number between 0 and 1 (inclusive), the average over these measures also results in a normalised measurement, satisfying Requirement CR6. Furthermore, by construction, recall/fitness is symmetric to precision (Requirement CR7), i.e. $recall(S, M, k) = precision(M, S, k)$.

Notice that we assume a closed world here, i.e. the alphabet Σ is assumed to be the same for S and M. If an activity is missing from M, we therefore consider M to express that the activity can never happen, by construction of the DFAs.

In this section, we introduced the PCC framework and explained its steps in detail for process trees and Petri nets. In the next section, we give an example and show how the intermediate results, i.e. the results for the individual k-subsets, might be used. In Section 7.3, we prove that the PCC framework guarantees reliable detection of language equivalence for certain classes of models.

7.2 An Example of Non-Conformance and Diagnostic Information

In this section, we illustrate the PCC framework using some examples, and illustrate how the intermediate measurements can be used to gain more detailed insight in the differences between model and log/model.

In this example, we will compare the process tree $M_{142} = $ $\rightarrow$ to an event

$$\overset{\displaystyle\rightarrow}{\underset{\displaystyle\underset{c\;\;d}{\overset{\wedge}{a\;b\;\circlearrowright}}}{\wedge}} \quad e$$

log L_{143} of 160 fitting traces of M_{142}, and the non-fitting trace $\langle a, b, c, c, e \rangle$. This non-fitness of the trace compared to the model can be explained in two ways: either an extra c was executed, or a d is missing. We illustrate how this case manifests in the PCC framework (for $k = 2$). The PCC framework computes fitness and precision of all subsets of 2 activities (as k is 2). We illustrate one of these computations, i.e. on the subset $\{c, d\}$.

- First, the log is projected to $\{c, d\}$ and transformed to a DFA. For instance, the trace $\langle a, b, c, c, e \rangle$ is projected to $\langle c, c \rangle$. The minimised DFA of the entire log is shown in Figure 7.8a. This DFA illustrates that in the event log, the loop of c and d was executed up to six times in the event log.

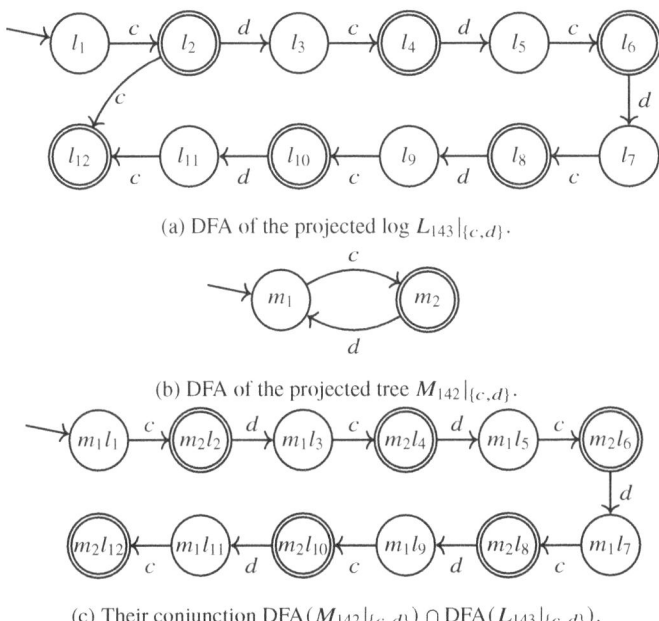

(a) DFA of the projected log $L_{143}|_{\{c,d\}}$.

(b) DFA of the projected tree $M_{142}|_{\{c,d\}}$.

(c) Their conjunction DFA$(M_{142}|_{\{c,d\}}) \cap$ DFA$(L_{143}|_{\{c,d\}})$.

Fig. 7.8: DFAs of an example run of the PCC framework. The labels of the states were added to keep track of corresponding states.

- Second, the process tree is projected and transformed into a DFA: projecting M_{142} to $\{c, d\}$ yields , which is reduced to ↺ using the reduction rules of Definition 5.1. This reduced tree is transformed to the minimal DFA shown in Figure 7.8b.
- Third, the minimal conjunction of both DFAs is computed: the result is shown in Figure 7.8c. This conjunction DFA accepts all languages that are in both DFAs, i.e. of the projected log and of the projected tree.
- Fourth, log precision is computed. For each combination of state in the DFA of the tree and state in the DFA of the conjunction, we count the outgoing edges:

state in tree-DFA	post-set size	state in conjunction	post-set size
m_1	1	$m_1 l_1$	1
m_1	1	$m_1 l_3$	1
m_1	1	$m_1 l_5$	1
m_1	1	$m_1 l_7$	1
m_1	1	$m_1 l_9$	1
m_1	1	$m_1 l_{11}$	1
m_2	2	$m_2 l_2$	2
m_2	2	$m_2 l_4$	2
m_2	2	$m_2 l_6$	2
m_2	2	$m_2 l_8$	2
m_2	2	$m_2 l_{10}$	2
m_2	2	$m_2 l_{12}$	1
Σ_1	18	Σ_2	17

- Fifth, the log precision for this subset is the sum of outgoing edges of the states of the conjunction (Σ_2) divided by the sum of outgoing edges of the states of the DFA of the tree (Σ_1): $17/18 = 0.9444\ldots$.
- Sixth, fitness is computed. That is, all traces are projected onto the activities $\{c, d\}$ and replayed onto the DFA of the model (Figure 7.8b). This results in 160 traces being accepted (the fitting traces) and one trace not being accepted (the unfitting trace $\langle c, c \rangle$), thus leading to a fitness of $\frac{160}{161} = 0.994$.

This procedure is repeated for all subsets of 2 activities. Table 7.3a shows the intermediate measures for all these subsets. As a final step, the PCC framework returns the average over these measures: fitness being 0.9993 and log precision being 0.9481. This matches our intuition, as 160 traces were fitting and 1 was not fitting, so we expect a fitness value being close to but not equal to 1. Furthermore, log precision matches intuition as well, as even though the model has an unbounded number of traces and the event log is bounded, the loop in the model was executed up to six times in the log[1], which intuitively leads to a log precision close to but not equal to 1, which is reflected by the measured log precision.

The intermediate computations provide insights into where the model and the event log deviate. For instance, the measurements show that for fitness, the only activity subset for which perfect fitness was not measured, was the pair $\{c, d\}$, which corresponds to the non-fitting trace we inserted, which has either an extra c or a missing d.

In our example, the subset $\{c, d\}$ stands out because it is the only one subset without a perfect fitness score. However, for real-life event logs and use cases, several pairs might have lower fitness scores, which might make detection of problematic *activities* more difficult. For instance, for the precision pairs it might be more challenging to spot the problematic parts of the model. To ease detection of problematic subsets of activities, consider the intermediate computations grouped by activity, i.e. for each activity, the average fitness/precision is computed over all k-subsets in which that activity is involved (see Table 7.3b).

---

[1] Log precision approaches one as the number of times the loop was taken in the log increases.

Table 7.3: Intermediate results of the PCC framework on M_{142} and L_{143}.

(a) As computed.

	fitness				log precision			
	a	b	c	$d\ e$	a	b	c	$d\ e$
a								
b	1.000				1.000			
c	1.000	1.000			0.926	0.926		
d	1.000	1.000	0.994		0.917	0.917	0.944	
e	1.000	1.000	1.000	1.000	1.000	1.000	0.929	0.923

(b) Averaged by activity.

	fitness	log precision
a	1.000	0.961
b	1.000	0.961
c	0.998	0.931
d	0.998	0.925
e	1.000	0.963

(c) Averaged.

fitness	log precision
0.999	0.948

In this table, the problematic activities in terms of fitness are clearly c and d, and it is also easier to see that c and d also have the lowest precision scores, which in this case indicates that they are involved in loop behaviour. In Section 7.4, we will describe the implemented visualisation of these results.

Example with edge cases and different k.

As a final example, we compare two process trees using several k's: $M_{144} =$

and $M_{145} =$  . Table 7.4 shows the results of applying the PCC framework to these trees, using every k from 1 to 6.

We illustrate edge cases using a few intermediate computations.

First, we consider $k = 1$. The $\{e\}$ represents an edge case as e does not appear in M_{145}. The corresponding projected trees are $M_{144}|_{\{e\}} = \times(\tau, e)$ and $M_{145}|_{\{e\}} = \tau$. Figures 7.9a and 7.9b show their DFAs and Figure 7.9c shows their conjunction. For $recall(M_{144}, M_{145}, \{e\})$, there are in total 3 outgoing edges in DFA$(M_{144}|_{\{e\}})$ and 1 outgoing edge in the conjunction, thus the mentioned recall is 0.333. With this recall measure, the PCC framework captures that in M_{144}, e can be executed but is not mandatory, which matches M_{145}, in which e is not present.

Table 7.4: Intermediate results for M_{144} and M_{145}.

(a) $k = 1$. Average recall is 0.889, average system precision is 0.889.

	recall	system precision		recall	system precision		recall	system precision
$\{a\}$	1.000	1.000	$\{b\}$	1.000	1.000	$\{c\}$	1.000	1.000
$\{d\}$	1.000	1.000	$\{e\}$	0.333	1.000	$\{f\}$	1.000	0.333

(b) $k = 2$. Average recall is 0.722, average system precision is 0.578.

	recall	system precision		recall	system precision		recall	system precision
$\{a,b\}$	1.000	1.000	$\{a,c\}$	1.000	0.750	$\{a,d\}$	1.000	0.750
$\{a,e\}$	0.667	0.667	$\{a,f\}$	0.333	0.167	$\{b,c\}$	1.000	0.750
$\{b,d\}$	1.000	0.750	$\{b,e\}$	0.667	0.667	$\{b,f\}$	0.333	0.167
$\{c,d\}$	1.000	0.500	$\{c,e\}$	0.250	0.333	$\{c,f\}$	1.000	0.750
$\{d,e\}$	0.250	0.333	$\{d,f\}$	1.000	0.750	$\{e,f\}$	0.333	0.333

(c) $k = 3$. Average recall is 0.556, average system precision is 0.388.

	recall	system precision		recall	system precision		recall	system precision
$\{a,b,c\}$	1.000	0.857	$\{a,b,d\}$	1.000	0.857	$\{a,b,e\}$	0.833	0.833
$\{a,b,f\}$	0.167	0.071	$\{a,c,d\}$	1.000	0.500	$\{a,c,e\}$	0.500	0.500
$\{a,c,f\}$	0.667	0.286	$\{a,d,e\}$	0.500	0.500	$\{a,d,f\}$	0.667	0.286
$\{a,e,f\}$	0.000	0.000	$\{b,c,d\}$	1.000	0.500	$\{b,c,e\}$	0.500	0.500
$\{b,c,f\}$	0.667	0.286	$\{b,d,e\}$	0.500	0.500	$\{b,d,f\}$	0.667	0.286
$\{b,e,f\}$	0.000	0.000	$\{c,d,e\}$	0.200	0.125	$\{c,d,f\}$	0.750	0.375
$\{c,e,f\}$	0.250	0.250	$\{d,e,f\}$	0.250	0.250			

(d) $k = 4$. Average recall is 0.360, average system precision is 0.225.

	recall	system precision		recall	system precision		recall	system precision
$\{a,b,c,d\}$	1.000	0.636	$\{a,b,c,e\}$	0.714	0.714	$\{a,b,c,f\}$	0.333	0.133
$\{a,b,d,e\}$	0.714	0.714	$\{a,b,d,f\}$	0.333	0.133	$\{a,b,e,f\}$	0.000	0.000
$\{a,c,d,e\}$	0.400	0.250	$\{a,c,d,f\}$	0.750	0.273	$\{a,c,e,f\}$	0.000	0.000
$\{a,d,e,f\}$	0.000	0.000	$\{b,c,d,e\}$	0.400	0.250	$\{b,c,d,f\}$	0.750	0.273
$\{b,c,e,f\}$	0.000	0.000	$\{b,d,e,f\}$	0.000	0.000	$\{c,d,e,f\}$	0.000	0.000

(e) $k = 5$. Average recall is 0.176, average system precision is 0.102.

	recall	system precision		recall	system precision		recall	system precision
$\{a,b,c,d,e\}$	0.625	0.455	$\{a,b,c,d,f\}$	0.429	0.158	$\{a,b,c,e,f\}$	0.000	0.000
$\{a,b,d,e,f\}$	0.000	0.000	$\{a,c,d,e,f\}$	0.000	0.000	$\{b,c,d,e,f\}$	0.000	0.000

(f) $k = 6$. Average recall is 0, average system precision is 0.

	recall	system precision
$\{a,b,c,d,e,f\}$	0.000	0.000

Reversely, when computing $precision(M_{144}, M_{145}, \{e\})$, the PCC framework considers $DFA(M_{145}|_{\{e\}})$ and the conjunction. As these DFAs are equivalent, precision is 1.

(a) $DFA(M_{144}|_{\{e\}})$ (b) $DFA(M_{145}|_{\{e\}})$ (c) $DFA(M_{144}|_{\{e\}}) \cap DFA(M_{145}|_{\{e\}})$

Fig. 7.9: DFAs computed by the PCC framework for $k = 1$ and $\{e\}$.

Second, we consider $k = 6$. As there are 6 activities in the two trees, projecting does not remove any behaviour and directly compares the DFAs shown in figures 7.10a and 7.10b. By not projecting, the PCC framework compares the two trees on their traces and as they have no traces in common, the conjunction 7.10c is empty. Therefore, recall and system precision are 0.

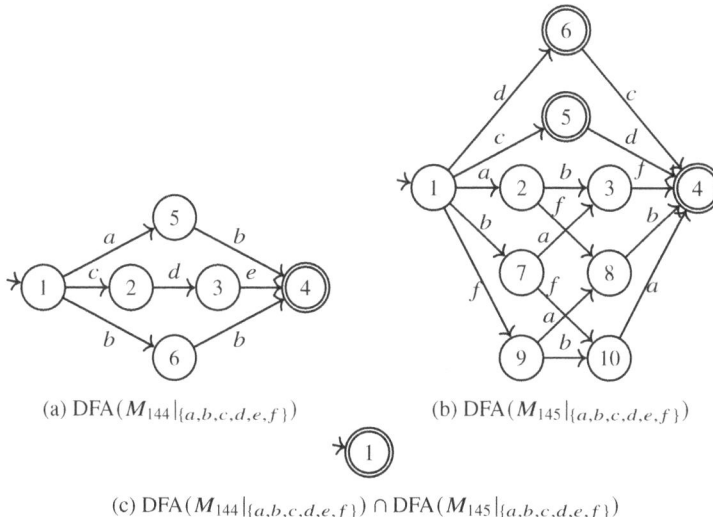

(a) $DFA(M_{144}|_{\{a,b,c,d,e,f\}})$ (b) $DFA(M_{145}|_{\{a,b,c,d,e,f\}})$

(c) $DFA(M_{144}|_{\{a,b,c,d,e,f\}}) \cap DFA(M_{145}|_{\{a,b,c,d,e,f\}})$

Fig. 7.10: DFAs computed by the PCC framework for $k = 6$ and $\{a, b, c, d, e, f\}$.

Finally, this example illustrates the influence of k: for $k = 1$, the PCC framework considers all activities in isolation and is unable to detect many behavioural differences between the process trees: only if an activity is absent, the measure drops below 1. In other examples, the measures might still be lower, e.g. if an activity is part of a loop in one model and not part of a loop in the system, then system precision will be below 1 as well. With the increase of k, PCC framework takes more and more inter-activity relations into account. Once k reaches its maximum, i.e. the size of the combined alphabets, the PCC framework requires that traces of the two models be equivalent: if there are no shared traces, recall and system precision will be 0.

7.3 Guarantees

So far, we introduced the PCC framework and showed that it accepts all bounded models, but not all weakly sound models (Requirement CR1), provides insights the on summarative and model but not on the log level (Requirement CR5), is normalisable (Requirement CR6), and symmetric (Requirement CR7). In this section, we prove that the PCC framework is able to reliably detect language equivalence between process trees of the class C_1, i.e. the class of process trees that may contain interleaved operators but not τ-leaves as defined in Section 5.4, using that for any two process trees of C_1 with a different language, the directly follows relation is different, and this difference is visible in projections if $k \geq 2$. This theorem will be useful in our evaluation in Chapter 8, where from recall and precision being 1 and the model being in C_1, we can conclude that the system was rediscovered. In these evaluations, we will also discuss the relation to other measuring techniques.

Theorem 7.1 (language decisive) *Let S and M be process trees of C_1. Then,* $recall(S, M, 2) = 1 \wedge precision(S, M, 2) = 1 \Leftrightarrow \mathcal{L}(S) = \mathcal{L}(M)$.

Proof As the PCC framework considers no structural properties of process trees but only languages, assume without loss of generality that S and M are in normal form according to Definition 5.1.

We prove the two cases $\Leftarrow$ and $\Rightarrow$ separately.

$\Leftarrow$ Assume $\mathcal{L}(S) = \mathcal{L}(M)$. By Lemma 5.13, $S = M$. By construction of PCC framework, $\forall_{a,b \in \Sigma(S)} recall(S, M, \{a, b\}) = precision(S, M, \{a, b\}) = 1$. Hence, recall and precision are 1.

$\Rightarrow$ Assume $recall(S, M, 2) = precision(S, M, 2) = 1$. Then, as any difference in activities would decrease recall or precision below 1, $\Sigma(S) = \Sigma(M)$. Take a set of activities $\{a, b\}$ with $\{a, b\} \subseteq \Sigma(M)$ and $a \neq b$. As $recall(S, M, \{a, b\}) = precision(S, M, \{a, b\}) = 1$, it holds that $\mathrm{DFA}(S|_{a,b}) = \mathrm{DFA}(M|_{a,b})$ and thus by definition $\mathcal{L}(S|_{a,b}) = \mathcal{L}(M|_{a,b})$. Then, by lemmas 5.11 and 5.12, the lowest common parents of a and b in S and M are equivalent, and the relative order of a and b matches (in case of $\circlearrowright$ or $\rightarrow$). This holds for all sets $\{a, b\}$.

Towards contradiction, assume that $S \neq M$. Without loss of generality, assume that S and M are reduced (Definition 5.1). Then, there must be a topmost different node $S' = \oplus(S'_1, \ldots S'_n)$ and $M' = \otimes(M'_1, \ldots M'_m)$ such that $\oplus \neq \otimes$ and/or $\exists_{1 \leq i \leq n,m} \Sigma(S'_i) \neq \Sigma(M'_i)$, although S and M are reduced and we know that for all pairs of activities, the lowest common parent is equivalent. By reasoning similar to lemmas 5.4 and 5.5, such a topmost different node cannot exist, and hence $S = M$ and therefore, $\mathcal{L}(S) = \mathcal{L}(M)$.

Hence, $recall(S, M, 2) = 1 \wedge precision(S, M, 2) = 1 \Leftrightarrow \mathcal{L}(S) = \mathcal{L}(M)$. $\square$

Unfortunately, this theorem does not hold for general process trees. For instance, take $S = \times(a, b, c, \tau)$ and $M = \times(a, b, c)$. For $k = 2$, the PCC framework will consider the subtrees $\times(a, b, \tau)$, $\times(a, c, \tau)$ and $\times(b, c, \tau)$ for both S and M, i.e. a τ will be introduced by projection, which hides the "real" tau from the measures. Hence, the PCC framework will not spot any difference: $recall = 1$ and $precision = 1$, even though the languages of S and M are clearly different. Only for $k = 3$, the PCC framework will detect the difference. A solution could be to treat projection-τ leafs and model-τ leafs separate to distinguish these cases.

7.4 Tool Support

As described in the previous sections, the PCC framework compares models and logs to models, in order to compute fitness, log precision, recall and system precision. The PCC framework has been implemented in the ProM framework and is accessible using several plug-ins. In this section, we describe these plug-ins. To use the PCC framework, choose one of the plug-ins "Compute projected recall and precision" or "Compute projected recall and fitness". Both of these plug-ins come in several variants, each for particular inputs: event logs, accepting Petri nets and process trees are supported. Figure 7.11 shows the settings available after selecting the plug-in. The first setting is the size of the subsets k ("size of projection"). Furthermore, a classifier can be chosen, which determines the activity corresponding to an event, there are options to not compute fitness or precision to save time, and there is a link to more information.

Once computations are finished, the results can be visualised in two ways:

- The results can be projected on the system, as shown in Figure 7.12. The model is visually laid out, and in each activity, the fitness/recall and precision measures are shown. To ease detection of deviating activities, both fitness/recall and precision influence the colouring of the activities: the red-most activities are the most deviating. Notice that this model projection shows only the first model, i.e. the event log and second model are not shown.
  The projected model can be exported as an image, its layout can be influenced and dragging/scrolling the mouse influences its position. For more information, move the mouse pointer to the question mark in the bottom right corner, and a popup with information will appear.
- The results can be visualised in a table, as shown in figures 7.13 and 7.14. These figures were obtained from applying the PCC framework to the BP11 log (see Section 8.3.1, [5]) and a model discovered by IMF (see Section 6.2). In this visualisation, first the average fitness/recall and precision values are shown as numbers, which shows that the model represents almost all behaviour of the event log, i.e. has a high fitness, however the rather low precision indicates that the model contains more behaviour. Second, fitness/recall is given, averaged for each activity. By the colouring, the more red an activity/pair is, the worse the measures fitness/recall/precision, which makes it easy to spot activities with a

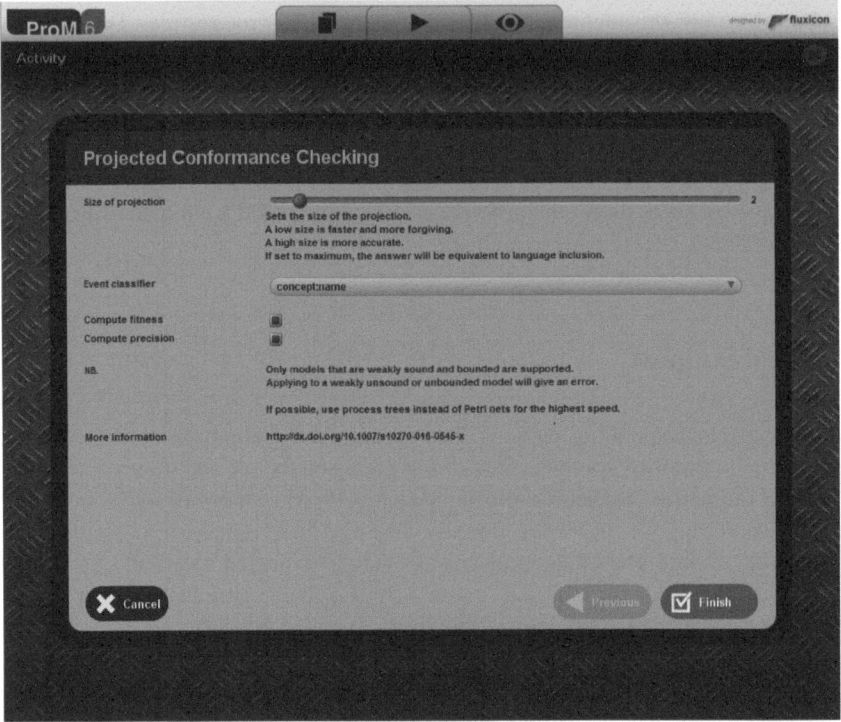

Fig. 7.11: The settings of the PCC framework plug-in in the ProM framework.

high or low fitness. In Figure 7.14, A_ACCEPTED had a particular low fitness. Third, fitness/recall is given for each pair of activities, for more insight into the precise measures. In our example figure, the pair of A_ACCEPTED and O_DECLINED was particulary problematic. Finally, precision is shown in a similar way.

Implementation.

The implementation uses adapted automata from [10], and is multithreaded. Developers that wish to call the PCC framework programmatically can freely choose k, using methods in one of the following classes:

- CompareLog2PetriNetPlugin,
- CompareLog2ProcessTreePlugin,
- ComparePetriNet2PetriNetPlugin or
- CompareProcessTree2ProcessTreePlugin

The code is available at https://svn.win.tue.nl/repos/prom/Packages/ ProjectedRecallAndPrecision/Trunk; for this evaluation, revision svn revision

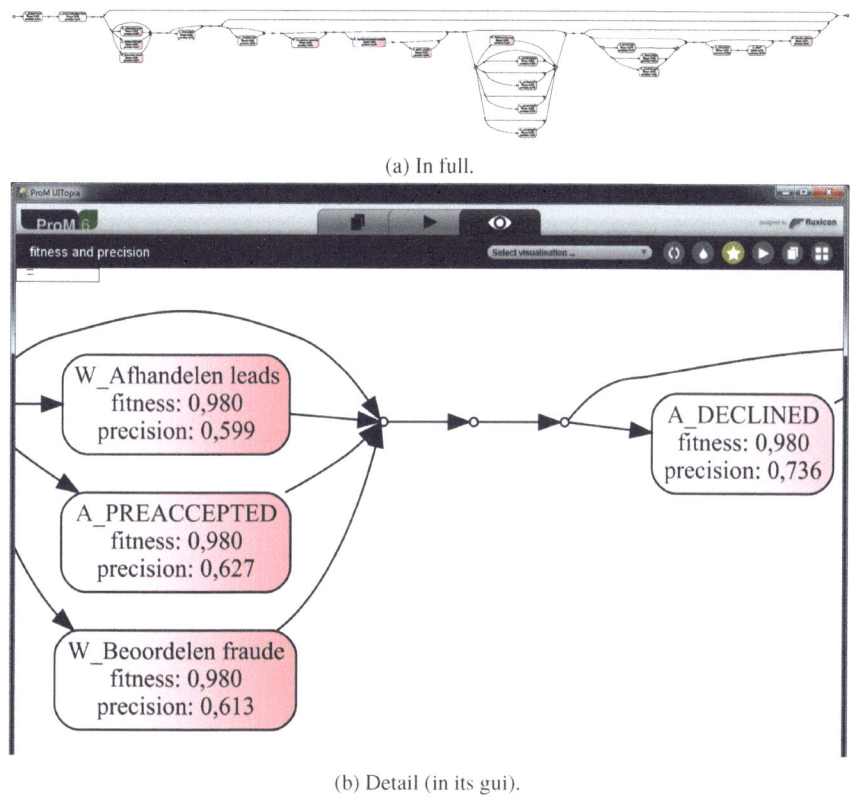

(a) In full.

(b) Detail (in its gui).

Fig. 7.12: Results of the PCC framework projected on a process model.

34642 was used. Furthermore, there is an option to not compute system precision or recall, as by default both measures are computed (similar for log precision and fitness). Choosing this option if one of the measures is not necessary for the use case at hand will save roughly half of the computation time.

Other formalisms can be supported, for which large parts of the framework can be reused. To support a new formalism, one should provide the framework with two pieces of information: the activities used in a model, and how to project and automatise the model, by extending either abstract class ModelModelFramework or LogModelFramework.

7.5 Conclusion

In this chapter, we introduced the PCC framework, a conformance checking technique that supports both log-conformance and model-conformance checking, both for

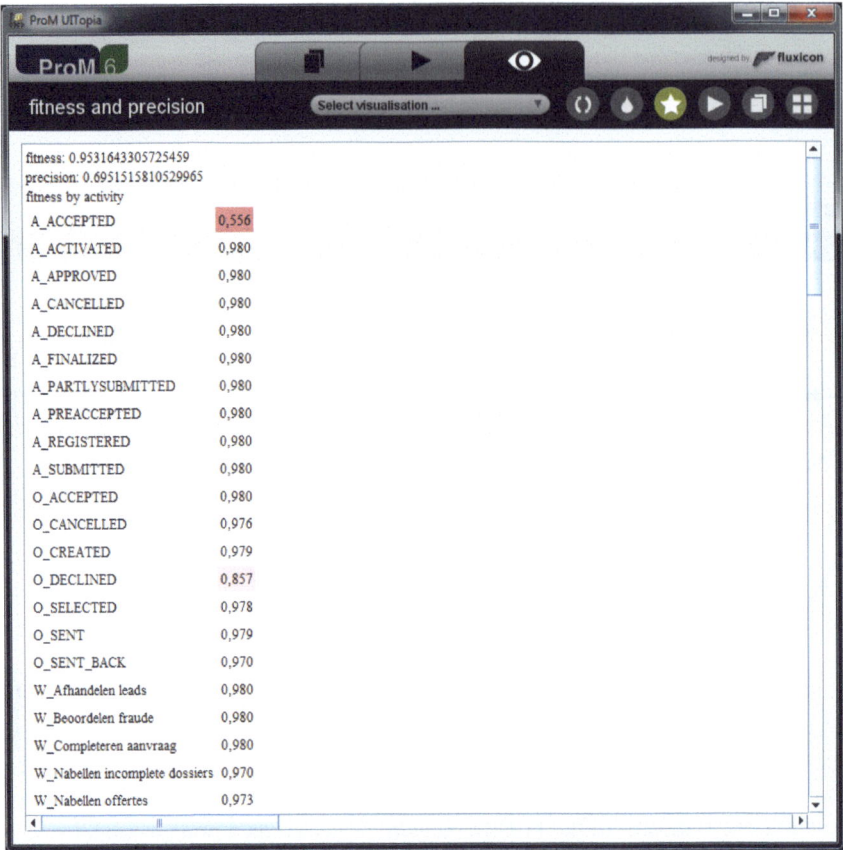

Fig. 7.13: Tabular results of the PCC framework (1).

arbitrary process model formalisms. Notice that the PCC framework could support log-log comparisons with minimal changes, however we did not implement this. It considers all subsets of activities of a user-choosable size k, and constructs DFAs from the behaviour of the model and the log/model projected on the activities. Then, the conjunction of these DFAs is used to compute recall and precision measures, by considering outgoing edges of states. Furthermore, fitness is computed by replaying the projected traces of the log on the projected model, and returning the fraction of traces that is accepted. These measures on k-subsets provide detailed insight into the location of deviations in the model, when averaged over activities. Furthermore, taking the average over all k-subsets provides an aggregated recall or precision measure.

We described how process trees and Petri nets fit in the framework, and that the PCC framework has been implemented in the ProM framework. Furthermore, we showed that if two process trees of C_1 are compared, choosing $k = 2$ will guarantee

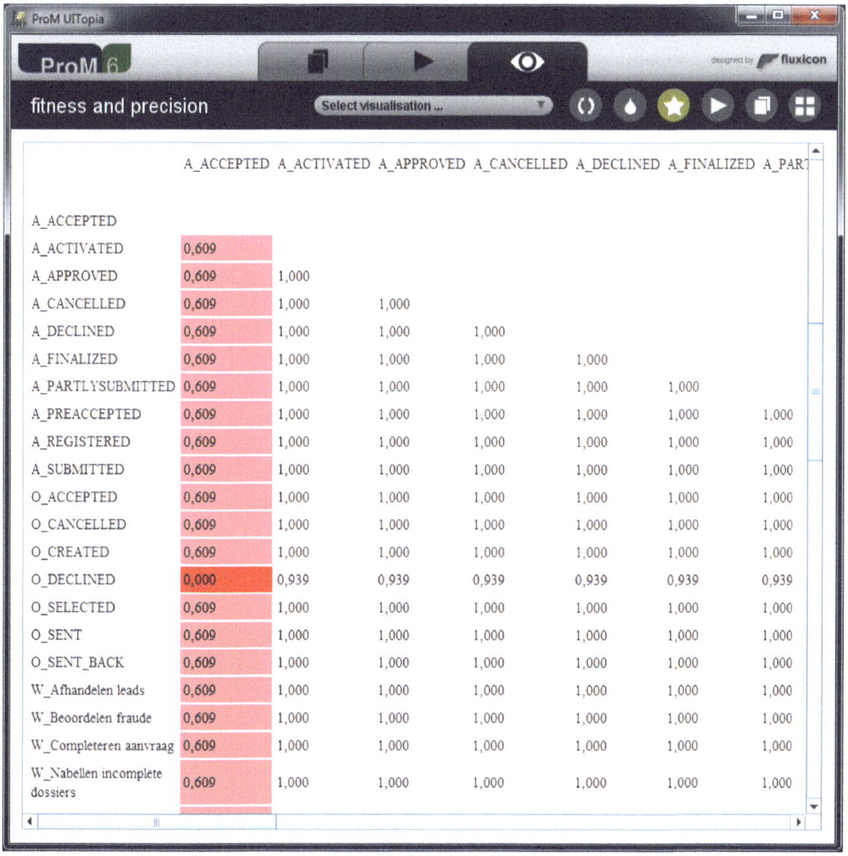

	A_ACCEPTED	A_ACTIVATED	A_APPROVED	A_CANCELLED	A_DECLINED	A_FINALIZED	A_PART
A_ACCEPTED							
A_ACTIVATED	0,609						
A_APPROVED	0,609	1,000					
A_CANCELLED	0,609	1,000	1,000				
A_DECLINED	0,609	1,000	1,000	1,000			
A_FINALIZED	0,609	1,000	1,000	1,000	1,000		
A_PARTLYSUBMITTED	0,609	1,000	1,000	1,000	1,000	1,000	
A_PREACCEPTED	0,609	1,000	1,000	1,000	1,000	1,000	1,000
A_REGISTERED	0,609	1,000	1,000	1,000	1,000	1,000	1,000
A_SUBMITTED	0,609	1,000	1,000	1,000	1,000	1,000	1,000
O_ACCEPTED	0,609	1,000	1,000	1,000	1,000	1,000	1,000
O_CANCELLED	0,609	1,000	1,000	1,000	1,000	1,000	1,000
O_CREATED	0,609	1,000	1,000	1,000	1,000	1,000	1,000
O_DECLINED	0,000	0,939	0,939	0,939	0,939	0,939	0,939
O_SELECTED	0,609	1,000	1,000	1,000	1,000	1,000	1,000
O_SENT	0,609	1,000	1,000	1,000	1,000	1,000	1,000
O_SENT_BACK	0,609	1,000	1,000	1,000	1,000	1,000	1,000
W_Afhandelen leads	0,609	1,000	1,000	1,000	1,000	1,000	1,000
W_Beoordelen fraude	0,609	1,000	1,000	1,000	1,000	1,000	1,000
W_Completeren aanvraag	0,609	1,000	1,000	1,000	1,000	1,000	1,000
W_Nabellen incomplete dossiers	0,609	1,000	1,000	1,000	1,000	1,000	1,000

Fig. 7.14: Tabular results of the PCC framework (2).

that the PCC framework correctly identifies language equivalence of the two trees, i.e. the two trees have the same language if and only if both recall and model-precision are 1. However, we did not prove tightness of the class C_1, i.e. there might be more process trees for which the PCC framework provides this guarantee. Further research should reveal this class of models.

Future work 7.2: Investigate properties of the PCC framework on models outside of C_1.

We finish this chapter with some ideas to extend the PCC framework to handle unbounded and weakly unsound Petri nets.

7.6 Ideas to Handle Unbounded & Weakly Unsound Petri Nets

As expressed by Requirement CR1, a conformance checking technique ideally deals with as many unsound Petri nets as possible. However, if the PCC framework could handle all Petri nets, it would be able to decide language inclusion, which is undecidable for unbounded models. Nevertheless, in this section we introduce several heuristics, while making sure that the consistency of the framework is preserved when the heuristics are applied to a sound workflow net, i.e. exact diagnostics should be returned. Notice that these heuristics have not been included in the PCC framework.

Making a Petri net Bounded.

To make the Petri net bounded, each place is given an artificial capacity. During state-space exploration, a transition is only enabled if firing it would not violate the bound of any place [3]. As sound workflow nets are bounded, this heuristic will not influence their semantics. However, if the capacity is chosen too low, not enough behaviour might be captured for a comparison, and language-preserving reduction rules might influence the result. This limitation is inherent to using DFAs and solving it would require other classes of models, for which the problem might be undecidable.

This heuristic presumably influences the conformance checking results, i.e. it might have influence on fitness or precision values. Therefore, if this heuristic would be used, its influence on these results should be studied further.

Heuristic for an Initial Marking.

Some algorithms, such as the Heuristics Miner, provide an initial marking. If no initial marking is present, we construct an initial marking by putting a token into each place without incoming transitions. In sound workflow nets, an initial marking is provided by the source place, which corresponds to our heuristic.

Heuristic for Final Markings.

Only few discovery algorithms are capable of producing Petri nets with a distinguished final marking. Some algorithms, such as ILP [15] and α [2, p.130], might benefit from the addition of artificial start and end events before discovery, which aids these algorithms in finding initial and final markings. In case no final marking is given, it can be obtained as follows:

1. Manually inspect the model and define a final marking. This is usually infeasible for complex models;

2. An approach taken in [4] is to consider each reachable marking to be a final marking. This heuristic increases the size of the language, as each prefix of a trace becomes a trace itself. In the PCC framework, this heuristic should be applied to both sides of the comparison, i.e. a model-with-heuristic should not be compared to a model-without-heuristic, as this would result in an overestimation of recall and an underestimation of precision (the model-with-heuristic obviously has a much larger language than the model-without-heuristic). Moreover, having each marking as a final marking is not what is *meant* by current discovery algorithms: corresponding to traces in the event log, algorithms aim to discover a model of a process with a clear start and a clear end. Nevertheless, in use cases in which such behaviour is intended [14], this strategy might be chosen (but must be chosen for both models to ensure comparability).

3. Another approach is to consider each reachable conditional livelock to be a final marking. A *conditional livelock* is a marking M in which for each enabled transition t, one of the following conditions holds:

   - t is not connected to any place, or
   - firing t leads to a marking M' that is equal to or strictly larger than M.

   Figure 7.15 shows an example of a Petri net in a conditional livelock: b is enabled but firing it would leave the net in a strictly larger marking (as b produces a token without consuming one), d is not connected to any place, and firing e would yield an equal marking. In sound workflow nets, the only conditional livelock is the one with a token in the sink place, which corresponds to this heuristic. A downside of this strategy is that nets without conditional livelocks (e.g. with livelocks) are considered to have the empty language.

   In case the Petri net is discovered by the ILP miner [15] with the empty-after-completion option enabled, the model can replay all traces of the event log and ends in an empty marking, which is a conditional livelock.

We prefer the second strategy as it keeps the framework consistent. Ideally, the discovery algorithm should provide an initial marking and final markings, and preferably a bounded net (as unbounded nets can inherently not be captured by DFAs).

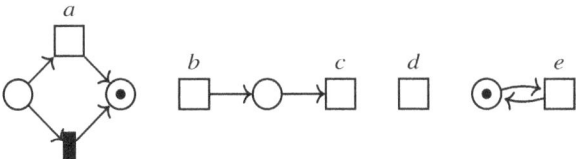

Fig. 7.15: An unbounded Petri net over transitions a, b, c, d and e in a conditional livelock.

In the next chapter, we will evaluate the process discovery and conformance checking techniques that were introduced before.

References

1. van der Aalst, W.M.P.: Verification of workflow nets. In: P. Azéma, G. Balbo (eds.) Application and Theory of Petri Nets 1997, 18th International Conference, ICATPN '97, Toulouse, France, June 23-27, 1997, Proceedings, *Lecture Notes in Computer Science*, vol. 1248, pp. 407–426. Springer (1997). DOI 10.1007/3-540-63139-9_48. URL http://dx.doi.org/10.1007/3-540-63139-9_48

2. van der Aalst, W.M.P.: Process mining - discovery, conformance and enhancement of business processes. Springer (2011). DOI 10.1007/978-3-642-19345-3. URL http://dx.doi.org/10.1007/978-3-642-19345-3

3. van der Aalst, W.M.P., Stahl, C.: Modeling Business Processes - A Petri Net-Oriented Approach. Cooperative Information Systems series. MIT Press (2011). URL http://mitpress.mit.edu/books/modeling-business-processes

4. Adriansyah, A.: Aligning Observed and Modeled Behavior. Ph.D. thesis, Eindhoven University of Technology (2014)

5. van Dongen, B.: BPI challenge 2011 dataset (2011). DOI doi:10.4121/uuid:d9769f3d-0ab0-4fb8-803b-0d1120ffcf54. URL http://dx.doi.org/10.4121/uuid:d9769f3d-0ab0-4fb8-803b-0d1120ffcf54

6. Esparza, J., Hoffmann, P.: Reduction rules for colored workflow nets. In: P. Stevens, A. Wasowski (eds.) Fundamental Approaches to Software Engineering - 19th International Conference, FASE 2016, Held as Part of the European Joint Conferences on Theory and Practice of Software, ETAPS 2016, Eindhoven, The Netherlands, April 2-8, 2016, Proceedings, *Lecture Notes in Computer Science*, vol. 9633, pp. 342–358. Springer (2016). DOI 10.1007/978-3-662-49665-7_20. URL http://dx.doi.org/10.1007/978-3-662-49665-7_20

7. Esparza, J., Nielsen, M.: Decidability issues for Petri nets - A survey. Bulletin of the EATCS **52**, 244–262 (1994)

8. Gelade, W.: Succinctness of regular expressions with interleaving, intersection and counting. Theor. Comput. Sci. **411**(31-33), 2987–2998 (2010). DOI 10.1016/j.tcs.2010.04.036. URL http://dx.doi.org/10.1016/j.tcs.2010.04.036

9. Linz, P.: An introduction to formal languages and automata (4. ed.). Jones and Bartlett Publishers (2006)

10. Møller, A.: dk.brics.automaton – finite-state automata and regular expressions for Java (2010). http://www.brics.dk/automaton/

11. Murata, T., Shenker, B., Shatz, S.M.: Detection of ada static deadlocks using Petri net invariants. IEEE Trans. Software Eng. **15**(3), 314–326 (1989). DOI 10.1109/32.21759. URL http://dx.doi.org/10.1109/32.21759

12. Rozinat, A., van der Aalst, W.M.P.: Conformance checking of processes based on monitoring real behavior. Inf. Syst. **33**(1), 64–95 (2008). DOI 10.1016/j.is.2007.07.001. URL http://dx.doi.org/10.1016/j.is.2007.07.001

13. de San Pedro, J., Cortadella, J.: Discovering duplicate tasks in transition systems for the simplification of process models. In: M.L. Rosa, P. Loos, O. Pastor (eds.) Business Process Management - 14th International Conference, BPM 2016, Rio de Janeiro, Brazil, September 18-22, 2016. Proceedings, *Lecture Notes in Computer Science*, vol. 9850, pp. 108–124. Springer (2016). DOI 10.1007/978-3-319-45348-4_7. URL http://dx.doi.org/10.1007/978-3-319-45348-4_7

14. Tapia-Flores, T., López-Mellado, E., Estrada-Vargas, A.P., Lesage, J.: Petri net discovery of discrete event processes by computing t-invariants. In: A. Grau, H. Martínez (eds.) Proceedings of the 2014 IEEE Emerging Technology and Factory Automation, ETFA 2014, Barcelona, Spain, September 16-19, 2014, pp. 1–8. IEEE (2014). DOI 10.1109/ETFA.2014.7005080. URL http://dx.doi.org/10.1109/ETFA.2014.7005080

15. van der Werf, J.M.E.M., van Dongen, B.F., Hurkens, C.A.J., Serebrenik, A.: Process discovery using integer linear programming. Fundam. Inform. **94**(3-4), 387–412 (2009). DOI 10.3233/FI-2009-136. URL http://dx.doi.org/10.3233/FI-2009-136

Chapter 8
Evaluation

Sander J. J. Leemans: Robust Process Mining with Guarantees, LNBIP 440, pp. 355–422, 2022
https://doi.org/10.1007/978-3-030-96655-3_8

Abstract In this chapter, we evaluate the introduced discovery techniques and conformance checking framework. We perform 5 experiments to study the following questions: (1) how scalable are the discovery algorithms? (2) how do the discovery algorithms compare to existing techniques in terms of log quality? (3) what are the boundaries of rediscoverability of the discovery algorithms in terms of incomplete, deviating and infrequent behaviour? (4) can the conformance checking framework provide similar results as existing techniques? and (5) how do discovery algorithms handle non-atomic behaviour?

In the previous chapters, we identified requirements for process discovery techniques and conformance checking techniques (Chapter 3), introduced process discovery techniques (Chapter 6) and introduced conformance checking techniques (Chapter 7).

The introduced process discovery algorithms guarantee several properties such as soundness, fitness and rediscoverability (requirements DR1, DR4 and DR2). In this chapter, we experimentally evaluate the other quality criteria described in Chapter 3: we evaluate scalability, log-conformance measures, soundness and handling of infrequent, incomplete and deviating behaviour.

We introduced conformance checking techniques that handle all bounded weakly sound models, guarantee returned measures being perfect bi-implies language equivalence, provide insight at two levels (summarative numbers and projections on models), and return normalised values that are symmetric and reflexive (requirements CR2, CR1 CR3, CR5, CR6 and CR7. In this chapter, we compare the PCC framework with existing approaches. Next to evaluating the quality of models, we focus on speed. (Requirement CR4).

In this chapter, we first evaluate process discovery techniques, after which we evaluate conformance checking techniques. We perform these evaluations to answer the following research questions:

RQ.1 What is the largest event log (number of events/traces or number of activities) that process discovery algorithms can handle?
This research question relates to Requirement DR5.

RQ.2 How do algorithms balance log-quality criteria? Can the user influence the balance between these criteria? How do the new algorithms of the IM framework and the IMD framework compare to existing discovery techniques?
These research questions relate to Requirement DR4. In the execution of the experiment to answer this research question, we will also check for soundness (Requirement CR1).

RQ.3 Can discovery algorithms rediscover the system if the event log contains deviating or infrequent behaviour? What is the influence of increasing levels of such behaviour on the quality of models discovered by discovery algorithms?
These research questions relate to Requirement DR3.

RQ.4 How is non-atomic behaviour in event logs handled by discovery algorithms? Do the newly introduced algorithms IMLC and IMFLC improve over existing techniques?
These research questions relate to requirements DR4 and CR1 for non-atomic event logs and process models.

RQ.5 How can the PCC framework be used for detailed analyses of the performance of discovery algorithms in the presence of infrequent, deviating and
incomplete behaviour? Can the PCC framework handle larger event logs and
models than existing conformance checking techniques? How do the reported
measures of the PCC framework compare to the measures returned by other
techniques, and can they be used interchangeably?

These research questions relate to Requirement CR4.

We start with a discussion of the selection of discovery algorithms that we will
evaluate (Section 8.1). The scalability of process discovery techniques (RQ.1) is
evaluated in Section 8.2, log-quality dimensions (RQ.2) in Section 8.3, the rediscoverability challenges (RQ.3) in Section 8.4. We evaluate conformance checking
techniques (RQ.5) in Section 8.5, non-atomic behaviour (RQ.4) in Section 8.6, and
Section 8.7 concludes the chapter.

8.1 Evaluated Process Discovery Algorithms

Table 8.1 shows the process discovery algorithms used in this evaluation. Unless
stated otherwise, all algorithms were applied with their default settings. We intended
to include all soundness guaranteeing discovery algorithms known to us (Evolutionary Tree Miner (ETM), Constructs Competition Miner (CCM), Maximal Pattern
Miner (MPM) and the algorithms described in previous chapters). However, there are
no public implementations of CCM and MPM available. Nevertheless, the authors
of CCM kindly performed some experiments for us.

As non-soundness guaranteeing algorithms, we included a broad selection of
other algorithms: the fitness and weak-soundness guaranteeing (Integer Linear Programming (ILP), the well-known α-algorithm (α), the filtering Flexible Heuristic
Miner (HM), and the recent Fodina (FO) and Structured Miner (SM). Finally, we
included the Tsinghua-α algorithm (Tα) as it handles non-atomic behaviour. As
commercial tools, we included Fluxicon Disco (FD) and Celonis Process Mining
(CPM), both of which were kindly provided for this experiment by Fluxicon and
Celonis.

As baselines, we included the Flower Miner (FM) and the Trace Model Miner
(TM): a flower miner returns a model that allows for any behaviour, while the
trace model miner returns a model that enumerates all traces. For instance, consider
the event log $\{\langle a, b, c \rangle, \langle b, c, a \rangle\}$. Then, the flower and trace models would be
respectively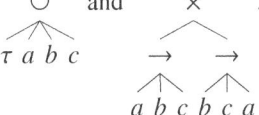

In all experiments, standalone versions of the implementations were used
wherever possible, e.g. all graphical user interfaces were circumvented. Furthermore, all algorithms were used with their default parameter settings.

Table 8.1: Discovery algorithms used in the evaluation.

ETM	Evolutionary Tree Miner [9] in ProM 6.6, random generator seed fixed to 1, standard termination condition with 30 minutes timeout.
CCM	Constructs Competition Miner [33]. Implementation not available: authors kindly performed some tests for us.
SM	Structured Miner [5]. Private version provided by authors.
IM	Inductive Miner in ProM 6.6 svn revision 34642.
IMF	Inductive Miner - infrequent in ProM 6.6 svn revision 34642, noise threshold 0.2.
IMA	Inductive Miner - all operators in ProM 6.6 svn revision 34642.
IMFA	Inductive Miner - infrequent - all operators in ProM 6.6 svn revision 34642, noise threshold 0.2.
IMC	Inductive Miner - incompleteness in ProM 6.6 svn revision 34642.
IMD	Inductive Miner - directly follows in ProM 6.6 svn revision 34642.
IMFD	Inductive Miner - infrequent - directly follows in ProM 6.6 svn revision 34642, noise threshold 0.2.
IMCD	Inductive Miner - incompleteness - directly follows in ProM 6.6 svn revision 34642.
IMLC	Inductive Miner - life cycle in ProM 6.6 svn revision 34642.
IMFLC	Inductive Miner - infrequent - life cycle in ProM 6.6 svn revision 34642, noise threshold 0.2.
FM	flower model in ProM 6.6 svn revision 34642.
TM	trace model in ProM 6.6 svn revision 34642.
ILP	Integer Linear Programming Miner [38] in ProM 6.6, empty-after-completion option set.
HM	Flexible Heuristic Miner [36] in ProM 6.6.
FO	Fodina [6] 20160706.
α	[1] in ProM 6.6.
Tα	Tsinghua-α [37] in ProM 5.2.
FD	Fluxicon Disco [17] 1.9.7. Fluxicon provided us an unlimited version for the purpose of these experiments.
CPM	Celonis Process Mining [12] 4.0.1. Celonis provided us a stand-alone version.

As a log importer, the OpenLogFileDiskImplWithoutCachePlugin-importer was used, as it caches the event log on disk (despite its name) and therefore has a small RAM overhead, while still enabling random access [28]. To create a directly-follows graph from a log and for FM and TM, the log was not imported or stored in memory at all, but incrementally parsed instead using a custom straightforward script.

8.2 Scalability of Discovery Algorithms

In Section 6.6, we introduced the IMD framework and several algorithms that are aimed at handling large event logs, i.e. logs with hundreds of activities and millions of events. In this section, we aim to answer RQ.1, i.e. we test existing and newly introduced algorithms on scalability. Event logs can be large on two dimensions: the number of activities and the number of events. In [23], we identified relevant gradations for these dimensions: for the number of activities, we identified COMPLEX logs, i.e. containing hundreds of activities, and MORE COMPLEX logs, i.e. containing thousands of activities. For the number of events we identified MEDIUM logs, i.e.

containing tens of thousands of events, LARGE logs, i.e. containing millions of events, and LARGER logs, i.e. containing billions of events.

In principle, complexity (number of activities) and size (number of events) are independent: event logs can be not COMPLEX and still LARGER, and can also be MORE COMPLEX and not MEDIUM. However, logs of higher complexity will need to contain more events to reach several notions of completeness. For instance, in [23] we showed that a system with 10^4 activities might need 10^5 traces to see all activities at least once, and many more to be directly-follows complete.

Such numbers of events and activities might seem large for a complaint-handling process in an airline, however processes of much larger complexity exist. For instance, even simple software tools contain hundreds or thousands of different methods. To study or reverse engineer such software, studies [20] have recorded method calls in event logs (at various levels of granularity), and process mining and software mining techniques have been used on small examples to perform the analyses [32, 20]. MEDIUM event logs can for instance be found in hospitals: BP11 was recorded in the emergency department of a Dutch hospital and contains over 600 activities [13]. Even though this log contains just 600 activities and 150,291 events, current discovery techniques have difficulties with this log (as we will show in this section). Other areas in which LARGER logs appear are click-stream data from web-sites, such as the web-site of a dot-com start-up, which produced an event log containing 3300 activities [19]. Even larger logs could be extracted from large machines, such as the Large Hadron Collider, in which over 25,000 distributed communicating components form just a part of the control systems [18], resulting in complicated behaviour that could be analysed using scalable process mining techniques.

Therefore, this experiment illustrates the feasibility of applying discovery algorithms to MORE COMPLEX (with up to 10^4 activities) and LARGER (with up to 10^{10} events) event logs, given limited computing resources (e.g. a limited amount of RAM).

8.2.1 Set-up

All discovery algorithms (Section 8.1) were tested on the same set of randomly generated logs and systems, which are iteratively increased in size and complexity (i.e. number of events and number of activities), until an algorithm is not able to return process models anymore.

For each algorithm, we first generate 10 random systems (process trees $\in C_B$) of a activities, starting a at 2. From these systems, we generate 10 event logs for each process tree, each containing t traces, starting t at 4. Second, we apply the discovery algorithm to these 10 logs with 2GB of RAM available for each run, and record how many of these applications are successful. If the discovery algorithm does not crash, does not run out of memory and returns a model within 5 hours, then we call the application successful. If successful on at least one event log, we multiply a by 2 and t by 4, and repeat the procedure (we refer to one such repetition as a *round*).

The number of successful applications per round, which is a number between 0 and 10, is recorded. Table 8.2 shows statistics describing the rounds, systems and event logs. For instance, the following tree illustrates a model used in round 5:

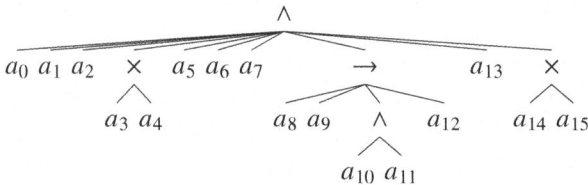

Notice that the generated event logs were not necessarily directly follows complete with respect to the randomly generated systems, however we verified that all were activity complete. For the purpose of this experiment, i.e. evaluating scalability and not model quality, stronger completeness notions than activity completeness were not necessary.

Table 8.2: Experiment rounds to test scalability.

round	activities	traces	events	
			μ	σ
1	2	4	7	2
2	4	16	60	30
3	8	64	502	277
4	16	256	3,283	1,342
5	32	1,024	26,316	18,852
6	64	4,096	154,509	66,567
7	128	16,384	1,387,570	609,401
8	256	65,536	10,456,186	4,890,888
9	512	262,144	69,510,326	43,495,480
10	1024	1,048,576	472,610,224	278,769,395

8.2.2 Results

Figure 8.1 shows the results. These have been split up in two graphs for readability reasons, i.e. each graph shows the results for 10-11 miners.

8.2.3 Discussion

Most algorithms handle event logs of up to 32 activities (not even COMPLEX) and on average 26,000 events (round 5, MEDIUM) without problems. We would argue

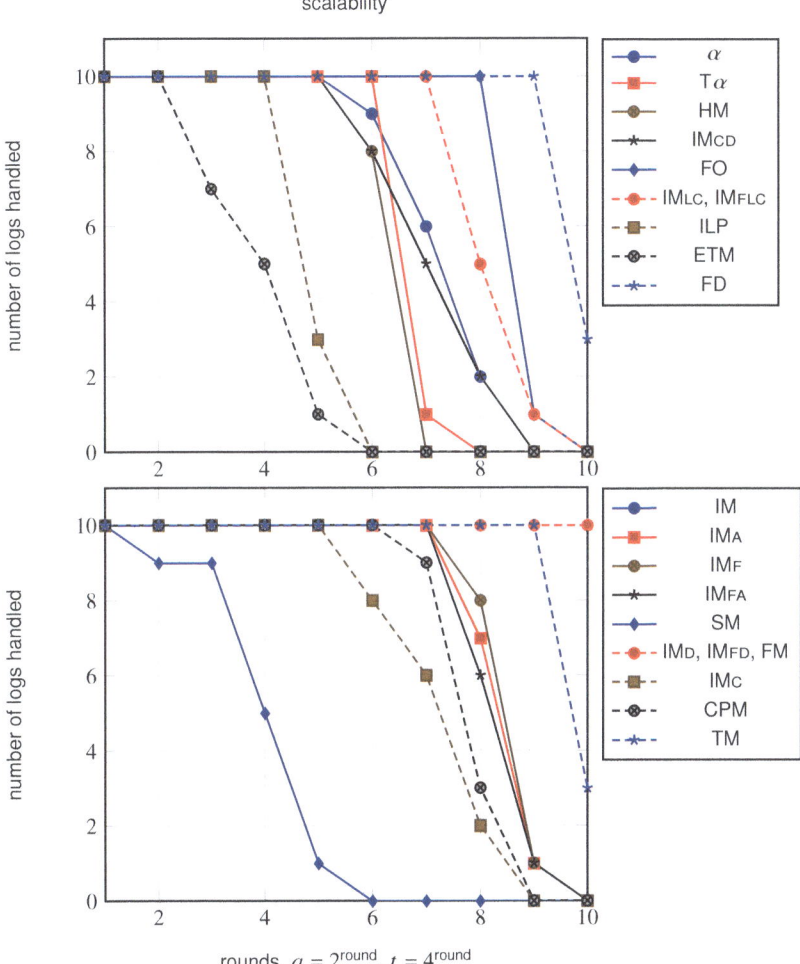

Fig. 8.1: Scalability results. Refer to Table 8.1 for information on the abbreviated algorithms.

that many real-life processes seem to be well-handled by these algorithms (see Section 3.1). Exceptions in this experiment are SM, for which we used a preliminary version, ETM, and ILP. The authors of SM kindly provided us several new versions while we were conducting the experiments. For ETM, we believe this is due to the repeated application of alignments to measure fitness, log-precision and generalisation. For ILP, we believe this is due to it constructing large ILP problems with constraints for every prefix of every trace in the event log.

Small event logs thus seem to pose little challenge for discovery algorithms. However, as discussed earlier in this section, much larger event logs exist in practise. A little bit larger event logs, i.e. with 64 activities, 4096 traces and on average 154,000 events (round 6, MEDIUM and not even COMPLEX) were not handled by α, HM, IMcD and IMc anymore: all these algorithms have an exponential run time in the number of activities. The single-pass algorithms HM, α and CCM (not tested) have been shown to be applicable in map-reduce [15] or streaming [11, 34] settings, thus one would expect that the used implementations would be scalable as well.

All remaining algorithms could handle the event logs of round 7, however in round 8 (256 activities, COMPLEX), the remaining algorithms of the IM framework start running out of memory: the main cause being the many copies of the event logs these algorithms have to keep in memory. Furthermore, the exponential Tα can handle only 1 of the 10 logs in round 7, and CPM still 9. The only algorithm remaining is Fodina (FO), which handles all 10 event logs in round 8, only 1 log in round 9, and no logs in round 10.

The trace model TM shows the challenge that algorithms face keeping the entire log in main memory: our implementation of TM keeps one integer per event and one integer per trace. Nevertheless, in round 10, 3 of the 10 event logs could not be handled by TM. Remarkably, FD could still handle 3 event logs in round 10, and could even visualise the results (partially), as FD caches the entire event log to disk.

The only algorithms that managed to handle all event logs in this experiment (MORE COMPLEX and LARGER) were the algorithms of the IMD framework, i.e. IMD and IMFD, as these do not keep the event log in main memory at all, but traverse it once to construct a directly follows graph. If we would have continued the experiment, these algorithms would fail at some point as well, however only when the directly follows graphs do not fit in main memory anymore. The other remaining algorithm is the baseline FM, that does not keep an event log in memory at all.

In answer of RQ.1, we conclude that all algorithms except SM are able to handle small event logs, i.e. logs of tens of activities and thousands of events (neither COMPLEX nor MEDIUM), and most algorithms of the IM framework handle all logs up to a hundred activities and a million events (COMPLES and LARGE). However, only the algorithms of the IMD framework and the baseline flower model FM are able to handle a thousand activities and hundreds of millions of events.

Notice that we evaluated the *implementations* of the techniques, rather than the techniques themselves, and these implementations might be optimised in the future. Nevertheless, this experiment illustrates the boundaries and potential of the techniques: if event logs get too large to keep in main memory, single-pass algorithms like the algorithms of the IMD framework might still be able to handle it.

In the future, we aim to extract such logs from real-life systems and apply our techniques to them, but currently, we would only discover a model but would not be able to process the discovered model further (no conformance checking and no visualisations are available on these scales).

In this experiment, we increased the number of activities a and the number of traces t synchronously, assuming that a more complex system (i.e. with a high a)

needs a log with more traces (i.e. a high t) to be represented well. In [23], we varied a and t independently, which led to similar conclusions.

In the remainder of this chapter, we will evaluate the quality of the models returned by the algorithms in more detail.

8.3 Log-Quality Dimensions

In Section 3.3, several challenges of process discovery were identified, one of them being that discovery algorithms inherently need to trade off log-quality criteria. Requirement DR10 states that discovery algorithms should provide a user-influenceable trade off between log-quality criteria. In this section, we perform two experiments to evaluate how process discovery algorithms balance the trade-off between log-quality criteria (RQ.2) on their default settings. In the first experiment, we apply these algorithms to several real-life logs using cross validation, and measure weak soundness, boundedness, fitness, log-precision and simplicity of the discovered models. In the second experiment, we apply the algorithms to the full event logs, and compare their results qualitatively.

We first describe the used event logs in Section 8.3.1, after which we describe the quantitative experiment (Section 8.3.2) and the qualitative experiment (Section 8.3.3).

8.3.1 Event Logs

In the log-quality experiments, we use several real life event logs. All event logs are publicly available.

BP11 The event log used in the Business Process Intelligence Challenge 2011 (BP11) [13] represents the gynaecology department of a Dutch academic hospital. The traces represent patients, for which medical activities are performed. This event log has been challenging for discovery algorithms [21, 23, 25, 27], given the large number of activities and the unstructured nature of the process. The event log contains much more data elements than just the activities used in this experiment: further research might reveal ways to use more data to aid discovery.
Traces: 1143, events: 150,291, activities: 624.
BP12 The event log used in the Business Process Intelligence Challenge 2012 (BP12) [14] is derived from a personal loan application process of a Dutch financial institution. The traces represent customers applying for loans, and the activities describe the steps taken for the customers.
The BP12 log contains three subprocesses: the activities are distinguishable by a prefix 'A_', 'O_' or 'W_'. Two of these three subprocesses (A and O) in isolation can be handled well by existing process discovery algorithms, as

they are relatively well structured. For instance, the A subprocess consists of 10 activities, is rather structured and contains only three activities in concurrency. However, the third subprocess (W) is more challenging: it contains start and completion events (which most discovery and conformance algorithms simply ignore), and is rather unstructured [21, 23, 29].

The entire event log contains the three subprocesses being executed. Not all traces contain every subprocess, and there is no sequential relation between the subprocesses. In this experiment, we use this full event log.

Traces: 13,087, events: 262,200, activities: 24.

RF   The road fines event log is a real-life event log of an information system managing Italian road traffic fines [24]. Each trace represents a fine, which can be objected, and should eventually be paid. Penalties are applied when payment is late.

Traces: 150,370, events: 561,470, activities: 11.

RPW   The WABO receipt phase log (RPW) [7] originates from a building permit application process in a Dutch municipality. This log describes the first phase in the process, in which the permit applications are received. Traces: 1,434, events: 8577, activities: 27.

WA1-5 The five WABO logs have been derived from a building permit application process in five Dutch municipality, each executing another 'flavour' of the same process. The differences between their processes have been studied in [8]. WA1 contains 434 traces, 13,571 events and 137 activities. WA2 contains 160 traces, 10,439 events and 160 activities. WA3 contains 481 traces, 16,566 events and 170 activities. WA4 contains 324 traces, 9,743 events and 133 activities. WA5 contains 432 traces, 13,370 events and 176 activities.

8.3.2 Quantitative

In the quantitative experiment, we compare the algorithms using cross validation.

8.3.2.1 Set-up

Each event log L is split randomly into v sublogs (we chose $v = 6$, as 3 would not yield cross validation and 9 would result in much more computations) and the following procedure is applied: $^2/_3$ of the sublogs form the *discovery log* L_d, while the other sublogs form the *test log* L_t. We did not use a default k-fold cross validation scheme, as the test log needs to be large enough to measure fitness as a generalisation measure. A process discovery algorithm is applied to L_d and a process model is obtained. First, simplicity is measured by taking the number of arcs, places and transitions in a Petri net (process trees and BPMN models are translated to Petri nets first for a sensible comparison). Second, weak soundness and boundedness are determined, as if a model is unbounded or weak sound, the PCC framework cannot handle it. Third, if the model is weak sound and bounded, then fitness and log precision are measured using the PCC framework with $k = 2$. Fitness is measured on the test log, while log precision is measured on the full unsplit log. Figure 8.2 illustrates this procedure, which is repeated for all combinations of discovery and test logs. Average scores of the measures are recorded and reported.

The above procedure is applied with each algorithm that was mentioned in Section 8.1 except CCM and MPM, for which we had no implementations available.

As discussed in Section 3.2.3, all three measures, i.e. fitness, log precision and simplicity, might be important in particular use cases of process mining. Therefore, we summarise the findings using pareto optimality: a result a *dominates* another result b if *all* measures of a are equal or better than the measures of b [8]. A result a is *pareto optimal* in a set of results A if it dominates all other measures in A. In our case, both the average fitness and log precision need to be equal or higher, and the average simplicity needs to be equal or lower for a model to be dominating.

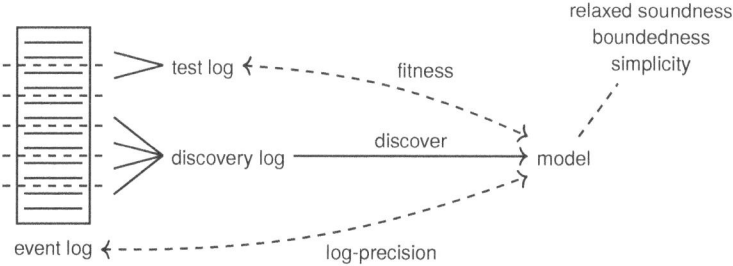

Fig. 8.2: Set-up of the log-quality experiment.

8.3.2.2 Checking Weak Soundness & Boundedness

In order to verify weak soundness and boundedness of models, as a first step the model was reduced using the Murata rules [31], similar to [16]. Second, weak soundness is checked by verifying that the final marking, i.e. a token in one of the places of the final marking, and no tokens elsewhere, is reachable. Boundedness is computed by verifying for each place that there is no reachable state with an unbounded number of tokens in that place. Both properties are checked using the Lola tool [39]. Some of the discovered models required Lola to traverse more than 200,000,000 markings and use more than 25GB of RAM, which would make performing any further computations on such a model infeasible. Therefore, even though such models might be bounded in theory, we consider them to be unbounded in practise, and we excluded them.

Models might be unbounded or weak unsound for several reasons: (1) the algorithm did not return a model within a reasonable time. For instance, IMc and IMcD are exponential in the number of activities and ran for over a week on the discovery logs with 133 activities or more. After the first failure of these algorithms on an event log, we sampled the remaining discovery logs: none of these samples succeeded and we did not perform the remaining runs. In this case, no simplicity could be reported and the model was left out. (2) the algorithm ran out of memory (we had 40GB of RAM available). This happened only for SM. In this case, no simplicity could be reported and the model was left out. (3) the algorithm produced models for which the Lola tool could not verify boundedness or weak soundness in at most 200,000,000 states. (4) the algorithm produced weak unsound or unbounded models. If any failure occurs, then the reported fitness and log precision are averages taken over a subset of the folds, which makes these measurements incomparable to measurements over all folds. Therefore, such measures have been ~~striked out~~ in the results.

8.3.2.3 Results

Tables 8.3 and 8.3 show the results: for each log and algorithm, it shows the average fitness (f), log precision (p), simplicity (s) and fraction of bounded weak-sound models (b) over the 15 folds. The pareto-optimal measures have been printed in a bold font. Notice that the measures have been rounded, but pareto optimality was computed before rounding.

The following table lists the number of times each algorithm was pareto optimal:

	pareto-optimal on event logs
FM, IMD	9
ETM, IMFD, IMA	8
IM	7
TM	5
IMF, IMLC	3
IMC, IMFA, SM	2
IMCD	1
α, FO, HM, ILP, IMFLC, Tα	0

8.3.2.4 Discussion

We first discuss boundedness & weak soundness, after which we discuss the baselines provided by this experiment. We finish the discussion with the performance of algorithms on the quality measures fitness, log precision and simplicity.

Boundedness & Weak Soundness.

In terms of returning a model, ETM, FM, FO, IM, IMA, IMD, IMF, IMFA, IMFD, IMFLC, IMLC and TM returned models for all event logs. For BP11, WA1, WA2, WA3, WA4 and WA5, IMC and IMCD ran out of time and did not return models, due

Table 8.3: Results of the log quality experiments, averaged over folds. f is fitness, p is log precision, s is simplicity and b is boundedness & weak soundness. Pareto-optimal results are shown in **bold**.

	BP12				BP11				RPW			
	f	p	s	b	f	p	s	b	f	p	s	b
α			70.0	0.0				0.0			195.3	0.0
ETM	**0.36**	**0.72**	92.6	1.0	**0.86**	**1.00**	13.8	1.0	**0.95**	**0.98**	59.1	1.0
FM	**1.00**	**0.61**	81.0	1.0	**1.00**	**0.65**	1672.4	1.0	**1.00**	**0.59**	88.6	1.0
FO			321.5	0.0			4576.3	0.0	0.73	0.60	156.0	1.0
HM			365.8	0.0				0.0			449.3	0.0
ILP			52.4	0.0				0.0			135.3	0.0
IM	**1.00**	**0.67**	166.7	1.0	1.00	0.65	1723.7	1.0	**1.00**	**0.63**	155.1	1.0
IM$_A$	**1.00**	**0.67**	166.7	1.0	1.00	0.65	1725.6	1.0	1.00	0.63	155.7	1.0
IM$_C$	1.00	0.67	313.0	1.0				0.0	**1.00**	**0.63**	290.1	1.0
IM$_{CD}$	0.99	0.67	401.7	1.0				0.0	1.00	0.63	238.5	1.0
IM$_D$	**1.00**	**0.67**	90.3	1.0	**1.00**	**0.65**	1681.8	1.0	**1.00**	**0.63**	101.4	1.0
IM$_F$	**0.96**	**0.69**	187.5	1.0	**0.98**	**0.84**	1038.7	1.0	1.00	0.69	163.7	1.0
IM$_{FA}$	**0.96**	**0.69**	195.8	1.0	**0.98**	**0.84**	1052.1	1.0	1.00	0.69	183.5	1.0
IM$_{FD}$	**1.00**	**0.67**	90.3	1.0	**1.00**	**0.79**	2402.2	1.0	**1.00**	**0.70**	118.7	1.0
IM$_{FLC}$	1.00	0.68	90.6	1.0	**0.98**	**0.84**	1038.7	1.0	1.00	0.69	164.5	1.0
IM$_{LC}$	**1.00**	**0.68**	80.0	1.0	1.00	0.65	1725.6	1.0	1.00	0.63	155.7	1.0
SM			672.0	0.0				0.0	0.97	0.71	214.9	0.2
Tα			173.9	0.0			611.8	0.0			42.1	0.0
TM	**0.98**	**0.93**	55654.1	1.0	**0.93**	**1.00**	5159.2	1.0	**0.99**	**0.95**	3579.6	1.0

	RF				WA1				WA2			
	f	p	s	b	f	p	s	b	f	p	s	b
α			60.0	0.0			1118.6	0.0			1383.3	0.0
ETM	0.85	0.90	115.4	1.0	**0.67**	**0.99**	59.5	1.0	**0.63**	**0.99**	72.5	1.0
FM	**1.00**	**0.59**	42.0	1.0	**1.00**	**0.68**	485.4	1.0	**1.00**	**0.66**	461.0	1.0
FO	1.00	0.69	146.4	1.0			857.2	0.0			844.9	0.0
HM	0.48	0.96	186.4	0.3			1623.3	0.0			1417.9	0.0
ILP			16.0	0.0				0.0			15815.5	0.0
IM	1.00	0.66	89.5	1.0	**1.00**	**0.70**	661.2	1.0	**1.00**	**0.67**	761.1	1.0
IM$_A$	1.00	0.66	162.4	1.0	1.00	0.70	671.4	1.0	1.00	0.67	767.7	1.0
IM$_C$	**1.00**	**0.93**	132.3	1.0				0.0				0.0
IM$_{CD}$	**0.76**	**0.93**	108.1	1.0				0.0				0.0
IM$_D$	**1.00**	**0.66**	50.0	1.0	**1.00**	**0.69**	532.1	1.0	**1.00**	**0.66**	483.1	1.0
IM$_F$	**0.99**	**0.92**	109.3	1.0	0.99	0.78	778.1	1.0	0.99	0.72	747.9	1.0
IM$_{FA}$	0.99	0.91	187.3	1.0	0.99	0.78	812.5	1.0	0.99	0.72	767.5	1.0
IM$_{FD}$	0.83	0.89	123.7	1.0	**1.00**	**0.79**	767.8	1.0	**1.00**	**0.76**	629.9	1.0
IM$_{FLC}$	0.99	0.92	192.5	1.0	0.99	0.78	778.8	1.0	0.99	0.72	748.3	1.0
IM$_{LC}$	1.00	0.66	122.5	1.0	**1.00**	**0.70**	661.2	1.0	**1.00**	**0.67**	761.1	1.0
SM				0.0	0.98	0.85	1546.5	0.5	0.99	0.76	781.5	1.0
Tα				0.0			203.1	0.0			184.9	0.0
TM	1.00	0.90	303056.4	1.0	**0.93**	**0.99**	2074.5	1.0	**0.88**	**0.99**	1326.3	1.0

Table 8.3: Results of the log quality experiments, averaged over folds. f is fitness, p is log precision, s is simplicity and b is boundedness & weak soundness. Pareto-optimal results are shown in **bold**.

	WA3				WA4				WA5			
	f	p	s	b	f	p	s	b	f	p	s	b
α			406.7	0.0			823.1	0.0			1601.9	0.0
ETM	**0.69**	**0.99**	**68.3**	1.0	**0.65**	**0.98**	**73.9**	1.0	**0.69**	**0.99**	**59.0**	1.0
FM	**1.00**	**0.67**	**498.8**	1.0	**1.00**	**0.66**	**395.4**	1.0	**1.00**	**0.67**	**498.2**	1.0
FO	~~1.00~~	~~0.78~~	973.6	0.1	~~0.38~~	~~0.53~~	699.3	0.5	~~1.00~~	~~0.82~~	857.1	0.5
HM			1670.3	0.0			1252.1	0.0			1546.2	0.0
ILP			16456.4	0.0			8789.6	0.0			26975.2	0.0
IM	**1.00**	**0.69**	1034.5	1.0	**1.00**	**0.68**	630.3	1.0	**1.00**	**0.68**	949.5	1.0
IM$_A$	1.00	0.69	1048.6	1.0	1.00	0.68	643.7	1.0	1.00	0.68	960.3	1.0
IM$_C$				0.0				0.0				0.0
IM$_{CD}$				0.0				0.0				0.0
IM$_D$	**1.00**	**0.69**	**606.9**	1.0	**1.00**	**0.68**	**447.1**	1.0	**1.00**	**0.68**	**531.5**	1.0
IM$_F$	**0.99**	**0.78**	**791.3**	1.0	0.99	0.76	682.9	1.0	0.99	0.82	844.5	1.0
IM$_{FA}$	**0.99**	**0.78**	**854.5**	1.0	0.99	0.76	691.2	1.0	0.99	0.82	852.0	1.0
IM$_{FD}$	**1.00**	**0.77**	1320.0	1.0	**1.00**	**0.78**	589.0	1.0	**1.00**	**0.83**	836.6	1.0
IM$_{FLC}$	0.99	0.78	794.5	1.0	0.99	0.76	683.1	1.0	0.99	0.82	847.4	1.0
IM$_{LC}$	1.00	0.69	1036.4	1.0	1.00	0.68	637.9	1.0	1.00	0.68	960.3	1.0
SM	~~0.99~~	~~0.78~~	1369.2	0.8	**1.00**	**0.82**	**901.6**	1.0	~~1.00~~	~~0.89~~	1397.6	0.9
Tα			206.5	0.0			163.2	0.0			197.3	0.0
TM	**0.94**	**0.99**	2395.2	1.0	**0.91**	**0.99**	1382.5	1.0	**0.94**	**0.99**	2078.8	1.0

to their exponential nature. α and HM did not return models for BP11, and ILP did not return all models for BP11 and WA1. SM did not return models for BP11 and RF. Tα did not return models for WA2 and RF. The inability of SM and Tα to discover models for RF is remarkable. Even though RF contains over 500,000 events, it only contains 11 activities, and the most time-consuming step of SM (after applying HM, which succeeded for some logs as shown by this experiment) is the structuring step, which only depends on the number of activities, and in contrast to Tα, α did return models.

In terms of boundedness and weak soundness, all models returned by ETM, FM, TM and the algorithms of the IM framework and IM$_D$ framework were bounded and weakly sound. HM, FO, and SM returned some bounded and weakly sound models: HM only for 5 logs of RF. FO returned bounded and weakly sound models for all logs of RPW, RF, 2 logs of WA3 and 8 logs of WA4 and WA5. SM returned bounded and weakly sound models for 3 logs of RPW, 8 logs of WA1, all logs of WA2, 12 logs of WA3, all logs of WA4 and 13 logs of WA5. These algorithms seem to be sensitive to variations in event logs, even if these logs are derived from the same system. The algorithms α, ILP and Tα did not return any bounded weakly sound model.

As discussed earlier, in the remainder of this discussion we will not consider incomplete measures, i.e. we will only consider combinations of algorithms and logs for which the algorithm discovered bounded weakly sound models for all discovery logs.

Baselines.

We first discuss these baseline algorithms. Due to the discovery and test log set-up, TM did not necessary reach perfect fitness. In our experiment, for only one log (RF), TM achieved perfect fitness, which indicates that the discovery and test log division made sense: the test logs contained behaviour different from the behaviour of the discovery logs. Thus, the cross validation and the measure on the trace miner (TM) provide insight into the completeness and *variety* of the event log: if the fitness of TM is high, the average test log contained many traces similar to traces in its corresponding discovery log, thus the variety in the event log was low. For instance, TM got low fitness for BP11 (0.40) and WA1-5 (0.56-0.70), hence it is to be expected that future behaviour has a low likelihood to resemble behaviour already seen in the event log, and variety is high. Given this insight, one might reconsider opting for a process model with a high log precision (and a bit lower fitness) in use cases that require a good generalisation, like prediction: on such logs, a high precision and low fitness will likely limit support of future behaviour.

This conclusion is strengthened by the fitness of FM (the all-behaviour-allowing flower model) for WA1, WA2 and WA5. On these logs, FM did not achieve perfect fitness, which implies that in some cases, not all activities were present in the discovery logs, which suggests that the log contains far from all behaviour.

In contrast, for RPW and RF, the TM fitness is 0.97 and 1 respectively, which is rather high and indicates that these log contain repetitive behaviour. Therefore, a high log precision would be desirable, as it is likely that such a model will represent all future behaviour. For RPW, one could even consider using the trace model itself, although that model contains 30 times more constructs than the models returned by all other algorithms.

The set-up of this experiment resembles a set-up commonly used in process discovery literature, for instance [9, 21, 22, 26, 33]. Often, these experiments measure fitness, precision, generalisation and simplicity on an event log. Generalisation "assesses the extent to which the resulting model will be able to reproduce future behavior of the process" [10]. By the separation of discovery and test data, our fitness measure already serves this role, thus we did not measure generalisation separately (as in [26]).

Fitness, Log Precision & Simplicity.

In terms of fitness, log precision and simplicity, algorithms need to strike a balance between FM (the all-behaviour-allowing flower model) with perfect fitness, low precision and high simplicity, and TM (the trace model) with perfect precision but high complexity.

Earlier in this section, we showed a frequency table that denotes how often each algorithm achieved pareto optimality. FM returned pareto-optimal models for all event logs, however the models returned by FM are inherently useless as they do not contain more information than which activities were contained in the discovery logs. IMD is also pareto optimal for all event logs, slightly improving over FM (up to 0.07 for RF) in terms of log precision, at the cost of simplicity (up to 108 extra constructs for WA3). IMD does not apply any infrequent behaviour filtering, and on the real-life event logs of this experiment, discovers models with large flower parts.

The algorithms ETM and IMFD were both pareto optimal for 8 event logs. For all these event logs, ETM used few Petri net elements), in some cases even less than FM, e.g. for RPW and WA1 - WA5, which implies that not all activities were present in the discovered models. While this might be a deliberate strategy, for WA1 - WA5, fitness was remarkably lower than other algorithms: less than 0.7, while all other non-baseline pareto-optimal algorithms got more than 0.9. Nevertheless, log precision was high: equal up to or exceeding the log precision of TM. IMFD is the variant of IMD that filters infrequent behaviour, which results in an improvement over IMD of 0.1 in log precision for BP11, RF, WA1, WA2, WA4 and WA5, at the expense of simplicity (up to 721 extra Petri net elements for BP11).

IM and IMA achieved pareto-optimality for 7 and 8 event logs, and both algorithms had almost equal fitness and log-precision characteristics, and IMA needing less Petri net constructs for all logs except RF. Their fitness and log-precision values also resembled IMD.

The baseline TM lists all traces of the discovery log, and discovered the most complex models, i.e. with the most Petri net constructs. Only for WA2 - WA5, ILP discovered more complex models. For 5 logs, TM achieved pareto optimality, due to the high log precision of its models. However, especially for the WA1 - WA5 logs, fitness was low, as TM does not attempt to generalise over the behaviour seen in the event log, which illustrates the need for discovery algorithms to generalise over the behaviour in the event log. This also illustrates the need for a generalisation measure: without cross validation, TM would have achieved perfect fitness and log precision for all event logs.

IMF and IMFA achieved similar fitness and log-precision values, with IMF discovering models with a lower complexity for all event logs. These algorithms achieved a higher log precision than IMFD for 2 event logs (BP12 and RF), and a higher fitness only for RF. IMC was pareto optimal for RPW and RF, with IMC achieving a perfect fitness and log precision of 0.93 for RF, which is even higher than TM. This illustrates the potential of the incompleteness handling despite exponential run times of this algorithm.

Of the remaining algorithms, SM was pareto optimal for all logs on which it completed the cross validation, i.e. WA2 and WA4, equalling (WA2) or outperforming (WA4) IMFD on log precision, equalling on fitness but being more complex for both event logs, which shows the potential of this algorithm. However, for two logs SM did not return any models, and for 5 logs the models were either unbounded or not weakly sound.

Conclusion.

In answer of RQ.2, FM and IM$_D$ were pareto optimal for 9 logs, ETM and IM$_{FD}$ for 8 logs, IM and IM$_A$ for 7 logs, TM for 5 logs, IM$_F$ and IM$_{LC}$ for 3 logs, IM$_C$, IM$_{FA}$ and SM for 2 logs, and IM$_{CD}$ for 1 log. The other algorithms did achieve pareto optimality for any log. For almost all event logs that were included in this experiment, the algorithms discovered different models, such that a human analyst could try several algorithms to see which model suits the use case at hand best. In the next section, we will study the discovered models qualitatively to provide more insight into the discovery algorithms.

8.3.3 Qualitative

Thus far we evaluated scalability and model quality in terms of fitness, log precision and simplicity. Now we evaluate the models qualitatively, i.e. we compare the models that are discovered by discovery algorithms on their visual appearance and their structure. As measures of log conformance express only how well the event log is represented and are only obtainable for bounded and weakly sound models, we do not use such measures in this experiment. Rather, we aim to illustrate the variety of process models that can be discovered from event logs and the quirks of discovery algorithms. Therefore, we apply the discovery algorithms to the event logs described in Section 8.3.1, and analyse the models manually.

We translate every model to a Petri net and visualise this Petri net using GraphViz, replacing the activity names by letters to save space. We use the Dot layout engine, however if this leads to unreadably small graphs, in which case we use the Sfdp or Neato engine. Some models by Tα(RF) could not be visualised. An exception to this are the models discovered by FD and CPM, which cannot be translated to Petri nets.

On these visualised Petri nets, we assess whether the model possess an unclear language. That is, whether the model, despite its Petri net semantics, does not give clear indications about their final markings, and/or allow for many executions of transitions without possibility to end in a final marking, which makes it harder for users to understand the model. Furthermore, we assess fitness and log precision manually (as automated measures are not available for weakly unsound or unbounded models), using the limitations the model puts on its behaviour. Finally, for one of the logs (RF) we assess the discovered models using information derived from the event log.

Notice that for this experiment, the authors of the CCM algorithm kindly provided us with the models discovered by the CCM algorithm. The only algorithm not included in this experiment is MPM, as its source code is not public and its authors chose not to participate in our experiment. Furthermore, the baseline algorithms FM and TM are not included in this experiment as the models they produce do not provide any insight into the process.

We limit this experiment to the two smallest event logs, BP12 and RF, as for the other event logs several issues challenged inclusion: not all algorithms succeeded in discovering a model, not all models were correctly visualised by the graph layout algorithm (GraphViz), and manual analysis on models containing over 100 activities is rather error-prone.

8.3.3.1 BP12

We start with BP12, the log of mortgage application process with 13,000 traces, 262,000 events and 24 activities. The activity names are replaced as follows:

a A_ACCEPTED	b A_ACTIVATED
c A_APPROVED	d A_CANCELLED
e A_DECLINED	f A_FINALIZED
g A_PARTLYSUBMITTED	h A_PREACCEPTED
i A_REGISTERED	j A_SUBMITTED
k O_ACCEPTED	l O_CANCELLED
m O_CREATED	n O_DECLINED
o O_SELECTED	p O_SENT
q O_SENT_BACK	r W_Afhandelen leads
y W_Afhandelen leads+SCHEDULE	s W_Beoordelen fraude
z W_Beoordelen fraude+SCHEDULE	t W_Completeren aanvraag
aa W_Completeren aanvraag+SCHEDULE	u W_Nabellen incomplete dossiers
ab W_Nabellen incomplete dossiers+SCHEDULE	v W_Nabellen offertes
ac W_Nabellen offertes+SCHEDULE	w W_Valideren aanvraag
ad W_Valideren aanvraag+SCHEDULE	x W_Wijzigen contractgegevens
ae W_Wijzigen contractgegevens+SCHEDULE	

Figures 8.3 to 8.18 show the results. We conclude that the BP12 log is rather challenging for current discovery algorithms: non-block structured algorithms return models and assessing their behaviour might be difficult (α, HM, FO), which despite their Petri net semantics do not give clear indications about their final markings, and/or allow for many executions of transitions without possibility to end in a

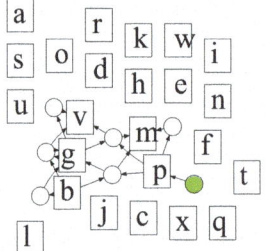

Fig. 8.3: ILP(BP12). The model returned by ILP expresses constraints for only 5 of the 24 activities. This part is not bounded however it is weakly sound, considering that the final marking is the empty marking. The other activities are not connected, i.e. have no input or output places, thus can be executed at any time.

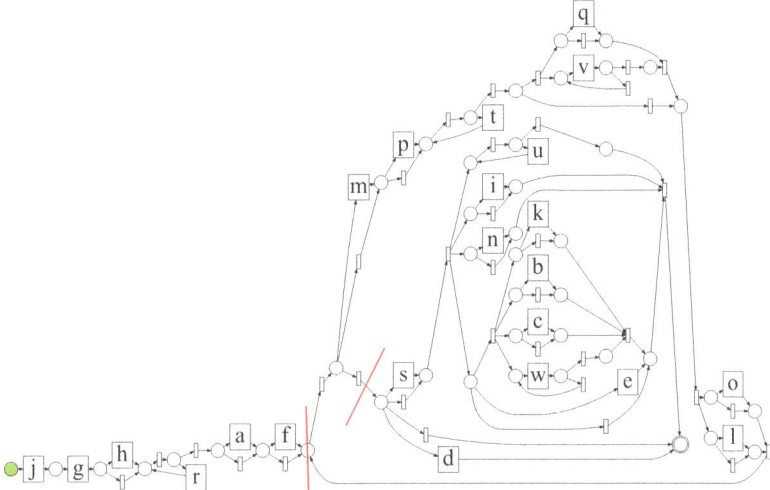

Fig. 8.4: CCM(BP12). The model returned by CCM is sound and consists of three sequential parts: the first part consists of 6 sequential activities that can each be skipped or repeated, which expresses a rather lot of structure. The second sequential part expresses little and is language equivalent to a flower model, except for a dependency between the topmost two activities. The third sequential part consists of a rather structured mix of choice, sequence and concurrency. There is 1 activity missing, and the 35 silent transitions in loop structures make the model difficult to understand.

final marking) or models that restrict behaviour little (ILP). Block-structured based algorithms improve over these algorithms by offering clear languages at all times. However, for BP12, block-structured based algorithms struggle to capture constraints on behaviour between activities and return flower-like models.

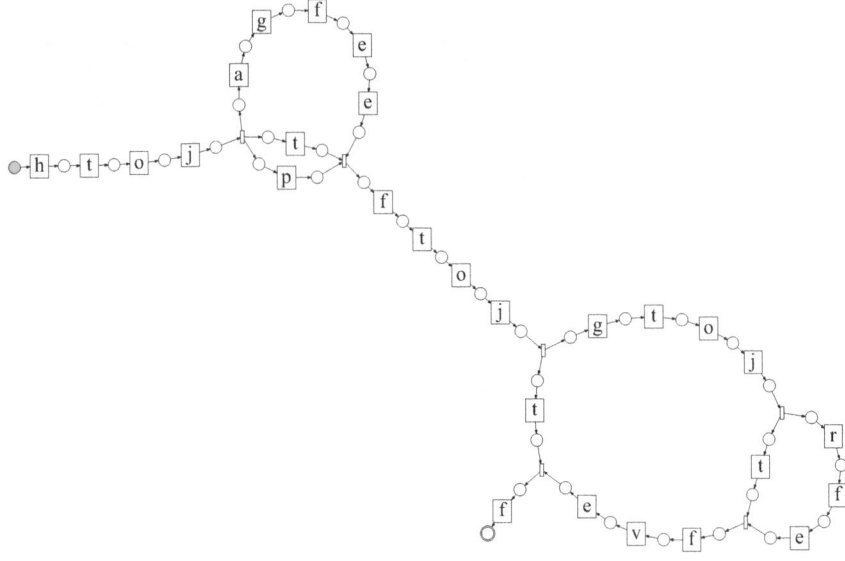

Fig. 8.5: ETM(BP12). The model returned by ETM consists of 28 transitions, but not all activities of the event log are present. This does not necessarily decrease model usability: an easy to understand model that does not contain a few infrequently executed activities might be preferable over a complex model that includes these activities. However, this depends on the use case of the analysis and analysts should be aware of this. The model is highly structured and sequential, i.e. its intention and behaviour are clearly visible. However, the model does not contain choices: all 28 transitions have to be executed precisely once (the only freedom in execution is that there is some concurrency), even though the number of events per trace in the model ranged from 3 to 175 with an average of 20, and which seems in contrast with the models discovered by all other algorithms.

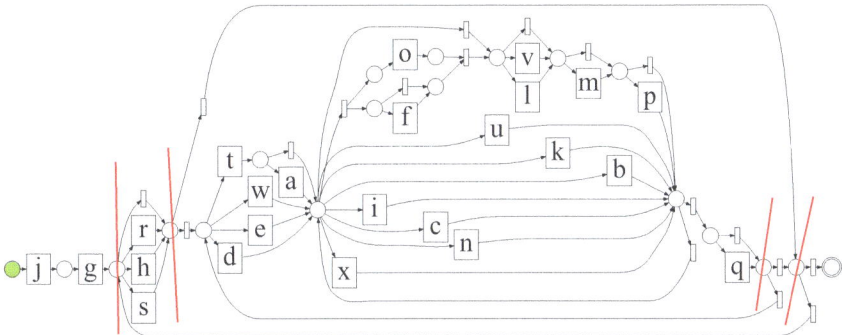

Fig. 8.6: IM(BP12) = IMₐ(BP12). The model returned by both IM and IMₐ contains all 24 activities and consists of several nested loops. The outer loop allows for any behaviour of 3 activities. The inner loop consists of 5 activities in sequence with a near-flower model and another activity. The behaviour of the outer loops are easily recognisable, however the inner flower model is hidden by a chain of 4 silent transitions.

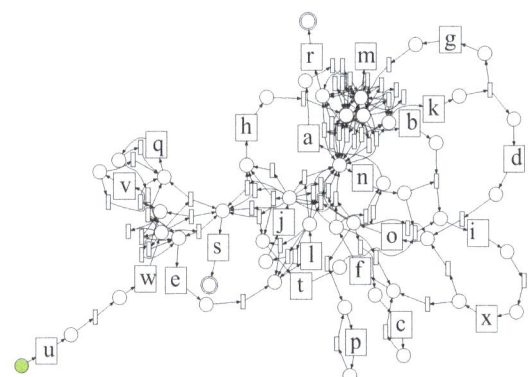

Fig. 8.7: HM(BP12). The model returned by HM contains all 24 activities, is not bounded and not weakly sound. Even though the behaviour of activities is seemingly restricted, it is difficult to truly grasp the behaviour of the model, due the to many groups of silent activities that enable several combinations of subsequent branches (this is not due to the layout). These branches again divert and enable subsequent branches, which introduces a lot of concurrency.

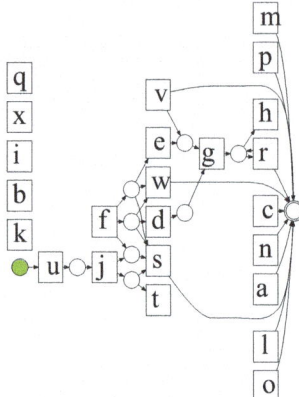

Fig. 8.8: α(BP12). The model returned by α has 5 unconnected activities, which means that the model poses no restrictions on these activities. Furthermore, the model contains many activities that only have output arcs, which can be arbitrarily executed as well, but produce tokens nevertheless. These token generators impose only weak restrictions on subsequent activities, as these become concurrent with the presence of multiple tokens in the net.

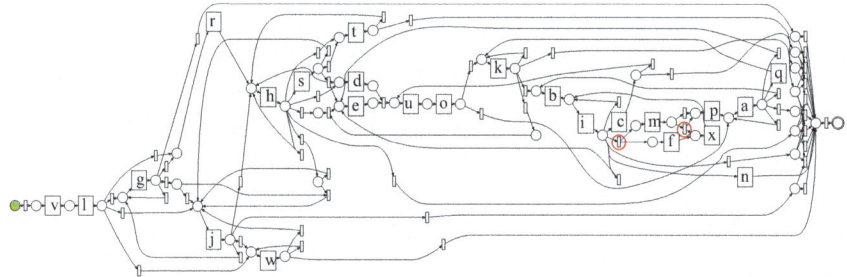

Fig. 8.9: FO(BP12). The model returned by FO contains all 24 activities and many silent activities. Most of the transitions in the net have a single input and output place, thus the net seems to have little concurrency at first sight. However, in the right half of the model, there are two silent transitions that have two output arcs. These transitions make the net unbounded. Nevertheless, the net is weakly sound and large parts are state machines, i.e. without concurrency, which makes its behaviour easily recognisable.

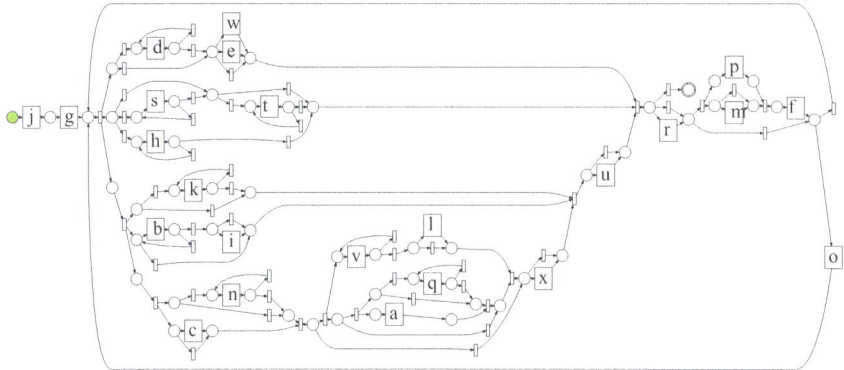

Fig. 8.10: IMc(BP12). The model returned by IMc contains all 24 activities and seems rather structured, containing structured loops, concurrency, sequence and choice, thus analysts might be able to gain insights from this model by manual inspection. However, due to the many silent transitions, many traces are possible, i.e. the model would have a low log precision (measured by the PCC framework: fitness is 1 and log precision is 0.67). Nevertheless, the model contains many inter-activity constraints, e.g. A_REGISTERED can only be executed after execution of W_Completeren aanvraag.

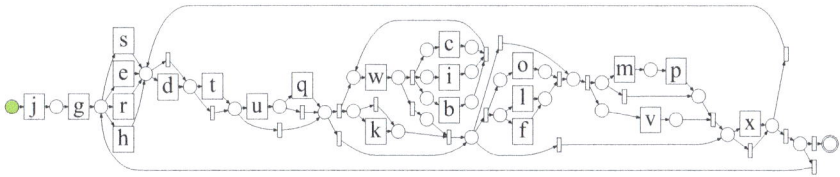

Fig. 8.11: IMFA(BP12). The model returned by IMFA is similar to the model returned by IMF, however poses a little more constraints as instead of 4 concurrent activities, expresses 3 concurrent activities in loop with another activity.

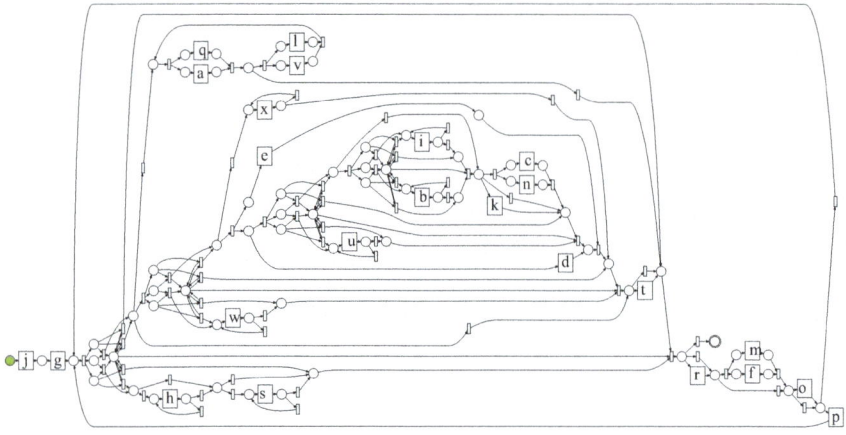

Fig. 8.12: IMcd(BP12). The model returned by IMcd contains all 24 activities. The process tree underlying the Petri net shown contains inclusive choice operators, which are translated using rather many silent activities to Petri nets, and which make the Petri net rather complicated. However, being aware of the underlying structure of these constructs, the model is fairly structured and contains many dependencies between activities.

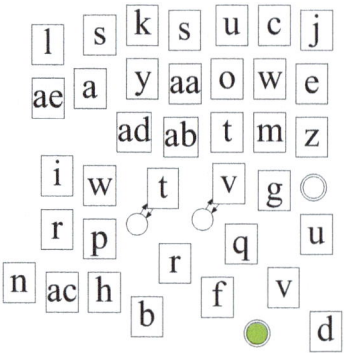

Fig. 8.13: SM(BP12). The model returned by SM contains little information, as no two tasks are connected. Moreover, the bottom-two activities cannot be executed (W_Completeren aanvraag and W_Nabellen offertes). As this model is different as the technique described in [5] would suggest, we suspect an inaccuracy in the implementation.

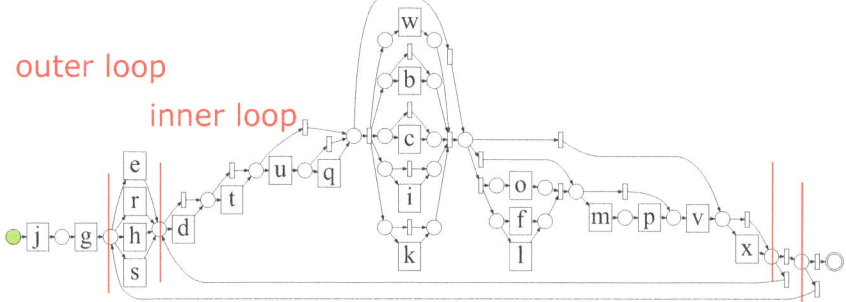

Fig. 8.14: IM_F(BP12). The model returned by IM_F contains 22 of the 24 activities. Depending on the use case, this might be desirable or problematic. The model consists of two nested loops, of which the outer one expresses that one of four activities needs to be executed before the remainder of the process (in subsequent iterations, all four activities can be executed arbitrarily). The inner loop structure consists of three parts in sequence, which can all be skipped, and due to the loop structure around it, arbitrarily executed. However, within the sequential parts lots of structure is expressed, e.g. a concurrency between 4 activities.

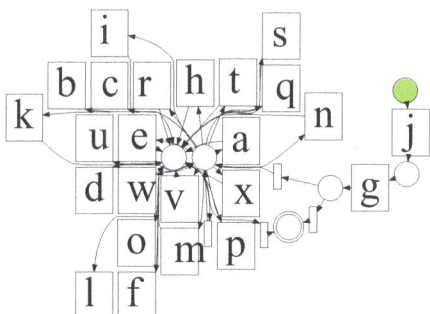

Fig. 8.15: IM_D(BP12) = IM_FD(BP12) = IM_FLC(BP12) = IM_LC(BP12). The model returned by IM_D, IM_FD and IM_FLC is almost a complete flower model, as only two activities are not arbitrarily executable.

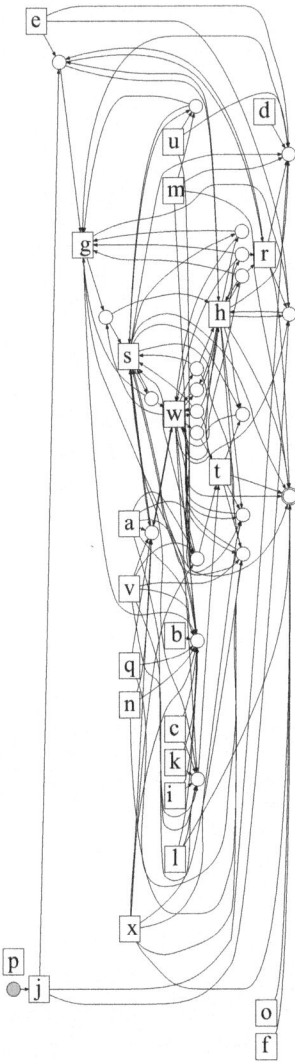

Fig. 8.16: Tα(BP12). Similar to the model returned by α, in the model returned by Tα many activities can be executed without restrictions. We argue that this model has a limited use for both human and machine analysis due to it containing unbounded behaviour, containing little structure and not containing a final marking.

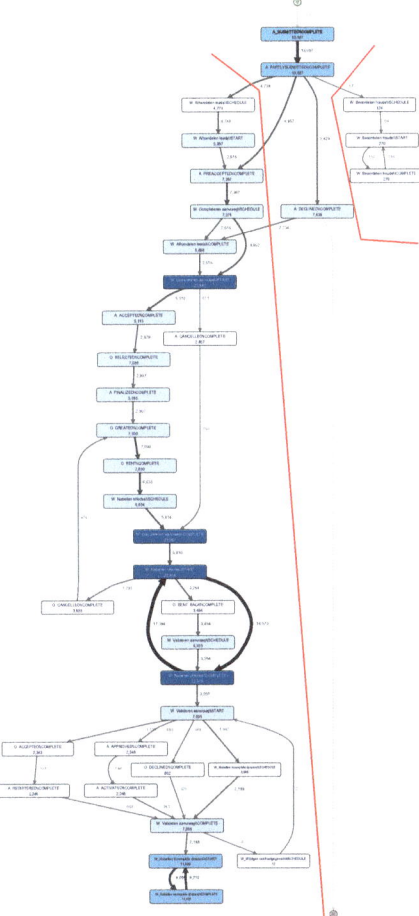

Fig. 8.17: FD(BP12). The model discovered by FD contains 36 activities as the start, completion and schedule life cycle transitions are included in the names of the activities. The model is highly understandable due its simplicity and lack of concurrency. This model also contains a soundness issue: only from the *three* activities in between the red lines, the end state is reachable.

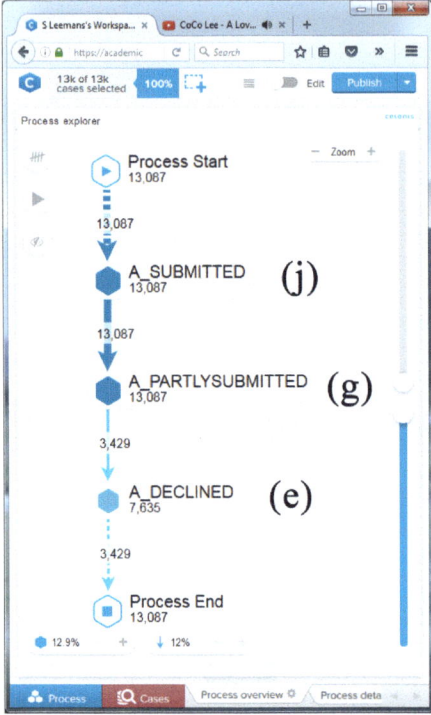

Fig. 8.18: CPM(BP12). Celonis discovers a model representing the most-occurring activities, and thereby this model represents the most-occurring trace, i.e. the trace which is rejected immediately. However, this leads to a model with only 3 activities.

8.3.3.2 RF

The event log RF was recorded in a road fines managing process and contains 150,000 traces, 5,600,000 events and 11 activities. The activity names are replaced as follows:

a Add penalty   b Appeal to Judge
c Create Fine   d Insert Date Appeal to Prefecture
e Insert Fine Notification   f Notify Result Appeal to Offender
g Payment   h Receive Result Appeal from Prefecture
i Send Appeal to Prefecture   j Send Fine
k Send for Credit Collection

Some domain knowledge about this process was obtained, which entails that the activity 'payment' ('g') may occur multiple times as people are allowed to pay their fines in chunks (in 7,500 cases, this happens), and payment can occur during a large part of the process, even before the fine is sent (we verified this in the event log). We focus our analysis on whether this knowledge is reflected in the model. In the models, the payment activity (g) are highlighted in blue, while loop patterns in which payment is involved are denoted in red. Furthermore, we assess the complexity of the models. Figures 8.19 to 8.32 show the discovered models.

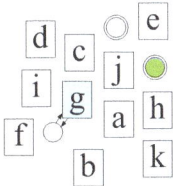

Fig. 8.19: SM(RF). SM returns a model without any useful connections or places. Payment cannot be executed.

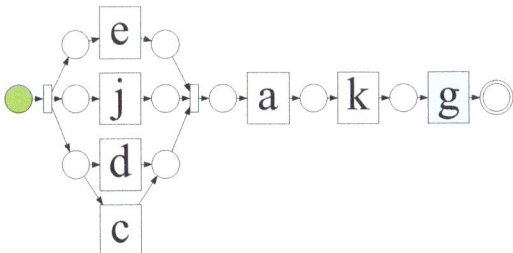

Fig. 8.20: ETM(RF). ETM returns a very simple model, which does not match intuition of a fine-handling process, e.g. in reality, not every case gets a penalty added. Furthermore, the model contains a choice between the creation of a fine and initiating an objection against a fine, which does not make much sense. This model is rather simple, as it contains only 7 of the 11 activities. We would argue that this is an oversimplification.

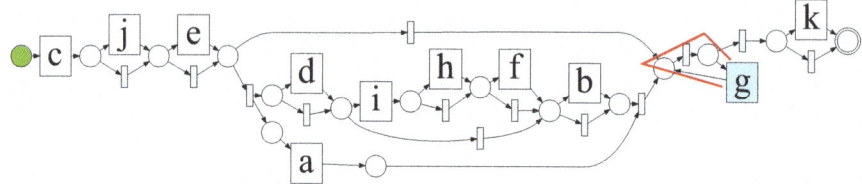

Fig. 8.21: CCM(RF). CCM returns a highly structured model that contains all activities, however without any loops, and payment only occurs at the end of the process. Even though this model contains 13 silent transitions, the block structure makes it easy to recognise the intention of the model: most of the silent transitions allow the skipping of parts of the model.

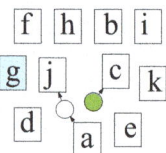

Fig. 8.22: ILP(RF). ILP returns a model without any useful connections or places.

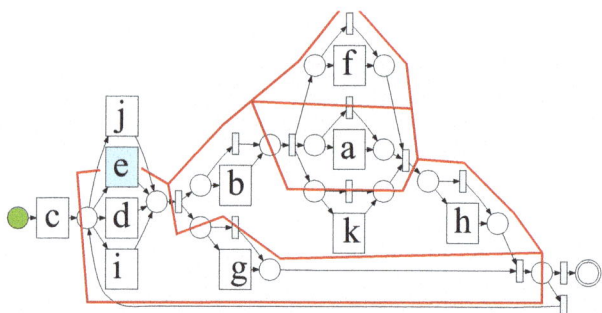

Fig. 8.23: IM(RF). IM returns a model with payment being part of a loop, however the loop spans all activities of the process except the first one, thereby lowering the log precision of the other activities. Furthermore, the loop obfuscates the behaviour of the model: the first part of the loop (including Payment) can be arbitrarily executed, but only by executing 11 silent transitions between each consecutive execution. It would have been better to put these four activities concurrent to the remaining part of the process.

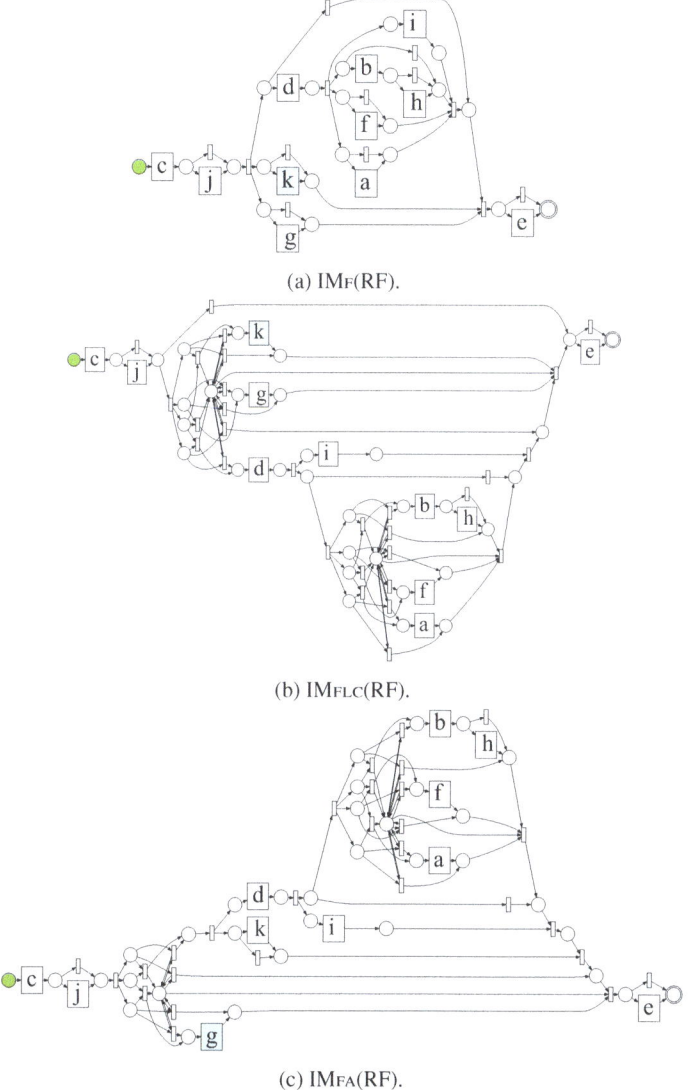

(a) IMF(RF).

(b) IMFLC(RF).

(c) IMFA(RF).

Fig. 8.24: IMF, IMFLC and IMFA put payment concurrently with the rest of the process, but not in a loop. The rather complicated inclusive choice structure, which was introduced by reduction rule C_V, is elegant in process trees, however rather many silent transitions are necessary in the corresponding Petri net representation. The model returned by IMF contains 13 silent transitions, however the block structure makes it easy to recognise the intention of the model.

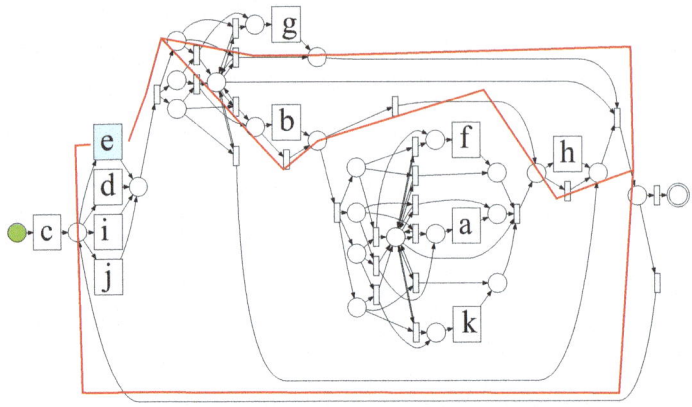

Fig. 8.25: IMA(RF). IMA discovers a model with several inclusive choice constructs, similar to IM with respect to payment, and similar in complexity.

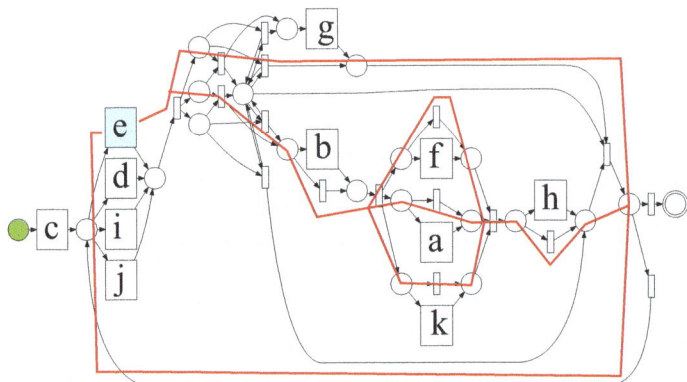

Fig. 8.26: IMLC(RF). IMLC discovers a model in which payment is part of a loop and can be executed throughout the process, however the loop spans all but one activities, which makes log precision lower. Complexity resembles the complexity of IM, however with an added inclusive choice construct.

We conclude that many discovery algorithms discovered models in which the log knowledge regarding payment is reflected, i.e. IM, IMA, IMC, IMCD, IMFD, IMLC and HM. For IM and IMLC, log precision could be higher as a lot of extra behaviour is included. Other algorithms discovered models reflecting our domain knowledge partially, i.e. CCM, IMFA, IMFLC and FO. The distinction between block-structured based algorithms and non-block-structured based algorithms can be made for the RF log as well: non-block-structured based models struggle to discover models with easy to spot behaviour, while block-structured based algorithms do not suffer from this. Exceptions are FO, which is simple due to the absence of concurrency, and IM which is not, due to the looping structure with many silent transitions. For block-structured

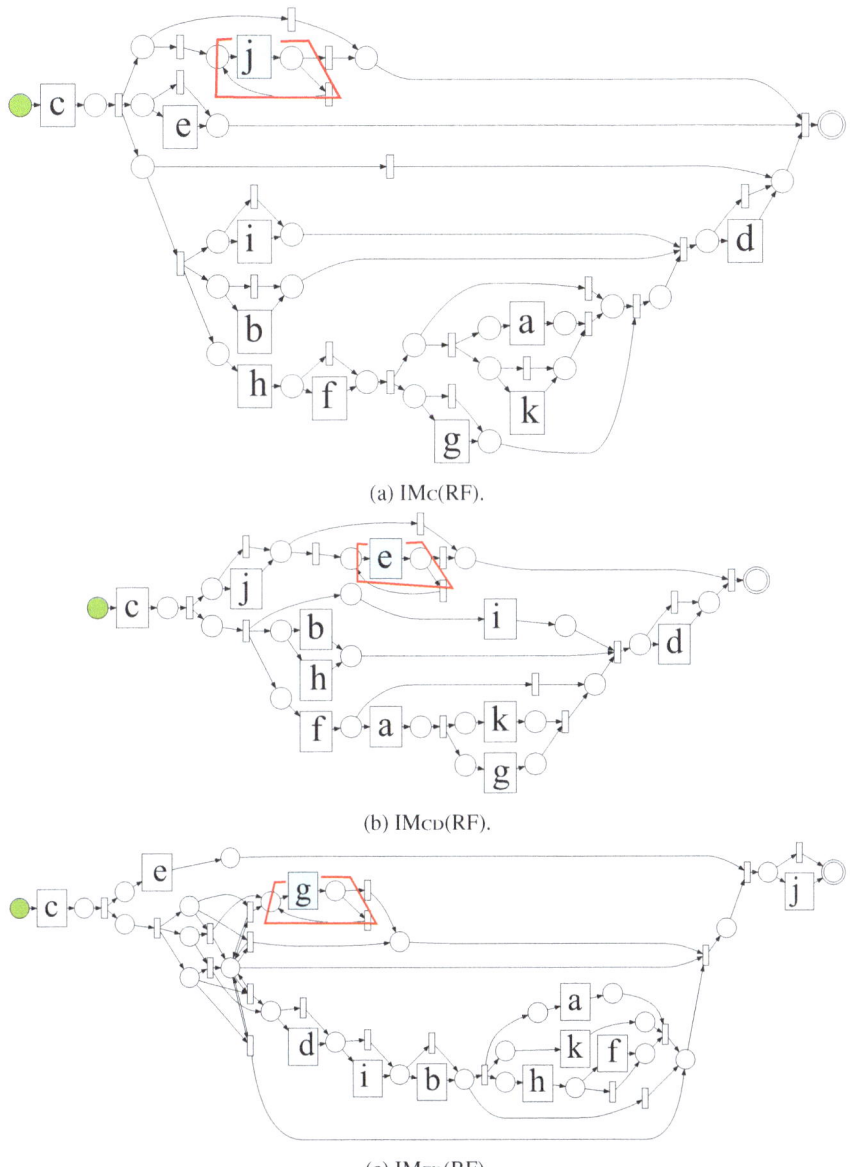

(a) IMc(RF).

(b) IMcD(RF).

(c) IMFD(RF).

Fig. 8.27: IMc, IMcD and IMFD discover a model in which payment can occur multiple times concurrently to the remaining parts of the process. IMcD contains the least silent transitions, followed by IMc and IMFD.

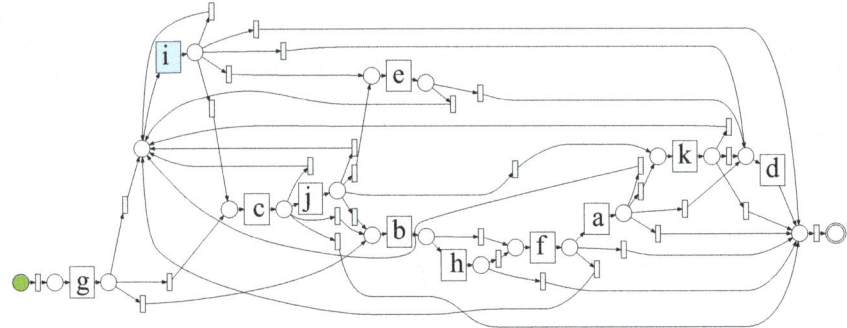

Fig. 8.28: FO(RF). FO discovers a model in which payment can be executed repeatedly and before/after many activities (i.e. concurrently), however this is modelled by loopbacks, which the side effects that at points in the model where such
a loopback is not present, payment cannot occur. Furthermore, as soon as payment
occurs, the 'state' of the remaining part of the model is lost and e.g. a new fine
notification can be initiated. This model, even though it is not block-structured and
has 31 silent transitions, is easy to grasp as it does not contain concurrency.

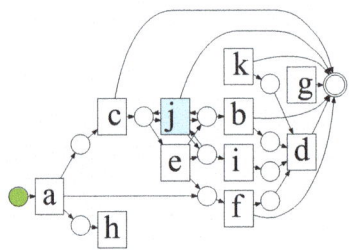

Fig. 8.29: α(RF). α discovers a model in which payment is part of a structural
loop, but nevertheless cannot be executed. The model seems simple, but grasping its
behaviour is challenging due to soundness issues.

models, simplicity is hindered by inclusive choice constructs, which are elegant in
process tree notation, but which need to be represented by several silent transitions
in a Petri net.

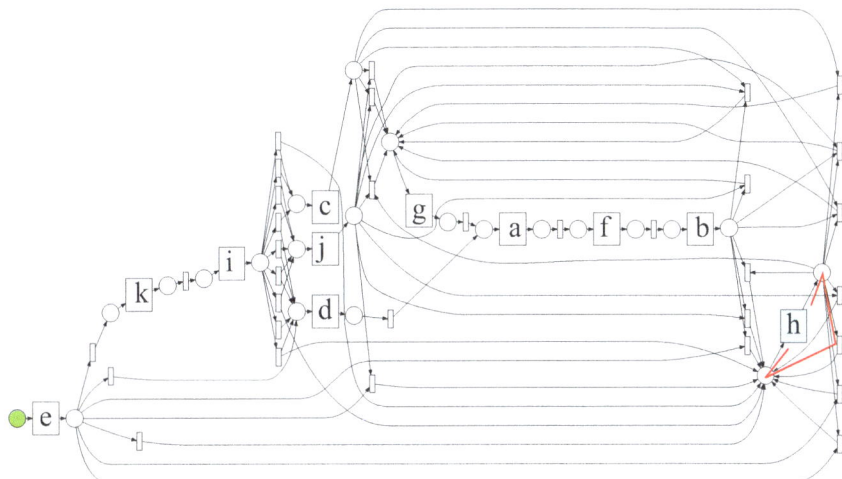

Fig. 8.30: HM(RF). HM discovers a model in which payment is part of a loop, and can be executed concurrently to some of the other activities. We consider this model more complex than the preceding models, due to the complex routing resulting from the 33 silent transitions: as the model is not block-structured and contains both dead silent transitions and silent transitions with duplicate effects, the behaviour and intention of the model are obfuscated.

Fig. 8.31: CPM(RF). CPM discovers an oversimplified model in which payment can be executed only once, in exclusive choice with sending the fine.

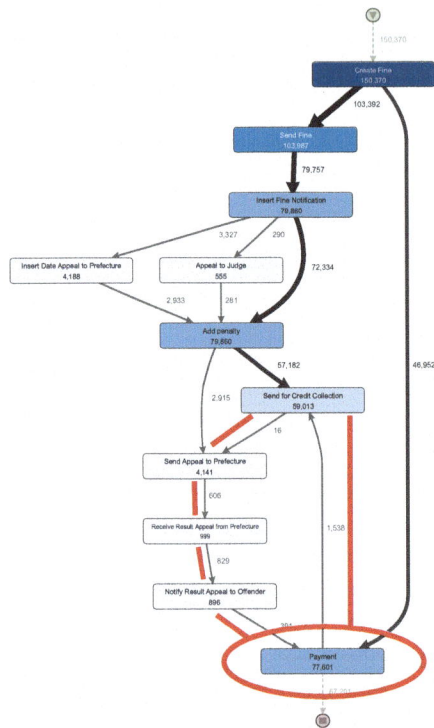

Fig. 8.32: FD(RF). FD discovers a model in which payment is part of a loop (FD models do not represent concurrency). Due to the absence of concurrency, the model seems easy to understand. However, conclusions should be drawn with care. For instance, the model suggests that after (partial) payment, no penalty can be added, as 'add penalty' is outside the loop of 'payment' denoted by red lines. However, in 4,112 of the 150,000 cases, penalties are added after payment. As conformance checking techniques cannot be applied, verifying the absence of this behaviour involves filtering the log and disable all other model filters.

8.3.4 Conclusion

In this section, we aimed to answer RQ.2, i.e. whether discovery algorithms are able to balance log-model quality criteria using their default parameters, and how discovery algorithms handle real-life event logs. We performed two experiments: a quantitative and a qualitative experiment.

In the quantitative experiment, we compared discovery algorithms using cross validation and measures fitness, log precision and simplicity. Of the 9 real-life event logs, the directly follows-based algorithm IMFD and IMD performed well, being two of the few algorithms that were pareto optimal for 9 and 8 logs, which might be due to these algorithms using *less* information of the event log, i.e. only the directly follows graph of the overall event log, than the algorithms of the IM framework. The best performing algorithms of the IM framework were IM and IMA, which were pareto optimal for 7 event logs. SM was pareto optimal for one log.

However, pareto optimality does not necessarily correspond to useful models. For instance, the trace model and flower model were pareto optimal in most cases as well, even though they do not provide any new information. Therefore, we conducted a second, qualitative, experiment, in which we applied the discovery algorithms to two of the nine event logs, and evaluated the discovered models manually. We found that on complicated real-life event logs, such as BP12, discovery algorithms struggle to discover process models that pose many restrictions on the behaviour that is expressed by the model, i.e. models with a high log precision, and are useful for human analysis. For instance, existing approaches that do not guarantee soundness, e.g. α, ILP, HM and SM, discover models that pose few constraints on their expressed behaviour (i.e. low log precision) and often lack easy to understand languages (i.e. not having a proper flow from initial to final marking, containing deadlocks, etc.). The algorithms of the IM framework and the IMD framework have difficulties with BP12 as well, however have the advantage of having well-defined and often easy to understand languages, i.e. clear initial and final markings, and being free of deadlocks. That is, some algorithms of these frameworks discover flower models, however every time that a flower model is avoided in recursion, the model restricts behaviour a little bit more (i.e. has a higher log precision), without losing the guarantees of a sound (and for some algorithms fitting) model. In contrast, non-soundness guaranteeing often result in unreadable unclear models in such cases.

The event log RF seems to be easier for discovery algorithms: more algorithms return models that restrict behaviour. The distinction between block-structured based algorithms and non-block-structured based algorithms can be made for this log as well: non-block-structured based models struggle to discover sensible models with easy to understand languages, while block-structured based algorithms do not suffer from this.

In both experiments, we found that some discovery algorithms did not return bounded and weakly sound models, or in some cases no models at all, e.g. the α algorithm did not return a single model that was bounded and weakly sound in the quantitative experiment, such that our measures could be applied.

In these experiments, we did not address human interaction with the algorithms, nor the influence of parameter settings on the discovered models. To evaluate how well algorithms are able to balance the log measures, a future experiment could include the algorithms at different parameter settings. Such an experiment would also provide insight in the variability in results depending on the parameter settings.

Future work 8.1: Perform experiment to investigate the influence of parameter settings on discovery algorithms.

In the next section, we will evaluate how discovery algorithms provide rediscoverability under log containing deviating, infrequent or incomplete behaviour.

8.4 Rediscoverability & its Challenges

A large part of this work addresses challenges of rediscoverability. That is, Chapter 4 introduced a formal framework for the influence of rediscoverability on abstractions, Chapter 5 studied these abstractions in more detail and in Chapter 6, we introduced discovery algorithms and showed their rediscoverability. However, all the given rediscoverability proofs assumed that the event log had the same abstraction as the system model that underlies the event log. For instance, for the basic IM, if the system is of C_B (i.e. consists of four basic operators and contains no duplicate activities, etc), and the event log fits the system has the same directly follows graph as the system, then IM will rediscover the system. Several factors might challenge these assumptions, as discussed in Section 3.4.2: the system might not be of C_B, and the event log might contain infrequent and deviating behaviour, or might not even be directly follows complete. In this section, we evaluate how these types of behaviour influence rediscovery (RQ.3).

In the previous section, we reported on experiments performed using single real-life logs with fixed characteristics whose underlying systems presumably are outside the class of C_B. In this section, we perform a more controlled experiment: we use multiple event logs generated from systems of C_B, in order to evaluate RQ.3. That is, we evaluate how discovery algorithms handle the following event log anomalies: incomplete behaviour, i.e. absence of information in the event log, deviating behaviour, i.e. presence of behaviour that is not described by the system, and infrequent behaviour, i.e. little-used parts of the system. We take systems of C_B and see how several discovery algorithms perform under these challenges.

In the execution of these experiments, we found that non-block-structured algorithms (e.g. α, Tα, SM, HM and FO) did not reliably return bounded weakly sound models, which corresponds to our findings of Section 8.3 (see tables 8.3 and 8.3). Therefore, we excluded these algorithms from this experiment. The models returned by FD and CPM lack clear semantics, thus these tools were excluded as well. Furthermore, due to the random nature of ETM and its limited scalability, this algorithm had to be excluded as well. No implementations of CCM and MPM were available to us, so, unfortunately, in this experiment we could only include the algorithms proposed.

We discuss the incompleteness handling experiment in Section 8.4.1, and in Section 8.4.2, we discuss the experiment of infrequent and deviating behaviour handling.

8.4.1 Incomplete Behaviour

To answer the first part of RQ.3, we investigate how incompleteness, i.e. the absence of behaviour that is necessary for algorithms to rediscover systems, influences rediscovery. To this end, we take a system, generate a base event log, add an increasing number of traces, and assess at which log size the system is rediscovered.

8.4.1.1 Set-up

The test procedure is illustrated in Figure 8.33: we reuse the 10 randomly generated systems having 32 activities of the scalability experiment (Section 8.2), as these trees can be handled well by all algorithms.

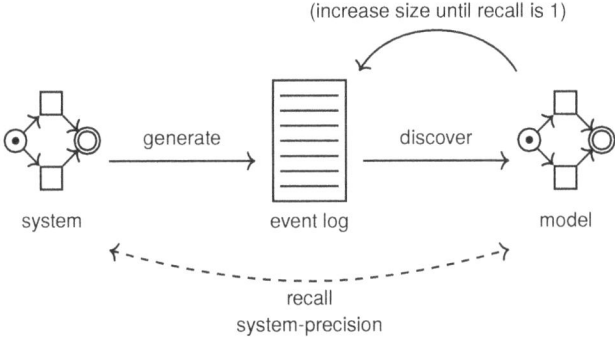

Fig. 8.33: Set-up of the incompleteness experiments.

Let n denote the number of traces, which initially is 1. That is, from each system we generate n traces and discover a model with all discovery algorithms under consideration. Second, we apply the PCC framework to measure recall and system precision (see Section 3.2.2.2) of the discovered models with respect to their randomly generated systems using a model-model comparison. The reported recall and system precision are the averages over all 10 logs and systems for a particular algorithm.

Finally, the experiment is repeated with n increasing 2-fold in each step, up to a maximum of $2^{13} = 8912$ traces. We chose this maximum as this is the first step in which all logs were directly follows complete and consequently all algorithms that rediscovered all systems managed to do so at this number of traces. We are particularly interested in the smallest n for which an algorithm scored 1 for both recall and system precision.

Figure 8.34 shows the completeness of the logs in terms of directly follows graphs. That is, the sizes of the directly follows graph of the systems were compared to the sizes of the directly follows graphs of the generated logs, in which the size of a directly follows graph is the count of edges, start activities and end activities, while disregarding edge/activity frequencies of the log. All event logs were directly follows complete to their systems at 8912 traces.

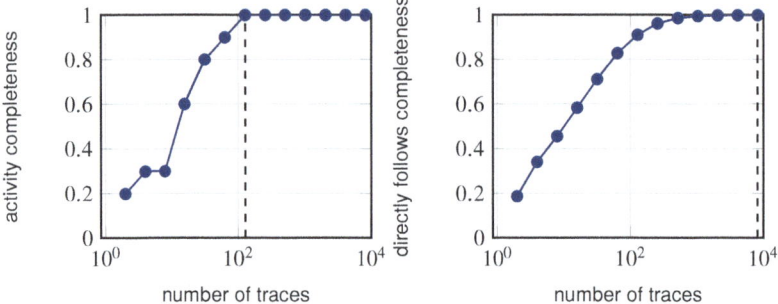

Fig. 8.34: Completeness of logs in the incompleteness experiment. The dashed line denotes where the measure hit 1.

8.4.1.2 Results

Figures 8.35 and 8.36 show the results of the incompleteness experiments. The results for IMc and IMcD could not be obtained, as they timed out. Each graph shows the influence of the number of traces on the discovered models, for a particular discovery algorithm. The first time an algorithm scored perfect is denoted with a vertical dashed line. Notice that the graphs have been truncated in the y-scale at 0.55 to better show differences.

8.4.1.3 Discussion

The flower model (FM) and trace model (TM) provide baselines, and illustrate the completeness and behaviour restrictions of the systems. For instance, the recall of FM provides a measure for the activity completeness of the generated logs. From

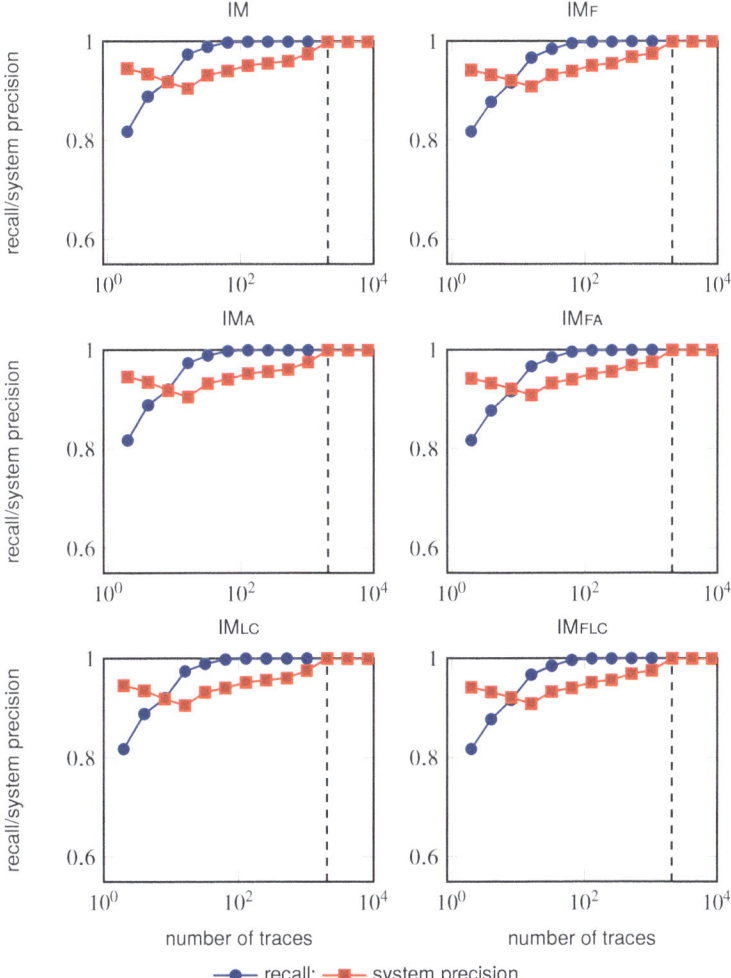

Fig. 8.35: Results of incompleteness experiments (1): recall/log precision vs increasing numbers of traces. A dashed line denotes reaching perfect fitness and log precision.

this, we derive that 128 traces were necessary for all logs to contain at least each activity once. Furthermore, FM provides a baseline for system precision, as a flower model allows for all behaviour, and therefore FM shows how much behaviour the systems actually restrict when compared to all behaviour: in this experiment, the systems on average allow for about 60% of all behaviour (as measured by the PCC framework).

The recall of the trace model (TM) shows what part of the behaviour of the models was present in the generated event logs, i.e. this indicates the completeness of the

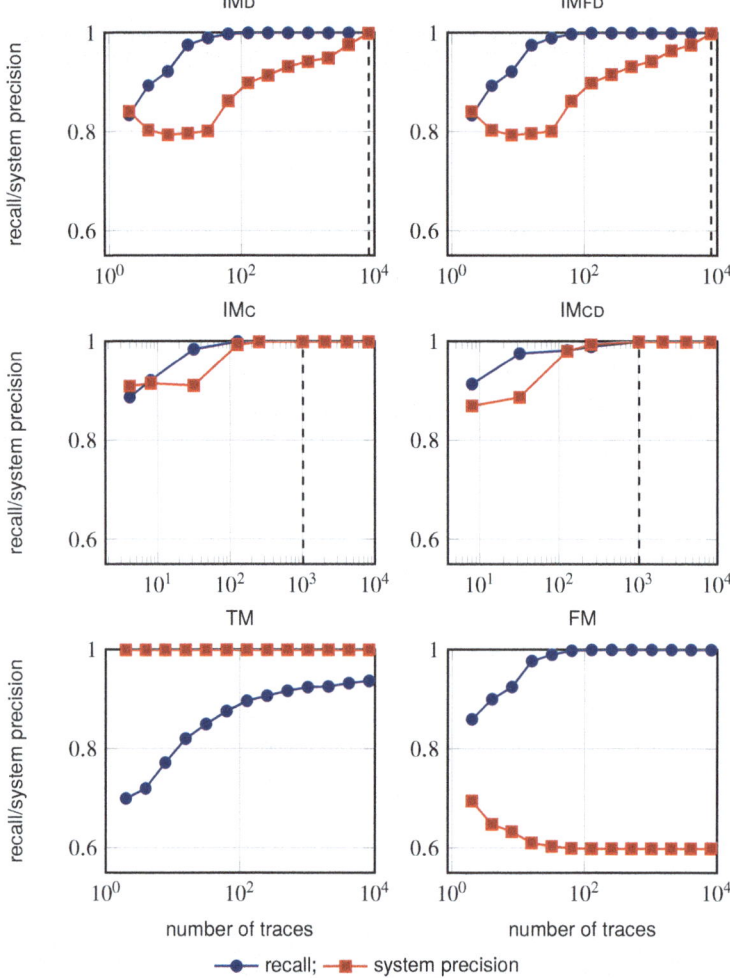

Fig. 8.36: Results of incompleteness experiments (1): recall/log precision vs in-creasing numbers of traces. A dashed line denotes reaching perfect fitness and log precision.

logs generated in this experiment, and would reach 1 in the limit. This measure illustrates the need and use cases for process discovery algorithms: choosing the trace model would yield a high system precision, but just describes the event log and makes no attempt to represent the system well. Consequently, comparing recall of other algorithms to TM provides an idea of their generalisation, i.e. that these other algorithms are able to rediscover the system without this behaviour. This illustrates the need for discovery algorithms to generalise and not simply represent the behaviour in the event log: for these systems of only 32 activities, event logs of even 10,000 traces do not contain all behaviour.

The results for IM, IMF, IMA, IMFA, IMLcand IMFLC are similar: fitness rises quickly (as over-estimations of behaviour, i.e. fall throughs are discovered) with the number of traces, and system precision increases until 2048, when the languages of all systems are rediscovered. In this experiment, the infrequent and deviating behaviour filtering of IMF, IMFA and IMFLC has little influence in presence of incomplete behaviour. That is, filtering seems not to influence fitness negatively, even though incomplete logs could contain behaviour that is prone to be filtered.

The directly follows based algorithms IMD and IMFD of the IMD framework need more traces to rediscover all systems (8912 vs 2048), and require the directly follows graphs to be complete before rediscovery is achieved. This highlights that these algorithms use less information, i.e. only the directly follows graph, than the algorithms of the IM framework, and hence have less chance to deal with incompleteness.

In sections 6.3 and 6.6.5, we introduced two algorithms that were focused on handling incomplete behaviour: IMc and IMcD. The results of this experiment shows both the strength and the weakness of these algorithms: both IMc and IMcD rediscover systems even in case of incomplete event logs and require less traces to rediscover all systems than the other algorithms (1024 vs 2048 traces), thus these algorithms are more robust to incompleteness. However, these algorithms did not discover models for all event logs. For instance, for the logs of 512 traces, both IMc and IMcD discovered models for 9 out of the 10 event logs, but could not discover a model for the remaining log in one *month* of running time, hence the gap in Figure 8.36 for log size 512. This illustrates their main weakness: both algorithms are exponential in the number of activities. A manual inspection revealed that IMc rediscovered the 9 systems, i.e. it might be that IMc would rediscover all systems at only 512 traces.

In answer of RQ.3, we conclude that all algorithms of the IM framework and IMD framework seem to provide rediscoverability, even if not all behaviour is present in the event log. The IMD framework algorithms (except IMcD) require the event log to contain the same directly follows graph as the system, which corresponds to theorems 6.16 and 6.17. The IM framework algorithms (except IMc) use entire event logs and provide rediscoverability on smaller logs than the algorithms of the IMD framework, which can only use directly follows graphs. That is, in addition to theorems 6.2 and 6.5, we have shown experimentally that these algorithms can provide rediscoverability, even if the log assumption that the log is directly follows complete from these theorems has not been met.

The algorithms that were introduced to handle incompleteness, i.e. IMc and IMcD, provide rediscoverability at even smaller event logs: in our experiment, the logs could be half the size, and presumably a quarter of the size.

8.4.2 Deviating & Infrequent Behaviour

To answer the second part of RQ.3, we investigate how infrequent and deviating behaviour, i.e. the presence of little-occurring behaviour and behaviour not according to the system, influences rediscovery. For the purposes of this experiment, we consider infrequent and deviating behaviour as one. To this end, we generate a base event log from each system, add an increasing number of deviating traces, and assess whether the system was rediscovered.

8.4.2.1 Set-up

The test procedure is illustrated in Figure 8.37. We reuse the 10 systems of the incompleteness experiment. To enable a fair comparison, we generate an event log (L_{146}) having 8192 traces from each of these 10 systems, such that this log is large enough to enable rediscovery for all algorithms (see the incompleteness experiment in Section 8.4.1).

To ensure that incompleteness does not influence the results of this experiment, we duplicate L_{146} by including each of the 8192 traces twice. During the experiment, we ensure that the used event log retains all information of L_{146} by only inserting deviations in the duplicated traces. In Figure 8.37 we would only add deviating behaviour below the dashed line.

The experiment consists of several rounds, in each of which a number n of deviations is inserted in each of the 10 base logs. A deviation is introduced by inserting an arbitrary event in an arbitrary trace at an arbitrary position. To these 10 deviating event logs, the discovery algorithms are applied and the resulting models are compared to the 10 systems using the PCC framework: recall and system precision are measured using $k = 2$.

In the first round, 1 deviation is inserted ($n = 1$). After each round, n is multiplied by 2, up to a maximum of $2^{13} = 8192$.

8.4.2.2 Results

Figures 8.38 and 8.39 show the results. Each graph shows the influence of the number of deviations on the discovered models, for a particular discovery algorithm. Notice that the graphs have been truncated in the y-scale at 0.55. We could not obtain measures for most of the logs for the incompleteness handling algorithms (IMc, IMcD), as on some deviating logs, the run time of these algorithms exceeded a week.

8.4.2.3 Discussion

The recall of the trace model (TM) shows the behavioural completeness of the generated event logs, i.e. the base logs contain on average 94% of the behaviour of

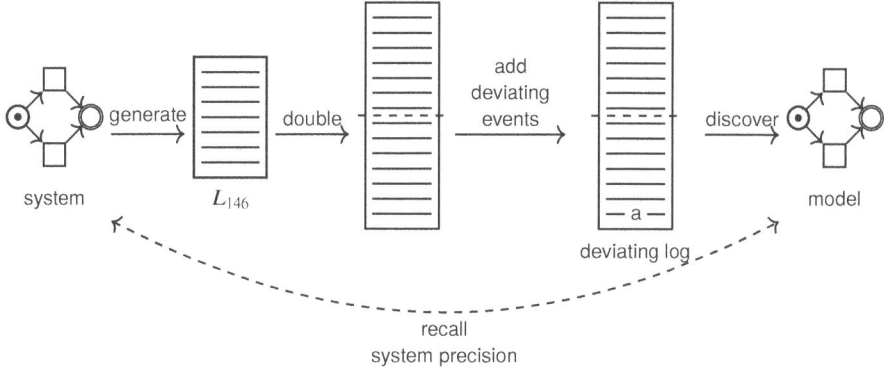

Fig. 8.37: Set-up of the deviating & infrequent behaviour experiment.

the systems (as measured by the PCC framework). Recall of TM is not constant but increases slightly with the addition of more arbitrary events, as it is not guaranteed that each inserted event is a deviation. However, the added events have a considerable influence on system precision, as witnessed by the system precision of TM, i.e. the addition of a deviation increases behaviour in the discovered trace models, which decreases precision of these discovered models with respect to the original systems. The flower model (FM) contains a baseline for system precision and a measure for the restrictions the 10 original systems pose on the event log, i.e. these systems allow for about 60% of the behaviour (according to the measure of the PCC framework) of all behaviour.

The graphs of most IM framework algorithms (i.e. IM, IMF, IMA, IMFA, IMLC, IMFLC) are of a similar structure and consist of two phases (see Figure 8.38): in the first phase system precision drops and recall is high and rather stable, while in the second phase, system precision increases a bit at the cost of recall for IMF, IMFA, IMD, IMFD and IMFLC. The first phase ends at around 100 - 1000 inserted deviations, which is substantial compared to the 26,000 events on average in the 10 event logs. In the second phase, the graphs show that behaviour is being filtered: recall drops below the recall of TM, so these algorithms are excluding non-deviating behaviour from the models they discover. Notice that this second phase is almost absent in IM, IMA and IMLC, and is most pronounced in IMF, IMFA and IMFLC. This shows that in these three last algorithms, the filtering of behaviour works as intended, i.e. to regain system precision, however at a loss of recall. All six algorithms find false structures and even though they are trading recall for system precision, it is clear that the deviating events are overwhelming the algorithms.

The algorithms of the IMD framework perform comparably to the algorithms of the IM framework in the first phase, i.e. have comparable recall and system precision levels, but in the second phase, i.e. after 100-1000 inserted deviations, IMFD does not exhibit the increase in system precision and drop in recall. We believe this

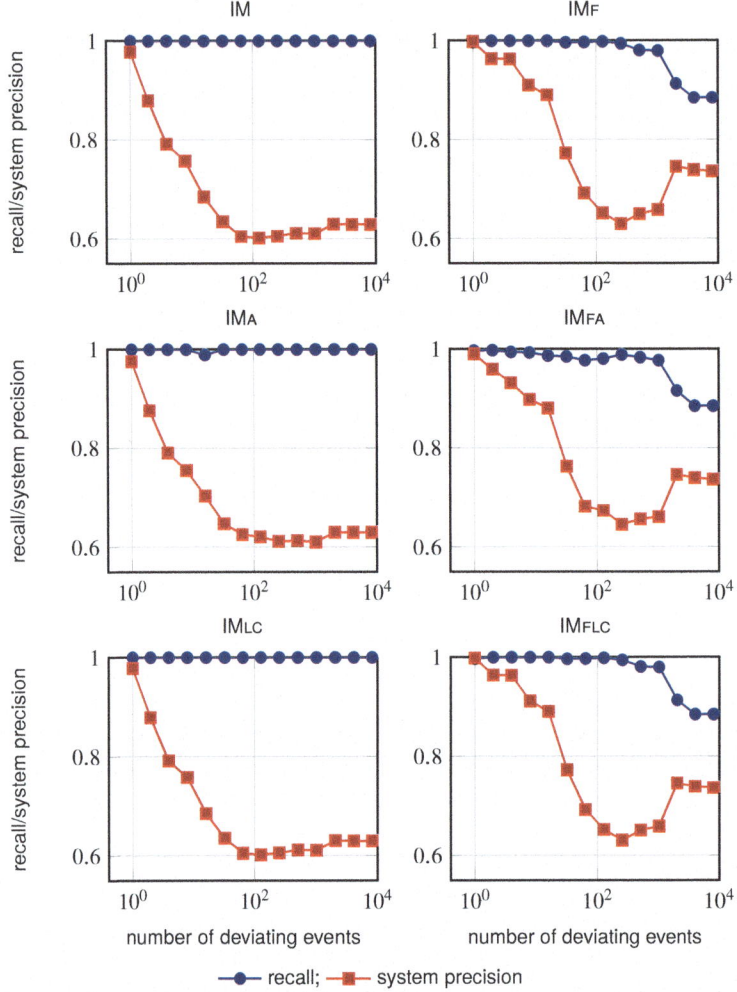

Fig. 8.38: Results of the deviating behaviour experiment (1): each graph shows how an algorithm handles deviating events on average over 10 models/logs.

is due to the directly follows graphs' ability to hide some deviating events, i.e. a deviating event does not necessarily introduce a deviating edge in a directly follows graph. However, in the second phase, the directly follows graphs contain too little information to allow for deviating and infrequent behaviour filtering.

The results show that the basic algorithms IM, IMA, IMD and IMLC are sensitive to the added deviations: they include all behaviour of the event log and try to generalise over this behaviour to discover a process tree (corollaries 6.1, 6.4 and 6.5), which manifests in the results as recall being higher than the recall of TM. However, this comes at the expense of a quickly dropping system precision, as the models allow

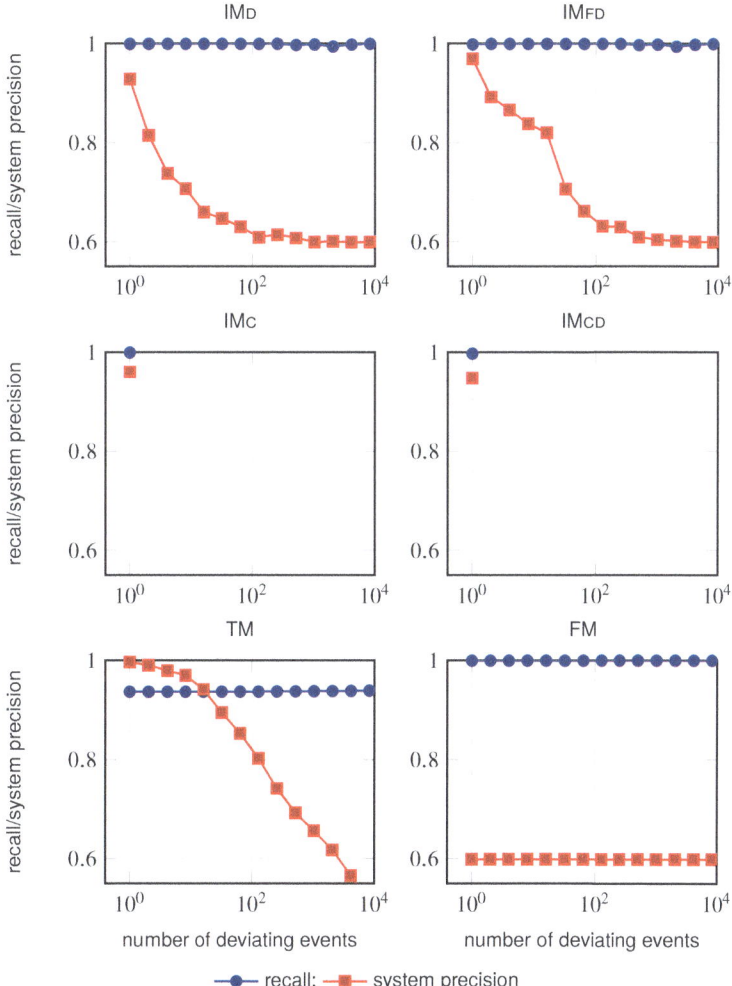

Fig. 8.39: Results of the deviating behaviour experiment (2): each graph shows how an algorithm handles deviating events on average over 10 models/logs.

for more and more behaviour. The infrequent and deviation filtering algorithms IMF, IMFA, IMFD and IMFLC perform better: they keep recall at comparable levels, while system precision is higher than system precision achieved by the basic algorithms.

Of the infrequent behaviour filtering algorithms, IMF, IMFLC and IMFA achieve the highest system precision. These three algorithms score equal on system precision, however IMFA achieves the highest system precision for 4 and 8 deviations, while IMF and IMFLC achieve the highest system precision for 128 and 256 deviations. In each model, IMFA discovers on average 1.0 ∨-node and 0.01 ↔-node, while the systems did not contain these constructs, while IMF does not discover these constructs. Next,

IMᶠᴅ follows as the miner with the highest average system precision. IMF and IMꜰʟᴄ are very similar algorithms as the event logs in this experiment do not trigger the life cycle handling capabilities of IMꜰʟᴄ and IMʟᴄ (IMᴄʟᴄ has not been implemented yet and thus not been included), which is reflected in the similar recall and system precision scores for these algorithms.

In answer of RQ.3 and in confirmation of Section 3.3, we conclude that infrequent and deviating behaviour indeed challenge rediscoverability: even a small number of deviations can prevent rediscoverability for all algorithms. Nevertheless, small numbers of deviations result in models that are close (in recall and system precision) to the systems, and large numbers of deviations result in models with low recall and system-precision values. Furthermore, in presence of large numbers of deviations, some algorithms (IMF, IMA, IMꜰA, IMꜰʟᴄ) seem to tend to not distinguish infrequent and deviating behaviour and erroneously filter the former. However, by applying filtering techniques in algorithms, models can be discovered that are closer to the original systems than the models discovered by non-filtering algorithms.

In this experiment, we included a single type of deviation, i.e. the addition of extra events, which, due to the representational bias of process trees in C_B, were deviations from the system. Other types of deviations that could occur in a system are for instance the removal of events or the swap of two events. We expect that performing the experiment described in this section using these deviations would lead to similar conclusions, however it would be interesting to study the influence of these types of deviations on rediscovery in more detail for particular algorithms.

Future work 8.2: Identify and analyse types of deviations and perform experiments to investigate the influence of these deviations on rediscovery.

8.5 Evaluation of Log-Conformance Checking

In the previous sections, we have evaluated several aspects of process discovery algorithms: log-quality measures, scalability, and handling infrequent, deviating and incomplete behaviour. In this section, we evaluate the conformance checking techniques introduced in Chapter 7 (RQ.5). That is, we evaluate the scalability of the PCC framework with respect to other techniques, and consider its results quantitatively. The research question is: could the PCC framework replace the existing techniques to determine which of two models has the highest log precision and fitness with respect to an event log?

8.5.1 Set-up

In this experiment, we take a selection of the real-life event logs described in Section 8.3.1: BP11 for its complexity in activities, RF for its many traces, and BP12 for the complexity of the models discovered by algorithms (see Section 8.3.3). To be able to perform a qualitative analysis between the conformance measures, we also included sublogs of BP12, i.e. the activities prefixed by A, O and W. To these event logs, we apply the top-performing discovery algorithms of Section 8.3.2, i.e. IM$_F$ and IM$_{FD}$. Furthermore, we include the flower model (FM) as a baseline.

On the combinations of event logs and discovered models, we apply the log-conformance techniques that measure fitness [2] and log precision [4] using alignments (we used the alignments from ProM 6.6). Furthermore, we apply the PCC framework, using both $k = 2$ and $k = 3$. We record the runtime of the computation and whether the log-conformance technique returns a result. Notice that for practical reasons (e.g. manually interfacing with the ProM GUI, non-exclusive use of hardware), the runtime measure is approximate: the results will show a clear difference nevertheless. The computations had 40GB of RAM available.

In the previous experiments (e.g. Section 8.3.3), we showed that FM could be used as a baseline for log precision, as FM returns models that allow for any behaviour. Therefore, in this section, we scale the log-precision measures using this baseline, such that the scaled log precision denotes the gain in precision compared to the flower model for the particular log:

$$\text{scaled log precision} = 1 - \frac{1 - \text{log precision of model}}{1 - \text{log precision of flower model}}$$

This scaling is performed for both the log-precision measure of the PCC framework and the existing alignment-based measure [4].

8.5.2 Results

Table 8.4 shows the results, extended with running times for the techniques. Figure 8.40 shows the results in a plotted form. Figure 8.41 illustrates the complexity of the logs and the scalability of the PCC framework. Some numbers are missing because the alignment-based approach could not handle WA3 and BP11.

8.5.3 Discussion

Fitness scores according to the PCC framework ($k = 2$) differ from the fitness scores by [2] by at most 0.05 (except for BP12|$_A$ and RF of IM$_{FD}$). For $k = 3$, the fitness measures differ more: some are higher, some are lower than for $k = 2$. Thus, this experiment suggests that the new fitness measurement could replace the alignment-based fitness [2] measure, while being generally faster on both smaller and larger logs, though additional experiments may be required to verify this hypothesis. More importantly, the PCC framework could handle logs (BP11, WA3) that the existing approach based on alignments could not handle.

Table 8.4: Log-conformance techniques compared on real-life logs.

		existing techniques			PCC framework, $k = 2$				PCC framework, $k = 3$			
		fitness [2]	log precision [4] scaled	time	fitness	log precision measure	scaled	time	fitness	log precision	scaled	time
BP11	IM$_F$	no measures obtained			0.627	0.764	0.472	25s	0.981	0.731	0.473	53h
	IM$_{FD}$	no measures obtained			0.997	0.766	0.477	1m	0.998	0.660	0.333	49h
	FM	1.000	0.002	0.000	5h	1.000	0.553	0.000	25s	1.000	0.490	0.000	54h
RF	IM$_F$	0.992	0.618	0.463	2.5m	0.991	0.963	0.909	≤1s	0.986	0.899	0.976	≤1s
	IM$_{FD}$	0.736	0.482	0.271	10s	0.803	0.918	0.798	≤1s	0.684	0.822	0.646	≤1s
	FM	1.000	0.289	0.000	≤5s	1.000	0.594	0.000	≤1s	1.000	0.497	0.000	≤1s
WA3	IM$_F$	0.952	-	-	5h+	0.992	0.729	0.203	≤1s	0.988	0.624	0.166	1m
	IM$_{FD}$	no measures obtained			0.999	0.769	0.321	5s	0.996	0.663	0.253	2m
	FM	1.000	-	-	5h+	1.000	0.660	0.000	10s	1.000	0.549	0.000	2m
BP12	IM$_F$	0.967	0.364	0.290	20m	0.978	0.668	0.092	≤1s	0.940	0.541	0.115	≤1s
	IM$_{FD}$	1.000	0.189	0.095	25m	1.000	0.693	0.161	≤1s	0.993	0.543	0.119	≤1s
	FM	1.000	0.104	0.000	30m	1.000	0.634	0.000	≤1s	1.000	0.481	0.000	≤1s
BP12$|_A$	IM$_F$	0.995	0.606	0.940	≤1s	0.999	0.967	0.931	≤1s	0.999	0.920	0.856	≤1s
	IM$_{FD}$	0.816	1.000	1.000	≤1s	0.700	1.000	1.000	≤1s	1.000	0.920	0.856	≤1s
	FM	1.000	0.227	0.000	≤1s	1.000	0.520	0.000	≤1s	1.000	0.445	0.000	≤1s
BP12$|_O$	IM$_F$	0.991	0.508	0.351	≤1s	0.981	0.809	0.407	≤1s	0.987	0.715	0.382	≤1s
	IM$_{FD}$	0.861	0.384	0.187	≤1s	0.862	0.794	0.360	≤1s	0.998	0.626	0.189	≤1s
	FM	1.000	0.242	0.000	≤1s	1.000	0.678	0.000	≤1s	1.000	0.539	0.000	≤1s
BP12$|_W$	IM$_F$	0.876	0.690	0.553	≤1s	0.875	0.836	0.611	≤1s	0.762	0.793	0.637	≤1s
	IM$_{FD}$	0.914	0.300	-0.010	≤1s	0.923	0.823	0.581	≤1s	0.963	0.699	0.473	≤1s
	FM	1.000	0.307	0.000	≤1s	1.000	0.578	0.000	≤1s	1.000	0.429	0.000	≤1s

Comparing the scaled log-precision measures, the PCC framework ($k = 2$) and the existing approach agree on the relative order of IM$_F$ and IM$_{FD}$ for BP12$|_A$, BP12$|_O$ and RF, disagree on BP12 and are incomparable on BP11 and WA3 (IM$_{FD}$) as the existing measure did not produce a result. For BP12$|_W$, IM$_{FD}$ performed *worse* than the flower model according to [4] but *better* according to our measure. This model, shown in Figure 8.42, is certainly more restrictive than a flower model, which is correctly reflected by our new log-precision measure. Therefore, likely the approach of [4] encounters an inaccuracy when computing the precision score. For BP12, precision [4] ranks IM$_F$ higher than IM$_{FD}$, whereas our precision ranks IM$_{FD}$ higher than IM$_F$. Inspecting the models, we found that IM$_F$ misses one activity from the log while IM$_{FD}$ has all activities. Apparently, our new measure penalises more for a missing activity, while the alignment-based existing measure penalises more for a missing structure.

Comparing the log-precision measures by $k = 2$ and $k = 3$ of the PCC framework, $k = 2$ measured a consistently higher log precision than $k = 3$. This is to be expected, as $k = 3$ takes more information into account and is therefore a 'stricter' measure (see Section 7.3). However, the $k = 2$ and $k = 3$ methods measure scaled log precision differently: in 10 cases, scaled log precision of $k = 2$ was higher than of $k = 3$ and 4 times the other way around. The $k = 2$ and $k = 3$ variants agree on the ranking of the discovered models for all logs except BP11. Thus, this experiment seems to suggest that it might not be necessary to opt for the more computationally expensive $k = 3$

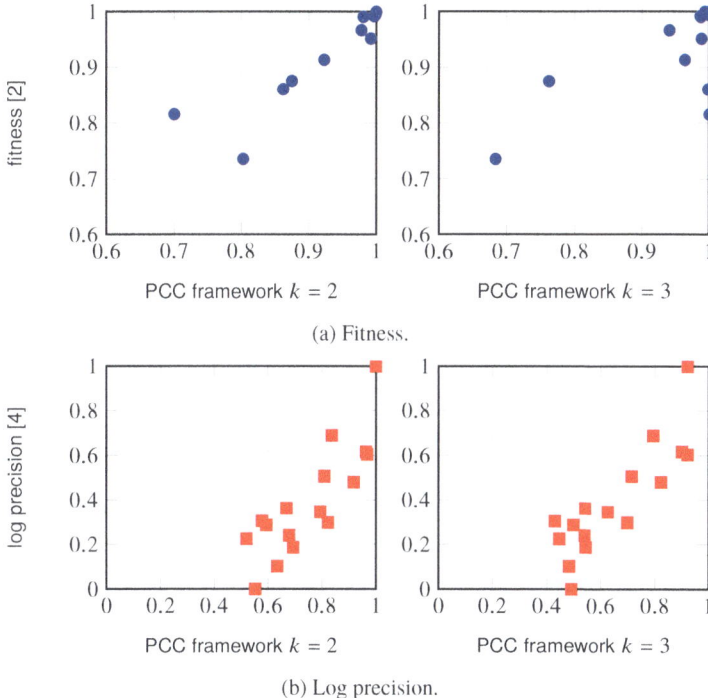

(a) Fitness.

(b) Log precision.

Fig. 8.40: Results of the conformance-checking experiment, showing how the PCC framework using $k = 2$ and $k = 3$ correlates to fitness [2] and log precision [4]. We scaled the fitness graphs to improve readability.

in all cases. For the BP11 log, the $k = 2$ and $k = 3$ disagreeing could indicate that the behaviour in the log is too complex to be captured well by projections of size 2 and 3, thus one might try a higher k.

This experiment does not suggest that our new measure can directly replace the existing measures in all cases, but log precision seems to be able to provide an intuitive categorisation, such as a good/bad precision, compared to the flower model. For instance, $IM_F(RF)$ and $IM_{FD}(BP12|_A)$ seem to have a rather good log precision, while $IM_F(BP12)$ and $IM_F(WA3)$ seem to have a bad log precision.

In answer of RQ.5, we showed that our new fitness and precision metrics are useful to quickly assess the quality of a discovered model and decide whether to continue analyses with it or not, in particular on event logs that are too large or complex for current techniques. Furthermore, the PCC framework provided measures much faster in this experiment.

The PCC framework-measured log precision can also be lower than the log precision of the flower model. The following example illustrates this: let $L = \{\langle a, b \rangle\}$ be a projected log and $M = a$ a projected model. Then, technically, their conjunction is empty and hence both precision and recall are 0. This matches intuition, as

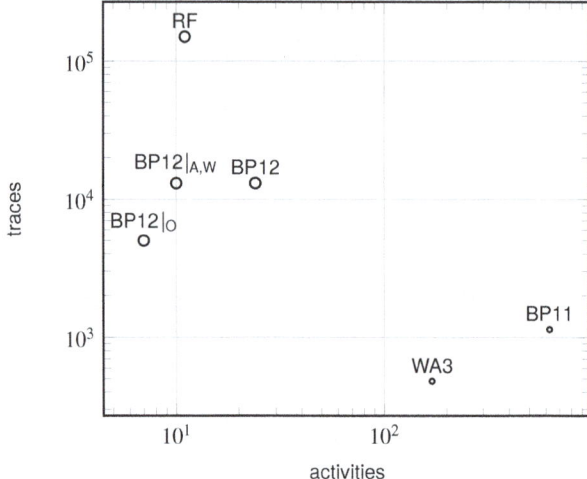

Fig. 8.41: Result of a scalability experiment for the PCC framework and [3]. The latter could not obtain a result on the logs WA3 and BP11, which have more activities than the other logs in this experiment.

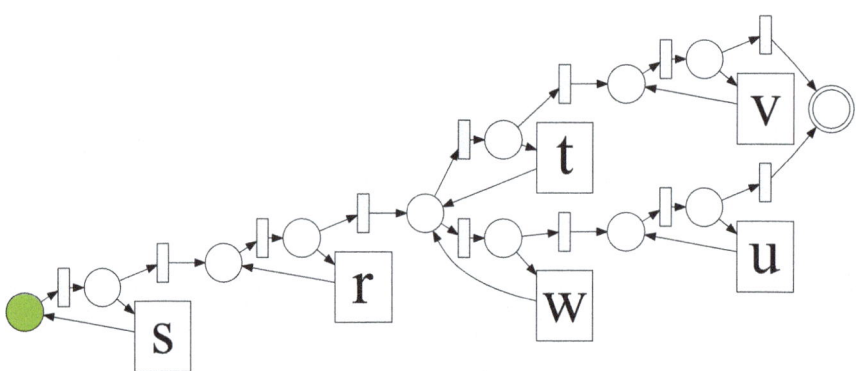

Fig. 8.42: IM_FD applied to BP12|_W. The activities have been encoded as letters for readability reasons: s = W_Beoordelen fraude, r = W_afhandelen leads, t = W_Completeren aanvraag, w = W_Valideren aanvraag, v = W_Nabellen offertes and u = W_Nabellen incomplete dossiers.

they have no trace in common. This sensitivity to missing activities is inherent to language-based measuring techniques, as no trace is in the languages of both L and M. Hence, any technique to measure this differently would need a different notion of language.

In addition to simply providing an aggregated fitness and precision value, both existing and our new technique allow for more fine-grained diagnostics of *where* in the model and event log fitness and precision are lost. For instance, by looking at the subsets $a_1 \ldots a_k$ of activities (in which k determines the size of the subset and consequently sets a trade-off between speed and accuracy; we limited our experiments to $k = 2$) with a low fitness or precision score, one can identify the activities that are not accurately represented by the model, and then refine the analysis of the event log accordingly. Figure 8.43 gives an impression how the results of both alignments and the PCC framework can be projected onto a process model, being $IM_F(BP12|_A)$: results are projected onto the activities. In Section 7.1.2, this projection notation was formally introduced.

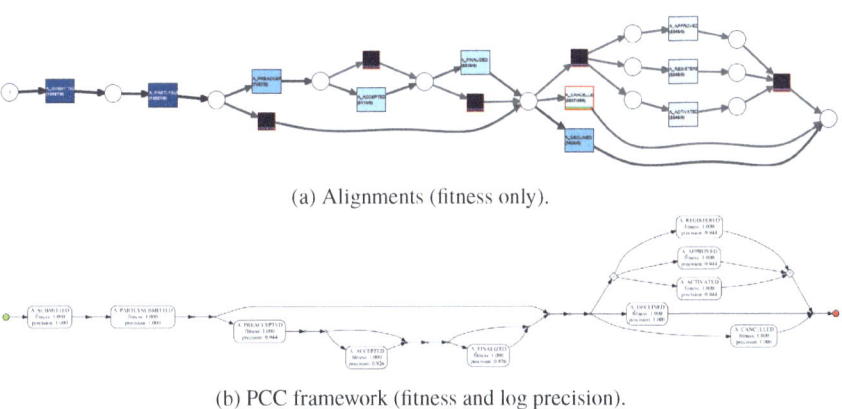

(a) Alignments (fitness only).

(b) PCC framework (fitness and log precision).

Fig. 8.43: Impression of log-conformance projected to a process model discovered from $BP12|_A$, as produced by the tools.

8.5.4 Evaluation Using the PCC framework

In the previous evaluations of this chapter, we have applied the PCC framework to perform our experiments. As the log-conformance experiment showed, executing these experiments with existing techniques would have been challenging, as in total over 2500 log-conformance checking computations were performed. Furthermore, we performed 3000 model-model conformance checking computations, which are not supported. Therefore, in support of RQ.5, we conclude that the PCC framework allows for detailed analyses of the model quality achieved by discovery algorithms in the presence of incomplete, infrequent and deviating behaviour, and for repeated measures such as required for cross validation.

8.6 Non-Atomic Behaviour

In the previous sections, we have evaluated several aspects of process discovery algorithms: log-quality measures, scalability, and handling infrequent, deviating and incomplete behaviour. In this section, we address RQ.4, i.e. how discovery algorithm handle non-atomic behaviour (i.e. life cycle information) in event logs.

In Section 3.4, we introduced several concepts to describe the relation between event log or system model and a discovered model: fitness, recall, system precision and log precision. However, these concepts have not been lifted yet to non-atomic behaviour, and consequently, the PCC framework has not been designed to handle adapted yet to handle non-atomic event logs and process models, thus we approach this experiment qualitatively. That is, we consider an artificial and a real-life event log, apply both atomic and non-atomic behaviour handling algorithms, and assess the discovered models manually.

The only other non-atomic behaviour handling algorithms known to us are Tα and the commercial FD and CPM, which are all included in this experiment. Furthermore, we include the newly introduced algorithms IMLC and IMFLC (see sections 6.5.2 and 6.5.3). The algorithm IMCLC (Section 6.5.3) has not been implemented yet. To illustrate the need for specialised algorithms to handle non-atomic event logs, we include HM, IMF and IMA as representatives.

8.6.1 Artificial Log

The artificial log we consider illustrates the distinction of interleaved and concurrent behaviour. The log, $\tilde{L}_{148}$, consists of 1000 traces generated from the process tree $M_{149} = $ 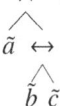 , which expresses that a can be executed concurrently to all activities, however b and c cannot overlap in time and are thus interleaved. We chose an event log with 1000 traces to take filtering steps of discovery algorithms out of the equation.

In this experiment, we consider two ways of handling non-atomic logs by transforming events into activities. The first way is to only consider the event name, e.g. "a", as the activity. Consequently, the life-cycle information in the trace is ignored, e.g. the non-atomic trace $\langle a_s, a_c \rangle$ will be interpreted as the atomic trace $\langle a, a \rangle$ by discovery techniques that are not aware of life-cycle information (see Section 6.5.1). This way is used by default in the algorithms that we introduced. The second way is to consider the combination of the life-cycle transition and the event name, e.g.

"a_c", as the activity. Consequently, discovery algorithms might produce inconsistent models, e.g. a_s and a_c might be unrelated in the model. This way is used by default in some other algorithm implementations, such as HM and α. Other ways to handle atomic event logs include the filtering of start or completion events, which is outside the scope of this experiment.

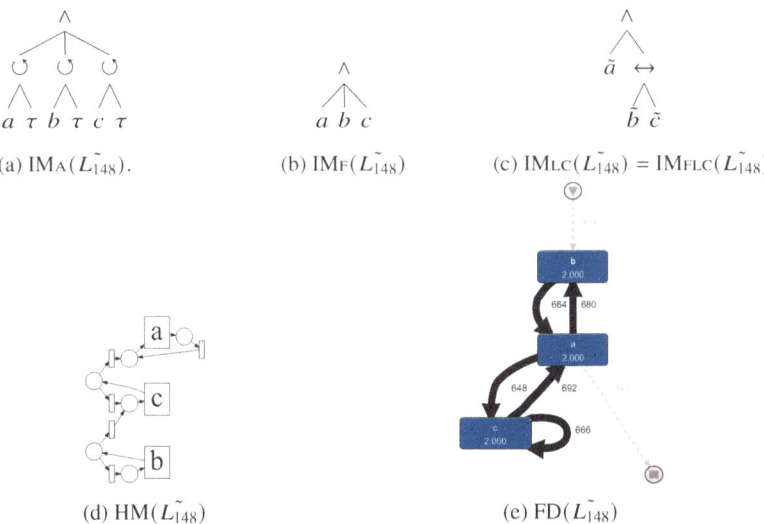

(a) $IM_A(L_{148}^{\sim})$. (b) $IM_F(L_{148}^{\sim})$ (c) $IM_{LC}(L_{148}^{\sim}) = IM_{FLC}(L_{148}^{\sim})$

(d) $HM(L_{148}^{\sim})$ (e) $FD(L_{148}^{\sim})$

Fig. 8.44: Results of the non-atomic artificial-log experiment, considering event names as activities.

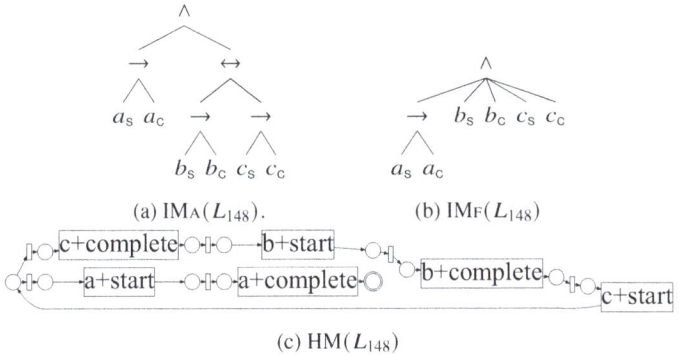

(a) $IM_A(L_{148})$. (b) $IM_F(L_{148})$

(c) $HM(L_{148})$

Fig. 8.45: Results of the non-atomic artificial-log experiment, considering life-cycle transitions and event names as activities.

We first consider the models discovered by algorithms on ($\tilde{L}_{148}$ using event names as activities. IMA discovers the model shown in Figure 8.44a. Even though this model is perfectly fitting, it misses the non-interleaved relation between b and c, thus system and log precision are lower. Furthermore, as each start event is considered an activity by these algorithms, the activities are all part of loop constructs. Notice that even though IMA supports the interleaved operator, the directly follows abstraction used by these algorithms does not distinguish concurrent from interleaved behaviour in case the behaviour consists of single activities. The filtering algorithm IMF filters behaviour and prevent these loops from appearing, however also does not discover the interleaved relation between b and c (see Figure 8.44b). FD discovers the model shown in Figure 8.44e, which suffers from overfiltering: a trace in this model can start with a b and end with a c only, while traces in the log start and end with all activities. Furthermore, c is put in a loop and it is not possible to execute c directly after b. HM discovers a model (Figure 8.44d) that is similar to the model discovered by FD, however a and b can be repeatedly executed. In contrast, the life cycle handling algorithms IMLC and IMFLC are able to distinguish this behaviour and rediscover the process tree M_{149} (see Figure 8.44c).

Second, we consider the models discovered by algorithms using life-cycle information and event names as activities. IMA discovers the model shown in Figure 8.45a, which corresponds to an expanded version of M_{149} perfectly. However, IMF Figure 8.45b is unable to discover the interleaved structure and discovers an inconsistent model, as e.g. b_s can be executed after b_c. This illustrates that even an algorithm that guarantees soundness and fitness can be challenged by non-atomic logs and the need for dedicated algorithms. HM (Figure 8.45c) discovers a model that could be consistent, depending on the position of the initial marking, e.g. when putting the initial token left of c+start. However, this model does not capture the concurrency between a and the combination of b and c, and does not capture the interleaving of b and c. Unfortunately, we were unable to load this artificial log into ProM 5, and therefore we could not apply $T\alpha$.

8.6.2 Real-Life Log

The real-life event log we consider is BP12 [14], which contains both atomic and non-atomic behaviour: the activities prefixed with A and O are atomic, while the activities prefixed with W are non-atomic. We consider the full log, and the projection onto the W activities.

Full Log.

The activity names have been replaced by letters as follows:

a A_ACCEPTED	b A_ACTIVATED
c A_APPROVED	d A_CANCELLED
e A_DECLINED	f A_FINALIZED
g A_PARTLYSUBMITTED	h A_PREACCEPTED
i A_REGISTERED	j A_SUBMITTED
k O_ACCEPTED	l O_CANCELLED
m O_CREATED	n O_DECLINED
o O_SELECTED	p O_SENT
q O_SENT_BACK	r W_Afhandelen leads
s W_Beoordelen fraude	t W_Completeren aanvraag
u W_Nabellen incomplete dossiers	v W_Nabellen offertes
w W_Valideren aanvraag	x W_Wijzigen contractgegevens

Figure 8.46 shows the result of applying IMF to BP12, while Figure 8.47 shows the result of IMFLC and IMLC, all of these as Petri nets.

A manual inspection revealed the differences between these algorithms: both first discover a sequential cut between two activities and the remainder of the activities. When recursing on the remainder, both algorithms use the *strictTauLoop* fall through, which splits traces on every occurrence of an end activity followed by a start activity. Both algorithms split traces, however IMFLC splits less traces than IMF, due to the requirement that the log needs to remain consistent. That is, in several splits performed by IMF, activity executions (i.e. combinations of corresponding start and completion events) are split, as start events get separated from their completion events. Furthermore, besides start and completion events, the event log contains "schedule" events, which are ignored by IMFLC but considered activity executions by IMF, as IMF is not aware of life-cycle information.

The model discovered by IMA (Figure 8.48) differs from the model returned by IMF in details.

We also applied HM, whose model is shown in Figure 8.49. This model is weakly sound, i.e. there is a path to the final marking, but all activities below the red dashed line cannot be part of any trace, as from this part there is no path to the final marking. Furthermore, the model contains concurrency splits, but no concurrency joins. Moreover, below the red dashed line there is no clear end place, thus even though some information can be derived from this part of the model

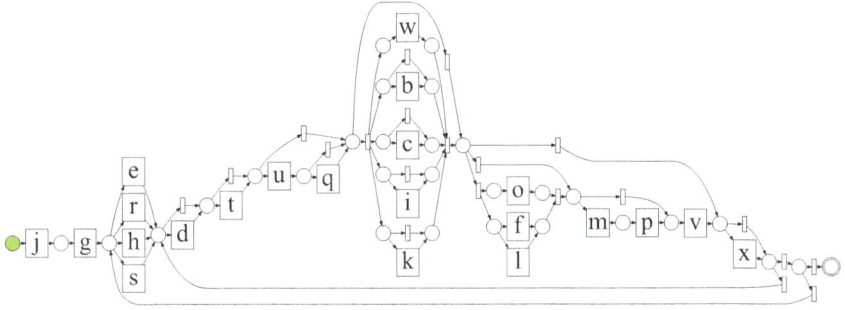

Fig. 8.46: IMF(BP12)

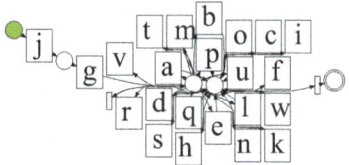

Fig. 8.47: IMʟᴄ(BP12) = IMꜰʟᴄ(BP12)

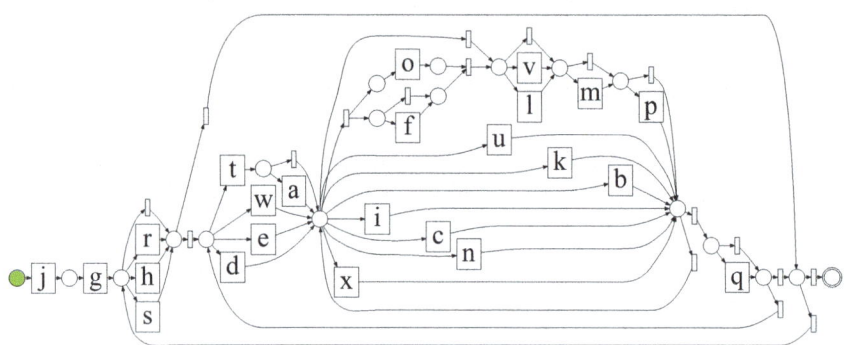

Fig. 8.48: IMᴀ(BP12)

manually, it is not part of the language of the model. Similarly, the model discovered by Tα (Figure 8.51) contains many features that make determining its language difficult: dead parts, token generators, unconnected places, etc. Furthermore, where all other algorithms discovered that the process always starts with A_SUBMITTED (j) followed by A_PARTLYSUBMITTED (g), Tα does not discover this structure and does not restrict execution of A_PARTLYSUBMITTED (g) at all.

However, without a proper framework to evaluate these models, we can only conclude that the model discovered by IMꜰ expresses more structure and limits behaviour more than the model discovered by IMꜰʟᴄ, which expresses a flower model, thus presumably the first has a higher log precision and is preferable.

To illustrate the drawbacks of using life-cycle information and event names as activities, we included the result of HM in Figure 8.50. Notice that the letters in this model differ from the letters in the other models. As this model contains over twice as many activities and does not discard the 'schedule' events, it is much more complex. Furthermore, manual analysis reveals that it is not consistent, which clearly shows the need for dedicated algorithms.

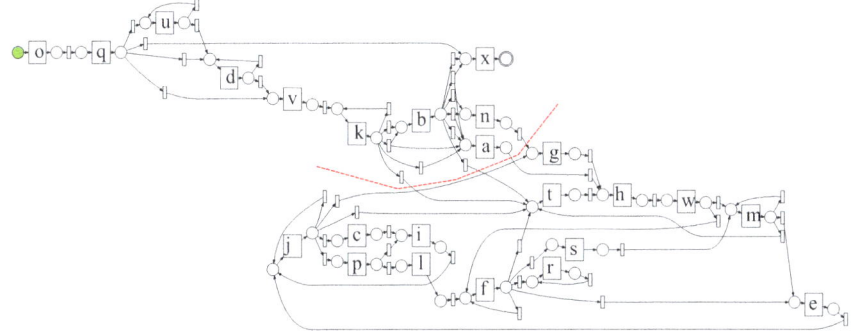

Fig. 8.49: HM(BP12)

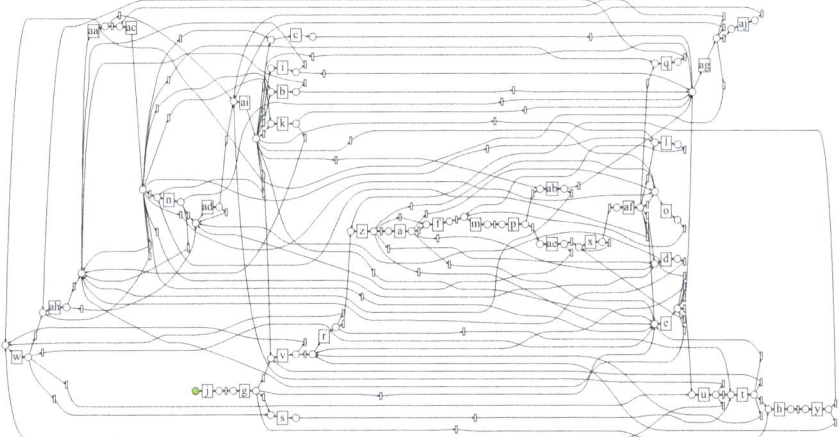

Fig. 8.50: HM(BP12) using life-cycle information and event names as activities.

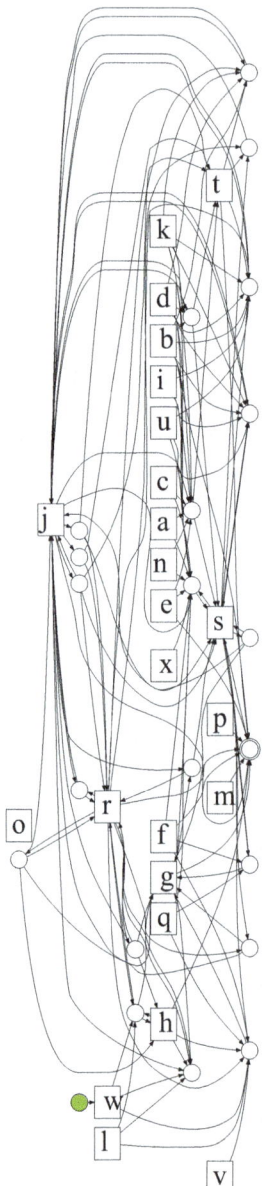

Fig. 8.51: Tα(BP12)

W Activities.

In real-life process mining projects, a next step could be to limit the scope of the analysis. For instance, focus could shift to the W-prefixed activities, as these appear in the event log as start, completion and schedule events. We filtered the log to only contain W-prefixed activities ($BP12|_W$).

The activity names have been replaced by letters as follows:

a W_Afhandelen leads   b W_Beoordelen fraude
c W_Completeren aanvraag   d W_Nabellen incomplete dossiers
e W_Nabellen offertes   f W_Valideren aanvraag
g W_Wijzigen contractgegevens

Figures 8.52 to 8.58 show the results.

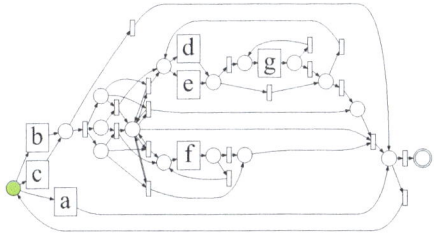

Fig. 8.52: $IM_A(BP12|_W)$. The IM_A algorithm clearly suffers from ignoring the life-cycle information, as this model expresses just a few constraints compared to a flower model. Notice that this model looks more complex than it would in process tree notation: the silent transitions in the left part of the model denote an inclusive choice split.

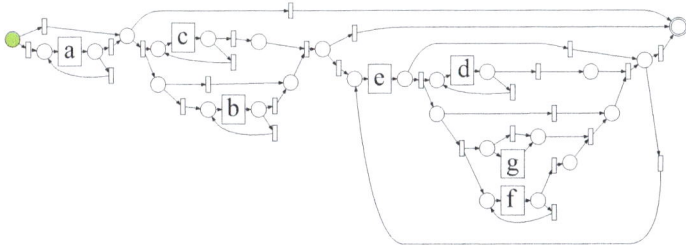

Fig. 8.53: $IM_F(BP12|_W)$. The model discovered by IM_F clearly suffer from ignoring the life cycle information, as this model express little more than a flower model, and even less than the model discovered by IM_A.

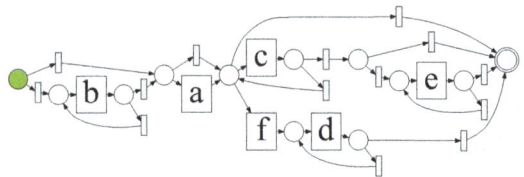

Fig. 8.54: IMFLC(BP12|w). The model discovered by IMFLC is highly structured and contains a lot of information, e.g. there are no loopbacks at all.

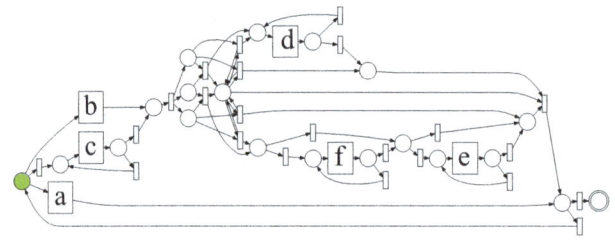

Fig. 8.55: IMLC(BP12|w). IMLC lacks the filtering of IMFLC and its model poses few constraints on its behaviour, i.e. closely resembles a flower model.

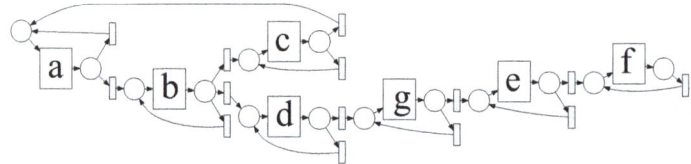

Fig. 8.56: HM(BP12|w). If the initial marking of the model discovered by HM lies somewhere in the loop of a, b and c, the model is likely weakly sound and exhibits lots of structure.

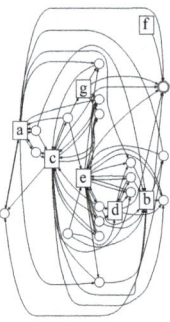

Fig. 8.57: Tα(BP12|w). This model discovered by Tα is dead, i.e. from the initial marking not a single transition can be fired.

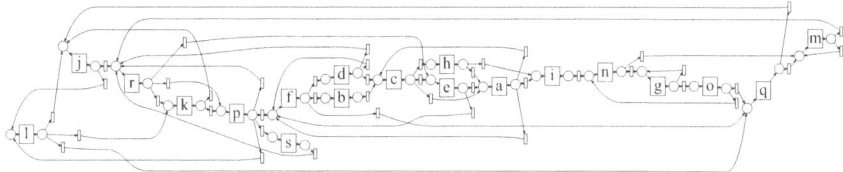

Fig. 8.58: HM(BP12|w) using life-cycle information and event names as activities. HM discovers a model that is close to useless: even though it seems to exhibit some structure and seems to be weakly sound, it has neither an initial nor a final marking. Guessing these markings has a large influence on the language of the model, thus the language of the model is completely unclear. Notice that the letters do not match the other models of this experiment.

Conclusion.

In answer of RQ.4, we conclude that the non-atomic behaviour handling algorithms have the potential to discover better models on event logs that contain non-atomic behaviour, however due to their stricter requirements on the event log, i.e. the log should be consistent in all iterations, might also overlook information in the event log that other algorithms can capture. A point of discussion that remains is the lack of quantitative methods to compare non-atomic event logs and non-atomic process models. Furthermore, in this experiment we applied process discovery techniques using their default settings to unfiltered event logs. In a process mining project, one would try different process discovery parameters and filters to achieve a satisfying model.

8.7 Conclusion

In this chapter, we evaluated existing and the newly introduced process discovery and conformance checking techniques.

RQ.1. In answer to RQ.1, i.e. the scalability of discovery techniques, we found that all algorithms of the IM framework except IMc handled all logs up to a hundred activities and a million events. Existing techniques handled logs up to 64 activities and 150,000 events. The algorithms of the IMD framework (except IMcD) and FD were able to handle a thousand activities and hundreds of millions of events, which were the largest logs we generated in our experiment.

RQ.2. Research question RQ.2 entails whether and how algorithms balance log-quality criteria, whether these balances are user-influenceable. In our cross validation experiment, we found that IMD, IMFD, IMA, IM, IMF, IMLC, IMC, IMFA and IMCD returned pareto optimal models, even though IMFD and IMD use less information than other algorithms and are much more scalable. All algorithms except ETM achieved a consistently high fitness, and a less high log precision, and different algorithms traded different amounts of fitness for log precision. Different trade-offs might be required by different use cases, as described in Section 3.1. In our qualitative experiments, we found that on complicated real-life event logs, such as BP12, discovery algorithms struggled to discover process models that pose many restrictions on the behaviour that is expressed by the model (the models have a low log precision, even though in this particular experiment we could not measure this), i.e. models that are useful for human analysis. The event logs RF and RPW seemed to be easier for discovery algorithms, as more restrictions to behaviour were visible in the discovered models.

RQ.3. In chapters 5 and 6, we proved rediscoverability, i.e. the ability of discovery algorithms to rediscover the language of a system, under laboratory conditions: the logs should be free of infrequent, deviating and incomplete behaviour. RQ.3 entails how close algorithms get to rediscovery under presence of such behaviour, i.e. what the influence of increasing levels of these three types of behaviour on the quality of models discovered by discovery algorithms is. We found that the IM framework algorithms, by their use of entire event logs, provided rediscoverability on smaller logs than the algorithms of the IMD framework, which can only use directly follows graphs. Furthermore, the algorithms that were introduced to handle incompleteness, i.e. IMc and IMcD, provided rediscoverability at even smaller event logs, i.e. requiring 1024 (presumably 512) traces instead of the 2048 by the IM framework and the 8912 by the IMD framework, in correspondence with the intention of these algorithms.

RQ.4. For non-atomic event logs and process models (RQ.4), we found that the newly introduced non-atomic behaviour handling algorithms have the potential to discover better models on event logs that contain non-atomic behaviour, however due to their stricter requirements on the event log, i.e. the log should be consistent in all iterations, might also overlook information in the event log that other algorithms can capture.

RQ.5. In most of the previous experiments, we used the PCC framework for its ability to handle large event logs and process models quickly. We quantified this

(RQ.5), and found that the PCC framework works faster and on larger event logs and process models than existing techniques. Furthermore, the measures of the PCC framework provided intuitive categorisations, such as good/bad precision, compared to the flower model, which in most cases corresponded to the results provided by the existing techniques, i.e. the relative order of models was preserved in many cases.

A similar set of experiments was described in [23], in which the PCC framework was used to assess incompleteness handling of algorithms, as well as the handling of both structured and random deviations from the model. We improved our deviation handling testing procedures by doubling the log, thereby assuring that inserting the deviations would not remove information from the event log. Even though the method of testing differed in details, similar conclusions were drawn.

A plethora of further experiments could have been performed to gain more insights into process discovery and conformance checking. For instance, we would welcome the opportunity to compare the algorithms we introduced to other soundness guaranteeing algorithms. Unfortunately, implementations of two out of the three other currently existing algorithms that have not been published. Furthermore, it would be interesting to repeat the rediscoverability experiments for other types of process models, outside the class of C_B. Besides the PCC framework, we compared one fitness and one log-precision measuring technique. However, many more such techniques exist, as well as model-model conformance checking techniques. An interesting field of further research would be to design test procedures for conformance checking techniques.

Another set of experiments that could be performed is the verification of requirements DR6, DR7, DR8 and DR9. These requirements all state that process discovery techniques should provide rediscoverability for particular types of constructs such as silent transitions, short loops, etc. Even though these constructs are out of the representational bias of most algorithms evaluated in this chapter, we believe that the experimental setups of this chapter could be reused for such experiments.

Finally, it would be interesting to investigate the influence of user-provided parameters on discovery algorithms and conformance checking techniques, e.g. what the influence of f is on the model discovered model by IM$_F$, or what the influence of k in the PCC framework would be on the measures provided by the framework.

References

1. van der Aalst, W.M.P.: Process mining - discovery, conformance and enhancement of business processes. Springer (2011). DOI 10.1007/978-3-642-19345-3. URL http://dx.doi.org/10.1007/978-3-642-19345-3
2. van der Aalst, W.M.P., Adriansyah, A., van Dongen, B.F.: Replaying history on process models for conformance checking and performance analysis. Wiley Interdisc. Rew.: Data Mining and Knowledge Discovery **2**(2), 182–192 (2012). DOI 10.1002/widm.1045. URL http://dx.doi.org/10.1002/widm.1045
3. Adriansyah, A.: Aligning Observed and Modeled Behavior. Ph.D. thesis, Eindhoven University of Technology (2014)

4. Adriansyah, A., Munoz-Gama, J., Carmona, J., van Dongen, B.F., van der Aalst, W.M.P.: Alignment based precision checking. In: Rosa and Soffer [35], pp. 137–149. DOI 10.1007/978-3-642-36285-9_15. URL http://dx.doi.org/10.1007/978-3-642-36285-9_15

5. Augusto, A., Conforti, R., Dumas, M., Rosa, M.L., Bruno, G.: Automated discovery of structured process models: Discover structured vs. discover and structure. In: I. Comyn-Wattiau, K. Tanaka, I. Song, S. Yamamoto, M. Saeki (eds.) Conceptual Modeling - 35th International Conference, ER 2016, Gifu, Japan, November 14-17, 2016, Proceedings, *Lecture Notes in Computer Science*, vol. 9974, pp. 313–329 (2016). DOI 10.1007/978-3-319-46397-1_25. URL http://dx.doi.org/10.1007/978-3-319-46397-1_25

6. vanden Broucke, S.K.L.M.: Advances in process mining: Artificial negative events and other techniques. Ph.D. thesis, KU Leuven (2014)

7. Buijs, J.: Receipt phase of an environmental permit application process (WABO), CoSeLoG project (2014). DOI doi:10.4121/uuid:a07386a5-7be3-4367-9535-70bc9e77dbe6. URL http://dx.doi.org/10.4121/uuid:a07386a5-7be3-4367-9535-70bc9e77dbe6

8. Buijs, J.C.A.M.: Flexible evolutionary algorithms for mining structured process models. Ph.D. thesis, Eindhoven University of Technology (2014)

9. Buijs, J.C.A.M., van Dongen, B.F., van der Aalst, W.M.P.: A genetic algorithm for discovering process trees. In: Proceedings of the IEEE Congress on Evolutionary Computation, CEC 2012, Brisbane, Australia, June 10-15, 2012, pp. 1–8. IEEE (2012). DOI 10.1109/CEC.2012.6256458. URL http://dx.doi.org/10.1109/CEC.2012.6256458

10. Buijs, J.C.A.M., van Dongen, B.F., van der Aalst, W.M.P.: On the role of fitness, precision, generalization and simplicity in process discovery. In: R. Meersman, H. Panetto, T.S. Dillon, S. Rinderle-Ma, P. Dadam, X. Zhou, S. Pearson, A. Ferscha, S. Bergamaschi, I.F. Cruz (eds.) On the Move to Meaningful Internet Systems: OTM 2012, Confederated International Conferences: CoopIS, DOA-SVI, and ODBASE 2012, Rome, Italy, September 10-14, 2012. Proceedings, Part I, *Lecture Notes in Computer Science*, vol. 7565, pp. 305–322. Springer (2012). DOI 10.1007/978-3-642-33606-5_19. URL http://dx.doi.org/10.1007/978-3-642-33606-5_19

11. Burattin, A., Sperduti, A., van der Aalst, W.M.P.: Control-flow discovery from event streams. In: Proceedings of the IEEE Congress on Evolutionary Computation, CEC 2014, Beijing, China, July 6-11, 2014, pp. 2420–2427. IEEE (2014). DOI 10.1109/CEC.2014.6900341. URL http://dx.doi.org/10.1109/CEC.2014.6900341

12. Celonis. https://www.celonis.com/. Accessed: 06-01-2017

13. van Dongen, B.: BPI challenge 2011 dataset (2011). DOI doi:10.4121/uuid:d9769f3d-0ab0-4fb8-803b-0d1120ffcf54. URL http://dx.doi.org/10.4121/uuid:d9769f3d-0ab0-4fb8-803b-0d1120ffcf54

14. van Dongen, B.: BPI challenge 2012 dataset (2012). DOI 10.4121/uuid:3926db30-f712-4394-aebc-75976070e91f. URL http://dx.doi.org/10.4121/uuid:3926db30-f712-4394-aebc-75976070e91f

15. Evermann, J.: Scalable process discovery using map-reduce. IEEE Transactions on Services Computing **9**(3), 469–481 (2016). DOI 10.1109/TSC.2014.2367525. URL http://dx.doi.org/10.1109/TSC.2014.2367525

16. Fahland, D., Favre, C., Koehler, J., Lohmann, N., Völzer, H., Wolf, K.: Analysis on demand: Instantaneous soundness checking of industrial business process models. Data Knowledge Engineering **70**(5), 448–466 (2011). DOI 10.1016/j.datak.2011.01.004. URL http://dx.doi.org/10.1016/j.datak.2011.01.004

17. Günther, C.W., Rozinat, A.: Disco: Discover your processes. In: N. Lohmann, S. Moser (eds.) Proceedings of the Demonstration Track of the 10th International Conference on Business Process Management (BPM 2012), Tallinn, Estonia, September 4, 2012, *CEUR Workshop Proceedings*, vol. 940, pp. 40–44. CEUR-WS.org (2012). URL http://ceur-ws.org/Vol-940/paper8.pdf

18. Hwong, Y., Keiren, J.J.A., Kusters, V.J.J., Leemans, S.J.J., Willemse, T.A.C.: Formalising and analysing the control software of the compact muon solenoid experiment at the large hadron collider. Sci. Comput. Program. **78**(12), 2435–2452 (2013). DOI 10.1016/j.scico.2012.11.009. URL http://dx.doi.org/10.1016/j.scico.2012.11.009

19. Kohavi, R., Brodley, C.E., Frasca, B., Mason, L., Zheng, Z.: Kdd-cup 2000 organizers' report: Peeling the onion. SIGKDD Explorations **2**(2), 86–98 (2000). DOI 10.1145/380995.381033. URL http://doi.acm.org/10.1145/380995.381033

20. Leemans, M., van der Aalst, W.M.P.: Process mining in software systems: Discovering real-life business transactions and process models from distributed systems. In: T. Lethbridge, J. Cabot, A. Egyed (eds.) 18th ACM/IEEE International Conference on Model Driven Engineering Languages and Systems, MoDELS 2015, Ottawa, ON, Canada, September 30 - October 2, 2015, pp. 44–53. IEEE (2015). DOI 10.1109/MODELS.2015.7338234. URL http://dx.doi.org/10.1109/MODELS.2015.7338234

21. Leemans, S.J.J., Fahland, D., van der Aalst, W.M.P.: Discovering block-structured process models from event logs containing infrequent behaviour. In: N. Lohmann, M. Song, P. Wohed (eds.) Business Process Management Workshops - BPM 2013 International Workshops, Beijing, China, August 26, 2013, Revised Papers, *Lecture Notes in Business Information Processing*, vol. 171, pp. 66–78. Springer (2013). DOI 10.1007/978-3-319-06257-0_6. URL http://dx.doi.org/10.1007/978-3-319-06257-0_6

22. Leemans, S.J.J., Fahland, D., van der Aalst, W.M.P.: Discovering block-structured process models from incomplete event logs. In: G. Ciardo, E. Kindler (eds.) Application and Theory of Petri Nets and Concurrency - 35th International Conference, PETRI NETS 2014, Tunis, Tunisia, June 23-27, 2014. Proceedings, *Lecture Notes in Computer Science*, vol. 8489, pp. 91–110. Springer (2014). DOI 10.1007/978-3-319-07734-5_6. URL http://dx.doi.org/10.1007/978-3-319-07734-5_6

23. Leemans, S.J.J., Fahland, D., van der Aalst, W.M.P.: Scalable process discovery and conformance checking. Software & Systems Modeling **special issue**, 1–33 (2016). DOI 10.1007/s10270-016-0545-x. URL http://dx.doi.org/10.1007/s10270-016-0545-x

24. de Leoni, M., Mannhardt, F.: Road traffic fine management process (2015). DOI dx.doi.org/10.1007/s00607-015-0441-1. URL http://dx.doi.org/10.1007/s00607-015-0441-1

25. Leontjeva, A., Conforti, R., Francescomarino, C.D., Dumas, M., Maggi, F.M.: Complex symbolic sequence encodings for predictive monitoring of business processes. In: Motahari-Nezhad et al. [30], pp. 297–313. DOI 10.1007/978-3-319-23063-4_21. URL http://dx.doi.org/10.1007/978-3-319-23063-4_21

26. Liesaputra, V., Yongchareon, S., Chaisiri, S.: Efficient process model discovery using maximal pattern mining. In: Motahari-Nezhad et al. [30], pp. 441–456. DOI 10.1007/978-3-319-23063-4_29. URL http://dx.doi.org/10.1007/978-3-319-23063-4_29

27. Maggi, F.M., Burattin, A., Cimitile, M., Sperduti, A.: Online process discovery to detect concept drifts in ltl-based declarative process models. In: R. Meersman, H. Panetto, T.S. Dillon, J. Eder, Z. Bellahsene, N. Ritter, P.D. Leenheer, D. Dou (eds.) On the Move to Meaningful Internet Systems: OTM 2013 Conferences - Confederated International Conferences: CoopIS, DOA-Trusted Cloud, and ODBASE 2013, Graz, Austria, September 9-13, 2013. Proceedings, *Lecture Notes in Computer Science*, vol. 8185, pp. 94–111. Springer (2013). DOI 10.1007/978-3-642-41030-7_7. URL http://dx.doi.org/10.1007/978-3-642-41030-7_7

28. Mannhardt, F.: Managing large XES event logs in ProM. BPM Center Report BPM-16-04, BPMcenter.org (2016)

29. Molka, T., Gilani, W., Zeng, X.: Dotted chart and control-flow analysis for a loan application process. In: Rosa and Soffer [35], pp. 223–224. DOI 10.1007/978-3-642-36285-9_26. URL http://dx.doi.org/10.1007/978-3-642-36285-9_26

30. Motahari-Nezhad, H.R., Recker, J., Weidlich, M. (eds.): Business Process Management - 13th International Conference, BPM 2015, Innsbruck, Austria, August 31 - September 3, 2015, Proceedings, *Lecture Notes in Computer Science*, vol. 9253. Springer (2015). DOI 10.1007/978-3-319-23063-4. URL http://dx.doi.org/10.1007/978-3-319-23063-4

31. Murata, T., Shenker, B., Shatz, S.M.: Detection of ada static deadlocks using Petri net invariants. IEEE Trans. Software Eng. **15**(3), 314–326 (1989). DOI 10.1109/32.21759. URL http://dx.doi.org/10.1109/32.21759

32. Pradel, M., Gross, T.R.: Automatic generation of object usage specifications from large method traces. In: ASE 2009, pp. 371–382. IEEE Computer Society (2009). DOI 10.1109/ASE.2009.60. URL http://dx.doi.org/10.1109/ASE.2009.60

33. Redlich, D., Molka, T., Gilani, W., Blair, G.S., Rashid, A.: Constructs competition miner: Process control-flow discovery of bp-domain constructs. In: S.W. Sadiq, P. Soffer, H. Völzer (eds.) Business Process Management - 12th International Conference, BPM 2014, Haifa, Israel, September 7-11, 2014. Proceedings, *Lecture Notes in Computer Science*, vol. 8659, pp. 134–150. Springer (2014). DOI 10.1007/978-3-319-10172-9_9. URL http://dx.doi.org/10.1007/978-3-319-10172-9_9

34. Redlich, D., Molka, T., Gilani, W., Blair, G.S., Rashid, A.: Scalable dynamic business process discovery with the constructs competition miner. In: R. Accorsi, P. Ceravolo, B. Russo (eds.) Proceedings of the 4th International Symposium on Data-driven Process Discovery and Analysis (SIMPDA 2014), Milan, Italy, November 19-21, 2014., *CEUR Workshop Proceedings*, vol. 1293, pp. 91–107. CEUR-WS.org (2014). URL http://ceur-ws.org/Vol-1293/paper7.pdf

35. Rosa, M.L., Soffer, P. (eds.): Business Process Management Workshops - BPM 2012 International Workshops, Tallinn, Estonia, September 3, 2012. Revised Papers, *Lecture Notes in Business Information Processing*, vol. 132. Springer (2013). DOI 10.1007/978-3-642-36285-9. URL http://dx.doi.org/10.1007/978-3-642-36285-9

36. Weijters, A.J.M.M., Ribeiro, J.T.S.: Flexible heuristics miner (FHM). In: Proceedings of the IEEE Symposium on Computational Intelligence and Data Mining, CIDM 2011, part of the IEEE Symposium Series on Computational Intelligence 2011, April 11-15, 2011, Paris, France, pp. 310–317. IEEE (2011). DOI 10.1109/CIDM.2011.5949453. URL http://dx.doi.org/10.1109/CIDM.2011.5949453

37. Wen, L., Wang, J., van der Aalst, W.M.P., Huang, B., Sun, J.: A novel approach for process mining based on event types. J. Intell. Inf. Syst. **32**(2), 163–190 (2009). DOI 10.1007/s10844-007-0052-1. URL http://dx.doi.org/10.1007/s10844-007-0052-1

38. van der Werf, J.M.E.M., van Dongen, B.F., Hurkens, C.A.J., Serebrenik, A.: Process discovery using integer linear programming. Fundam. Inform. **94**(3-4), 387–412 (2009). DOI 10.3233/FI-2009-136. URL http://dx.doi.org/10.3233/FI-2009-136

39. Wolf, K.: Generating Petri net state spaces. In: J. Kleijn, A. Yakovlev (eds.) Petri Nets and Other Models of Concurrency - ICATPN 2007, 28th International Conference on Applications and Theory of Petri Nets and Other Models of Concurrency, ICATPN 2007, Siedlce, Poland, June 25-29, 2007, Proceedings, *Lecture Notes in Computer Science*, vol. 4546, pp. 29–42. Springer (2007). DOI 10.1007/978-3-540-73094-1_5. URL http://dx.doi.org/10.1007/978-3-540-73094-1_5

Chapter 9
Enhancement & Inductive visual Miner

Sander J. J. Leemans: Robust Process Mining with Guarantees, LNBIP 440, pp. 423–447, 2022
https://doi.org/10.1007/978-3-030-96655-3_9

Abstract In this section, we discuss enhancement strategies using a end-user focused process mining tool that provides process discovery, conformance checking and enhancement: the Inductive visual Miner. We describe its architecture, introduce its user-focused process tree and highlight some of its features. The Inductive visual Miner supports several enhancements: we describe how deviations, frequency, performance information and animation are computed from conformance checking results and visualised on the model.

In Chapter 6, we introduced several new process discovery techniques of the IM framework, and these techniques were evaluated in Chapter 8. We showed that discovery algorithms offer several guarantees and compare favourably to existing algorithms. However, different algorithms strike different balances of log-quality criteria (e.g. fitness, log precision and simplicity) and including or excluding infrequent, deviating and incomplete behaviour. As described in Section 3.1, different event logs and use cases might require different parameter settings and different algorithms, and it might be challenging to choose an algorithm and its parameter settings for a given event log. Furthermore, discovered models should be evaluated, and if the model does not suit the use case at hand, a new model should be discovered using different parameters or a different algorithm. Therefore, typical process mining projects contain an explorative phase, in which the analyst interactively and repeatedly discovers models and evaluates these models.

For instance, for a case study we performed (Section 3.1, [6]), during the explorative part, human understanding of the business process was important, so the model parameters were chosen to result in a simple model. In later parts of the project, results and insights were evaluated, to ensure the drawn conclusions were not artifacts of process discovery.

Commercial process mining software, such as Fluxicon Disco [9] (FD, we used version 1.9.7) and Celonis Process Mining [5] (CPM), make iteration in process mining projects easy, e.g. discovering a new model is as easy as dragging a slider. Furthermore, such commercial tools typically offer several options to enhance the log and model by visualising extra data on it. However, even though these tools filter behaviour from their models, they offer no way to evaluate the discovered models using conformance checking, i.e. to visualise the behaviour that was excluded. Furthermore, as described in Section 3.3.2, their discovered models might contain ambiguities, which challenges evaluation.

This chapter serves two purposes: we introduce a process mining tool (the Inductive visual Miner (IvM) that combines process discovery (i.e. the algorithms introduced in Chapter 6) with conformance checking to provide users with an easy-to-use package. Second, we use IvM to discuss key challenges and limitations of log and model enhancement concepts, and how they were implemented in IvM.

In this chapter, we will describe several types of enhancements:

- **deviations**. To enable users to evaluate discovered models, we apply a conformance checking technique, alignments, whose results are to be visualised on log and model to provide maximum insight into deviations between the discovered model and the event log.

- **frequency**. If the use case of the analysis is to identify the most frequently used paths through or parts of the model, the model can be enhanced with frequency information. Furthermore, frequency information enables users to assess log precision manually: if parts of the model are not or little used in the event log, the model is not very precise.
- **performance**. An aim of the analysis might be to improve the process in terms of time, such as decreasing employee time spent on traces or improving the user experience of customers by eliminating time spent waiting. In such cases, enhancing the log and model with performance information provides an overview of the parts that take the most time.
- **animation**. Deviations, frequency and performance might vary over time, e.g. the process might change, or seasonal factors influence measurements. Enhancing the model with animation, i.e. visually replaying every trace on the model, highlights several potential issues: waiting customers, slowly moving cases, little-used parts of the model, etc. Furthermore, in our experience, animation increases the confidence of process owners and other stakeholders in the discovered model and increases understandability of the model.

We start with the introduction of IvM in Section 9.1. We describe the capabilities of IvM to diagnose deviations in Section 9.2, show its abilities to filter and visualise frequencies in Section 9.3, discuss how IvM projects performance diagnostics on process trees in Section 9.4 and show IvM's animation capabilities in Section 9.5. Section 9.6 concludes the chapter.

9.1 Inductive visual Miner (IvM)

In the previous section, we described several types of enhancements for event logs and process models. Further on in this chapter, we describe these enhancements in more detail. However, to ease their explanations, we first introduce the software tool that we developed to make process mining more accessible to end users. Using the architecture and notation of this tool, we will explain the enhancement techniques in subsequent sections.

Inductive visual Miner (IvM) is a process exploration tool: it discovers a process model, aligns it to the event log and enhances the resulting model [10]. It is a plug-in ("mine with Inductive visual Miner") of the ProM framework, and interacts with many other plug-ins.

To use IvM, load an event log in the full version of ProM and apply the plug-in "Mine with Inductive visual Miner". Alternatively, IvM can also visualise a log and an existing process tree. Use the plug-in "Visualise deviations on process tree" to start IvM without the mining controls & options, but with alignments, deviations, animation and highlighting filters.

In the remainder of this section, we first describe the steps that are taken by IvM and its architecture. Second, we explain the visualisation of the model. Third, we explain the options and controls, after which we finish the section with a discussion of the extension points of the IvM.

9.1.1 Steps & Architecture

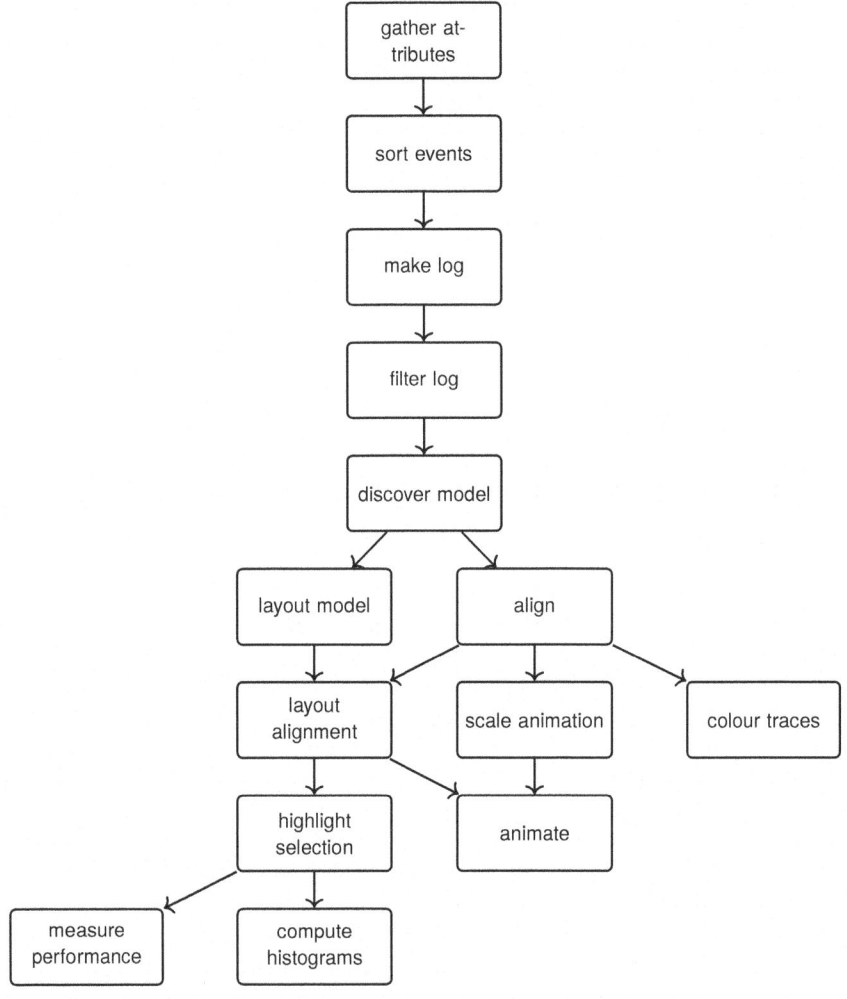

Fig. 9.1: The architecture of IvM. The arrows denote constraints: a task is started as soon as its preceding tasks are finished.

The IvM performs several steps automatically. The computations steps can be interrupted by the user at any time, and IvM will automatically redo steps on user input. Figure 9.1 shows these steps and their dependencies. The main steps are:

- Sort events.
  Some event logs contain traces in which the timestamps are out of order. For instance, in the trace $\langle a^{14:00}, b^{13:00} \rangle$ a happened first according to the order of the events in the trace, but b occurs first according to the timestamps of the events. Such anomalies make animation and performance measures unreliable, so IvM offers the user the choice to either sort the events (in our example, IvM would continue as if $\langle b^{13:00}, a^{14:00} \rangle$ was given) or disable the animation and performance measures.
- Filter log.
  Let L be the event log. This step will remove events of which the activities do not occur enough. See the activities slider and the pre-mining filters in the Controls & Parameters settings for more information. This step is also available (with even more fine-grained options) as a separate plug-in of ProM ("Filter events").
- Discover a process model from log L_1,
  which is done using either the algorithm IMF, IMFLC or IMFA (depending on the miner selector).
- Align the model and the log L.
  The alignment is based on work described in [4]. Before aligning, the discovered model is expanded (as described in Section 5.7), i.e. each activity a is transformed into a nested process tree $\rightarrow(a_E, a_S, a_C)$. This expansion is used at all times, i.e. the alignment *always* takes enqueue, start and completion events into account.
- Visualise the model and the alignment.
- Animate the alignment.
- Compute performance measures and visualise them.

9.1.2 Model Visualisation

A key aspect of IvM is its visualisation of the discovered process model: on this model, all further enhancements, which will be described later on this section, will be visualised. In this section, we discuss our choice for this visualisation.

In Section 8.3.3, we showed several Petri nets that were translated from the process trees returned by the discovery algorithms introduced in earlier chapters. Some of these Petri nets contained many silent transitions. Some of these silent transitions originated from $\times(\tau, \ldots)$ constructs of the discovered process trees, while others were introduced by translating $\vee$ or $\leftrightarrow$ constructs. We believe that such silent transitions are confusing and make models hard to read, and that they therefore best be avoided.

Therefore, IvM shows models in an intuitive formalism that closely resembles Petri nets, process trees and BPMN models. Figure 9.2 shows the constructs of these models. In such a model, each trace traverses edges from the source to the sink, thereby executing each activity on its path. Figure 9.3 shows an example, which corresponds to the process tree $\rightarrow(\times(a,b), \wedge(c,d), \leftrightarrow(e,f))$. In case of concurrency, the path is "split" in multiple branches, e.g. in our example c and d are both executed, and these paths are merged again at a concurrency join. Inclusive choice and interleaving are similar, corresponding to their process tree semantics, i.e. inclusive choice ($\vee$) splits the path into one or more subsequent branches, while interleaving ($\leftrightarrow$) splits the paths but allows only one to be "active" at the same time.

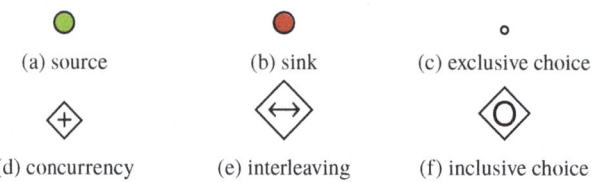

| (a) source | (b) sink | (c) exclusive choice |
| (d) concurrency | (e) interleaving | (f) inclusive choice |

Fig. 9.2: IvM model constructs.

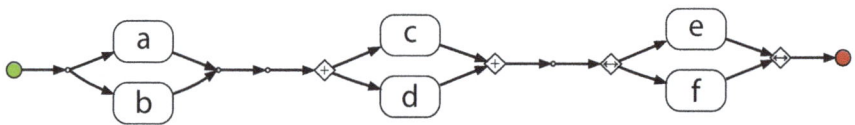

Fig. 9.3: A model in IvM.

9.1.3 Controls & Parameters

As described, IvM will perform the steps described in Section 9.1.1, and show intermediate results. It is not necessary to wait for IvM to complete these steps; users can change parameters any time, and IvM will automatically recompute the necessary steps. Figure 9.4 shows these parameters, and we will explain them in more detail in this section.

9.1.3.1 Activities Slider

The activities slider controls the fraction of activities that is included in the event log on which a discovery algorithm is applied. That is, before discovery, the event log is filtered. The position of the slider (between 0 and 1) determines how many of the activities remain in the filtered event log. For instance, the log $[\langle a,b,c \rangle, \langle a,b \rangle, \langle a \rangle]$,

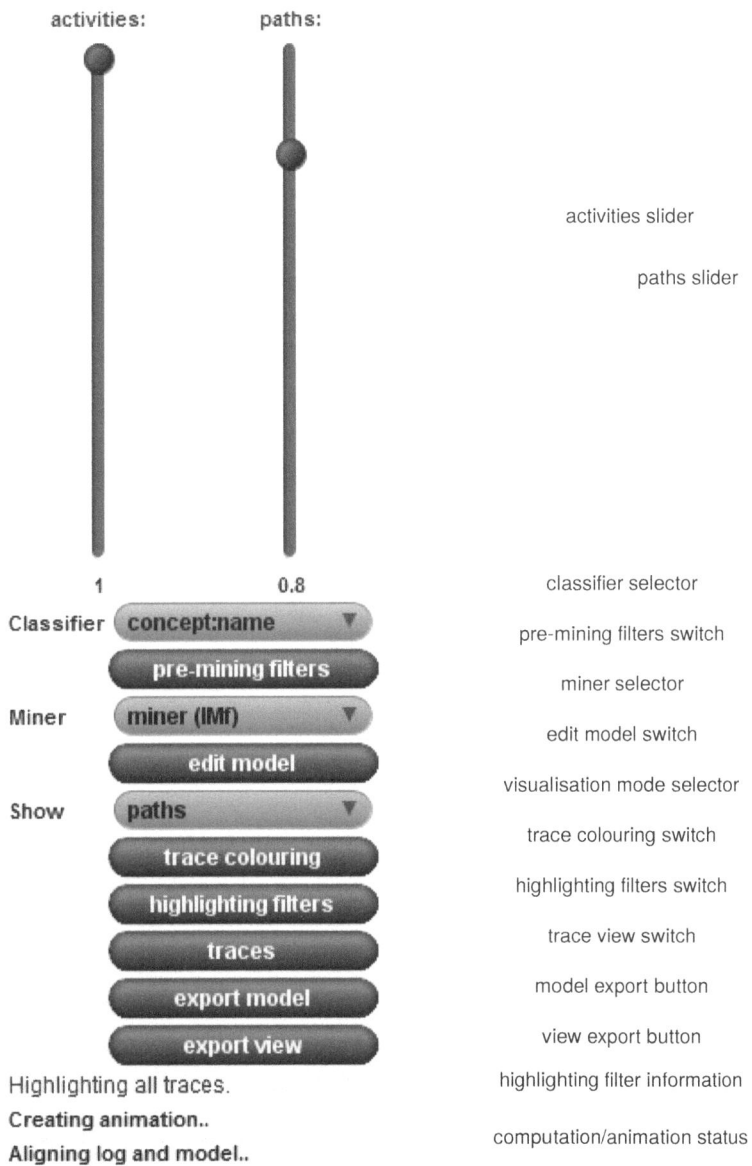

activities slider

paths slider

classifier selector

pre-mining filters switch

miner selector

edit model switch

visualisation mode selector

trace colouring switch

highlighting filters switch

trace view switch

model export button

view export button

highlighting filter information

computation/animation status

Fig. 9.4: Controls of IvM.

has the frequency table $[a^3, b^2, c]$, and if the activities slider would be set to 0.4, then all events corresponding to the activities that occur more than 0.4 times the occurrence of the most-occurring activity would be included. In out example, the filtered event log would be $[\langle a, b\rangle^2, \langle a\rangle]$, and to this filtered event log, the discovery algorithm is applied. Notice that this only affects the discovery, i.e. all other parts of the IvM including alignments and animation are not affected by this slider.

Putting this and the paths slider (described next) all the way up to 1.0 and setting the miner selector to IMᴘ guarantees fitness. However, if the event log contains life-cycle transitions besides complete, deviations might be shown.

9.1.3.2 Paths Slider

The paths slider controls the amount of noise filtering applied: if set to 1, then no noise filtering is applied, while set at 0, maximum noise filtering is applied. Technically, the slider sets the input for the discovery algorithm to 1 - the value of the slider. Please refer to Section 6.2 or 6.5 for more information on the mining algorithms. The default is 0.8, which corresponds to $1 - 0.8 = 0.2$ noise filtering in IMᴘ, IMꜰʟᴄ and IMꜰᴀ.

Putting both sliders all the way up to 1.0 and setting the miner selector to IMᴘ guarantees fitness. However, the alignment of IvM always takes life-cycle information into account, thus deviations might still be present.

9.1.3.3 Classifier Selector

The classifier selector controls what determines the activities of events: events in XES-logs can have several data attributes [7], and this selector determines which one of these data attributes determines the types of activities. As described in Section 6.7, any combination of event attributes can be chosen.

9.1.3.4 Pre-Mining Filters Switch

The pre-mining filters switch opens a panel to set pre-mining filters. A pre-mining filter does not alter the alignment, the performance measures or the animation, but filters the log that is used to discover a model. To activate a pre-mining filter, check its checkbox.

For instance, the pre-mining filter 'Trace filter' allows to discover a model using only the customers who spent more than €10,000.

9.1.3.5 Miner Selector

The miner selector allows to select which mining algorithm is to be used. Default is IM$_F$, other included options are the life-cycle algorithm IM$_{FLC}$ and the more-operators algorithm IM$_{FA}$. We limit the choice to ease the users: these algorithms were shown to be the most applicable to real-life event logs in Chapter 8.

9.1.3.6 Edit Model Switch

The edit model switch opens a panel to manually edit the discovered model, as explained below. This allows users to correct the discovery algorithm if its result is not satisfactory, and to try the effect of a different, custom, model on the same event log. In this panel, the currently discovered process tree is displayed in a custom notation, and can be edited. While typing, the IvM redoes computations automatically. A screenshot is shown in Figure 9.5.

The notation is as follows: each process tree node should be on its own line. The white space preceding the node declaration matters, i.e. a child should be more indented than its parent. Reserved keywords are *xor*, *sequence*, *concurrent*, *interleaved*, *or*, *loop* and *tau*. Loops should be given in an unary ($\circlearrowright(a)$), binary ($\circlearrowright(a, b)$) or ternary ($\circlearrowright(a, b, c)$) form, in which the c denotes the loop exit, i.e. $\circlearrowright(a, b, c) = \rightarrow(\circlearrowright(a, b), c)$. This guarantees compatibility with process trees of the process tree package in ProM, even though this syntax differs from the process trees introduced in Section 2.2.5. Any other text is interpreted as an activity name. In case a keyword is used as an activity name, it should be put in between double quotes (e.g. "sequence").

In case the edited process tree contains a syntactical error, this will be shown at the bottom of the panel, and an approximate location of the error will be highlighted. The manual changes are overwritten if the automatic discovery is triggered, however, *ctrl z* reverts the edit model view to a previous state.

9.1.3.7 Visualisation Mode Selector

The visualisation mode selector allows user to choose between several information to be added to the model. There are four options:

- **paths** This is the default mode, showing the model; the numbers on the activities and edges denote the total number of executions of each of them. Figure 9.6a shows an example: activity b was executed 3952 times, just as the incoming edge to the left of it. In Section 9.3, we will elaborate on frequencies.
- **paths and deviations** shows the model; the numbers in the activities denote the total number of executions of each of them. Moreover, red-dashed edges denote the results of alignments (which were discussed in Section 3.4.1.1): Figure 9.6b shows a model move (see Section 3.4.1.1), indicating that activity d was skipped once in the event log, while the model said it should have been

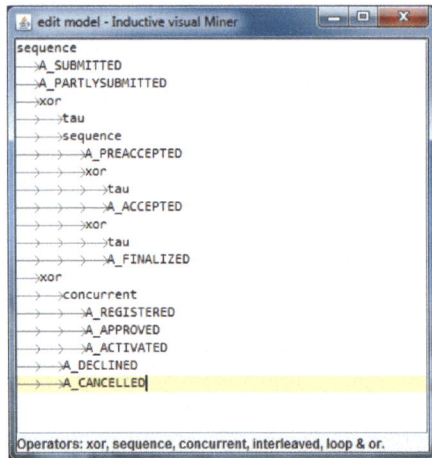

Fig. 9.5: In the edit model panel, the model can be edited.

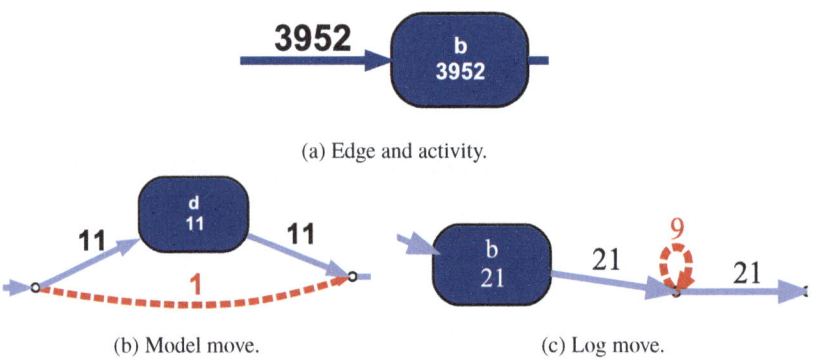

(a) Edge and activity.

(b) Model move.   (c) Log move.

Fig. 9.6: IvM visualisation mode concepts.

executed. Figure 9.6c shows a log move, indicating that 9 times in the event log, after the execution of activity b, an event happened in the event log while this should not happen according to the model. In Section 9.2, we will elaborate on deviation enhancements.

- **paths and queue lengths** shows the model, and denotes each activity with the queue length in front of it, i.e. the number of cases waiting for this activity to start. If the event log contains both events with enqueue and start life-cycle information, this queue length is accurate. Otherwise, it is estimated using the method described in [12]. This queue size is updated as the animation progresses. In Section 9.4, we will elaborate on performance enhancements.
- **paths and sojourn times** shows the model, and denotes each activity with the average sojourn time for that activity. Sojourn times are computed using

completion events, as described in Section 9. The sojourn times are not estimated, i.e. if not both necessary completion events are present and have timestamps, the activity instance is excluded from the average. Performance measures can also be inspected by putting the mouse cursor on an activity: a pop-up will show the performance measures and a histogram. In Section 9.4, we will elaborate on performance enhancements. Furthermore, performance measures are updated when any log filtering is applied.

• **paths and service times** shows the model, and denotes each activity with the average service time for that activity. Service times are computed using start and completion events, as described in Section 9. The service times are not estimated, i.e. if for an activity instance not both start and completion events are present and have timestamps, that activity instance is not considered in the average.

9.1.3.8 Trace Colouring Switch

The IvM can colour traces in the animation and the trace view. Using this colouring, different categories of traces can be easily distinguished. For instance, Figure 9.7 contains a screenshot of coloured traces in the animation and the trace view (which we will explain later). This event log represents an ore mining process, and the traces have been coloured with the hardness of the rock that is being processed.

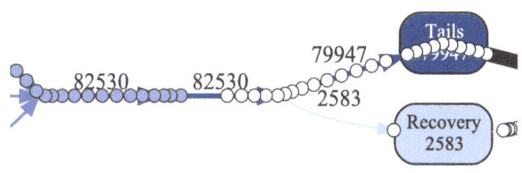

(a) In the animation.

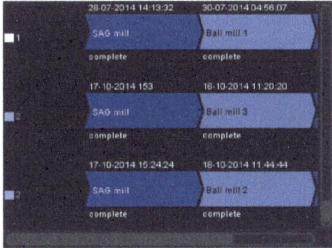

(b) In the trace view. The little blocks on the left denote the category of rock hardness (in the log, this was decoded with a number).

Fig. 9.7: Coloured traces in IvM.

The trace colouring switch opens a panel to set up the trace colouring. In this panel, a trace attribute can be chosen, as well as the derived properties 'duration' and 'number of events'. IvM supports up to 7 colours, and if the attribute is numeric, the domain of the numbers is split into 7 even parts automatically. Date and time attributes are handled similarly. If the attribute is literal and there are more than 7 different values, the colouring will remain disabled.

To enable quick enabling and disabling of the trace colouring, a checkbox has been added to the left side of the panel, which should be checked to enable trace colouring.

9.1.3.9 Highlighting Filters Switch

The highlighting filters switch opens a panel to set highlighting filters. A highlighting filter does not alter the model or the alignment, but filters the log that is shown in the animation and the information projected on the activities and edges of the model. If a highlighting filter is enabled, the highlighting filter information will show this.

A highlighting filter can also be applied to an activity in the model: by clicking on an activity (i.e. selecting it), the event log is filtered to only contain traces for which this activity was executed in accordance with the model, i.e. log-moves and model-moves are excluded. Hold the control-key to select multiple activities; edges can be selected as well. The highlighting filter information will textually show these click-highlighting filters as well.

To enable quick enabling and disabling of highlighting filters, a checkbox has been added to the left side of each filter, which should be checked to enable filtering.

9.1.3.10 Trace View Switch

To allow inspection of the traces, and to provide insight to the deviations between model and log on the log-level (Requirement CR5), IvM offers a trace view. The trace view switch enables or disables the trace view.

Figure 9.8 shows a trace in this trace view: the name of the trace (i.e. the *concept:name* extension) is displayed to the left of the events, which are the coloured wedges to the right. Above the wedges, time stamps are displayed in day-month-year hour:minute:second:millisecond. The wedge itself shows the activity (depending on the classifier selector) of the event. Below the wedge, the first line shows the life-cycle transition information (if that is not present, it shows *complete*). Second, below the wedge the alignment information is shown: in Figure 9.8, the first event is a synchronous event, the second is a model move ("only in model") and the third one is a log move ("only in log").

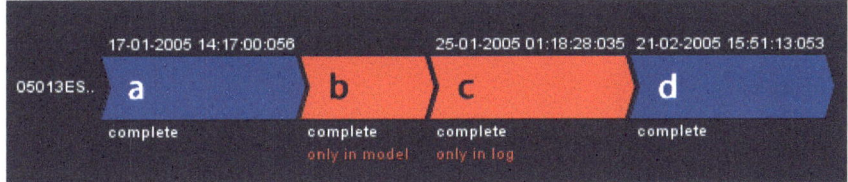

Fig. 9.8: Trace view.

9.1.3.11 Model Export Button

The model export button allows the current model to be exported as a Petri net or process tree to the workbench of ProM.

9.1.3.12 View Export Button

The view export button exports the current image to an image file. Moreover, the animation can be exported (rendered) as a movie, and some statistics about the activities can be exported as a comma-separated-value (csv) file.

9.1.3.13 Changing the View

The model can be moved by dragging it, or by using the arrow keys. Zooming in and out can be done with a scroll wheel, or with the key combination *ctrl =* or *ctrl -*. *Ctrl 0* (zero) resets the model to its initial position.

Once zoomed in, a navigation image will appear in the upper left corner. A click on this navigation image will move the model to that position, and scrolling while the mouse pointer is in the navigation image will zoom the navigation image.

The graph direction, i.e. the position of the green and red start and end places, can be changed by pressing *ctrl d*. The distance between activities and edges can be altered using the key combinations *ctrl q* and *ctrl w*.

9.1.4 Adding Extensions

The IvM can be extended in several ways without changing the source code: miners, pre-mining filters and highlighting filters can be added. A developer should simply add a class that extends the abstract class VisualMinerWrapper, PreMiningEvent-Filter, PreMiningTraceFilter or HighlightingFilter. ProM will automatically detect these classes once on the classpath, and they will automatically be added to IvM. For more information, please refer to the documentation of these classes.

Other extensions to the architecture-chain are possible. Ideally, each step of the chain should be cancellable and applicable to all process trees.

9.2 Deviations

As described in Chapter 3 and shown in Chapter 8, process discovery algorithms might leave behaviour that was recorded in the event log out of the discovered model, and might include behaviour in the model that is not recorded in the event log, as discovery algorithms try to represent the behaviour of event logs into a certain representational bias. Therefore, the discovered model should be evaluated before reliable conclusions can be drawn from such models. Conformance checking techniques, as described in Section 3.4, enable the evaluation of models on three levels: summarative measures, projections on models and projections on event logs. Furthermore, the process model can be entered by hand (using the "edit model" function) to circumvent process discovery and assess an idealised or normative model. (See also the plug-in "Visualise deviations on process tree".)

In this section, we show the output and intermediate computation results of two conformance checking techniques that can be used to evaluate models. That is, we show how the intermediate results of the PCC framework (see Chapter 7) can be projected onto a process model, and we show how alignments ([3], see Section 3.4.1.1) can be projected on process models and event logs.

9.2.1 Deviations and the PCC framework

The PCC framework computes fitness and log precision of subsets of activities of the model and the event log. These fitness and log-precision measures can be averaged over activities to provide insight into the location of deviations in the process. For screenshots and an explanation of this visualisation, please refer to Section 7.4. In the future, we intend to integrate (an extension of) the PCC framework into IvM.

9.2.2 Deviations and Alignments

The PCC framework provides insight into deviations on the level of summarised measures, and on the model. However, currently it does not provide insight on the level of the event log, i.e. given an event log, it currently cannot determine whether an event occurred according to the model. As described in Section 3.4.1.1, alignments can provide this information, as it considers two types of steps: log moves, i.e. events not represented in the model, and model moves, i.e. model steps not represented by an event in the log.

A key property of alignments is that they provide a path through the model that is most similar to the trace in the log. For instance, consider the following alignment of the trace $t = \langle b, c \rangle$ and the model $M = \to$:

$$\begin{array}{c} \wedge \\ a \quad \times \\ \wedge \\ b \quad c \end{array}$$

trace	b	-	c
model	-	a	c

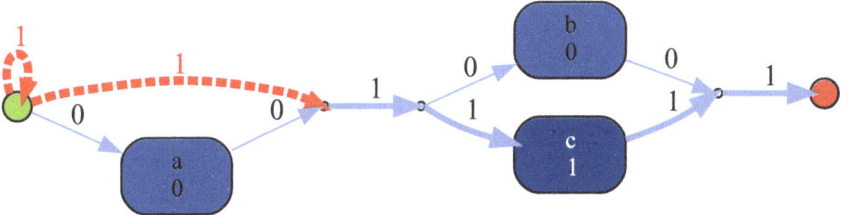

Fig. 9.9: Visualisation of moves.

This alignment provides several pieces of useful information about deviations:

- According to the model, a should have been executed, however no such event was found in the trace. This model move can be considered as a 'skip' of a. In Figure 9.9, this concept is visualised on the model as a red-dashed edge that bypasses a.
- According to the event log, b should have been executed, however the model did not support this. This log move denotes that the event is considered to be superfluous. In Figure 9.9, we show this concept being visualised on a model: the red dashed self-loop indicates this log move.

Notice that these concepts are applicable to a variety of process model notations, such as BPMN, Petri nets and YAWL.

Pitfalls.

Despite the intuitiveness of the visualisations, these concepts should be interpreted with care, as they might convey more information than is actually available. For instance, the alignment shown is not the only 'optimal' alignment. There may be other alignments with the same number of log and model moves. For instance, the following alignment has the same number of deviations:

trace	-	b	c
model	a	b	-

In this alignment, c is a log move instead of b.

Another example is the model $\rightarrow(a, b)$ and the trace $\langle a, c \rangle$. There two optimal alignments:

trace	a - c
model	a b -

trace	a c -
model	a - b

Notice the difference in order between log move c and model move b: this order is arbitrary. Nevertheless, in the visualisation, a choice had to be made to position the log move before or after activity b.

The current alignment implementations traverse the state space defined by log and model, and deterministically choose one option in case there are multiple optimal possibilities. Which possibility is chosen is not always easily determinable by the user, as it depends on internal ordering and sorting. Here, we do not describe these computations in more detail. For more details, we refer to [3].

Furthermore, notice that the different alignments discussed here are all 'optimal'. However, there is no guarantee that an optimal alignment reflects reality, i.e. a non-optimal alignment (i.e. with more deviations than an optimal alignment) could also explain the deviations between event log and model, and might explain this better if domain knowledge is taken into account. Thus, one should be careful interpreting alignments.

9.3 Frequency Information

Given a control flow model of the process described in the event log, a question could be what the important paths through the process are, i.e. which paths are used often. This allows analysts to focus on the main behaviour of the process, or study the little-used parts that represent the exceptional behaviour in the process. For instance, Figure 9.10 shows a model representing the event log $[\langle a, b \rangle^{290}, \langle a, c \rangle^{10}]$, enhanced with frequency information. That is, activity a was executed 300 times, b 290 times and c 10 times, which is indicated by the shade of blue of the activities. Furthermore, the edge leaving a has been used 300 times, the edge arriving at b 290 times and the edge arriving at c 10 times, which is reflected in the thickness of the edges.

The combination of activity colouring and edge thickening makes spotting frequent parts easy: everything that stands out is frequent. For instance, Figure 9.11 shows a model that was discovered from a real-life event log (BP12, see Chapter 8). Even though the scale of the figure makes it difficult to read the activity labels and see the precise control flow, it is obvious that the most frequent part of the process is the looping behaviour in the upper right corner. In this case, the looping behaviour indicates that a more suitable discovery algorithm could have been used, i.e. on

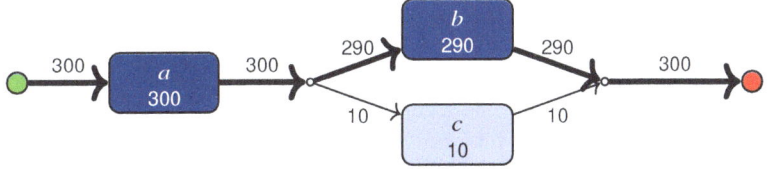

Fig. 9.10: A model enhanced with frequency information.

closer inspection, the event log contains start and completion events, thus one could try to apply one of the non-atomic algorithms, e.g. IM$_{\text{FLC}}$ (Section 6.5.3).

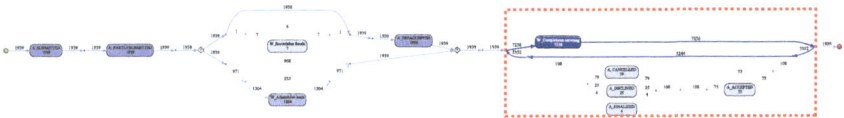

Fig. 9.11: An example of a model, discovered from a real-life event log, and enhanced with frequency information. The most activities are executed in the upper right part of the process. The red box contains looping behaviour which is discovered because the discovery algorithm ignores life-cycle information.

As the edges of the visualisation are annotated with frequency information, frequencies need not only be computed for activities, but for every node of the process tree. To compute these frequencies, we chose to let IvM use the computed alignments, as alignments provide a consistent view on the paths that were (likely) taken through the model. Therefore, IvM counts occurrences of process tree nodes by walking over the aligned traces, keeping track of which process tree nodes are being used.

9.4 Projecting Performance Information on Process Trees

Using an event log, an alignment and a process model, the performance of a process can be computed, i.e. for each activity, time measures such as service time can be computed. In this section, we explore these performance measures and how they can be computed. We first extend the life cycle transitions beyond the start and completion notions that are present in non-atomic event logs. Second, we introduce the four performance measures that are considered in this work: queueing time, waiting time, service time and sojourn time. Third, we show that for reliable measures, a process model and alignment should be taken into account.

In Section 3.5.1.1, we showed that events in event logs might be annotated with a life cycle transition, which denotes the life cycle the activity execution enters with the execution of that event. We introduced the start and completion life cycle transitions, which denote the start and end of activity executions, thereby making the activity execution and event log non-atomic. Many more life cycle transitions have been defined, e.g. in [7, 1, 12], and the techniques described in this section are mostly agnostic the the precise life cycle transition model. However, a precise definition of the transitions is essential for reliable performance measures. For instance, if an activity execution can be started twice, e.g. $\langle a_s, a_s, a_c \rangle$, then it should be clear what this means both behaviourally and in terms of performance.

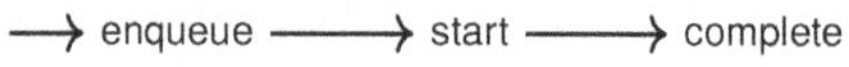

$$\longrightarrow \text{enqueue} \longrightarrow \text{start} \longrightarrow \text{complete}$$

Fig. 9.12: The life cycle model used in this work.

To illustrate the possibility of adding more life cycle transitions to the model, we introduce a third life cycle transition: *enqueue*, which happens before start and denotes that the activity is ready to be executed, but for some reason is delayed (see Figure 9.12). For instance, consider a call center receiving calls from customers. After an initial computer voice menu, a customer enters a queue. After a while the customer is connected to an agent, who services the customer, until the connection is terminated. The events of this execution of 'service customer' are enqueue, start and completion [12]. Another example is the flow of patients through a hospital, specifically for a visit to a doctor. The patient arrives, reports at a desk and waits until a doctor is available for treatment. At the desk queueing starts, upon entering the treatment room service starts, and upon leaving the treatment room service completes.

Using the life cycle model, we define several performance measures, as shown in Figure 9.13. The measure *queueing time* denotes how long the activity execution was queued, and *service time* denotes how long execution of the activity took. The remaining measures, waiting and sojourn time, express how long it took before the activity was respectively started and completed. These measures are taken from the moment the activity *could have* started until it started or completed.

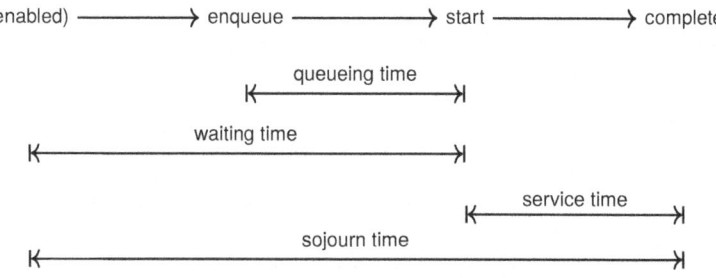

Fig. 9.13: Performance measures.

The computation of the above times depends on the time that an activity execution could have started, in which obviously control flow restrictions, captured in the process model, play a major role.

We illustrate the importance of process models to compute precise performance measures using an example. Consider the trace $t = \langle a_{\text{S}}^{11:\,25}, a_{\text{C}}^{11:\,26}, b_{\text{E}}^{11:\,30}, b_{\text{S}}^{11:\,34},$ $b_{\text{C}}^{11:\,36}, c_{\text{S}}^{11:\,40}, c_{\text{C}}^{11:\,50}\rangle$, in which the events are annotated with time stamps. In this trace, we consider the sojourn and waiting time of c, which both ended at $11:50$. We show that two process models lead to two different values for these measures. That is, in the process tree $M_{150} = $  , c could have been executed at $11:36$,

$$\begin{array}{c} \rightarrow \\ \nearrow\uparrow\nwarrow \\ a \;\; b \;\; c \end{array}$$

i.e. immediately after b_{C}. Thus, for this process tree M_{150} the sojourn time of c is $11:50 - 11:36 = 0:14$ and the waiting time of c is $11:40 - 11:36 = 0:04$. However, in the process tree $M_{151} = \;\rightarrow\;$, c could have been executed at $11:26$, i.e.

$$\begin{array}{c} \rightarrow \\ \nearrow\;\nwarrow \\ a \;\;\; \wedge \\ \nearrow\nwarrow \\ b \;\; c \end{array}$$

immediately after a_{C}, as b and c are concurrent and can hence start independently of each other. For this process tree M_{151}, the sojourn time of c is $11:50 - 11:26 = 0:24$ and the waiting time of c is $11:40 - 11:26 = 0:14$, which are both more than the $0:14$ and $0:04$ for M_{150}. Hence, whenever sojourn time and waiting time are measured, a process model should be taken into account. Queueing time is not affected by the process model.

The first moment an event could have been executed corresponds with an event of another activity execution. In order to link events to the execution of activities in a process model, an alignment is used. In case the event log and process model deviate from one another, we argue that the measures should not be taken. For instance, consider our example tree M_{150} and the trace $\langle a_{\text{S}}^{11:25}, a_{\text{C}}^{11:26}, c_{\text{S}}^{11:40}, c_{\text{C}}^{11:50}\rangle$. For this trace, an alignment computation will deduct that b is not executed while it should according to the model. We argue that as it is unknown when c could have started, c has neither a sojourn time nor a waiting time (or they are unknown). Similarly, if the enabling event is present but has no time stamp, we argue that no sojourn or waiting time exists.

This procedure is repeated for all executions of the activity (i.e. leaf in the process tree) and all results, in which all necessary time stamps and events are present in the log, are averaged. That is, log moves, model moves and events without time stamps are excluded. This yields performance measures for each activity, and these measures can be visualised using colouring in the model or using histograms. Furthermore, besides performance measures, using the enqueue events, queue properties can be computed and visualised. In [12], queue properties, e.g. the number of cases in the queue at any given moment, is estimated in absence of enqueue events by considering throughput, however discussing these estimations in detail is outside the scope of this work.

Many more performance measures could be useful to identify bottlenecks in business processes, such as cost and resource utilisation performance measures [11].

Such measures rely on other information from the event log and it would be interesting to investigate these measures further, eventually standardising such information in event logs and adding these measures to process mining tools such as IvM.

Future work 9.1: Explore cost, resource utilisation and other performance measures.

9.5 Animation

In the previous section, we introduced several performance measures. However, activities, bottlenecks, busy periods, waiting times and other performance measures might vary over time and influence each other. Animation might be suitable to visualise the interplay of these concepts. Animation represents the control-flow state of each trace with one or multiple tokens, by letting tokens flow along the arcs/through activities as events occur for the trace.

Furthermore, animation might highlight changes in the process (*concept drift*). For instance, Figure 9.14 contains two screenshots of an event log animated over a process map, each represent a different time of the event log. The animation shows that the rightmost part of the process was not used in the first part of the event log, but only in the second part, which indicates that the process changed. An analyst could try to explain this change and e.g. filter the log to zoom in on the process before or after the change.

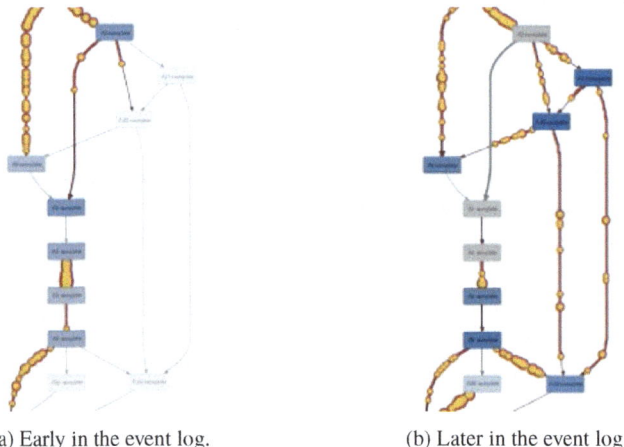

(a) Early in the event log.   (b) Later in the event log.

Fig. 9.14: Concept drift shown by animation in Fluxicon Disco: the right part of the process is only used later in the time span of the event log.

Figure 9.15 contains the principles of animation on an activity (notice that this concept is applicable to a variety of process model notations, such as BPMN, Petri nets and YAWL). In this figure, activity a is represented by a rectangle, and the flows to and from the activity are represented by edges. Each trace is represented by tokens (a thick circle) that move over the edges to and from activities. The execution of an activity is represented by a token being inside the rectangle of the activity. Before the token reaches an activity, it is waiting to be executed, and this gives a clear visual indication that the activity might be a bottleneck (notice that the difference between queueing and waiting is not visualised, see Figure 9.12). Once the start event of the execution occurs, the token flows into the box of the activity, and execution starts. During execution (service) of the activity, the token moves over the activity, such that the speed with which tokens move over an activity is an indication for the service times. Once the completion event of the execution of the activity occurs, the token reaches the end of the activity, and moves towards (i.e. waits for) the next activity.

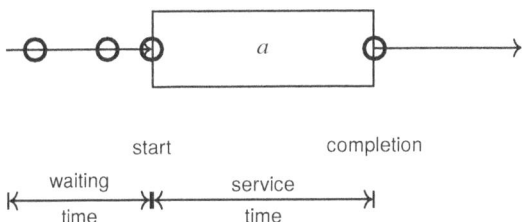

Fig. 9.15: Animation principles on activities.

For splits and joins, there's not much information to be conveyed, as typically, splits and joins have no associated time stamp information in the event log, thus all token movement must be interpolated. Nevertheless, to make the animation smooth, easily interpretable and to convey the message of control flow, token flow over splits and joins should be smooth. The animation strategy depends on whether the split or join entails concurrency. If the split or join is exclusive, then the token takes its intended outgoing or incoming edge and continues. However, if the split or join is non-exclusive, e.g. in a concurrent split or join, then the tokens should split up or merge, as illustrated in Figure 9.16: the split on the left shows a token just before splitting up. Then, the token splits up and the two tokens leave the split in different directions as shown in the figure. At the corresponding join, both tokens arrive at the merge point *simultaneously*, such that they smoothly merge and move on. This implies that the tokens might move at different speeds to arrive at the same time, due to different execution times of the previous activities, or the length of the path from that activity to the join.

In case of missing information, for the performance measures, we argued that performance measures should not be interpolated if time stamps are missing or behaviour in event log and process model does not match. Instead, such cases should be ignored to guarantee reliability of the measures. For animation, this is clearly undesirable, as it makes tokens appearing and disappearing seemingly random, which makes it hard to track cases. Therefore, we argue that intuitiveness and clarity of the visual appearance of the token flow should have priority over a 100% accurate

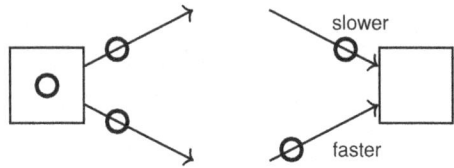

Fig. 9.16: Animation principles for splits and joins: tokens split and join from and into single tokens. Therefore, tokens should arrive at the same time and might depart the split and arrive at the join at different speeds.

visualisation of the underlying data in situations of incomplete event data, and thus the tokens should flow smoothly through the process. However, we make an exception for missing start events, i.e. if the activity execution is atomic, then we choose to reflect this by making the time in the activity zero, i.e. the token jumps from the left to the right of the activity instantly.

In case of tools that provide animations on top of process maps, e.g. in the Fuzzy Miner (FM) [8] and in commercial tools such as Fluxicon Disco (FD) [9] and Celonis Process Mining (CPM), a challenge is that some edges are filtered out. That is, tokens ideally flow from activity to activity via an edge, however this edge might have been removed by filtering. To solve this, in FD, tokens appear at the start of an edge and disappear again at the end of the edge, and do not flow over activities. Therefore, waiting time and bottlenecks are visualised by tokens on incoming edges of activities, but service times cannot be visualised by the tokens, and tokens have to 'jump' over the missing edges. Instead, in FD, activity executions are summarised by the activity changing colour if it is being executed. This approach shows that the activity is executing, but not how many cases are being executed or how long a particular execution takes if multiple executions are being performed at the same time. In CPM, this is not an issue, as traces are filtered instead of edges, which ensures that if a trace is to be animated, all the edges on its flow path are present.

9.6 Conclusion

In this chapter, we showed several ways in which extra information in event logs can be used to enhance event logs and corresponding process models. That is, we showed how highlighting frequencies helps to reveal the busy parts of the process, we showed how deviations highlight where model and log deviate, we showed how performance quantifies busy parts of the process, and we showed how animation conveys concept drift and the interplay of activities, bottlenecks, queue sizes and throughput.

We introduced the Inductive visual Miner (IvM), of which a screenshot is shown in Figure 9.17. With the IvM, we aim to offer as reliable visualisations as possible, such that even if information is missing from the event log, conclusions can still be drawn

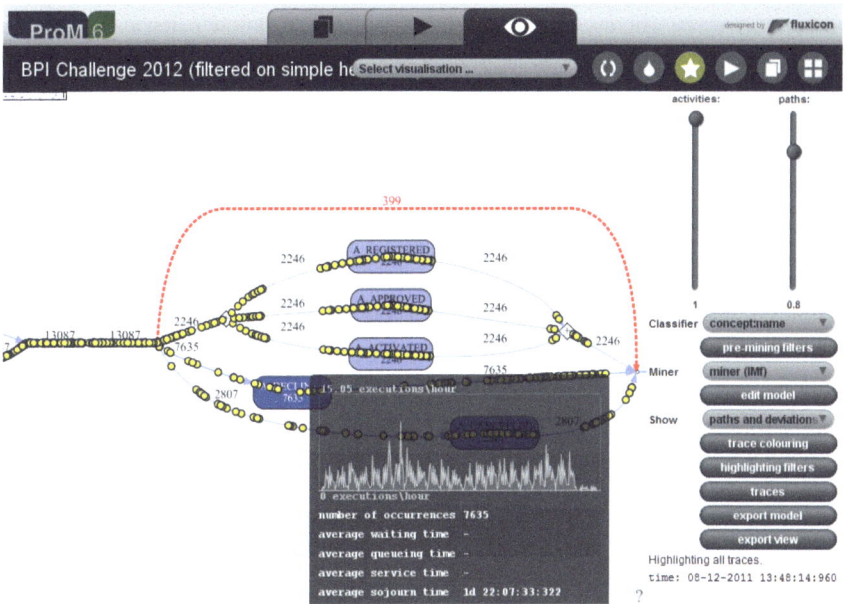

Fig. 9.17: A screenshot of the Inductive visual Miner.

reliably. For instance, if an event in the log lacks a time stamp, we disregards that event in performance computations and histograms. In contrast, to keep animation smooth, we interpolate missing time stamps and users should be aware of this and validate conclusions drawn from animations in IvM.

Even though the provided filters of IvM are not as extensive as in commercial tools such as Fluxicon Disco and Celonis Process Mining, new and even custom filters can be added without much effort. IvM supports log filtering at two stages: before discovering a model, and after the alignment. Regardless of filter settings, the alignment is always computed over the entire unfiltered event log, and deviations can be visualised (taking the limitations of alignments as described in Section 9.2 into account). This is an improvement over commercial tools such as FD and CPM, as these simply leave out filtered behaviour, such that users might have to choose between a readable model and visualising all behaviour, while IvM, log-model quality can be assessed in depth.

Concurrency, interleaving, milestones and other advanced workflow patterns (see [2] for more details) are essential for process mining: without these constructs, only the most simple processes can be described. Existing commercial tools such as Fluxicon Disco and Celonis Process Mining are limited to directly follows relations, and do not support these constructs. IvM is not limited to a particular discovery algorithm: other discovery techniques can be added without much effort, as long as these algorithms return process trees. Nevertheless, the algorithms introduced in Chapter 6 return sound process models fast, such that IvM can be used in interactive and iterative ways. The use of process trees limits the workflow patterns that can be supported by IvM, however enable to not consider unsound models or models with deadlocks, and process trees make the models easy to read.

Due to the automation and interactivity of steps, and the guarantees provided by the steps of IvM, extensions are easier to implement and to use. For instance, queue estimations, performance computations and histograms were added as subsequent steps to IvM. However, even though the steps are executed concurrently by IvM as much as possible, they are not independent. That is, each step has to be aware of and work with every other step in IvM to work properly, and thus adding features becomes more complex with the addition of more features.

References

1. van der Aalst, W.M.P.: Process mining - discovery, conformance and enhancement of business processes. Springer (2011). DOI 10.1007/978-3-642-19345-3. URL http://dx.doi.org/10.1007/978-3-642-19345-3
2. van der Aalst, W.M.P., ter Hofstede, A.H.M., Kiepuszewski, B., Barros, A.P.: Workflow patterns. Distributed and Parallel Databases **14**(1), 5–51 (2003). DOI 10.1023/A:1022883727209. URL http://dx.doi.org/10.1023/A:1022883727209
3. Adriansyah, A.: Aligning Observed and Modeled Behavior. Ph.D. thesis, Eindhoven University of Technology (2014)
4. Buijs, J.C.A.M.: Flexible evolutionary algorithms for mining structured process models. Ph.D. thesis, Eindhoven University of Technology (2014)
5. Celonis. https://www.celonis.com/. Accessed: 06-01-2017
6. van Eck, M.L., Lu, X., Leemans, S.J.J., van der Aalst, W.M.P.: PM^2 : A process mining project methodology. In: J. Zdravkovic, M. Kirikova, P. Johannesson (eds.) Advanced Information Systems Engineering - 27th International Conference, CAiSE 2015, Stockholm, Sweden, June 8-12, 2015, Proceedings, *Lecture Notes in Computer Science*, vol. 9097, pp. 297–313. Springer (2015). DOI 10.1007/978-3-319-19069-3_19. URL http://dx.doi.org/10.1007/978-3-319-19069-3_19
7. Günther, C., Verbeek, H.: XES v2.0 (2014). URL http://www.xes-standard.org/
8. Günther, C.W., van der Aalst, W.M.P.: Fuzzy mining - adaptive process simplification based on multi-perspective metrics. In: G. Alonso, P. Dadam, M. Rosemann (eds.) Business Process Management, 5th International Conference, BPM 2007, Brisbane, Australia, September 24-28, 2007, Proceedings, *Lecture Notes in Computer Science*, vol. 4714, pp. 328–343. Springer (2007). DOI 10.1007/978-3-540-75183-0_24. URL http://dx.doi.org/10.1007/978-3-540-75183-0_24
9. Günther, C.W., Rozinat, A.: Disco: Discover your processes. In: N. Lohmann, S. Moser (eds.) Proceedings of the Demonstration Track of the 10th International Conference on Business Process Management (BPM 2012), Tallinn, Estonia, September 4, 2012, *CEUR Workshop Proceedings*, vol. 940, pp. 40–44. CEUR-WS.org (2012). URL http://ceur-ws.org/Vol-940/paper8.pdf
10. Leemans, S.J.J., Fahland, D., van der Aalst, W.M.P.: Process and deviation exploration with Inductive visual Miner. In: L. Limonad, B. Weber (eds.) Proceedings of the BPM Demo Sessions 2014 Co-located with the 12th International Conference on Business Process Management (BPM 2014), Eindhoven, The Netherlands, September 10, 2014., *CEUR Workshop Proceedings*, vol. 1295, p. 46. CEUR-WS.org (2014). URL http://ceur-ws.org/Vol-1295/paper19.pdf
11. Rozinat, A.: Process Mining: Conformance and Extension. Ph.D. thesis, Eindhoven University of Technology (2010)

12. Senderovich, A., Leemans, S.J.J., Harel, S., Gal, A., Mandelbaum, A., van der Aalst, W.M.P.:
    Discovering queues from event logs with varying levels of information. In: M. Reichert,
    H.A. Reijers (eds.) Business Process Management Workshops - BPM 2015, 13th International
    Workshops, Innsbruck, Austria, August 31 - September 3, 2015, Revised Papers, *Lecture Notes
    in Business Information Processing*, vol. 256, pp. 154–166. Springer (2015). DOI 10.1007/
    978-3-319-42887-1_13. URL http://dx.doi.org/10.1007/978-3-319-42887-1_13

Chapter 10
Conclusion

Sander J. J. Leemans: Robust Process Mining with Guarantees, LNBIP 440, pp. 449–459, 2022
https://doi.org/10.1007/978-3-030-96655-3_10

Process mining aims to extract information from event logs, which are recorded from running business processes. Process mining projects may go through multiple phases in which different process mining techniques are used: process discovery, conformance checking and model enhancement, to all of which we contributed concepts and techniques. In this chapter, we summarise the previous chapters and reflect on how well these techniques address the challenges of process mining identified in 3, and point out open problems and future work.

10.1 Process Discovery

After an event log is obtained, a first step in a typical process mining project is to discover a process model from the event log by applying a process discovery algorithm. In Chapter 3, we discussed several properties discovery algorithms need to possess: (1) the returned model needs to be free of deadlocks and other anomalies, i.e. be sound, (2) the algorithm needs to be able to balance fitness, log precision and simplicity (how these are best balanced might depend on the use case), and (3) the algorithm needs to be able to rediscover the entire language of the underlying real-life system (possess *rediscoverability*), even if the log contains only a small subset of the possible behaviour. Many existing process discovery techniques do not guarantee soundness, do not allow users to adjust the balance of fitness, log precision and simplicity, or do not possess rediscoverability. As some of the identified requirements involve contradicting trade-offs and different use cases might require different discovery algorithms, we conjecture that no single discovery strategy can satisfy all identified requirements.

Use cases might require several different process discovery techniques, but these techniques might offer similar guarantees that can be proven using similar proofs. Therefore, in Chapter 4, we introduced the Inductive Miner framework (IM framework), which aids algorithms in providing several guarantees and enables algorithm designers to consider only the *most important* behaviour in an event log, instead of *all behaviour*. The IM framework searches for the most important behaviour in an event log. That is, the framework searches for the combination of the root process tree operator and a proper division of the activities in the event log (a *cut*). If such a cut can be found, the event log is split into several sublogs and the IM framework is applied to it recursively until a base case (e.g. an event log containing only a single activity) is encountered. If no such cut can be found, a fall through (e.g. a model allowing for a superset of the behaviour in the event log) is returned. The guarantees aided by the IM framework are soundness, which is guaranteed for any algorithm by the use of process trees, and fitness, log-precision and rediscoverability, for which proof obligations have been expressed as properties that are local in the IM framework.

In order to ease generalising over the behaviour in an event log, many process discovery algorithms use an abstraction of the behaviour in the event log, e.g. a directly follows graph. We used these abstractions to establish proof obligations for rediscoverability, i.e. we expressed a set of requirements on such abstractions that guarantee rediscoverability. Furthermore, we expressed these requirements in the parameter functions of the IM framework (Theorem 4.2).

In Chapter 5, we performed a systematic study towards these abstractions: we analysed what classes of languages have equivalent abstractions, and therefore cannot be distinguished reliably by algorithms that use these abstractions (and hence cannot be uniquely discovered). For several abstractions, we identified classes of languages such that no two models of a class with different languages have the same abstraction. Furthermore, we addressed the mismatch between semantics and syntax of process trees: there can be many process trees with the same language, and we are not interested in the difference between two process models if they have the same language. Therefore, we introduced a set of reduction rules for process trees, such that applying these rules exhaustively yields a normal form. Ideally, these normal forms have one-to-one mappings to languages, i.e. for each language there is precisely one process tree in normal form and vice versa. We proved this property for several classes of process trees using abstractions, i.e. we proved that two process trees in normal form have different abstractions. This establishes the close relation between the syntax of process trees in normal form, the abstraction under consideration and the semantics (i.e. the language) of process trees (as the abstractions are language based, obviously two different abstractions represent different languages). Studying these abstractions gives a better understanding of how these abstractions influence process discovery and conformance checking, enables the comparison of these techniques and sketches the formal boundaries of process discovery techniques.

Using the IM framework and the studied abstractions, we introduced several discovery algorithms in Chapter 6, which are summarised in Table 10.1: we introduced a fitness-guaranteeing basic algorithm IM, a deviating- and infrequent-behaviour handling algorithm IM_F and an incomplete-behaviour handling algorithm IM_C. Furthermore, we introduced algorithms to handle more process tree constructs: a fitness guaranteeing IM_A and a deviating- and infrequent-behaviour handling IM_{FA}. These algorithms highlight the flexibility of the IM framework: one can take an existing algorithm and improve it locally with little-impacting changes, and likely rediscoverability and perhaps fitness guarantees will be preserved.

For all these algorithms, we proved rediscoverability, using the proof obligations identified in Chapter 4. Furthermore, we evaluated these algorithms in Chapter 8 and found that they handle deviating, incomplete and infrequent behaviour well: in our experiment of 9 real-life logs, the IM_D algorithm was pareto optimal for all event logs, and IM_{FD} and IM_A were pareto optimal for 8 logs. Existing techniques achieving pareto optimality were the Evolutionary Tree Miner [4] (8 times pareto optimal, however many activities were left out of the models) and the Structured Miner [2] (2 times).

Some event logs contain non-atomic executions of activities, i.e. activities take time, which is denoted by the presence of events denoting the start and end of

---

[1] Future work.

[2] We chose not to guarantee fitness for IM_D (see Section 6.6.6).

Table 10.1: The family of discovery algorithms that implement the IM framework, and their guarantees and purposes. Due to the frameworks, all algorithms guarantee soundness, termination and rediscoverability. The algorithms will be introduced in Chapter 6.

use cases	framework	fitness guaranteed	infrequent & deviating behaviour	incomplete behaviour
	IM framework	IM	IM_F	IM_C
discover more behaviour	IM framework	IM_A	IM_{FA}	-[1]
handle non-atomic event logs	IM framework	IM_{LC}	IM_{FLC}	IM_{CLC}
handle larger logs	IM_D framework	IM_D[2]	IM_{FD}	IM_{CD}

executions. To handle such event logs, we introduced a family of algorithms that constructs a non-atomic directly follows graph as a first step, but further resembles algorithms mentioned earlier: a fitness-guaranteeing basic IM_{LC}, a deviating and infrequent behaviour handling IM_{FLC} and an incomplete behaviour handling IM_{CLC}. Also for these algorithms, we proved rediscoverability.

Most of the algorithms mentioned have a run time that is polynomial in the number of activities and run quick on real-life event logs, and, as shown in Chapter 8, can be applied on normal (2GB RAM) hardware to event logs containing millions of events and hundreds of activities. However, the IM framework requires the event log to be copied for every recursion, thus even larger event logs might be problematic. To handle larger event logs, i.e. containing tens of millions of events and thousands of activities, we adapted the IM framework to recurse on a directly follows abstraction instead of on event logs, such that in each recursion, only a directly follows relation needs to be copied instead of an event log. We introduced three algorithms that use the adapted framework: the basic IM_D, the infrequent and deviating behaviour handling IM_{FD} and the incompleteness handling IM_{CD}. In our evaluation (Chapter 8), we found that the algorithms of the IM_D framework handle event logs of tens of millions of events and thousands of activities, while sacrificing little fitness, log-precision and simplicity over (and sometimes even surpassing) the IM framework algorithms.

Even though the algorithms presented in this chapter apply different strategies, instead of searching for the *entire* behaviour while worrying about soundness, the IM framework allowed us to focus on strategies to find the *most important* behaviour in an event log (i.e. the root operator and root activity partition) and have soundness guaranteed. By using the IM framework in different settings and for different algorithms, we have shown that it, and the proofs for guarantees, can be reused. Therefore, the IM framework can be seen as a starting point for more algorithms that leverage the algorithms and proofs we provided. That is, in future work, many advanced techniques might be designed to handle specific events and use cases. All of these techniques might benefit from the ideas of the IM framework.

10.2 Conformance Checking

While discovering a model, process discovery algorithms might need to exclude behaviour of the event log from the model, or include behaviour that is not in the event log into the model, in order to obtain a model with the "right" balance. Therefore, discovered models should be evaluated before further usage, for which a conformance checking technique could be used. Two types of conformance checking techniques were addressed in previous chapters: log-conformance checking, which compares a process model to an event log and advises on their differences, and model-conformance checking, which compares two process models. For instance, the discovered model and another model representing a reference implementation, representing a different geographical area or representing a different time period could be compared. We identified three levels on which conformance checking techniques provide information about the correspondence between logs and models: a summarised measure (e.g. a fitness or precision number), information on the model level, and information on the log level (see Chapter 3). Many existing conformance checking techniques are either unable to deal with the complexity of real-life event logs and the models discovered from these logs, do not support all features of such discovered models, or use an abstraction that is too coarse to capture the behaviour of logs and models well.

In Chapter 7, we presented our approach to conformance checking: the PCC framework. This framework is applicable to compare event logs to models and models to models, and checks for conformance by constructing the language of these logs and models explicitly in DFAs and compares their behaviour to measure fitness and precision. Thus, the PCC framework supports all regular languages, regardless of the model formalism used. To avoid constructing the entire state space and consequently take a lot of time, the PCC framework constructs DFAs of all *subsets* of activities of a user-specified length in the logs and the model. This allows the PCC framework to consider behaviour on a scale from fine-grained to very coarse, depending on the size of the subsets. These partial measures provide insight in the locations of deviations between the logs and models, i.e. average fitness and precision can be computed for each activity and this can be visualised on the model. Furthermore, the partial measures can be averaged over all subsets of activities to provide a summarised fitness and precision measure.

We evaluated the PCC framework on real-life event logs and discovered models in Chapter 8, and found that it is applicable to the large event logs that cannot be handled by current conformance checking techniques. Furthermore, we found that in many cases, the measures of the PCC framework rank process models discovered by discovery techniques similar to existing alignment-based techniques.

10.3 Enhancement & Tool Support

Given a discovered process model and the result of a log-conformance checking technique, a process model and an event log can be enhanced with additional information. In Chapter 9, we described four types of information to enhance models and event logs: deviations, frequency, performance and animation. Performance information, e.g. the sojourn, waiting, queueing or service time, enables analysts to discover time-consuming activities in the process. For log animation, the event log is visually replayed on the model: each case can be visually tracked as it traverses the process, which enables the detection of changes in the process (concept drift) and bottlenecks. Queues might determine the majority of waiting times in a business process, so analysing queues might reveal bottlenecks. Deviations between log and model, i.e. log moves and model moves, are essential to evaluate a model and should be considered before drawing conclusions about a process using a model.

We described a software tool, the Inductive visual Miner (IvM), which performs several steps, all fully automated. First, Inductive visual Miner discovers a process model using several of the algorithms described in Chapter 6. Second, it performs conformance checking, i.e. computes an alignment, between the event log and the discovered model. Third, it enhances the model based on this alignment (we described how the IvM supports the four types of information). Options for enhancement include performance information on the model and the event log, animation on the model, and deviations projected on both event log and model.

The Inductive visual Miner combines the strong points of commercial products with strong points of academic software. For instance, commercial products offer ease-of-use and practical applicability, while academic software provides reliability and semantic results that allow an analyst to validate any gained insights. Using Inductive visual Miner, analysts can explore the event log by repeatedly discovering a model (which is guaranteed sound, and potentially is fitting and language equivalent to the system), evaluate this model to ensure its validity, filter the event log and enhance it with performance information. Inductive visual Miner shows that it is possible to use powerful techniques with formal guarantees in a user-friendly package. We hope that the Inductive visual Miner will inspire commercial vendors to consider models with executable semantics and support deviation analysis.

10.4 Remaining Challenges

In this section, we elaborate on remaining challenges and future work. We first reiterate detailed identified areas of potential future work. Second, we elaborate on future challenges that lie beyond the scope of this work.

10.4.1 Detailed

In the discussions, several detailed areas of future research have been identified. Here, we reiterate these areas:

3.1   Investigate semantics for arbitrary life cycle models.   92

3.2   Use other information next to event logs in process discovery and conformance checking, and apply ideas of PCC framework to similarity measures stronger than language-equivalence. 108

3.3   Investigate whether it's possible to extend the PCC framework to provide information on the log level (Requirement CR5). 108

3.4   Obtain and visualise deviations and performance measures without alignments. 108

3.5   Study what enqueue events can contribute to process discovery. 108

5.1   Extend reduction rules to reduce trees with τ leafs as non-first children and duplicate activities. 144

5.2   Extend C_B with non-arbitrarily nestable trees. 151

5.3   Study the influence of τ, $\leftrightarrow$ and $\vee$ on activity relations. 162

5.4   Find or disprove a footprint LC-property of $\wp$-graphs to distinguish all trees of C_M. 176

5.5   Identify requirements such that nested $\vee$ and $\wedge$ can be handled without coo abstractions, and identify an abstraction to identify nested $\leftrightarrow$. . . 200

6.1   Explore other techniques as fall throughs. 233

6.3   Research more elegant locally log-precision preserving IM framework functions. 237

6.4   Consider other deviation-filtering techniques to distinguish concurrency and deviating/infrequent behaviour. 240

6.6   Prove rediscoverability of IM_F for logs with deviating and infrequent behaviour. 249

6.7   Consider other deviation-filtering techniques to distinguish concurrency and deviating/infrequent behaviour. 249

6.10   Extend IM_{FA} to apply deviation filtering to the detection of $\wedge$ and $\vee$. . 286

6.12   Implement and evaluate IM_{CLC}. 300

6.13   Combine life cycle handling capabilities with the minimum self-distance relation $\wp$. 301

6.14   Include support for $\leftrightarrow$ in IM_{LC} and IM_{FLC}. 301

6.18   Develop cut detection, log splitting, base case detection and fall through techniques further. 324

6.19   Engineer a do-it-yourself graphical user interface to compose an algorithm in the IM framework. 324

7.2   Investigate properties of the PCC framework on models outside of C_I. 351

8.1   Perform experiment to investigate the influence of parameter settings on discovery algorithms. 392

8.2   Identify and analyse types of deviations and perform experiments to investigate the influence of these deviations on rediscovery. 402

9.1   Explore cost, resource utilisation and other performance measures. . . 442

10.4.2 Future Work

Beyond our scope, we identified several areas of future research: (1) considering stronger notions than language equivalence, (2) rediscoverability on more constructs, (3) guaranteeing soundness without the use of process trees, (4) distinguishing infrequent and deviating behaviour and (5) performing usability experiments on IvM.

10.4.2.1 Beyond Languages

In the lion's share of this work, we assumed that event logs contain only the order of activities. Therefore, we limited ourselves to languages, i.e. we focused on process discovery techniques being able to rediscover the language of an underlying real-life system and conformance checking techniques that verify whether the language of a model corresponds to an event log. However, event logs might contain more information that enables techniques to consider stronger equivalence notions.

For instance, the life cycle transition might reveal the moment of choice. Consider the event log $[\langle b_E, a, b_S, b_C\rangle, \langle c_E, a, c_S, c_C\rangle]$. Figure 10.1a shows a model that would be returned by e.g. IM$_{LC}$, which ignores the enqueue event of b and c. For this event log, one could argue that the choice between b and c is made before the execution of a, as the enqueue events happen before a happens. Therefore, even though the language of this model corresponds to the event log, the moment of choice is captured incorrectly, as the model puts this choice after a. In contrast, the model of Figure 10.1b puts this choice before a. Therefore, this model is "closer" to the event log in the sense of bisimilarity, or even bisimilar to it. Further research needs to be performed how process discovery and conformance techniques can incorporate bisimilarity by using extra data from the event log.

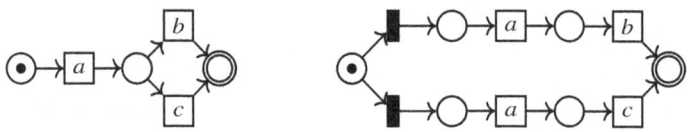

(a) A model in which the choice for b or c is made after a.   (b) A model in which the choice for b or c is made before a.

Fig. 10.1: Two language-equivalent but not (weakly) bisimilar models.

10.4.2.2 Extended Rediscoverability

We proved rediscoverability for arbitrarily nestable process tree operators $\times$, $\rightarrow$, $\wedge$, $\circlearrowleft$, $\vee$, and, with certain nesting restrictions, $\leftrightarrow$ and τ constructs. Even though

$\leftrightarrow$ and $\vee$ are translated to Petri nets using some non-free-choice elements, we did not cover general non-free-choice constructs, duplicate activities, long-distance dependencies (Requirement DR6), milestones, parallel interleaved routing, arbitrary cycles (Requirement DR9) and many more workflow patterns [1].

Some of these constructs can be incorporated incrementally in the IM framework, such as duplicate activities (see [3] and the discussion on Heuristic Miner in Section 3.3.2). Other constructs would be discoverable by introducing new process tree operators, such as certain types of milestones and parallel interleaved routings. These constructs might require post-processing steps outside the IM framework. Finally, certain constructs might be challenging to represent in the process tree formalism, such as general non-free-choice constructs, long-distance dependencies and arbitrary cycles.

A solution to represent these constructs could be the use of hybrid models, in which some of the nodes in a process tree are Declare models, which in turn might contain activities that represent process trees. Thus, Declare and process trees form a hierarchy [7]. The process tree parts of such models might for instance be discovered by using the IM framework until a fall through would be necessary. Then, instead of choosing a fall through, patterns of a few activities could be identified (e.g. one could detect that two activities are sequential), and continue the recursion bottom-up instead of top-down. The final "glue" between the top-down and the bottom-up part could then be expressed using Declare.

Two major challenges of process discovery that arise with the addition of such powerful constructs are (1) the potential lack of generalisation and the risk of overfitting, as for each log there might be a perfectly fitting and perfectly log-precise construct, and (2) proving rediscoverability might require infeasibly strict log completeness requirements (see Corollary 5.10). Nevertheless, future research might explore the boundaries of discovering such constructs.

10.4.2.3 Soundness

In Section 3.2.1, we showed that process discovery algorithms should guarantee to return sound models at all times. In our techniques, we guaranteed soundness by the use of process trees, which, by their block structure, are inherently sound. In essence, we use the representational bias of the discovery algorithms to guarantee soundness. However, this representational bias of process trees inherently brings some challenges and restrictions.

There might be other approaches to guarantee soundness. That is, other algorithms have been proposed that guarantee soundness without using process trees, such as Maximal Pattern Mining ([6], see Section 3.3), which iteratively adjusts block-structured Petri nets to guarantee sound models. Furthermore, quick soundness checks, e.g. [5], might be used to steer genetic algorithms to sound models. Future research could reveal further approaches to guarantee soundness in process discovery.

10.4.2.4 Infrequent and Deviating Behaviour

In Section 3.2.2.3, we discussed infrequent and deviating behaviour. Infrequent behaviour is behaviour according to the system that occurs little, while deviating behaviour is in violation of the system. That is, both types of behaviour typically occur rarely, however we do not fully understand the difference between deviating and infrequent behaviour yet. Further study might reveal ways to distinguish these types of behaviour and, consequently, process discovery algorithms might be improved.

10.4.2.5 Usability Experiments

In our evaluation section, we evaluated the introduced discovery algorithms and conformance checking techniques. However, we did not evaluate the Inductive visual Miner (IvM). Such an evaluation would entail the usability of IvM, which would be best evaluated using systematic usability tests with real-life users. However, such tests would require a different scientific approach than we performed. Therefore, we suggest such an experiment to be performed in the future.

References

1. van der Aalst, W.M.P., ter Hofstede, A.H.M., Kiepuszewski, B., Barros, A.P.: Workflow patterns. Distributed and Parallel Databases **14**(1), 5–51 (2003). DOI 10.1023/A:1022883727209. URL http://dx.doi.org/10.1023/A:1022883727209
2. Augusto, A., Conforti, R., Dumas, M., Rosa, M.L., Bruno, G.: Automated discovery of structured process models: Discover structured vs. discover and structure. In: I. Comyn-Wattiau, K. Tanaka, I. Song, S. Yamamoto, M. Saeki (eds.) Conceptual Modeling - 35th International Conference, ER 2016, Gifu, Japan, November 14-17, 2016, Proceedings, *Lecture Notes in Computer Science*, vol. 9974, pp. 313–329 (2016). DOI 10.1007/978-3-319-46397-1_25. URL http://dx.doi.org/10.1007/978-3-319-46397-1_25
3. vanden Broucke, S.K.L.M.: Advances in process mining: Artificial negative events and other techniques. Ph.D. thesis, KU Leuven (2014)
4. Buijs, J.C.A.M.: Flexible evolutionary algorithms for mining structured process models. Ph.D. thesis, Eindhoven University of Technology (2014)
5. Fahland, D., Favre, C., Koehler, J., Lohmann, N., Völzer, H., Wolf, K.: Analysis on demand: Instantaneous soundness checking of industrial business process models. Data Knowledge Engineering **70**(5), 448–466 (2011). DOI 10.1016/j.datak.2011.01.004. URL http://dx.doi.org/10.1016/j.datak.2011.01.004
6. Liesaputra, V., Yongchareon, S., Chaisiri, S.: Efficient process model discovery using maximal pattern mining. In: H.R. Motahari-Nezhad, J. Recker, M. Weidlich (eds.) Business Process Management - 13th International Conference, BPM 2015, Innsbruck, Austria, August 31 - September 3, 2015, Proceedings, *Lecture Notes in Computer Science*, vol. 9253, pp. 441–456. Springer (2015). DOI 10.1007/978-3-319-23063-4_29. URL http://dx.doi.org/10.1007/978-3-319-23063-4_29
7. Slaats, T., Schunselaar, D.M.M., Maggi, F.M., Reijers, H.A.: The semantics of hybrid process models. In: C. Debruyne, H. Panetto, R. Meersman, T.S. Dillon, eva Kühn, D. O'Sullivan, C.A. Ardagna (eds.) On the Move to Meaningful Internet Systems: OTM 2016 Conferences - Confederated International Conferences: CoopIS, C&TC, and ODBASE 2016, Rhodes, Greece,

October 24-28, 2016, Proceedings, *Lecture Notes in Computer Science*, vol. 10033, pp. 531–551 (2016). DOI 10.1007/978-3-319-48472-3_32. URL http://dx.doi.org/10.1007/978-3-319-48472-3_32

Index

α (α-algorithm), 73, 357
α-algorithm, 73, 357
$*$ Kleene star (regular expression), 24
$\square$ nothing, 123
End end activities (language), 44
Σ activities (process tree), 40
Σ alphabet of activities, 40
Σ alphabet of directly follows relation, 44
$\Sigma^{\oslash}$ activity sets of non-coo subtrees, 190
Start start activities (language), 44
$\underline{\wedge}$ concurrent (activity relation), 159
$\underline{\leftrightarrow}$ interleaved (activity relation), 162
$\underline{\circlearrowleft}_i$ loop indirect (activity relation), 159
$\underline{\circlearrowleft}_s$ loop single (activity relation), 159
$\underline{\rightarrow}$ sequence (activity relation), 159
$\underline{\times}$ exclusive choice (activity relation), 159
$\bar{?}$ optionality (process tree), 177
$\cdot$ sequence (regular expression), 24
$\cdot$ trace concatenation, 24
$\wedge_{\mathcal{L}}$ concurrent join (process tree), 36
$\wedge$ concurrent operator (process tree), 35

$\twoheadrightarrow(L)$ directly follows relation (log), 44
$\twoheadrightarrow(M)$ directly follows relation (model), 44
$\twoheadrightarrow$ directly follows relation, 44
$\twoheadrightarrow^+$ transitive closure of $\twoheadrightarrow$, 45
ϵ empty trace, 24
$\in$ element of multiset, 24
$\leftrightarrow_{\mathcal{L}}$ interleaved join (process tree), 36
$\leftrightarrow$ interleaved operator (process tree), 36
$\circlearrowleft_{\mathcal{L}}$ loop join (process tree), 37
$\circlearrowleft$ loop operator (process tree), 37
$\mathbb{L}$ all regular languages, 26
$\mathbb{M}^{\Sigma}$ merge superset of coo subtrees, 192
$\mathbb{M}$ unbounded multiset, 24
$\mathbb{T}$ all process trees, 40
$\mathcal{L}^{\Sigma}$ activity set language, 190
$\mathcal{L}$ language (Petri net), 28
$\mathcal{L}$ language (process tree), 34
$|$ choice (regular expression), 24
$\oslash$ minimum self-distance, 169
$\dashrightarrow$ non-atomic directly follows relation, 204
$\oplus$ operator (process tree), 34
$\oplus_{\mathcal{L}}$ language-join (process tree), 34
$\vee_{\mathcal{L}}$ inclusive choice join (process tree), 38

∨ inclusive choice (process tree), 38

⤳ firing sequence (Petri net), 28

→_L sequence join (process tree), 35

→ sequence operator (process tree), 35

set set of multiset, 24

⊔⊔ shuffle traces, 36

⊆ subset of multiset, 24

τ silent activity (process tree), 34

$\tilde{L}$ non-atomic event log, 41

$\tilde{L}$ non-atomic language, 204

$\tilde{a}$ non-atomic activity, 41

$\tilde{a}$ non-atomic activity (process tree), 203

⊎ multiset difference, 24

⊎ multiset sum, 24

×_L exclusive choice join (process tree), 35

× exclusive choice operator (process tree), 34

$\xrightarrow{a}$ edge executing *a* (DFA), 25

$e^{\bullet}$ post set of *e* (Petri net), 28

a_c completion event, 41, 202

a_E enqueue event, 92

a_s start event, 41, 202

$^{\bullet}e$ pre set of *e* (Petri net), 28

PCC framework (Projected Conformance Checking framework), 328

so stem (sequence-optional stem), 185

∧.1-∧.1
    concurrency footprint, 206

∧.1-∧.2
    footprint, 152

∧↔.1-∧↔.1
    minimum self-distance footprint, 171

↔.1-↔.1
    footprint, 164

↔↺.1-↔↺.1
    concurrency footprint, 207

↺.1-↺.4

↺.1-↺.4
    footprint, 153

→.1-→.1
    footprint, 152

×.1-×.1
    footprint, 152

×(τ).1-×(τ).1
    footprint, 178

×(τ→(...)).1-×(τ→(...)).5
    footprint, 186

AP.1-AP.6
    abstraction preservation, 133

DAP.1-DAP.6
    directly follows abstraction preservation, 316

RF.1-RF.5
    rediscoverability framework, 130

RQ.1-RQ.5
    research question, 356

C_B.1-C_B.5, 148

C_{coo}.1-C_{coo}.4, 184

C_I.1-C_I.4, 164

80% model, 54

abstraction
    preserving, 133, 316
    rediscoverability, 131
    rediscovery, 131

activities, alphabet of (Σ), 40

activity, 2

activity instance, 91

activity partitions, 154, 166

activity relations, 158, 159
    concurrent (∧), 159
    exclusive choice (×), 159
    interleaved (↔), 162
    loop indirect (↺ᵢ), 159
    loop single (↺ₛ), 159
    sequence (→), 159

activity set log, 269

alignment, 16, 83
    log move, 83

model move, 83
  optimal, 83
  synchronous move, 83
atomic, 9, 201
atomic event log, 40

behavioural appropriateness, 86
bisimilarity, 58
  branching, 59
  weak bisimilarity, 59
body end activity, 254
BPMN (Business Process Model and Notation), 33
Business Process Model and Notation, 33
  gateway, 33

canonicity, 139, 144
causal nets, 75
C_B (rediscoverable process trees), 148
CCM (Constructs Competition Miner), 71, 357
C_{coo} (rediscoverable process trees), 184
Celonis Process Mining, 98, 357, 444
C_I (rediscoverable process trees), 164
C_{LC} (rediscoverable process trees), 207
C_M (rediscoverable process trees), 170
completeness, 64
completion event (a_C), 202
concatenation of traces ($\cdot$), 24
concept drift, 442
concurrency graph, 206
concurrent-optional-or, 188
  abstraction, 192
  activity set language ($\mathcal{L}^\Sigma$), 190
  activity set trace, 191
  merge superset of coo subtrees ($\mathbb{M}^\Sigma$), 192
  non-coo subtrees, 189
    activity sets of ($\Sigma^\oslash$), 190
  operators, 189
  relations, 177, 192
  stem, 189
conditional livelock, 353

confluency
  local, 145
conformance checking, 2, 11, 50, 55
connected component, 44
Constructs Competition Miner, 71, 357
coo
  see concurrent-optional-or, 188
correct behaviour, 63
correctness, 64
CPM (Celonis Process Mining), 98, 357, 444
cut, 121, 123, 304
  concurrent, 152, 206
  conforms to process tree, 160
  cross a, 161
  exclusive choice, 152
  interleaved, 164
  loop, 153
  non-trivial, 152
  sequence, 152

Declare, 79
deterministic finite automaton, 25, 104
  conjunction, 335
  edge executing a ($\xrightarrow{a}$), 25
  minimal, 26
  post set, 336
deviating behaviour, 63
DFA (deterministic finite automaton), 25
directed graph, 43
directly follows graph, 44, 148
directly follows relation
  alphabet (Σ), 44
  complete, 250
  directly follows relation ($\twoheadrightarrow$), 44
  of log ($\twoheadrightarrow(L)$), 44
  of model ($\twoheadrightarrow(M)$), 44
  transitive closure ($\twoheadrightarrow^+$), 45
discovery log, 364
dominates, 365
duplicate activities, 72

edge weight, 43

EM (Evolutionary Miner), 71
enactment, 54
enhancement, 2, 5
    model enhancement, 51
ETConformance, 86
ETM (Evolutionary Tree Miner), 71, 357
event, 2, 40
event log, 2, 40
eventually follows graph, 71
Evolutionary Miner, 71
Evolutionary Tree Miner, 71, 357

fall through, 122, 229, 304
    activity concurrent, 229
    activity per trace, 229
    flower model, 230
    flower model with epsilon, 231
    strict tau loop, 230
    tau loop, 230
    trace model, 231
FD (Fluxicon Disco), 78, 95, 357, 444
fitness, 7, 54, 61, 85, 328
fitting behaviour, 60
Flexible Heuristic Miner, 75, 357
Flower Miner, 357
flower model, 68, 229
Fluxicon Disco, 78, 95, 357, 444
FM (Flower Miner), 357
FM (Fuzzy Miner), 78, 444
FO (Fodina), 76, 139, 148, 357
Fodina, 76, 139, 148, 357
footprint, 152
    concurrency graph
        concurrent, 206
        interleaved, 207
        loop, 207
    concurrent-optional-or relations
        concurrent, 194
        inclusive choice, 194
    directly follows relation
        concurrent, 152
        exclusive choice, 152
        loop, 153
        optionality, 178

    sequence, 152, 164
    sequence (optionality), 187
minimum self-distance
    concurrent, 172
    interleaved, 172
    loop, 172
footprint matrices, 86
Fuzzy Miner, 78, 444

generalisation, 62, 86
graph, 43

happy flow models, 54
Heuristic Miner, 139, 148
HM (Flexible Heuristic Miner), 75, 357
HM (Heuristic Miner), 139, 148
hybrid process model, 324

ILP (Integer Linear Programming Miner), 78
ILP (Integer Linear Programming), 357
incomplete, 65
Inductive Miner, 124
    - all operators (IMA), 268, 272, 278
    - directly follows (IMD), 301, 304
    - directly follows based framework (IMD framework), 301
    - incompleteness (IMC), 249, 251, 258, 267
    - incompleteness - directly follows (IMCD), 301, 312
    - incompleteness - life cycle (IMCLC), 286, 294
    - infrequent (IMF), 246
    - infrequent - all operators (IMFA), 268, 279, 281
    - infrequent - directly follows (IMFD), 301, 310
    - infrequent - life cycle (IMFLC), 286, 294
    - life cycle (IMLC), 286

(IM), 218, 232
- infrequent (IM_F), 240
framework, 9, 120, 123
framework (IM framework), 102
Inductive visual Miner, 16, 105, 424,
        425, 458
    activities, 428
    animation export, 435
    change view, 435
    classifier, 430
    deviations, 431
    edit model, 431
    filter before discovery, 430
    filters after discovery, 434
    highlighting filters, 434
    image export, 435
    mining algorithm, 431
    model export, 435
    paths, 430, 431
    pre-mining filters, 430
    queue length, 431
    service time, 431
    sojourn time, 431
    statistics export, 435
    trace colouring, 433
    trace view, 434
infrequent behaviour, 65
Integer Linear Programming, 357
Integer Linear Programming Miner,
        78
IvM (Inductive visual Miner), 16,
        105, 424, 425, 458

language, 26
    end activities (End), 44
    start activities (Start), 44
language complete, 64
language equivalent, 58
language unique, 88, 138, 144, 148
language uniqueness, 130, 216
language-class preserving, 135, 317
LC-property   (loop-concurrent-
        property), 173
life cycle, 91
life-cycle transition

complete, 92
enqueue, 92
start, 92
Little Thumb, 75
locally fitness preserving, 127
log conformance, 60
log precision, 8, 54, 61, 85, 328
    log-precise behaviour, 60
log quality, 63
log-conformance checking, 2, 5, 11,
        51
long-distance dependency, 72, 222
loop body, 37
loop redo, 37
loop-concurrent-property   (LC-
        property), 173
LT (Little Thumb), 75

Maximal Pattern Miner, 72, 357
minimum self-distance, 169
    graph, 170
    relation ($\circleddash$), 169
model enhancement, 50
model-conformance checking, 3, 5,
        11, 51
model-model comparison, 87
MPM (Maximal Pattern Miner), 72,
        357
multiset, 24
    corresponding set (*set*), 24
    difference ($\uplus$), 24
    element of ($\in$), 24
    subset ($\subseteq$), 24
    sum ($\uplus$), 24
    unbounded multiset ($\mathbb{M}$), 24

negative events, 79
Newman's Lemma, 147
non-atomic, 9, 201
    activity ($\tilde{a}$), 41, 203
    activity (process tree), 202
    completion event (a_c), 41
    consistent trace, 42, 202
    directly follows graph, 204
    directly follows relation ($\twoheadrightarrow$),
        204

end activity, 204
end activity instance, 204
enqueue event (a_E), 92
event log, 10
event log ($\tilde{L}$), 41
expanded (Petri net), 202
language, 41
language ($\tilde{L}$), 204
leaf (process tree), 202
process models, 201
process tree, 203
start activity instance, 204
start event (a_s), 41
transition (Petri net), 202
nothing (□), 123

observed behaviour, 63

pareto optimal, 365
parsing measure, 86
path, 43
PCC framework (Projected Conformance Checking framework), 13, 104
Petri net, 27
    expanded, 202
    firing sequence ($\rightsquigarrow$), 28
    free choice, 30
    inhibitor arc, 31
    language ($\mathcal{L}$), 28
    non-free choice constructs, 72
    post set ($e^\bullet$), 28
    pre set ($^\bullet e$), 28
    reset arc, 31
    silent transition, 28
    unlabelled, 27
pivot, 186
    scope, 186, 187
plug-ins
    compute projected fitness and precision, 347
    compute projected recall and precision, 347
    convert log to directly follows graph, 320

expand collapsed process tree, 298
filter events, 427
mine Petri net with Inductive Miner, 318
mine Petri net with Inductive Miner - directly follows, 319
mine process tree with Inductive Miner, 318
mine process tree with Inductive Miner - directly follows, 319
mine with Inductive visual Miner, 425
visualise deviations on process tree, 425
PM (Process Miner), 73
principal transition sequences, 89
process discovery, 2, 50, 51, 55
Process Miner, 73
process mining, 2
process tree, 33, 34
    activities (Σ), 40
    all process trees ($\mathbb{T}$), 40
    concurrent join ($\wedge_{\mathcal{L}}$), 36
    concurrent operator ($\wedge$), 35
    conforms (binary tree), 160
    exclusive choice join ($\times_{\mathcal{L}}$), 35
    exclusive choice operator ($\times$), 34
    inclusive choice ($\vee$), 38
    inclusive choice join ($\vee_{\mathcal{L}}$), 38
    interleaved join ($\leftrightarrow_{\mathcal{L}}$), 36
    interleaved operator ($\leftrightarrow$), 36
    language ($\mathcal{L}$), 34
    language-join function ($\oplus_{\mathcal{L}}$), 34
    loop join ($\circlearrowleft_{\mathcal{L}}$), 37
    loop operator ($\circlearrowleft$), 37
    lowest common ancestor, 40
    non-atomic, 203
    normal form, 141, 148
    operator ($\oplus$), 34
    optionality property ($\bar{?}$), 177
    reduced, 141
    sequence join ($\rightarrow_{\mathcal{L}}$), 35

sequence operator (→), 35
silent activity (τ), 34
Projected Conformance Checking
      framework, 13, 104, 328

queueing time, 92

recall, 8, 54, 61, 62, 88, 328
rediscoverability, 7, 54, 63, 129
redo start activity, 254
regular expression
    choice (|), 24
    Kleene star (∗), 24
    sequence (·), 24
regular language, 26
    all ($\mathbb{L}$), 26
rewriting rules
    system of, 147

sequence-optional stem (so stem),
    185
service time, 92
short loops, 74
simplicity, 54, 66, 67
SM (Structured Miner), 76, 357
sociometry, 94
sojourn time, 92
soundness, 29
    weak soundness, 58
split point (sequence split), 242
start activity, 204
start event (a_s), 202
strongly connected component, 44
Structured Miner, 76, 357

system, 2
system conformance, 61
system model, 5
system precision, 8, 54, 62, 88, 328
    system-precise behaviour, 61

Tα (Tsinghua-α algorithm), 75, 357
test log, 364
TM (Trace Model Miner), 357
token, 27
token-based replay, 86
trace, 2, 24, 26
    concatenation, 24
    empty (ϵ), 24
    shuffle (⊔), 36
trace model, 68, 231
Trace Model Miner, 357
Tsinghua-α algorithm, 75, 357

undirected graph, 43
undirected path, 43

variety, 369

waiting time, 92
workflow net, 28
    block structured, 29
workflow patterns, 32

YAWL (Yet Another Workflow Language), 32
Yet Another Workflow Language, 32
    cancellation region, 32
    multiple instance, 32

The manufacturer's authorised representative in the EU is Springer
Nature Customer Service Centre GmbH, Europaplatz 3, 69115 Heidelberg,
Germany. If you have any concerns regarding our products, please
contact ProductSafety@springernature.com

Printed and bound by CPI Group (UK) Ltd, Croydon, CR0 4YY
29/04/2026
02099470-0010